# Basic Topology 1

**Professor Mahima Ranjan Adhikari (1944–2021)**

Avishek Adhikari · Mahima Ranjan Adhikari

# Basic Topology 1

## Metric Spaces and General Topology

 Springer

Avishek Adhikari 🄳
Department of Mathematics
Presidency University
Kolkata, West Bengal, India

Mahima Ranjan Adhikari
Institute for Mathematics, Bioinformatics,
Information Technology and Computer
Science (IMBIC)
Kolkata, West Bengal, India

Professor Mahima Ranjan Adhikari is deceased

ISBN 978-981-16-6511-0       ISBN 978-981-16-6509-7   (eBook)
https://doi.org/10.1007/978-981-16-6509-7

Mathematics Subject Classification: 54-XX, 46A11, 18-XX, 46Axx, 46Bxx

This Springer imprint is published by the registered company Springer Nature Singapore Pte Ltd.
The registered company address is: 152 Beach Road, #21-01/04 Gateway East, Singapore 189721,
Singapore

*Dedicated to*

*my grandparents Naba Kumar and Snehalata Adhikari,*

*my mother Minati Adhikari*

*who created my interest in mathematics*

*at my very early childhood.*

*Avishek Adhikari*

*Dedicated to*

*Prof. A. C. Chaudhuri and Prof. B. C. Chatterjee*

*who gave me the first lesson of topology*

*at the University of Calcutta, India,*

*in the session 1965–1966.*

*Mahima Ranjan Adhikari*

# Preface

This volume studies metric spaces and general topology. It considers the general properties of topological spaces and their mappings. The special structure of a metric space induces a topology having many applications of topology in modern analysis, geometry and algebra. Contents of Volume 1 are expanded in eight chapters. The chapterwise texts run as follows:

Chapter 1 assembles together some basic concepts and results of set theory, algebraic systems, analysis, Euclidean spaces and category theory through the concepts of categories, functors and natural transformations with the standard notations for smooth reading of the book.

Chapter 2 starts with the concept of the metrics, which is an abstraction of distance in the Euclidean space and conveys an axiomatic framework for this abstraction with a systemic study of elementary basic properties of metric spaces with a view to define open sets and hence to study continuity of functions. Urysohn lemma for metric spaces facilitates to provide a vast number of continuous functions while metric spaces provide a rich supply of topological spaces. In fact, most of the applications of topology to analysis arise through metric spaces. Normed linear spaces form a special class of metric spaces which provide Banach and Hilbert spaces. Metric spaces give the simplest setting for the study of certain problems arising in analysis. The framework for topology begins with an introduction to metric spaces. The special structure of a metric space induces a topology having many applications of topology in modern analysis and modern algebra.

Chapter 3 conveys the basic concepts of topological spaces. In fact, their continuous mappings in an axiomatic framework by introducing the concept of open sets without the notion of distance function or metric, where the basic objects are topological spaces and basic functions between them are continuous maps. The basic motivation of framing the axioms is to introduce the notion of continuity, which is the central concept in topology. It provides a convenient language to study when different points in a space come near to each other, and hence, the subject plays an important role in science and technology.

Chapter 4 studies topological spaces by imposing certain conditions, called *separation axioms*, as the defining properties of a topological are weak to study most of the topological spaces of our interest, which carry more structure (not like a metric). These axioms initially used by P. S. Alexandroff (1896–1982) and H. Hopf (1894–1971) facilitate to classify topological spaces and provide enough supply of continuous functions which are linked to open sets. Separation properties provide enough supply of continuous functions which are linked to open sets. Many important topological properties can be characterized with the help of separation axioms by distributing the open sets in the space $X$ and imposing natural conditions on $X$ such that $X$ behaves like a metric space.

Chapter 5 studies compactness and connectedness in topological settings, which are two important topological properties. A compact space is a natural generalization of closed and bounded sets in the Euclidean space $\mathbf{R}^n$. On the other hand, the concept of connectedness as a single piece generalizes the intuitive idea of nonseparateness of a geometric object. These two topological concepts are utilized to solve many problems in topology, mainly classification of topological spaces up to homeomorphism, and are fundamental in the study of modern analysis, geometry, topology, algebra and many other areas. Moreover, this chapter studies compactification, which is a process or result of making a topological space into a compact space. There are many noncompact spaces. Considering the importance of compactness in mathematics, this study includes Stone–Čech compactification and Alexandroff one-point compactification.

Chapter 6 presents more results on continuous functions from a topological space to the real line space, called *real-valued continuous functions*, which plays a central role in topology and studies uniform convergence and normal spaces through separation by such functions. This chapter proves Urysohn lemma for normal spaces with the help of dyadic rational numbers, which is a surprising result and applies it to Tietze extension theorem and ring theory.

Chapter 7 studies certain class of topological spaces satisfying the two axioms of countability formulated by F. Hausdorff in 1914 and establishes a connection between compactness and the Bolzano–Weierstrass property (B-W property) and proves some embedding theorem. Motivation for the study of the concepts of countability and separability of topological spaces comes from some natural problems discussed in this chapter.

Chapter 8 conveys the history of emergence of the concepts leading to the development of general topology as a subject with their motivations.

The book is a clear exposition of the basic ideas of topology and conveys a straightforward discussion of the basic topics of topology and avoids unnecessary definitions and terminologies. Each chapter starts with highlighting the main results of the chapter with motivation and is split into several sections which discuss related topics with some degree of thoroughness and ends with exercises of varying degrees of difficulties, which not only impart an additional information about the text covered previously but also introduce a variety of ideas not treated in the earlier texts with certain references to the interested readers for more study. All these constitute the basic organizational units of the book.

This three-volume book together with the authors' two other Springer books *Basic Modern Algebra with Applications* (Springer, 2014) and *Basic Algebraic Topology and its Applications* (Springer, 2016) will form a unitary module for the study of modern algebra, general and algebraic topology with applications in several areas.

The authors acknowledge the Higher Education Department of the Government of West Bengal for sanctioning the financial support to the "Institute for Mathematics, Bioinformatics and Computer Science (IMBIC)" toward writing this book vide order no 432 (Sanc)/ EH/P/SE/ SE/ 1G-17/07 dated August 29, 2017, and also to IMBIC, University of Calcutta, Presidency University, Kolkata, India, and Moulana Abul Kalam Azad University of Technology, West Bengal, for providing the infrastructure toward implementing the scheme.

The authors are indebted to the authors of the books and research papers listed in the bibliography at the end of each chapter and are very thankful to Profs. P. Stavrions (Greece), Constantine Udriste (Romania), and Akira Asada (Japan) and to the reviewers of the manuscript for their scholarly suggestions for improvements. We are thankful to Md. Kutubuddin Sardar for his cooperation towards the typesetting of the manuscript and to many UG and PG students of Presidency University and Calcutta University, and to many other individuals who have helped in proofreading the book. Authors apologize to those whose names have been inadvertently not entered. Finally, the authors acknowledge, with heartfelt thanks, the patience and sacrifice of long-suffering family of the authors, especially Dr. Shibopriya Mitra Adhikari and Master Avipriyo Adhikari.

Kolkata, India
June 2021

Avishek Adhikari
Mahima Ranjan Adhikari

# A Note on Basic Topology—Volumes 1–3

The topic "**Topology**" has become one of the most exciting and influential fields of study in modern mathematics, because of its beauty and scope. The aim of this subject is to make a qualitative study of geometry in the sense that if one geometric object is continuously deformed into another geometrical object, then these two geometric objects are considered topologically equivalent, called *homeomorphic*. Topology starts where sets have some cohesive properties, leading to define continuity of functions.

The series of three books on *Basic Topology* is a project book funded by the Government of West Bengal, which is designed to introduce many variants of a basic course in topology through the study of point set topology, topological groups, topological vector spaces, manifolds, Lie groups, homotopy and homology theories with an emphasis of their applications in modern analysis, geometry, algebra and theory of numbers. Topics in topology are vast. The range of its basic topics is distributed among different topological subfields such as general topology, topological algebra, differential topology, combinatorial topology, algebraic topology and geometric topology. Each volume of the present book is considered as a separate textbook that promotes active learning of the subject highlighting elegance, beauty, scope and power of topology.

## Basic Topology—Volume 1: Metric Spaces and General Topology

This volume majorly studies metric spaces and general topology. It considers the general properties of topological spaces and their mappings. The special structure of a metric space induces a topology having many applications of topology in modern analysis, geometry and algebra. The texts of Volume 1 are expanded in eight chapters.

## Basic Topology—Volume 2: Topological Groups, Topology of Manifolds and Lie Groups

This volume considers additional structures other than topological structures studied in Volume 1, links topological structure with other structures in a compatible way to study topological groups, topological vector spaces, topological and smooth manifolds, Lie groups and Lie algebra and also gives a complete classification of closed surfaces without using the formal techniques of homology theory. Volume 2 contains five chapters.

## Basic Topology—Volume 3: Algebraic Topology and Topology of Fiber Bundles

This volume mainly discusses algebraic topology and topology of fiber bundles. The main aim of topology is to classify topological spaces up to homeomorphism. To achieve this goal, algebraic topology constructs algebraic invariants and studies topological problems by using these algebraic invariants. Because of its beauty and scope, algebraic topology has become an essential branch of topology. Algebraic topology is an important branch of topology that utilizes algebraic tools to study topological problems. Its basic aim is to construct algebraic invariants that classify topological spaces up to homeomorphism. It is found that this classification, usually in most cases, is up to homotopy equivalence.

This volume conveys a coherent introduction to algebraic topology formally inaugurated by H. Poincaré (1854–1912) in his land-marking *Analysis situs*, Paris, 1895, through his invention of fundamental group and homology theory, which are topological invariants. It studies Euler characteristic, the Betti number and also certain classic problems such as the Jordan curve theorem. It considers higher homotopy groups and establishes links between homotopy and homology theories, axiomatic approach to homology and cohomology inaugurated by Eilenberg and Steenrod. It studies the problems of converting topological and geometrical problems to algebraic one in a functorial way for better chance for solution.

This volume also studies geometric topology and manifolds by using algebraic topology. Contents of Volume 3 are expanded in seven chapters.

Just after the concept of homeomorphisms is clearly defined, the subject of topology begins to study those properties of geometric figures which are preserved by homeomorphisms with an eye to classify topological spaces up to homeomorphism, which stands the ultimate problem in topology, where a geometric figure is considered to be a point set in the Euclidean space $R^n$. But this undertaking becomes hopeless, when there exists no homeomorphism between two given topological spaces. The concepts of topological properties and topological invariants play key tools in such problems:

(a)     The concept of "topological property," such as compactness and connectedness, is introduced in general topology, which solves this problem in a very few cases, which are studied in Volume 1. A study of the subspaces of the Euclidean plane $\mathbf{R}^2$ gives an obvious example.

(b)     On the other hand, the subjects algebraic topology and differential topology (studied in Volume 2) were born to solve the problems of impossibility in many cases with a shift of the problem by associating *invariant objects* in the sense that homeomorphic spaces have the same object (up to equivalence), called *topological invariants*. Initially these objects were integers, and subsequent research reveals that more fruitful and interesting results can be obtained from the algebraic invariant structures such as groups and rings. For example, homology and homotopy groups are very important algebraic invariants which provide strong tools to study the structure of topological spaces.

# Contents

Contents

# About the Authors

**Avishek Adhikari**, Ph.D., M.Sc. (Gold Medalist), is a Professor at the Department of Mathematics, Presidency University, Kolkata, India. A recipient of the President of India Medal, ISCA Young Scientist Award, and the NANUM Fund by International Mathematical Union, Prof. Adhikari did his Ph.D. from the Indian Statistical Institute, India, under the guidance of Prof. Bimal Roy, the former Director of the Indian Statistical Institute. Earlier, Prof. Adhikari was a faculty member at the Department of Pure Mathematics, University of Calcutta, Kolkata, from July 2006–January 2019. He is the founder Secretary of the Institute for Mathematics, Bioinformatics, Information Technology and Computer Science (IMBIC), India, having branches in Sweden and Japan, and the Treasurer of the Cryptology Research Society of India. He was the former Eastern Zonal coordinator of the M.Sc. and Ph.D. scholarship examinations by the National Board of Higher Mathematics (NBHM). He was a Post-Doctoral fellow at INRIA-Rocquencourt, France, and Visiting Scientist at Linkoping University, Sweden; Indian Statistical Institute, Kolkata. Professor Adhikari delivered invited talks at various universities and institutions abroad.

He has authored four textbooks on mathematics including *Basic Modern Algebra with Applications* (Springer, 2014) and edited two volumes including *Mathematical and Statistical Applications in Life Sciences and Engineering* (Springer, 2017). His research papers have been published in reputed international journals, conference proceedings and contributed volumes. Professor Adhikari is one of the investigators of the twelve sponsored research projects funded by agencies like DRDO, WESEE (the Ministry of Defense), DST, DIT, NBHM of the Government of India, including two international collaborative projects supported by DST-JSPS and DST-JST (both Indo-Japan projects). Five of his Ph.D. students have already been awarded degrees, and two of the students have submitted their theses and, currently, four Ph.D. scholars are conducting research under his supervision.

**Mahima Ranjan Adhikari**, Ph.D., M.Sc. (Gold Medalist), is the founder president of the Institute for Mathematics, Bioinformatics, Information Technology and Computer Science (IMBIC), Kolkata, India. He is a former professor at the Department of Pure Mathematics, University of Calcutta, India. His research papers are published in national and international journals of repute, including the *Proceedings of American Mathematical Society*. He has authored nine textbooks and is the editor of two, including: *Basic Modern Algebra with Applications* (Springer, 2014), *Basic Algebraic Topology and Applications* (Springer, 2016), and *Mathematical and Statistical Applications in Life Sciences and Engineering* (Springer, 2017).

Twelve students have been awarded Ph.D. degree under his guidance on various topics such as algebra, algebraic topology, category theory, geometry, analysis, graph theory, knot theory and history of mathematics. He has visited several universities and research institutions in India, USA, UK, Japan, China, Greece, Sweden, Switzerland, Italy, and many other counties on invitation. A member of the American Mathematical Society, Prof. Adhikari is on the editorial board of several journals of repute. He was elected as the president of the Mathematical Sciences Section (including Statistics) of the 95th Indian Science Congress, 2008. He has successfully completed research projects funded by the Government of India.

# Chapter 1
# Prerequisites: Sets, Algebraic Systems and Classical Analysis

This chapter assembles together some basic concepts and results of **set theory, modern algebra, classical analysis and also of category theory by using a natural language** for smooth reading of the book. It is assumed that the readers are familiar with these basic concepts. In mathematical problems, subspaces of the $n$-dimensional Euclidean spaces arise frequently. Such spaces are used both in theory and application of topology. Some standard notations used throughout the book are also given in this chapter.

The notion of functions is the most important concept used in all branches of mathematics. Sets and functions are closely related. They have the vast potential to enrich mathematical sciences and form the basic concepts of set theory. There are two different approaches to the theory of sets: one is known as **naive set theory** created by the German mathematician G. Cantor (1845–1918) in 1870s and the other one is known as **axiomatic set theory** born through the work of Zermelo, Frankel, Skolem and others. D. Hilbert (1862–1943) wrote in 1910 "set theory is that mathematical discipline which today occupies an outstanding role in our science and radiates its powerful influence in all branches of mathematics." The basic concepts of naive set theory as well as Cantor set constructed by Cantor himself in 1883 are also given in this chapter.

For detailed study of the concepts and results given in this chapter, the books Adhikari and Adhikari (2014), Adhikari (2016), Conway (1914), Alexandrov (1979), Simmons (1963) and some other references are given in Bibliography.

## 1.1 Sets and Set Operations

This section presents the meanings of sets and subsets from the set-theoretic viewpoint of **naive set theory**. There are certain terms, called undefined terms whose meanings require no explanation. A set is such an undefined term. But mathematicians

© The Author(s), under exclusive license to Springer Nature Singapore Pte Ltd. 2022
A. Adhikari and M. R. Adhikari, *Basic Topology 1*,
https://doi.org/10.1007/978-981-16-6509-7_1

have accepted in "naive set theory" the Cantor's concept "a set is any collection of definite, distinguishable objects of our intuition or of our intellect to be conceived as a whole," which we use throughout this book. An object of a set $X$ is called an element or member of the set $X$. If $x$ is an element of the set $X$, then it is symbolized by $x \in X$. On the other hand, if $x$ is not an element of $X$, it is expressed by the symbol $x \notin X$. If one defines a set as a well-defined collection of objects, then the meaning of "collection of objects" is not clear. Again if one defines a "collection of objects," as an "aggregate of objects," then a natural question arises about the meaning of "aggregate." As other synonyms like "class," "family," etc. are limited, they will also exhaust soon. But an operational and intuitive approach is taken throughout the book to define a set as a well-defined collection of distinguishable objects.

**Example 1.1.1** Mathematicians accept the following familiar notations of the sets:

- (i)  $\mathbf{N} = \{1, 2, 3, \dots\}$: the set of all natural numbers (positive integers);
- (ii)  $\mathbf{Z}$: the set of all integers;
- (iii)  $\mathbf{Q}$: the set of all rational numbers;
- (iv)  $\mathbf{Q}^+$: the set of all positive rational numbers;
- (v)  $\mathbf{R}$: the set of all real numbers;
- (vi)  $\mathbf{I}$: the closed unit interval $[0, 1]$ in $\mathbf{R}$;
- (vii)  $\mathbf{R} - \mathbf{Q}$: the set of all irrational numbers;
- (viii)  $\mathbf{R}^+$: the set of all positive real numbers;
- (ix)  $\mathbf{C}$: the set of all complex numbers.
- (x)  $\mathbf{H}$: the set of all quaternionic numbers (quaternions).

**Definition 1.1.2** A set having no element is called the **empty set, or the void set, or the null set**, abbreviated $\emptyset$, which is unique and is considered a subset of every set.

**Definition 1.1.3** Given two sets $X$ and $A$, if every element of $A$ is an element of $X$, then $A$ is said to be a **subset** of $X$, symbolized $A \subset X$ or $X \supset A$. The symbol $\subset$ is called the inclusion relation for sets with the possibility of equality. If $X \supset A$, then $X$ is said to contain $A$ with the possibility of equality.

**Example 1.1.4** $\mathbf{N} \subset \mathbf{Z} \subset \mathbf{Q} \subset \mathbf{R}$.

**Proposition 1.1.5**   (i) *Two sets $X$ and $Y$ are equal $(X = Y)$ iff $X \subset Y$ and $Y \subset X$.*
  (ii) *Given three sets $X$, $Y$ and $Z$, the inclusion relations $X \subset Y$, $Y \subset Z$ imply $X \subset Z$.*
 (iii) *Every set $X$ is a subset of itself, written as $X \subset X$.*

**Proof** Left as an exercise.                                    □

### 1.1.1   Indexed Family of Sets

This subsection conveys the concept of an indexed family of subsets of a fixed set.

**Definition 1.1.6** Let $\mathbf{A}$ be a nonempty set. If for each $a \in \mathbf{A}$, $X_a$ is a subset of a given set $X$, then $\mathbf{A}$ is called an **indexing set**, and the family of subsets of $X$, denoted by $\{X_a\}_{a \in \mathbf{A}}$ or simply by $\{X_a\}$ is called **an indexed family of subsets of** $X$. The definition of indexed family $\{X_a\}_{a \in \mathbf{A}}$ of sets (not necessarily subsets of a given set) with $\mathbf{A}$ indexing set is similar.

**Definition 1.1.7** If $\mathbf{A}$ is a finite set given by $\mathbf{A} = \{1, 2, \ldots, n\}$, then $\{X_1, X_2, \ldots, X_n\}$ is said to be a finite collection of sets.

## 1.1.2  Set Operations: Union, Intersection and Complement

This subsection addresses set operations such as the intersection, union and complement of subset, which are the main set operations. Cartesian product of sets is also a basic set operation discussed in Sect. 1.4.4. Let $\{X_a\}$ be an indexed family of sets with an indexing set $\mathbf{A}$ (finite or infinite).

**Definition 1.1.8** **The join or union of the indexed family of sets** $\{X_a\}_{a \in \mathbf{A}}$ (written also as $\{X_a : a \in \mathbf{A}\}$) is the set $X$, defined by the property that an element $x$ is in $X$ if $x$ is in at least one member of the given family $\{X_a : a \in \mathbf{A}\}$, it is symbolized as

$$X = \bigcup \{X_a : a \in \mathbf{A}\}.$$

It is also said that $X$ is covered by $\{X_a : a \in \mathbf{A}\}$. The union of a family of sets is empty if each member of the family is empty. The join of two sets $X$ and $Y$ is denoted by $X \cup Y$.

**Definition 1.1.9** **The meet or intersection of the indexed family of sets** $\{X_a : a \in \mathbf{A}\}$ or $\{X_a\}_{a \in \mathbf{A}}$ is the set $M$ defined by the property that an element $x \in M$ if $x \in X_a$ for each $a \in \mathbf{A}$, it is symbolized as

$$M = \bigcap \{X_a : a \in \mathbf{A}\}.$$

The intersection of two sets $A$ and $B$ is denoted by $A \cap B$. The sets $A$ and $B$ are said to be disjoint if $A \cap B = \emptyset$.

**Proposition 1.1.10** *The set operations $\cup$ and $\cap$ satisfy the following properties:*

(i) $X \cap X = X$ *and* $X \cup X = X$; *(idempotent properties)*

(ii) $X \cup Y = Y \cup X$ *and* $X \cap Y = Y \cap X$; *(commutative properties)*

(iii) $(X \cup Y) \cup Z = X \cup (Y \cup Z)$ *and* $(X \cap Y) \cap Z = X \cap (Y \cap Z)$; *(associative properties)*

(iv) $X \cap (Y \cup Z) = (X \cap Y) \cup (X \cap Z)$ *and* $X \cup (Y \cap Z) = (X \cup Y) \cap (X \cup Z)$; *(distributive properties)*

(v) $X \cap (X \cup Y) = X$ *and* $X \cup (X \cap Y) = X$; *(absorptive properties)*

*(vi) Any one of the three properties:*

$$X \subset Y, \ X \cap Y = X \text{ and } X \cup Y = Y$$

*implies the other two (consistency property).*

**Remark 1.1.11** Since $\emptyset$ is a subset of every set $X$, then $\emptyset \subset X$. This asserts by part *(vi)* of Proposition 1.1.10 that $X \cap \emptyset = \emptyset$ and $X \cup \emptyset = X$, for every set $X$.

**Definition 1.1.12** Given two sets $X$ and $Y$, their **difference set** $X - Y$ (or $X \setminus Y$) is defined to be the set, consisting of all those elements of $X$, which are not elements of $Y$, in notation:

$$X - Y = \{x \in X : x \notin Y\}.$$

Clearly, the difference set $X - Y = \emptyset$ iff $X \subset Y$ and $X - Y = X$, iff $X \cap Y = \emptyset$.

**Definition 1.1.13** Given a subset $A$ of a set $X$, the difference set $X - A$ is called the **complement** of $A$ in $X$ and is written as $X - A$ or $A^c$.

**Remark 1.1.14** The difference set $X - A$ is a subset of $X$. Again, a subset $B$ of $X$ is the complement of $A$ in $X$ iff $A \cup B = X$ and $A \cap B = \emptyset$. Let $X$ be a fixed set, called the **universal set** and $A, B, C, \ldots$ be subsets of $X$. In considering the intersection, union, complements in $X$ of these subsets, the set $X$ is usually taken as the universal set.

**Proposition 1.1.15** *(De Morgan's Rules) Let $\{A_\alpha\}$ be a family of subsets of a fixed set $X$. Then the following dualization properties hold:*

$$X - \bigcap\{A_\alpha\} = \bigcup\{X - A_\alpha\}, \text{ and } X - \bigcup\{A_\alpha\} = \bigcap\{X - A_\alpha\}$$

**Proof** Left as an exercise.      □

**Corollary 1.1.16** *Given two subsets $A, B$ of $X$,*

$$X - (A \cap B) = (X - A) \cup (X - B) \text{ and } X - (A \cup B) = (X - A) \cap (X - B).$$

**Definition 1.1.17** The **union of an indexed family of subsets** $\{X_a\}_{a \in \mathbf{A}}$ of a fixed set $X$, denoted by $\bigcup X_{a \in \mathbf{A}}$, is the set defined by

$$\bigcup X_{a \in \mathbf{A}} = \{x \in X : x \in X_a \text{ for at least one index } a \in \mathbf{A}\}.$$

**Definition 1.1.18** The **intersection of an indexed family of subsets** $\{X_a\}_{a \in \mathbf{A}}$ of a fixed set $X$, denoted by $\cap X_{a \in \mathbf{A}}$, is the set defined by

$$\bigcap X_{a \in \mathbf{A}} = \{x \in X : x \in X_a \text{ for each index } a \in \mathbf{A}\}.$$

## 1.2  Basic Properties of Real Numbers

This section conveys some basic properties of the set $\mathbf{R}$ of real numbers, which play a key role in mathematics, in particular, in algebra, geometry and analysis. The theory of real analysis of a single variable is mainly based on some basic properties enjoyed by real numbers. One of them is that $\mathbf{R}$ is a complete Archimedean ordered field. The geometrical representation of $\mathbf{R}$ asserts that each point on the real line represents a unique point, and conversely, each real number represents a unique point. Moreover, $\mathbf{R}$ has its natural or usual order "$\leq$." The natural order relation on $\mathbf{R}$ defines the usual topology on $\mathbf{R}$ (see Chap. 3).

**Definition 1.2.1**  A subset $X$ of $\mathbf{R}$ is said to be **bounded above** if there is a real number $M$ such that $x \leq M, \forall x \in X$. The number $M$ is called an upper bound of $X$. A lower bound of $X$ is defined in a similar way. The subset $X$ is said to be **bounded** if it is bounded both above and below.

**Definition 1.2.2**  If $X \subset \mathbf{R}$ is bounded above, then a **least upper bound (lub) or supremum** of $X$ is a real number $M_0$, symbolized $M_0 = \sup X$ such that

(i)  $M_0$ is an upper bound for $X$;
(ii)  $M_0 \leq M$ for any upper bound $M$ for $X$.

**Example 1.2.3**  For the subset $X = \{1/n : n = 1, 2, \ldots\} \subset \mathbf{R}$, $sup\ X = 1$. The set $X \subset [0, 1]$ is bounded and its $sup\ X \in X$.

**Definition 1.2.4**  If $X \subset \mathbf{R}$ is bounded below, then a **greatest lower bound (glb) or infimum** of $X$ is a real number $m_0$, symbolized $m_0 = \inf X$ such that

(i)  $m_0$ is a lower bound for $X$;
(ii)  $m_0 \geq m$ for any lower bound $m$ for $X$.

**Example 1.2.5**

(i)  For the subset $X = (4, 5) \subset \mathbf{R}$, $\inf X = 4$, and $\sup X = 5$ and both of them are not in $X$.
(ii)  The set of all subsets of a given nonempty set $X$ is accepted to be the **power set of** $X$ and is denoted by $\mathcal{P}(X)$. Then for any family $\mathcal{C}$ of subsets of $X$, the union $\cup\{S : S \in \mathcal{C}\}$ is an upper bound and the intersection $\cap\{S : S \in \mathcal{C}\}$ is a lower bound for $\mathcal{C}$ with respect to set inclusion.
(iii)  The set $X = \{\frac{1}{x} : x \in (0, \infty)\}$ is bounded below but it has neither minimum nor maximum.
(iv)  The set $X = \{x \in \mathbf{Q} : x > 0 \text{ and } 2 < x^2 < 3\}$ has an infinite number of both upper and lower bounds but it has neither infimum nor supremum present within the set $X$.

**Remark 1.2.6**  Example 1.2.5 shows that the concepts of maximum, minimum, infimum or supremum of a nonempty set are independent. Given a nonempty subset $X \subset \mathbf{R}$, if supremum of $X$ or infimum of $X$ exists, then it must be unique.

The following fundamental properties of **R** are taken as axioms:

(i) (**Least Upper Bound Axiom**): If $X$ is any nonempty subset of **R** bounded above, then $X$ has a lub.

(ii) (**Density Axiom**): Let $r, t$ be two rational numbers such that $r < t$. Then there is an irrational number $x$ with $r < x < t$. If $r, t$ are two irrational numbers such that $r < t$, then there is a rational number $x$ with $r < x < t$.

(iii) (**Completeness Property**): Given a nonempty subset $X \subset$ **R**, if $X$ is bounded above, then it has a least upper bound lub, and hence, sup $X$ exists. Its dualization asserts that if given a nonempty subset $X \subset$ **R**, if $X$ is bounded below, then it has a greatest lower bound glb and hence inf **X** exists.

(iv) (**Archimedean Order Axiom on R**): The set **N** $= \{1, 2, \ldots\}$ of natural numbers is not bounded above in the sense that there exists no real number $r \in$ **R** such that $r > n$, $\forall n \in$ **N**.

## 1.2.1 Least Upper Bound or Completeness Property of R

This property of **R** satisfying the least upper bound axiom solves many problems such as the problem whether any nonempty subset of **R** bounded above has the lub. Least upper bound property of **R** asserts that if $X$ is any nonempty subset of **R** bounded above, then $X$ has a lub, which implies that sup **X** exists.

**Remark 1.2.7** Completeness property of **R** is the same as the least upper bound of **R** and it asserts that if a nonempty subset $X \subset$ **R** has an upper bound , then $X$ has a supremum.

**Proposition 1.2.8** *Let $X$ be a nonempty subset of* **R***, which is bounded below. Then $X$ has the greatest lower bound, i.e., inf* **X** *exists.*

**Proof** It follows by applying completeness property of **R** (see Remark 1.2.7) to the set $X$.                                                                          □

**Remark 1.2.9** Though **R** has the completeness property, **Q** does not enjoy the completeness property. Because, if $X = \{x \in$ **Q** $: x^2 < 2\}$, then $X$ has no supremum in **Q**, but it has the supremum in **R**, which is $\sqrt{2}$.

## 1.2.2 Archimedean Property of R

This subsection proves Archimedean property of **R**, which is used subsequently.

**Theorem 1.2.10** (*Archimedean property*) *Given two real numbers $x, y$ with $y > 0$, there exists a positive integer $n$ such that $ny > x$.*

**Proof** Suppose the theorem is not true. Then $ny \leq x$, $\forall n \in \mathbf{N}$. Consider the set $X = \{u \in \mathbf{R} : u = ny, \ n \in \mathbf{N}\} \subset \mathbf{R}$. Hence, $x$ is an upper bound of $X$. Then by completeness property of $\mathbf{R}$ (see Remark 1.2.7), it follows that $X$ has a least upper bound $M$ say. Then $M - y < M$, since by hypothesis $y > 0$. Hence, $M - y$ cannot be an upper bound of the set $X$. This shows that there is a real number $ny$ in $X$ such that $ny > M - y$. This implies that $(n + 1)y > M$ which is not possible, since $(n + 1)y \in X$. This contradiction proves the theorem. $\qquad \square$

**Corollary 1.2.11** *For every $x \in \mathbf{R}^+$, there exists an $n \in \mathbf{N}$ such that $n > x$.*

**Corollary 1.2.12** *For every $x \in \mathbf{R}$, there exists an $n \in \mathbf{Z}$ such that $n < x$.*

**Corollary 1.2.13** *For every $x \in \mathbf{R}^+$, there exists an $n \in \mathbf{N}$ such that $\frac{1}{n} < x$.*

**Corollary 1.2.14** *For every $x \in \mathbf{R}^+$, there exists an $n \in \mathbf{N}$ such that $n - 1 \leq x < n$.*

**Corollary 1.2.15** *Given $x, y \in \mathbf{R}^+$ with $x < y$, there exists a number $q \in \mathbf{Q}$ such that $x < q < y$.*

### 1.2.3  Denseness Property of Q in R

This subsection conveys "denseness property" of $\mathbf{Q}$ in $\mathbf{R}$, which is a fundamental property of the set $\mathbf{R}$ of real numbers.

**Theorem 1.2.16  (Denseness Property of Q)**

(i) *For any two distinct real numbers $x$ and $y$ with $x < y$, there exists a rational number $\alpha$ such that*

$$x < \alpha < y.$$

(ii) *Given any irrational number $x$, there exist rational numbers $x_n$ and $y_n$ such that*

$$x_n < x < y_n \text{ and } x_n - y_n \to 0, \text{ as } n \to \infty.$$

**Proof** Left as an exercise. $\qquad \square$

**Remark 1.2.17** Denseness property of $\mathbf{Q}$ in $\mathbf{R}$ given in Theorem 1.2.16 asserts that any real number $x$ can be approximated as close as we please by a rational number. Moreover, as there exists at least one rational number between any two distinct real numbers by this theorem, it follows that there exist infinitely many rational numbers between any two distinct real numbers. This property of $\mathbf{Q}$ is called the denseness property of $\mathbf{Q}$ in $\mathbf{R}$.

### 1.2.4 Complex Numbers: Ordered Pairs of Real Numbers

This subsection defines a complex number as an ordered pair of real numbers with certain conditions.

**Definition 1.2.18** $\mathbf{C}$ is defined as $\mathbf{R} \times \mathbf{R}$ such that

(i)
$$(a, b) + (c, d) = (a + c, \ b + d), \quad \forall a, b, c, d \in \mathbf{R};$$

and

(ii)
$$(a, b) \cdot (c, d) = (ac - bd, \ ad + bc), \quad \forall a, b, c, d \in \mathbf{R};$$

Using usual convention, $(a, 0)$ is identified with the real number $a$, because,

$$(a, 0) + (c, 0) = (a + c, 0), \text{ and } (a, 0) \cdot (c, 0) = (ac, 0),$$

and $i$ stands for $(0, 1)$. Then $i^2 = i \cdot i = -1$, and hence if $z = (a, b) \in \mathbf{C}$, then the complex number $z$ takes the form $z = a + ib$, where $a, b \in \mathbf{R}$.

## 1.3 Binary Relations on Sets

This section conveys some basic binary relations such as equivalence relation, partial ordered relation and ordered relation, which are important binary relations in mathematics.

**Definition 1.3.1** A binary relation $\rho$ on a nonempty set $X$ is a subset $\rho \subset X \times X$.

**Example 1.3.2** The diagonal $\triangle = \{(x, x) : x \in X\} \subset X \times X$ is a binary relation on a nonempty set $X$, called the **relation of equality**.

### 1.3.1 Equivalence Relation

This subsection presents the concept of an equivalence relation, which is very important to form new sets, called quotient sets, from the old one.

**Definition 1.3.3** A binary relation $\rho$ on a nonempty set $X$ is said to be an **equivalence relation** if

(i) $(x, x) \in \rho, \ \forall x \in X$ (reflexive property);
(ii) $(x, y) \in \rho$ implies $(y, x) \in \rho$ (symmetric property);
(iii) $(x, y) \in \rho$ and $(y, z) \in \rho$ imply $(x, z) \in \rho$ (transitive property).

For an equivalence relation $\rho$ on $X$, if $(x, y) \in \rho$, then the elements $x$ and $y$ are said to be $\rho$-equivalent, sometimes denoted by $x \sim y$.

**Example 1.3.4**  Define a binary relation $\rho$ on $\mathbf{Z}$ by the rule: $(x, y) \in \rho$ iff the integer $x - y$ is divisible by a fixed positive integer $n > 1$. Then $\rho$ is an equivalence relation on $\mathbf{Z}$.

**Definition 1.3.5**  Given a nonempty set $X$, the family $\mathcal{P} = \{X_i : i \in \mathbf{A}\}$ of subsets of $X$ is said to be a partition of $X$, if

(i)  each $X_i$ is nonempty;
(ii)  $X_i \cap X_j = \emptyset, \forall i, j \in \mathbf{A}$ with $i \neq j$;
(iii)  $X = \cup\{X_i : X_i \in \mathcal{P}\}$.

**Example 1.3.6**  For the set $X = \{1, 2, 5, 6, 9, 10\}$, the subsets $X_1 = \{1, 2, 5\}$, and $X_2 = \{6, 9, 10\}$, form a partition of $X$. Again $X_3 = \{1, 5, 9, 10\}$ and $X_4 = \{2, 6\}$ also form a different partition of $X$. This shows that partition of $X$ is not unique.

Theorem 1.3.7 asserts that the set of all equivalence classes on a nonempty set $X$ and the set of all partitions of $X$ are closely related.

**Theorem 1.3.7**  *Let $\rho$ be an equivalence relation on a nonempty set $X$, and for each $x \in X$, the class $[x]$ be defined by $[x] = \{y \in X : (x, y) \in \rho\}$. Then*

*(i)  $x \in [x], \forall x \in X$;*
*(ii)  for $x, y \in X$, either $[x] = [y]$ or $[x] \cap [y] = \emptyset$:*
*(iii)  $X = \cup\{[x] : x \in X\}$:*
*(iv)  if $\mathcal{P}_\rho = \{[x] : x \in X\}$, then $\mathcal{P}_\rho$ is a partition of $X$, called the partition induced by $\rho$, denoted by $X/\rho$.*

*There exists a bijective correspondence $\psi$ between the set $\mathcal{E}$ of all equivalence classes and the set of all partitions $\mathcal{P}$ on a nonempty set $X$ defined by*

$$\psi : \mathcal{E} \rightarrow \mathcal{P}, \rho \mapsto \mathcal{P}_\rho.$$

**Proof**  Left as an exercise.  □

**Definition 1.3.8**  The disjoint classes $[x]$ into which a nonempty set $X$ is partitioned by an equivalence relation "$\sim$" form a new set, called the quotient set of $X$ by $\sim$, written as $X/\sim$, where $[x]$ denotes the class containing the element $x \in X$. The element $x$ is called a representative of the class $[x]$.

**Example 1.3.9**  The quotient set $\mathbf{Z}/\sim$ defined by an equivalence relation "$\sim$" in Example 1.3.4 consists of the $n$ distinct classes $[0], [1], \ldots, [n - 1]$, denoted by $\mathbf{Z}_n$. It is called the set of residue classes of $\mathbf{Z}$ modulo $n$. It admits different algebraic structures depending on $n$.

## 1.3.2 Order Relation

The ordered sets form an important class of sets in mathematics, specially, for the study of analysis, topology and algebra. An ordered set is a nonempty set with an order relation.

**Definition 1.3.10** An order relation, abbreviated "$<$" on a nonempty set $X$, is a binary relation such that

(i) if $x, y \in X$, then one and only one of the relations holds

$$x < y, \; x = y, \; y < x$$

(ii) if $x, y, z \in X$ are the three elements with $x < y$ and $y < z$, then $x < z$.

**Remark 1.3.11** If $x < y$, then it is read as "$x$ is less than $y$" or "$x$ is smaller than $y$." Sometimes, it is convenient to write $y > x$ in place of $x < y$. The symbol $x \le y$ is used to indicate

$$x < y \text{ or } x = y.$$

**Definition 1.3.12** An ordered set $X$ is a nonempty set with an order relation $<$ defined on $X$.

**Example 1.3.13** The set $\mathbf{R}$ with an order relation $<$ defined by the rule $x < y$ if $y - x$ is a positive real number is an ordered set. The ordering $\le$ on $\mathbf{R}$ is called the natural ordering on $\mathbf{R}$.

**Definition 1.3.14** Let $X$ be an ordered set with ordering $\le$ and $A \subset X$. If there exists an element $\mu \in X$ such that $x \le \mu$, $\forall x \in A$, then $A$ is said to be bounded above and $\mu$ is said to be an upper bound of $A$. Lower bound is defined in a similar way.

**Definition 1.3.15** Let $X$ be an ordered set with order relation $<$ and $A \subset X$ be bounded above. If there exists an element $\mu \in X$ such that

(i) $\mu$ is an upper bound of $A$;
(ii) for any $\alpha < \mu$, $\alpha$ is not an upper bound of $A$. Then $\mu$ is called the least upper bound (lub) of $A$ or supremum of $A$ , denoted by $\mu = \sup A$.
(iii) The greatest lower bound (glb) or infimum $\beta$ of $A$ is defined in a similar way, and it is denoted by $\beta = \inf A$.

**Example 1.3.16** Let $A = \{1, 1/2, 1/3, 1/4, \ldots\} \subset \mathbf{R}$. Then $\sup A = 1$ and $\inf A = 0$.

**Definition 1.3.17** An ordered set $X$ is said to have the least upper bound property: if $A \subset X$, $A \ne \emptyset$ and $A$ is bounded above, then $\sup A$ exists in $X$. The greatest lower bound property is similar.

**Example 1.3.18** $\mathbf{Q}$ has no least upper bound property.

### 1.3.3  Partial Order Relation and Zorn's Lemma

Zorn's lemma supplies a very powerful tool in mathematics. Partial order relation is essential in Zorn's lemma, which is important concept to prove many results in mathematics.

**Definition 1.3.19**  A reflexive, antisymmetric and transitive relation $\rho$ on a nonempty set $P$ is called a partial order relation, usually written "$\leq$" in place of $\rho$ and the pair $(P, \leq)$ is called a **partial order set or a poset**.

**Definition 1.3.20**  Let $(P, \leq)$ be a poset such that for every pair of elements $x, y \in P$, exactly one of the relations holds:

$$x < y, \ x = y, \ y < x.$$

Then $P$ is said to be a **totally (fully or linearly) ordered set or a chain**.

**Example 1.3.21**  With natural ordering, the poset $(\mathbf{R}, \leq)$ is a totally ordered set.

**Lemma 1.3.22**  *(Zorn) Let $(P, \leq)$ be a nonempty partial order set. If every subset $X \subset P$, which is totally ordered by $\leq$, has an upper bound in $P$, then $P$ has at least one maximal element.*

**Remark 1.3.23**  Zorn's lemma is logically equivalent to the axiom of choice and well-ordering principle.

**Example 1.3.24**  With natural ordering the poset $(\mathbf{R}, \leq)$ is a totally ordered set.

**Definition 1.3.25**  Let $(P, \leq)$ be a poset and $a, b \in X$. Then an element $x \in X$ is said to be a lower bound of $a$ and $b$ if $x \leq a$ and $x \leq b$. If $x$ is a lower bound of $a, b \in X$ and $y \leq x$ for any lower bound $y$ of $a$ and $b$, then $x$ is uniquely determined and is called the greatest lower bound (glb or infimum or meet) of $a$ and $b$, denoted by $a \wedge b$. Similarly, the least upper bound (lub or supremum or join) of $a$ and $b$ denoted by $a \vee b$.

## 1.4  Functions or Mappings

This section gives the concept of functions (mappings), which is perhaps the single most important and universal notion used in all branches of mathematics. Different types of functions are studied in different areas of mathematics. For example, a function $f$ in elementary calculus is a correspondence that assigns to a real number $x$, a real number $f(x)$. The functions in algebra are polynomials, permutations, homomorphisms; in linear algebra, they are linear transformations, matrix multiplications, and in analysis, they are continuous, differentiable, integrable functions, etc. An extended version of a function is closely related to sets.

**Definition 1.4.1** Let $X$ and $Y$ be two nonempty sets. A function or a mapping $f$ from $X$ to $Y$ is a correspondence which assigns to each element $x \in X$, a unique element $y \in Y$, written, $f : X \to Y, x \mapsto y$ ( or simply, $y = f(x)$). The set $X$ is called the domain and $Y$ is called the codomain of $f$. The set $\{y \in Y : y = f(x)$, for some $x \in X\}$ is called the image or range of $f$, denoted by Im $f$ or rang $f$. If $X \subset Y$, then the map $i : X \to Y$ is said be an **inclusion map** if $i(x) = x$ for all $x \in X$, and it is symbolized as $i : X \hookrightarrow Y$.

*Example 1.4.2*

(i) $f : \mathbf{R} \to \mathbf{R}, x \mapsto x^2$ is a function, usually written in elementary calculus as $f(x) = x^2$, whose graph $G_f = \{(x, x^2)\}$ represents the parabola $y = x^2$ on the Euclidean plane.

(ii) $\det : M(n, \mathbf{R}) \to \mathbf{R}, A \mapsto \det A$ is a function, known as determinant function on the set of square matrices of order $n$ over $\mathbf{R}$.

(iii) $t : M(n, \mathbf{R}) \to M(n, \mathbf{R}), A \mapsto A^t$ is a function, known as transpose function on the set of square matrices of order $n$ over $\mathbf{R}$.

(iv) $i : \mathbf{N} \hookrightarrow \mathbf{R}, n \mapsto n$ is an inclusion function or map.

*Remark 1.4.3* All functions in elementary calculus are taken to have the same range (the real numbers, represented geometrically as $y$-axis). On the other hand, there are different ranges in algebra.

**Definition 1.4.4** A map $f : X \to Y$ is called

(i) an injective map if $x \neq x'$ in $X$ implies $f(x) \neq f(x')$ in $Y$, i.e., different elements in $X$ have different images in $Y$;

(ii) a surjective map if $f(X) = Y$, i.e., every element in $Y$ is the image of some element in $X$;

(iii) a bijective map if it is both injective and surjective.

Injective, surjective and bijective maps or functions are sometimes called injections, surjections and bijections, respectively.

*Example 1.4.5*

(i) The map $f : \to \mathbf{R} \to \mathbf{R}, x \mapsto x^2$ is neither injective nor surjective;

(ii) The determinant function $\det : M(n, \mathbf{R}) \to \mathbf{R}, A \mapsto \det A$ is not injective but surjective.

(iii) The function $f : (0, 1) \to \mathbf{R}, x \mapsto \log \frac{x}{1-x}$ is a bijection.

(iv) If $m$ and $n$ are two distinct natural numbers with $m < n$, then there exists no injective mapping

$$f : \{1, 2, \ldots, n\} \to \{1, 2, \ldots, m\} \text{ (Pigeonhole Principle)}.$$

**Definition 1.4.6 Inverse function**: Let $f : X \to Y$ be a function and $B \subset Y$ be nonempty. Then the inverse image of $B$ under $f$, abbreviated $f^{-1}(B)$, is defined to be the subset

$$f^{-1}(B) = \{x \in X : f(x) \in B\} \subset X.$$

**Remark 1.4.7** Given a subset $B \subset Y$, the subset $f^{-1}(B) \subset X$ is uniquely determined, even when $B$ is a singleton set (i.e., consists of one element only) but $f^{-1}$ cannot be considered as a function on $Y$ into $X$. Because, for an element $y \in Y$, there may not exist any element $x \in X$ such that $f(x) = y$, unless $f$ is, in particular, surjective. Again if $f$ is surjective, then the function $h$ on $Y$ onto $X$, such that $h(y) = \{x \in X : f(x) = y\}$, is said to be the inverse of $f$ and is denoted by $f^{-1}$. The function $f^{-1}$ may not however be single-valued, unless $f$ is injective. If $f$ is bijective, then $f^{-1}$ is also bijective and every element $x \in X$ has a unique image $y \in Y$, and every $y \in Y$ is the image of a unique element $x \in X$. If $f$ is bijective, then it establishes a $(1, 1)$-correspondence between the elements of $X$ and $Y$.

**Definition 1.4.8** Two sets $X$ and $Y$ are said to be equivalent denoted by $X \sim Y$, if there exists a bijection $f : X \to Y$.

**Example 1.4.9** The open intervals $(a, b)$ and $(c, d)$ are equivalent, since the mapping

$$f : (a, b) \to (c, d), x \mapsto c + \frac{d - c}{b - a}(x - a)$$

is a bijection.

**Theorem 1.4.10** *Every map $f : X \to Y$ induces an equivalence relation $\sigma(\sim)$ on $X$, such that $x\sigma x'$ holds iff $f(x) = f(x')$; and produces a bijective map*

$$\psi : X/\sigma \to f(X) \subset Y, C_x \mapsto f(x),$$

*where $C_x \in X/\sigma$ denotes the particular class to which the element $x$ belongs.*

**Proof** Left as an exercise.  □

## 1.4.1  Geometrical Examples of Functions

This subsection presents some geometrical examples of functions, which are used throughout the book. We are interested in geometrical configurations, which are subsets of Euclidean line $\mathbf{R}$ (or $\mathbf{R}^1$), Euclidean plane $\mathbf{R}^2$ or in general, Euclidean $n$-space $\mathbf{R}^n$, where there are Cartesian coordinates having each point is expressed uniquely by coordinates of an ordered $n$-tuple of real numbers $(x_1, x_2, \ldots, x_n)$. In particular, it represents a point in the Euclidean line $\mathbf{R}$ (or $\mathbf{R}^1$) for $n = 1$; a point in the Euclidean plane $\mathbf{R}^2$ for $n = 2$ or in general, a point in the Euclidean $n$-space $\mathbf{R}^n$.

**Example 1.4.11**  (i) (**Translation**) A translation $T_a$ by $a \in \mathbf{R}$ on the Euclidean line $\mathbf{R}$ is a function

$$T_a : \mathbf{R} \to \mathbf{R}, x \mapsto x + a.$$

This translation carries a point $x \in \mathbf{R}$ to the point $x + a \in \mathbf{R}$. Similarly, given $a, b \in \mathbf{R}$, a translation $T : \mathbf{R}^2 \to \mathbf{R}^2$, $(x, y) \mapsto (x + a, y + b)$ carries the point $(x, y) \in \mathbf{R}^2$ to the point $(x + a, y + b) \in \mathbf{R}^2$.

(ii) **(Rotation)** A rotation of the plane $\mathbf{R}^2$ is described by specifying its center and the angle of rotation. For example,

$$R_\theta : \mathbf{R}^2 \to \mathbf{R}^2, \ (x, y) \mapsto (x \cos\theta - y \sin\theta, \ x \sin\theta + y \cos\theta)$$

represents a rotation of the plane $\mathbf{R}^2$ about its origin through an angle $\theta$.

(iii) **(Reflection)** Reflection on a line $\mathbf{l}$ in $\mathbf{R}^2$ is a continuous function $R_1 : \mathbf{R}^2 \to \mathbf{R}^2$, which keeps every point of the line $\mathbf{l}$ fixed and takes the mirror image of points of the two sides of $\mathbf{l}$, i.e., interchanges the the two sides of the line $\mathbf{l}$.

***Remark 1.4.12*** Any translation, rotation, reflection or any reflection followed by a translation described in Example 1.4.11 keeps the shape and size of the configurations in the plane unchanged but their positions and sizes may be altered. Such functions are the **congruences** in elementary geometry.

**Definition 1.4.13** A **rigid motion** on $\mathbf{R}^n$ with Euclidean distance function $d$ is a continuous function $f : \mathbf{R}^n \to \mathbf{R}^n$ such the distance between every pair of points is preserved under $f$ in the sense that

$$d(f(x), f(y)) = d(x, y), \ \forall x, y \in \mathbf{R}^n$$

(see Chap. 2).

***Example 1.4.14*** Every translation, rotation, reflection or any reflection followed by a translation described in Example 1.4.11 is a rigid motion.

### 1.4.2 Geometrical Examples of Bijections

This subsection presents some geometrical examples of bijections.

***Example 1.4.15***    (i) Every translation, rotation, reflection or any reflection followed by a translation described in Example 1.4.11 is a bijection. The translation

$$T_a : \mathbf{R} \to \mathbf{R}, x \mapsto x + a$$

has its inverse which is $T_{-a}$. The rotation

$$R_\theta : \mathbf{R}^2 \to \mathbf{R}^2, \ (x, y) \mapsto (x \cos\theta - y \sin\theta, \ x \sin\theta + y \cos\theta)$$

has its inverse $R_{-\theta}$.

(ii) Every rigid motion of the plane is a bijection.

(iii)  Let $S^2 = \{(x_1, x_2, x_3) \in \mathbf{R}^3 : x_1^2 + x_2^2 + x_3^2 = 1\}$ be the unit sphere in $\mathbf{R}^3$ and $\mathbf{N} = (0, 0, 1)$ be the north pole of $S^2$. Then $S^2 - \mathbf{N}$ and $\mathbf{R}^2$ are equivalent sets, since the map

$$f : S^2 - \mathbf{N} \to \mathbf{R}^2, (x_1, x_2, x_3) \mapsto \frac{1}{1 - x_3}(x_1, x_2)$$

is a bijection, called **stereographic projection**.

## 1.4.3   Restriction, Extension and Composition of Functions

This subsection gives the concepts of restriction, extension and usual composition of functions.

**Definition 1.4.16**  Let $f : X \to Y$ be a function and $A \subset X$ be nonempty. Then $f(A) \subset f(X) \subset Y$. The function

$$r : A \to f(A), \ x \mapsto f(x), \ \forall x \in A$$

is called the **restriction** of $f$ to $A$, denoted by $f|_A$.

**Definition 1.4.17**  Let $A \subset X$ and $f : A \to Y$ be a given function. Then a function $\tilde{f} : X \to Y$ is said to be an **extension** of $f$ over $X$ if

$$\tilde{f} : X \to A : \tilde{f}(x) = f(x), \ \forall x \in A.$$

An extension $\tilde{f}$ of a function $f : A \subset X \to Y$ is not unique. Because, the values of $\tilde{f}(x)$ may be arbitrarily chosen for $x \in X - A$. On the other hand, the restriction of a function to subset is unique.

**Definition 1.4.18**  (**Composition of Functions**): Let $f : X \to Y$, $g : Y \to Z$ be two functions. Then the function $h : X \to Z$, $x \mapsto g(f(x))$ is called the composite of the maps $f$ and $g$, denoted by $g \circ f$.

**Proposition 1.4.19** (*Associativity of Composition of Functions*): *Let* $f : X \to Y$, $g : Y \to Z$ *and* $h : Z \to W$ *be three functions. Then their composition maps* $(h \circ g) \circ f$ *and* $h \circ (g \circ f)$ *are well defined and they are equal.*

**Proof**  It follows from the Definition 1.4.18 of the composition of functions.   □

### *1.4.4   Cartesian Product of Any Collection of Sets*

This subsection conveys the concept of Cartesian product of sets born through the coordinate plane of the analytical geometry along with its generalization for any family of sets.

**Definition 1.4.20**  Given two nonempty sets $X$ and $Y$,

$$X \times Y = \{(x, y) : x \in X, y \in Y\}$$

is called the **Cartesian product** of $X$ and $Y$.

***Example 1.4.21***   (i) $\mathbf{I} \times \mathbf{I}$ represents geometrically the set of all points in the unit square $\mathbf{I}^2$.
(ii) Let $S^1$ be the unit circle. Then $S^1 \times \mathbf{I}$ represents geometrically the set of all points in the hollow unit cylinder over $S^1$.
(iii) $\mathbf{R} \times \mathbf{R}$ represents the points in the Euclidean plane with usual distance function (see Chap. 2).

Definition 1.4.20 is now extended to the finite product of $n$ $(n > 2)$ nonempty sets $X_1, X_2, \ldots, X_n$.

**Definition 1.4.22**  Given a finite collection of nonempty sets $X_1, X_2, \ldots, X_n$, their Cartesian product (or combinatorial product), denoted by $X_1 \times X_2 \times \cdots \times X_n$, is the set defined by

$$X_1 \times X_2 \times \cdots \times X_n = \{(x_1, x_2, \ldots, x_n) : x_i \in X_i, \ i = 1, 2, \ldots, n\}.$$

In particular, for $X_1 = X_2 = \cdots = X_n = X$, their Cartesian product symbolized by $X^n$, yields well-known sets such as $\mathbf{R}^n$ and $\mathbf{C}^n$.

The Definition 1.4.22 of Cartesian product can be extended in Definition 1.4.23 for any collection of sets.

**Definition 1.4.23**  Let $\{X_a\}_{a \in \mathbf{A}}$ be an arbitrary collection of sets (the sets $X_a$ may not be different). The product of the sets denoted by $\Pi_{a \in \mathbf{A}} X_a$ is defined to be the set of all functions $\psi$ with domain $\mathbf{A}$ such that $\psi(a)$ is an element in $X_a$ for each index $a \in \mathbf{A}$. The set $X_a$ is called the $a$-th coordinate set. It is assumed that if any set $X_a = \emptyset$, then the product set $\Pi_{a \in \mathbf{A}} X_a$ is also $\emptyset$. So we take each coordinate set $X_a$ is nonempty for nonempty product. If the indexing set $\mathbf{A} = \emptyset$, then the Cartesian product of the sets $\{X_a\}$ is taken to be empty.

### 1.4.5 Choice Function and Axiom of Choice

This subsection defines choice function and describes the axiom of choice, which has wide applications in different areas of mathematics such as in set theory, analysis, topology and algebra.

**Definition 1.4.24** (**Choice Function**) Let $\{X_a : a \in \mathbf{A}\}$ be any nonempty family of nonempty sets. Then there exists a function $\psi : \mathbf{A} \to \cup_{a \in \mathbf{A}} X_a$, with the property that $\psi(a) \in X_a$, $\forall a \in \mathbf{A}$. The map $\psi$ is called a choice function for the family $\{X_a : a \in \mathbf{A}\}$ of sets.

**Remark 1.4.25** One cannot prove the existence of a choice function for an arbitrary family of sets. To avoid this difficulty, we take recourse to an axiom, known as the axiom of choice. The axiom of choice has wide applications in different areas of mathematics. **The axiom of choice** formulated by E. Zermelo in 1904 asserts that for any family of sets $\{X_a : a \in \mathbf{A}\}$, such that $\mathbf{A} \neq \emptyset$ and $X_a \neq \emptyset$ for each $a \in \mathbf{A}$, there always exists at least one choice function. If the indexing set $\mathbf{A} = \emptyset$, then the null function is the only possible choice function. On the other hand, if one of the sets $X_a$ is empty, then the Cartesian product of the sets $\{X_a\}$ is empty. G. Peano (1858–1932) introduced this axiom in 1889 for natural number (in another form), known as the **principle of mathematical induction**.

## 1.5 Cantor Set

This section constructs the **Cantor set** $C$ obtained by successively deleting the open middle one-third of the closed interval $\mathbf{I} = [0, 1] \subset \mathbf{R}$, defined by G. Cantor in 1883. It is named after him. It is a basic uncountable set used in analysis, topology and geometry. It conveys the richness of the structure of real number system. Its length is zero, because its complement in the unit interval is of length unity. The importance of the set can be realized in the study of metric spaces and in illustration of several concepts discussed in subsequent chapters. For example, the Cantor set $C$ is totally disconnected, and it is compact (see Chap. 5).

### 1.5.1 Construction of the Cantor Set

This subsection describes construction of the Cantor set. Denote the closed unit interval $\mathbf{I} = [0, 1] \subset \mathbf{R}$ by $\mathbf{I}_0$. Delete from $\mathbf{I}$ the middle third $(1/3, 2/3)$ and denote by $\mathbf{I}_1$ the remaining closed set, i.e., $\mathbf{I}_1 = [0, 1] - (1/3, 2/3) = [0, 1/3] \cup [2/3, 1]$. Again delete from $\mathbf{I}_1$ the middle thirds of its two parts $(1/9, 2/9)$ and $(7/9, 8/9)$ and denote by $\mathbf{I}_2$, the remaining closed set, i.e., $\mathbf{I}_2 = ([0, 1/3] - (1/9, 2/9)) \cup ([2/3, 1] - (7/9, 8/9)) = [0, 1/9] \cup [2/9, 1/3] \cup [2/3, 7/9] \cup [8/9, 1]$. If we go on continuing this process as shown in Fig. 1.1 by deleting at every stage the open middle third of every closed interval remaining from the previous stage, a sequence of closed sets $\{I_n\}$ is obtained such that each $I_n$ contains all its successors.

**Definition 1.5.1**  The set $C$ obtained by

$$C = \bigcap_{n=1}^{\infty} I_n$$

is called the **Cantor set**.

**Remark 1.5.2**  The Cantor set $C$ consists of those points in $I = [0, 1]$ which ultimately survive after deleting all the open intervals $(1/3, 2/3)$, $(1/9, 2/9)$, $(7/9, 8/9)$, . . . . This asserts that $C$ contains the end points of the closed intervals which constitute each set $I_n$:

$$0, 1, 1/3, 2/3, 1/9, 2/9, 7/9, 8/9, \ldots .$$

The point 1/4 is also in $C$ but not an end point of any one of the closed intervals which build up $I_n$. The cardinal number $|C|$ of $C$ is $c$, which is the cardinal number of the continuum.

## 1.5.2   Geometrical Method of Construction of the Cantor Set

**The Cantor's ternary set** may be constructed by an interesting geometrical method as described in Fig. 1.1. The closed unit interval [0, 1] is divided into the three equal parts, and the middle open segment $(1/3, 2/3)$ is deleted. The remaining closed subsegments [0, 1/3] and [2/3, 1] are again trisected, and their middle open segments, viz. $(1/3^2, 2/3^2)$ and $(2/3 + 1/3^2, 2/3 + 2/3^2)$, are deleted. The remaining 4 closed subsegments are then trisected, and their middle open segments are deleted. This process is continued ad infinitum, and the points, which survive, constitute Cantor's ternary set, denoted by $C$.

**Definition 1.5.3  The Cantor's ternary set** $C$ is defined to be the set of all those real numbers $x$ which are expressible in the form,

$$x = x_1 + x_2/3^2 + \cdots + x_n/3^n + \ldots \tag{1.1}$$

where $x_1, x_2, \ldots, x_n, \ldots$ assume the value 0 and 2 only.

**Remark 1.5.4 (Geometrical Representation of Cantor Set)** The closed unit interval [0, 1] is divided into the three equal parts, and the middle open segment $(1/3, 2/3)$ is deleted. The remaining closed subsegments [0, 1/3] and [2/3, 1] are again trisected, and their middle open segments, viz. $(1/3^2, 2/3^2)$ and $(2/3 + 1/3^2, 2/3 + 2/3^2)$, are deleted . The remaining 4 closed subsegments are then trisected, and their

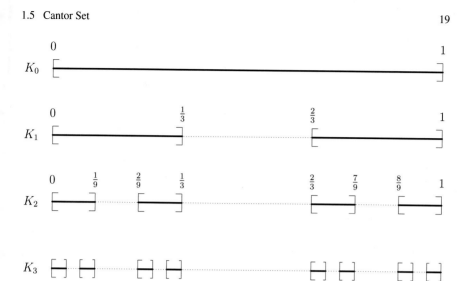

**Fig. 1.1** Geometrical construction of the Cantor set

middle open segments are deleted. This process is continued ad infinitum as shown in Fig. 1.1 and the points, which survive, constitute the Cantor's ternary set, denoted by $C$. The Cantor's ternary set $C$ is also defined to be the set of all those real numbers $x$ which are expressible in the form,

$$x = x_1 + x_2/3^2 + \cdots + x_n/3^n + \cdots \tag{1.2}$$

where $x_1, x_2, \ldots, x_n, \ldots$ assume the value 0 and 2 only.

**Example 1.5.5** Let $C$ be the Cantor set. Then the set

$$A = \{x + y : x, y \in C\} = [0, 2].$$

Because, $C \subset [0, 1]$ and hence $A \subset [0, 2]$. Conversely, for any element $a \in A$, there exist two elements $x, y \in C$ such that $x + y = a$.

## 1.6 Countability and Cardinality of Sets

This section conveys the concepts of countability and cardinality of sets, which play a key role in the study of mathematics.

### 1.6.1  Countability of Sets

We use positive integers $1, 2, \ldots$, for counting in daily life that are the elements of a finite set but there are many sets which are not finite (called infinite sets) such as $\mathbf{N}, \mathbf{Z}, \mathbf{Q}, \mathbf{R}$, etc. For example, the set $\mathbf{R}$ is infinite in the sense that the set of points in the real line $\mathbf{R}$ is not finite. Cantor described a method for counting infinite sets by introducing the concepts of cardinality of sets, which plays a key role in the study of algebra, analysis, topology and in many areas, and this concept carries a great aesthetic appeal developed into a natural structure of thought.

**Definition 1.6.1** Two sets $X$ and $Y$ are said to be equivalent, denoted $X \sim Y$, if there exists a bijection $f : X \to Y$.

**Example 1.6.2** Let $S^2 = \{(x, y, z) \in \mathbf{R}^3 : x^2 + y^2 + z^2 = 1\}$ be the unit sphere in the Euclidean space $\mathbf{R}^3$ and $\mathbf{C}_\infty = \mathbf{C} \cup \{\infty\}$ be the extended complex plane. Then the stereographic projection $f : S^2 \to \mathbf{C}_\infty$ defined in Example 1.4.15 is a bijection. Thus as sets $\mathbf{C}_\infty$ are equivalent to the sphere $S^2$, called the **Riemann sphere**.

**Definition 1.6.3** A set $X$ which is equivalent to the set $\mathbf{N}$ of natural numbers is said to be an **enumerable (or denumerable)** set. The set $X$ is said to be **countable**, if it is either finite or enumerable.

**Example 1.6.4** The set of rational numbers $\mathbf{Q}$ is countable.

**Example 1.6.5** If $\mathbf{A}$ is a countable indexing set given by $\mathbf{A} = \mathbf{N} = \{1, 2, 3, \ldots\}$, then $\{X_1, X_2, X_3, \ldots\}$ is said to be a countable collection (family) of sets.

**Definition 1.6.6** Given a nonempty set $X$, a map $f : \mathbf{N} \to X$ is said to be a **sequence** in $X$. It is usually written as $f(n) = x_n \in X, \forall n \in N$, and the sequence $f$ is symbolized by $\{x_n\}$.

### 1.6.2  Cardinality of Sets

The concept of counting for sets is introduced by assigning to every set $X$ (finite or infinite), an object, called the cardinal number or cardinality of $X$, denoted by $|X|$ with the property that $|X| = |Y|$ iff $X \sim Y$. Cardinality of a set plays a key role in mathematics, and it extends the concept of counting.

**Example 1.6.7** If $X$ is a finite set of $n$ distinct elements, then $|X| = n$.

**Definition 1.6.8** If $X$ is an infinite set, then $|X|$ is called a **transfinite cardinal number**. In particular, $|\mathbf{N}| = d$ and $|\mathbf{R}| = c$.

**Example 1.6.9** An infinite set $X$ has the cardinal number $d$ or $c$ according as $X$ is infinitely countable or not so. As $|\mathbf{R}| \neq d$, and $|\mathbf{R}| = c$, the cardinal number $c$ stands for the power of continuum. Thus the cardinal number of an infinite set $X$ is $d$ or $c$ asserts that $X$ is infinite countable or not.

**Definition 1.6.10** Two sets $X$ and $Y$ are said to have the same cardinality if there exists a bijection $f : X \to Y$.

**Example 1.6.11** (i) The sets $\mathbf{N} \times \mathbf{N}(= \mathbf{N}^2)$ and $\mathbf{N}$ have the same cardinality.
(ii) Any two concentric circles in the Euclidean plane have the same cardinality.
(iii) The sets $S^2 - \mathbf{N}$ and $\mathbf{R}^2$ have the same cardinality, because the map $f : S^2 - \mathbf{N} \to \mathbf{R}^2$, $(x_1, x_2, x_3) \mapsto \frac{1}{1-x_3}(x_1, x_2)$ is a bijection.
(iv) The open interval $(0, 1)$ and the real line $\mathbf{R}$ have the same cardinality, because the function $f : (0, 1) \to \mathbf{R}$, $x \mapsto \log \frac{x}{1-x}$ is a bijection.

A cardinal number is assigned to every set and signifies a property which is common to all sets which are equivalent with each other.

**Definition 1.6.12** The **cardinal number** of a set $X$ is denoted by $|X|$ or $\text{card}(X)$ and has the characteristic property: for two sets $X$ and $Y$, the equality $\text{card}(X) = \text{card}(Y)$ holds if and only if $X \sim Y$.

**Remark 1.6.13** Two finite sets are evidently equivalent if and only if they have the same number of elements. A finite set, consisting of $n$ elements, is assigned the number $n$ as its cardinal number (for $n = 1, 2, \ldots$). The cardinal number of the null set $\emptyset$ is taken to be the number 0. The cardinal number of an infinite set is sometimes called a **transfinite cardinal number**. In particular, the cardinal number assigned to the set of natural numbers $\mathbf{N}$ (a denumerable set) is denoted by $d$. The cardinal number assigned to the set of real numbers $\mathbf{R}$, called the **power (or potency) of the continuum**, is denoted by $c$.

**Definition 1.6.14** A set $X$ is said to be **nondegenerative** if $\text{card}(X) > 1$.

**Example 1.6.15** $\mathbf{N}, \mathbf{Q}$ and $\mathbf{R}$ are important examples of nondegenerative sets.

**Theorem 1.6.16** *Every infinite set has infinitely many countable subsets.*

**Proof** Let $X$ be an infinite set. We claim that, for each natural number $n$, there exists a subset $X_n \subset X$, such that $|X_n| = n$. As $X \neq \emptyset$, there exists a one-pointic subset $X_1 \subset X$. Let $X_i \subset X$, where $|X_i| = i$. As $X$ is an infinite set, $X - X_i \neq \emptyset$; and let $x \in X - X_i$. Then $X_{i+1} = X_i \cup \{x\} \subset X$, and $|(X_{i+1})| = i + 1$. Hence, by the principle of mathematical induction, there exists a subset $X_n \subset X$, with $|(X_n)| = n$, for each natural number $n$. Define a family of sets $Y_n$ by

$$Y_n = X_{2^n} - (X_1 \cup X_2 \cup X_{2^2} \cup \cdots \cup X_{2^{n-1}}), \text{ for } n = 1, 2, \ldots.$$

Then $Y_1 = X_2 - X_1$, $Y_2 = X_4 - (X_1 \cup X_2)$, .... This produces a family $\{Y_n : n \in \mathbf{N}\}$ of pairwise disjoint nonempty subsets of $X$ such that $|Y_n| \geq 2^n - (2^n - 1) = 1$. Let $f$ be a choice function for the nonempty family of nonempty sets $\{Y_n : n \in \mathbf{N}\}$. Then $f : \mathbf{N} \to X$ is injective, which asserts that $\text{Im } f$ (Image of $f$) is an enumerable subset of $X$. $\qquad\square$

**Theorem 1.6.17** *Any subset of a countable set is countable.*

**Proof** Let $A$ be a subset of $X$ of a countable set $X$. If $A$ is finite; then it is obviously countable. If $A$ is an infinite subset of $X$, then as $X$ is infinitely countable, there exists a function $f : \mathbf{N} \to A$ defined recursively:

$f(1) = x_{n_1}$, where $n_1$ is the smallest positive integer $n$ such that $a_n \in A$,
$f(m + 1) = x_{n_{m+1}}$, where $n_{m+1}$ is the smallest positive integer $n$ such that $a_n \in A - \{a_1, a_2, \ldots, a_{n_m}\}$,

This shows that $f$ is a bijection and hence $A$ is countable.                                 □

**Theorem 1.6.18** *The Cartesian product of two countable sets is countable.*

**Proof** Let $X$ and $Y$ be two countable sets, and let $P$ denote their Cartesian product. Now, if both the sets $X$ and $Y$ are finite, then $P$ is also a finite set and hence also countable. Next, let $X = \{x_1, \ldots, x_r\}$ be a finite set, and $Y$ a denumerable set with $Y = \{y_1, y_2, \ldots, y_n, \ldots\}$ an enumeration of $Y$. Then the elements of the Cartesian product $P$ can be arranged as a sequence

$$(x_1, y_1), \ldots, (x_r, y_1); (x_1, y_2), \ldots, (x_r, y_2); \ldots, (x_1, y_n), \ldots, (x_r, y_n); \ldots;$$

and $P$ is therefore a denumerable set (and hence also countable). Finally, let both $X$ and $Y$ be denumerable, and $X = \{x_1, x_2, \ldots\}$ and $Y = \{y_1, y_2, \ldots\}$ be their enumerations. Then the elements of their Cartesian product $P$ can be arranged as a sequence,

$$(x_1, y_1); (x_1, y_2), (x_2, y_1); \ldots; (x_1, y_n), (x_2, y_{n-1}), \ldots, (x_n, y_1); \ldots,$$

where the element $(x_i, y_j)$ precedes the element $(x_m, y_n)$ iff either $i + j < m + n$, or $i < m$ when $i + j = m + n$. This defines a bijection $f : P \to \mathbf{R}$ (for its explicit expression proceed as in the proof of Theorem 1.6.19).                         □

**Theorem 1.6.19** *The union of any countable family of countable sets is countable.*

**Proof** First, let $X_1, X_2, \ldots, X_n$ be a finite collection of denumerable sets, and let $\{x_{r1}, x_{r2}, \ldots, x_{rn}, \ldots\}$ be an enumeration of the set $X_r$, for $r = 1, 2, \ldots, n$. Then their union consists of the elements $\{x_{ij}\}$, for $i = 1, 2, \ldots, n$ and $j = 1, 2, \ldots, n, \ldots$ (or a subset of this collection, in case the given sets are not pairwise disjoint). The elements $x_{ij}$ can be arranged as a sequence:

$$x_{11}, x_{21}, \ldots, x_{n1}; x_{12}, x_{22}, \ldots, x_{n2}; \ldots; x_{m1}, x_{m2}, \ldots, x_{mn}; \ldots,$$

where $x_{ij}$ precedes $x_{rs}$ if, either $j < s$, or $i < r$ when $j = s$. In fact, $f(x_{ij}) = n(j - 1) + i$ is a collection of the $\{x_{ij}\}$ onto $\mathbf{N}$. Next, let $X_1, X_2, \ldots, X_n, \ldots$ be a denumerable family of denumerable sets. Let $\{x_{r1}, x_{r2}, \ldots, x_{rn}\}$ be an enumeration of the elements of the set $X_r$, for $r = 1, 2, \ldots$. Then the elements $\{x_{ij}\}$, for $i = 1, 2, \ldots$, and $j = 1, 2, \ldots$, can be arranged as a sequence,

$$x_{11}; x_{12}, x_{21}; x_{13}, x_{22}, x_{31}; \ldots; x_{1n}, x_{2,n-1}, \ldots, x_{n1}; \ldots,$$

where $x_{ij}$ precedes $x_{mn}$ iff either $i + j$, $m + n$, or $i < m$ when $i + j = m + n$. Then $f(x_{ij}) = [1 + 2 + \cdots + (i + j - 2)] + i$ is a bijection of the set of elements $\{x_{ij} : i = 1, 2, \ldots, j = 1, 2, \ldots\}$ onto $\mathbf{N}$. The union of the given family of sets is given by the denumerable set of elements $\{x_{ij} : i = 1, 2, \ldots, j = 1, 2, \ldots\}$ or a subset of this set; hence the union is countable.                                                        □

**Theorem 1.6.20**  *Each of the following sets is countable:*

(i)   *the set of all integers $\mathbf{Z}$ is countable;*
(ii)  *the set of all rational numbers $\mathbf{Q}$ is countable;*
(iii) *the set of all polynomials with rational coefficients is countable;*
(iv)  *A real number is said to be a (real) **algebraic number**, if it is a root of some (nonnull) polynomial with rational coefficients (or, equivalently, with integral coefficients). The set of all (real) algebraic numbers is countable.*

***Proof***  (i) $\mathbf{Z} = \mathbf{N} \cup \{0\} \cup \{-n : n \in \mathbf{N}\}$; hence $\mathbf{Z}$ is countable.

(ii) The set $\mathbf{Q}_r^+ = \{1/r, 2/r, \ldots, n/r, \ldots\}$ is denumerable, for $r = 1, 2, \ldots$. So, $\mathbf{Q}_r = \mathbf{Q}_r^+ \cup \{0\} \cup \{-n/r : n \in N\}$ is also a denumerable set, for $r = 1, 2, \ldots$. Hence, $\mathbf{Q}_1 \cup \mathbf{Q}_2 \cup \cdots \cup \mathbf{Q}_n \cup \ldots$ is also denumerable and is therefore countable. The set of rational numbers $\mathbf{Q}$ is a subset of the set $\mathbf{Q}_1 \cup \mathbf{Q}_2 \cup \ldots$; hence the set $\mathbf{Q}$ is also countable.

(iii) Let $P_n$ denote the set of all polynomials of degree $n$ and with rational coefficients (of which the leading coefficient $\neq 0$). Hence $P_n$ is equipotent with the Cartesian product $\mathbf{Q}_1 \times \mathbf{Q}_2 \times \cdots \times \mathbf{Q}_n \times [\mathbf{Q} - \{0\}]$, where $\mathbf{Q}_1 = \mathbf{Q}_2 = \cdots = \mathbf{Q}_n = \mathbf{Q}$ denotes the set of all rational numbers. $P_n$ is therefore denumerable (by generalization of Theorem 1.6.18). Hence the union, $P_0 \cup P_1 \cup P_2 \cup \cdots \cup P_n \cup$, is also denumerable; in other words, the set of all polynomials with rational coefficients is countable.

(iv) Let $P$ denote the set of all polynomials with rational coefficients. Then $P$ is a countable set, by (iii). Again every polynomial belonging to $P$ has only a finite number of real roots, not exceeding the degree of the polynomial. Hence, the set of all real algebraic numbers being a subset of the denumerable set formed by the union of a denumerable collection of finite sets is necessarily a countable set.

□

**Definition 1.6.21**  An infinite set, which is not denumerable, is called an **uncountable set**.

**Example 1.6.22**  (i) The set of all real numbers $\mathbf{R}$ is uncountable.
(ii)  A (real) number, which is not algebraic, i.e., which is not a root of any (nonnull) polynomial with rational coefficients, is called a (real) **transcendental number**. The set of all (real) transcendental numbers is uncountable.
(iii) The Cantor set $C$ is uncountable.

### 1.6.3 Order Relations on Cardinal Numbers

The concept of natural ordering of positive integers is now extended to cardinal numbers.

**Definition 1.6.23** Let $\alpha$ and $\beta$ be two cardinal numbers and $A$ and $B$ be two sets, such that $|A| = \mathrm{card}(A) = \alpha$ and $|B| = \mathrm{card}(B) = \beta$. A binary relation "$\leq$" (equivalently, $\geq$) is defined on the set of cardinal numbers by the rule that $\alpha \leq \beta$ (or $\beta \geq \alpha$) if the set $A$ is equipotent to a subset of $B$. Again $\alpha < \beta$ ($\beta > \alpha$), if, $\alpha \leq \beta$ and $\alpha \neq \beta$.

**Theorem 1.6.24** *Let* $\alpha$, $\beta$ *and* $\gamma$ *be cardinal numbers. Then*

(i) $\alpha \leq \alpha$;
(ii) $\alpha \leq \beta$ *and* $\beta \leq \gamma$ *imply* $\alpha \leq \gamma$.

**Proof** Let $A, B$ and $C$ be three sets, such that $\mathrm{card}(A) = \alpha$, $\mathrm{card}(B) = \beta$ and $\mathrm{card}(C) = \gamma$.

(i) Since the identity mapping of $A$ onto $A$ is a bijection (and $A$ is a subset of itself), it follows that $\alpha \leq \alpha$.
(ii) By hypothesis, there exists a bijection $f$ of $A$ onto a subset $B_1 \subset B$ and also a bijection $g$ of $B$ onto a subset $C_1 \subset C$. Let $g(B_1) = C_2$, then $C_2 \subset C_1 \subset C$, and the composite $g \circ f$ is a bijection of $A$ onto the subset $C_2 \subset C$. Hence $\alpha \leq \gamma$.

$\square$

**Corollary 1.6.25** "$\leq$" *is an order relation on the set of cardinal numbers.*

**Proof** The relation "$\leq$" is both reflexive and transitive by Theorem 1.6.24. It is not symmetric, since if $A$ and $B$ are two finite sets with 3 and 5 elements, respectively, then $\mathrm{card}(A) = 3$ and $\mathrm{card}(B) = 5$, where $3 \leq 5$, but $5 \leq 3$ does not hold. $\square$

**Theorem 1.6.26** *(Schroeder and Bernstein) If $A$ and $B$ are two sets, such that $A \sim B_1 \subset B$ and $B \sim A_1 \subset A$, then $A \sim B$.*

**Proof** To prove the theorem use the result which asserts that if $A_2 \subset A_1 \subset A$, and $A \sim A_2$, then $A \sim A_1$. $\square$

**Corollary 1.6.27** *The relation "$\leq$" is antisymmetric.*

**Proof** It follows from Schroeder and Bernstein Theorem 1.6.26. $\square$

**Corollary 1.6.28** Let $\mathbf{I} = [0, 1]$. Then $\mathbf{I}^n = \{(x_1, x_2, \ldots x_n) : x_i \in \mathbf{I}, \ i = 1, 2, \ldots, n\}$ is called the n-**dimensional unit cube**. Then $\mathbf{I} \sim \mathbf{I}^n$.

### 1.6.4  Sum, Product and Powers of Cardinal Numbers

This subsection defines sum, product and power of cardinal numbers.

**Definition 1.6.29**  Let $\alpha$ and $\beta$ be two cardinal numbers, and let $X$ and $Y$ be two sets such that card$(X) = \alpha$ and card$(Y) = \beta$. If then their sum $\alpha + \beta$, product $\alpha \cdot \beta$ and power $\alpha^\beta$ are defined by the rules:

$$\alpha + \beta = \text{card}(X \cup Y), \ if \ X \cap Y = \emptyset;$$
$$\alpha \cdot \beta = \text{card}(X \times Y), \ if \ X \neq \emptyset \ and \ Y \neq \emptyset;$$
$$\alpha^\beta = \text{card}(X^Y),$$

where $X^Y$ denotes the set of all functions $f : Y \to X$.

## 1.7  Well-Ordered Sets and Ordinal Numbers

This section introduces the concept of ordinal numbers of a well-ordered set, which is a concept different cardinal numbers, because, every set (not necessarily, ordered set) has a unique cardinal number; on the other hand, only well-ordered sets have ordinal numbers. Since a nonempty set can be well-ordered in essentially different ways, and hence correspondingly different ordinal numbers are assigned to the same set; on the other hand a set has a unique cardinal number.

**Definition 1.7.1**  Let $(X, <)$ be an ordered set. An element $a \in X$ is called the first element in $(X, <)$, if $a < b$ for every element $b (\neq a) \in X$. An element $c \in X$ is said to be last element in $(X, <)$, if $b < c$ for every element $b (\neq c) \in X$. If they exist, they are unique.

**Definition 1.7.2**  An ordered set $X$ is said to be well-ordered if every nonempty subset of $X$ has a first element.

**Definition 1.7.3**  Let $(X, <_X)$ and $(Y, <_Y)$ be two ordered sets with their respective order relations $<_X$ and $<_Y$. A bijective mapping

$$f : X \to Y$$

is said to an order-isomorphism, if

$$x_1 <_X x_2 (\text{in } X) \implies f(x_1) <_Y f(x_2)(\text{in } Y).$$

**Remark 1.7.4**  As the relation of an order-isomorphism is an equivalence relation on any collection of ordered sets, this collection is accordingly partitioned into disjoint classes, and each such class is called an **order type**. It asserts that every ordered set belongs to an order type or it is usually said that every ordered set has an ordered type. Two sets have the same order type if, and only if, they are order-isomorphic. This leads to the definition of an ordinal number.

**Definition 1.7.5** The ordered type of a well-ordered set is called an **ordinal number**.

*Example 1.7.6* Let $N$ denote the set of natural numbers. Then $(N, <)$ is an ordered set $\{1 < 2 < 3 \ldots\}$. Let $N^*$ denote the same set $N$, but with the inverse order relation $>$; i.e., $(N^*, >)$ is the ordered set $\{\cdots > 3 > 2 > 1\}$. The two sets are equivalent by identity map but they are not order-isomorphic, since $(N, <)$ has 1 as its first element, but $(N^*, >)$ has no first element.

*Remark 1.7.7* (**Notations of Ordered Types**)

(i) If $\alpha$ be the order type of an ordered set $(X, <)$, then the same set with the inverse order relation will be denoted by $(X^*, >)$, and its order type will be denoted by $\alpha^*$.

(ii) The order type of the ordered set $(N, <)$ of natural numbers is denoted by $\omega$, whereas that of $(N^*, >)$ is denoted by $\omega^*$.

(iii) The order type of the set of rational numbers $Q$, ordered by the usual inequality relation $<$, i.e., $\mathrm{Ord}(Q, <)$, is denoted by $\eta$.

(iv) The order type of the set of real numbers, ordered by the usual inequality relation $<$, i.e., $(R, <)$, is denoted by $\lambda$.

*Remark 1.7.8* Two finite sets $X$ and $Y$, ordered by their respective order relations $<_X$ and $<_Y$, are order-isomorphic, iff they are equivalent as sets. Hence the order type of a finite set with $n$ (distinct) elements is denoted by the natural number $n$. In particular, we take $\mathrm{Ord}(\emptyset) = 0$.

**Theorem 1.7.9** *Let $(X, <)$ be a well-ordered set and $f$ be an order-isomorphism of $(X, <)$ onto a subset $Y \subset X$. Then*

$$x < f(x), \ \forall x \in X.$$

*Proof* If possible, there exists an element $x \in X$ such that $f(x) < x$. Then the set of all such elements $x$ forms a nonempty subset $U$ of $X$, and hence $U$ has a first element $u$ (say). If $f(u) = y$, then $y < u$. Since $f$ is an order-isomorphism by hypothesis, $y < u$ asserts that

$$f(y) < f(u), \text{ i.e., } f(y) < y.$$

But this contradicts the defining property of $U$. Hence it follows that

$$x < f(x), \ \forall x \in X.$$

$\square$

**Definition 1.7.10**  Let $x$ be an element of a well-ordered set $(X, <)$, and let $X_x$ be a subset of $X$, defined by $X_x = \{y \in X : y < x\}$. Then $X_x$ is called the **initial segment** of $X$ determined by $x$.

**Theorem 1.7.11**    (i)  *A well-ordered set cannot be ordered-isomorphic to any of its initial segments.*

(ii) *Initial segments determined by two distinct elements cannot be order-isomorphic.*

**Proof**    (i) If possible, $(X, <)$ is order-isomorphic to an initial segment $X_x$ determined by $x \in X$ under an order-isomorphism $f$. Then $f(x) \in X_x$ and hence $f(x) < x$. But it contradicts Theorem 1.7.9.

(ii) Let $x$, $y$ be two distinct elements of $X$ with $y < x$. Then the initial segment $X_x$ determined by $x$ is a well-ordered set, and $y \in X_x$. This asserts that the initial segment $X_y$ determined by $y$ is also an initial segment of the well-ordered $(X_x, <)$. Hence it follows that $X_x$ cannot be order-isomorphic to $X_y$, by the first part of the theorem.

$\square$

**Definition 1.7.12**  Let $\alpha$ and $\beta$ be the ordered types of two well-ordered sets $X$ and $Y$ respectively. Then the ordinal number $\alpha$ is said to be less than the ordinal number $\beta$, abbreviated $\alpha < \beta$ if $X$ is order-isomorphic to an initial segment of $Y$.

Theorem 1.7.13 generalizes the well-known principle of mathematical induction for any well-ordered set.

**Theorem 1.7.13  (Principle of Transfinite Induction)** *Let $(X_x, <)$ be a well-ordered set and $Y$ be a subset of $X$ such that the initial segment $X_x$ determined by $x \in X$ has the property*

$$X_x \subset Y \implies x \in Y.$$

*Then $Y = X$.*

**Proof**  Suppose $Y \neq X$. Then $X - Y \neq \emptyset$. Hence it has a first element $x$ (say), since $X$ is well-ordered. Since any element of $X_x$ precedes $x$, it follows that it does not belong to $X - Y$. This shows that it belongs to $Y$. Thus $X_x \subset Y$. Hence, by hypothesis, $x \in Y$, which is not possible. This concludes that $Y = X$.  $\square$

## 1.8   Rings and Ideals

This section recalls some concepts of ring theory needed for our future study. A group is an algebraic system having with only one binary operation; a ring is also an algebraic system having two binary operations which are connected by some inter-relations. A field is an important special ring. The set of integers $\mathbf{Z}$ is the prototype of rings and $\mathbf{R}$ is the prototype of fields.

**Definition 1.8.1** A ring is an algebraic system $(R, +, \cdot)$ consisting of a nonempty set $R$ together with two binary operations "$+$" (called *addition*) and "$\cdot$" (called *multiplication*) such that

(i) $(R, +)$ is an abelian group, i.e., it is an additively an abelian group;
(ii) $(R, \cdot)$ is semigroup, i.e., $(x \cdot y) \cdot z = x \cdot (y \cdot z)$ for all $x, y, z \in R$; and
(iii) the operation "$\cdot$" is *distributive* (on both sides) over the operation "$+$," i.e.,
$x \cdot (y + z) = x \cdot y + x \cdot z$ and $(y + z) \cdot x = y \cdot x + z \cdot x$ for all $x, y, z \in R$.
(*distributive laws*).

A ring $(R, +, \cdot)$ is conventionally written by the set symbol $R$ and $x \cdot y$ by $xy$ simply.

**Definition 1.8.2** A ring $R$ is said to be commutative if its multiplication is commutative, i.e., $xy = yx$, $\forall x, y \in R$. It is said to have an identity element denoted by 1 if $1x = x1 = x \, \forall x \in R$.

*Example 1.8.3*    (i)   $\mathbf{Z}, \mathbf{Q}, \mathbf{R}, \mathbf{C}$ are important commutative rings under usual compositions of addition and multiplication.
(ii) $\mathbf{Z}_n (n > 1)$ is commutative ring under usual compositions of addition and multiplication of classes.
(iii) Each of the set of all square matrices $M(n, \mathbf{Z}), M(n, \mathbf{Q}), M(n, \mathbf{R}), M(n, \mathbf{C})$ of order $n$ is a noncommutative ring under usual addition and multiplication of matrices.
(iv) The set of all polynomials $R[x]$ over a ring $R$ is a ring, called the polynomial ring over $R$ in indeterminate $x$.

**Definition 1.8.4** Let $R$ be a ring. An element $a \in R$ is said to be a divisor of zero, if there exists a nonzero element $b \in R$ such that $ab = 0$ and $ba = 0$.

*Example 1.8.5* In the ring $\mathbf{Z}_{12}$ the element $[3]$ is a zero divisor of the element $[4]$.

**Definition 1.8.6** A commutative ring with identity element having no zero divisor is called an integral domain.

*Example 1.8.7* $\mathbf{Z}$ is an integral domain.

**Definition 1.8.8** An element $x$ in a ring $R$ with identity element 1 is said to be invertible if there exists an element $y \in R$ such that $xy = 1 = yx$. The element $y$ (if it exists) is unique and said to be an inverse element of $x$, abbreviated $x^{-1}$. An invertible element is sometimes called a unit.

*Example 1.8.9* The set of all units in a ring $R$ with 1 forms a (multiplicative) group.

*Remark 1.8.10* The integral domain $\mathbf{Q}$ and the integral domain $\mathbf{R}$ enjoy an algebraic advantage over the integral domain $\mathbf{Z}$: every equations $ax = b$ ($a$ is not zero) has a unique solution in them. Integral domains with this property are called fields.

**Definition 1.8.11** A ring $R$ with 1 is said to a **division ring or a skew field** if every nonzero element of $R$ is invertible, i.e., if its nonzero elements form a multiplicative group. A commutative division ring is said to be a **field**.

**Example 1.8.12** Every finite integral domain is a field.

**Example 1.8.13**    (i)  All square matrices of order 2 of the form

$$\begin{pmatrix} x_1 + x_2 i & x_3 + x_4 i \\ -x_3 + x_4 i & x_1 - x_2 i \end{pmatrix}$$

where $i^2 = -1$ and $x_i$ are real numbers ($i = 1, 2, 3, 4$) is a division ring under usual addition and multiplication of matrices, but it not a field.

(ii)  (**Quaternion Ring**) The real quaternion ring **H** introduced by Wiliam R. Hamilton (1805–1865) is a division ring but it is not a field. It consists of all ordered 4-tuples of real numbers such that if $\mathbf{H} = \{(x, y, z, t) : x, y, z, t \in \mathbf{R}\}$, then **H** forms a division ring under the composition

$$(a, b, c, d) + (x, y, z, t) = (a + x, b + y, c + z, d + t),$$
$$(a, b, c, d) \cdot (x, y, z, t) = (ax - by - cz - dt, ay + bx + ct - dz,$$
$$az - bt + cx + dy, at + bz - cy + dz$$

The symbols

$$1 = (1, 0, 0, 0), \; i = (0, 1, 0, 0), \; j = (0, 0, 1, 0), k = (0, 0, 0, 1),$$

used in **H** are such that

$$i^2 = j^2 = k^2 = -1, i \cdot j = k, j \cdot k = i, k \cdot i = j, j \cdot i = -k, k \cdot j = -i, i \cdot$$

$$k = -j.$$

In **H**, the element 1 is the multiplicative identity and an element $(x, y, z, t) \in \mathbf{H}$, called a real quaternion takes the form

$$(x, y, z, t) = x + yi + zj + tk.$$

The subset $K \subset \mathbf{H}$ defined by $K = \{(x, y, 0, 0) : (x, y, 0, 0) \in \mathbf{H}\}$ forms a subring of **H**, isomorphic to **C**. So real quaternions may be considered as generalization of complex numbers and an element $k = (x, y, 0, 0) \in K$ takes the form $k = x + yi$.

The **integral quaternion ring and rational quaternion ring** are defined in an analogous way by taking, respectively, integral and rational coefficients.

(iii)  The commutative rings **R**, **Q** and **C** are important examples of fields, called the field of real numbers, the field of rational number and the field of complex numbers, respectively.

**Definition 1.8.14** Let $R$ be a ring and $A$ be an additive subgroup of $R$. Then $A$ is said to an ideal of $R$ if $ra, ar \in A$, $\forall r \in R$, i,e., the subgroup $A$ swallows up multiplication by elements of $R$ from sides. Every ring $R$ has at least two ideals viz. $\{0\}$ and $R$. The two ideals such as $\{0\}$ and $R$ are called trivial ideals, and any other ideals (if there exist) are called nontrivial ideals. An ideal different from $R$ is said to be proper ideal. A proper ideal $I$ of $R$ is said to be **prime** if for $x, y \in R$, $xy \in I$ implies either $x \in I$ or $y \in I$ and it said to be **maximal** if there exists no ideal $K$ of $R$ such that $I \subset K \subset R$ satisfying $I \neq K \neq R$.

**Proposition 1.8.15** *A proper ideal $I$ of a commutative ring $R$ with 1 is prime iff its quotient ring $R/I$ is an integral domain, and it is maximal iff $R/I$ is field.*

**Example 1.8.16** (i)  A division ring has no nontrivial ideals.

(ii)  For every nonnegative integer $n$, the subgroup $A = n\mathbf{Z} = \{nm : m \in \mathbf{Z}\}$ is an ideal of the ring of integers $\mathbf{Z}$. Conversely, for any ideal $A$ of the ring $\mathbf{R}$, there is a unique nonnegative integer $n$ such that $A = n\mathbf{Z}$.

(iii)  Every maximal ideal of a commutative ring is a prime ideal but its converse is not true. For example, $\langle x \rangle$ is a prime ideal but it is not a maximal ideal in the ring of polynomials $\mathbf{Z}[x]$ over $\mathbf{Z}$.

## Ring of Real-Valued Continuous Functions

The ring $\mathcal{C}([0, 1])$ of real-valued continuous functions on the closed interval $[0, 1]$ is an important ring in mathematics. This ring gives an interplay between real analysis and algebra.

**Definition 1.8.17** Let $\mathcal{C}([0, 1])$ be the set of all real-valued continuous functions on the closed interval $[0, 1]$. i.e., $\mathcal{C}([0, 1]) = \{f : [0, 1] \to \mathbf{R} \text{ such that } f \text{ is continuous}\}$. Then $\mathcal{C}([0, 1])$ forms a commutative ring under usual pointwise addition and multiplication:

$$(f + g)(x) = f(x) + g(x), \quad (f \cdot g)(x) = f(x)g(x),$$

for all $f, g \in \mathcal{C}([0, 1])$ and for all $x \in [0, 1]$, where the right-hand addition and multiplication are the usual addition and multiplication of real numbers. The ring $\mathcal{C}([0, 1])$ is called the **ring of real-valued continuous functions on** $[0, 1]$.

**Example 1.8.18** The ring $\mathcal{C}([0, 1])$ contains divisors of zero and hence it is not an integral domain. Consider

$$f : [0, 1] \to \mathbf{R} : f(x) = \begin{cases} 0, & \text{if } 0 \leq x \leq 1/2 \\ x - \frac{1}{2}, & \text{if } 1/2 \leq x \leq 1 \end{cases}$$

$$g : [0, 1] \to \mathbf{R} : g(x) = \begin{cases} \frac{1}{2} - x, & \text{if } 0 \leq x \leq 1/2 \\ 0, & \text{if } 1/2 \leq x \leq 1 \end{cases}.$$

Then $f, g \in C([0, 1])$ are nonzero elements such that their product $f \cdot g = 0$. This implies that the ring $C([0, 1])$ cannot be an integral domain.

**Theorem 1.8.19** *For each $x \in I = [0, 1]$ the ideal $M_x = \{f \in R : f(x) = 0\}$ is a maximal ideal of the ring $C([0, 1])$. Moreover, if $\mathcal{M}$ is the set of all maximal ideals of $C([0, 1])$, then the map $\psi : I \rightarrow \mathcal{M}, x \mapsto M_x$ is a bijection.*

**Proof** Left as an exercise. □

## 1.9 Vector Space and Linear Transformations

This section recalls the concept of vector spaces (linear spaces) which has a very strong algebraic structure to solve many specific problems. Let $F$ be the field $F = \mathbf{R}$ or $\mathbf{C}$. For an arbitrary field $F$ the definition of a vector space is similar.

**Definition 1.9.1** A **vector space or a linear space** over a field $F$ ( whose elements are called scalars) with identity element $e$, is an additive abelian group $V$ (whose elements are called vectors) together with an external law of composition, called scalar multiplication) $m : F \times V \rightarrow V$, the image of $(\alpha, v)$ under $m$ abbreviated $\alpha v$, if the following conditions are satisfied:

(i) $ex = x$;
(ii) $\alpha(x + y) = \alpha x + \alpha y$;
(iii) $(\alpha + \beta)x = \alpha x + \beta x$;
(iv) $(\alpha\beta)x = \alpha(\beta x)$
$\quad \forall x, y \in V, \alpha, \beta \in F$.

**Example 1.9.2** For a given field $F$, $F^n$ is a vector space over $F$. In particular, the $n$-dimensional Euclidean space $\mathbf{R}^n$ is a vector space over the field $\mathbf{R}$.

### 1.9.1 Basis for a Vector Space

This subsection studies basis of vector spaces. The $n$ vectors such as $e_1 = (1, 0, \ldots, 0)$, $\ldots, e_n = (0, 0, \ldots, 1) \in \mathbf{R}^n$ determine every vector of $\mathbf{R}^n$ uniquely. This leads to the concept of basis (finite) of a vector space.

**Definition 1.9.3** Let $B$ be a nonempty subset (finite or infinite) of a vector space $V$ over $F$. A vector $v \in V$ is called a linear combinations of vectors of $B$ over $F$ if it can be expressed as

$$v = a_1 v_1 + \cdots + a_n v_n, a_i \in F; v_i \in B; i = 1, 2, \ldots, n.$$

**Proposition 1.9.4** *Let B be a nonempty subset (finite or infinite) of a vector space V over F. Then the set* $\mathbf{L}(B)$ *of all linear combinations of vectors of B is a subspace of the vector space V, which is the smallest subspace of V containing B.*

**Definition 1.9.5** $\mathbf{L}(B)$ is called the subspace generated or spanned by $B$ in $V$. In particular, if $V = \mathbf{L}(B)$ for some $B \subset V$, then $B$ is said to be a set of generators for $V$. It is said to finitely or infinitely generated according as card $B$ is finite or infinite.

*Example 1.9.6* The vector space $\mathbf{C}^n$ is finitely generated. On the other hand, the polynomial ring $\mathbf{R}[x]$ with real coefficients is not finitely generated. Because, any linear combination of a finite set of polynomials is a polynomial whose degree does not exceed the maximum degree say $k$ of the set of polynomials but $\mathbf{R}[x]$ has polynomials having degree $> k$.

**Definition 1.9.7** A nonempty subset $B$ of a vector $V$ is said to be a basis of $V$ over $F$ if

 (i)  $V$ is generated by the set $B$; and
 (ii)  $B$ is linearly independent over $F$.

If a vector space $V$ is trivial, i.e., $V = \{0\}$, then the empty set $\emptyset$ is conventionally taken as its basis.

**Theorem 1.9.8** *(Existence of a Basis) Every vector space has a basis. The cardinality of every basis of a vector space is the same.*

**Definition 1.9.9** Given a nonzero vector space $V$ over $F$, the cardinality of every basis of $V$ being the same, this common value is said to **dimension** of $V$, abbreviated as dim $V$. The vector space is said to be finite or infinite dimensional if $V$ has a finite or an infinite basis $B$, i.e., if card $V$ is finite or not. If dim $B$ is $n$, then $V$ is said to be an $n$-dimensional vector space over $F$.

*Example 1.9.10* $\mathbf{R}^n$ is an $n$-dimensional vector space over $\mathbf{R}$. On the other hand, $\mathbf{R}[x]$ is an infinite dimensional vector space over $\mathbf{R}$.

*Example 1.9.11* The real quaternions form a four-dimensional vector space $\mathcal{H}$ over $\mathbf{R}$, with a basis $\mathbf{B} = \{1, i, j, k\}$.

## 1.9.2  Linear Transformations

A linear transformation between vector spaces is an analogue concept between groups. Many mathematical problems, when properly posed, may be solved with the help of linear transformations. The concept of linear transformation is used throughout **Basic Topology, Volumes I, II, III** of the present book series.

**Definition 1.9.12** Given two vector spaces $V$ and $U$ over the same field $F$, a **linear transformation** is a map $T : V \to U$ such that for all $u, v \in V$ and for all $a, b \in F$,

(i) $T(u + v) = T(u) + T(v)$ (**additivity law**);

(ii) $T(av) = aT(v)$ (**homogeneity law**).

The above two conditions can be combined together to have an equivalent condition:

(iii) $T(au + bv) = aT(u) + bT(v)$.

In particular, for $U = V$, a linear transformation $T : V \to V$ is said to be a **linear operator**.

A linear transformation is also a group homomorphism between the corresponding additive groups.

**Definition 1.9.13** A linear transformation $T : V \to U$ is said to be

(i) a monomorphism if $T$ is injective;

(ii) an epimorphism if $T$ is surjective;

(iii) an isomorphism of $T$ is bijective.

### 1.9.3  Dual Space of a Vector Space

**Definition 1.9.14** Let $V$ be a vector space over a field $F$. Then the set $L(V, F)$ of all linear transformations $T : V \to F$ forms a vector space over $F$, called the dual space of $V$, denoted by $V^d$. An element $T \in V^d$ is called a linear functional of $V$ into $F$. A linear functional $T \in V^d$ transforms a vector to a scalar.

### 1.9.4  $\mathcal{C}([0, 1])$ as a Vector Space

This subsection defines linear functional for the vector space $\mathcal{C}([0, 1])$. Clearly, $\mathcal{C}([0, 1])$ is a commutative ring under the operations

$$(f + g)(x) = f(x) + g(x), \ (f \cdot g)(x) = f(x) \cdot g(x), \ \forall x \in [0, 1].$$

Again $V = \mathcal{C}([0, 1])$ is also a vector space over the field $\mathbf{R}$ under pointwise addition and scalar multiplication. The function $\psi : V \to \mathbf{R}$ defined by $f(x) \mapsto \int_0^1 f(x)dx$ is a linear function such that $\psi \in V^d = L(V, \mathbf{R})$, called the dual space of $V$.

## 1.10  Euclidean Spaces and Related Spaces with Standard Notations

In mathematical problems, subspaces of an $n$-dimensional Euclidean space arise frequently. Such spaces are used both in theory and application of topology. Some standard notations used throughout the book are given.

| | |
|---|---|
| $\emptyset$: | empty set |
| $\mathbf{Z}$: | ring of integers (or set of integers) |
| $\mathbf{Z}_n$: | ring of integers modulo $n$ |
| $\mathbf{R}$: | field of real numbers |
| $\mathbf{C}$: | field of complex numbers |
| $\mathbf{Q}$: | field of rational numbers |
| $\mathbf{H}$: | division ring of quaternions |
| $\mathbf{R}^n$: | Euclidean $n$-space, with $\|x\| = \sqrt{\sum_{i=1}^{n} x_i^2}$ and $\langle x, y \rangle$ $= \sum_{iz=1}^{n} x_i y_i$ for $x = (x_1, x_2, \ldots, x_n)$ and $y = (y_1, y_2, \ldots, y_n)$ $\in \mathbf{R}^n$ |
| $\mathbf{C}^n$: | complex $n$-space |
| $\mathbf{I}$: | $[0, 1]$ |
| $\dot{\mathbf{I}}$: | $\{0, 1\} \subset \mathbf{I}$ |
| $\mathbf{I}^n$: | $n$-cube=$\{x \in \mathbf{R}^n : 0 \le x_i \le 1$ for $1 \le i \le n\}$ for $x = (x_1, x_2, \ldots, x_n)$ |
| $\mathbf{D}^n$: | $n$-disk or $n$-ball=$\{x \in \mathbf{R}^n : \|x\| \le 1\}$ |
| $S^n$: | $n$-sphere $= \{x \in \mathbf{R}^{n+1} : \|x\| = 1\} = \partial \, \mathbf{D}^{n+1}$ (the boundary of the $(n+1)$-disk $D^{n+1}$) |
| $\mathbf{R}P^n$: | real projective space $=$ quotient space of $S^n$ with $x$ and $-x$ identified for all $x \in S^n$ |
| $\mathbf{C}P^n$: | complex projective space $=$ space of all complex lines through the origin in the complex space $\mathbf{C}^{n+1}$ |
| $\sqcup$: | disjoint union of sets or spaces |
| $\times, \Pi$: | product of sets, groups, modules, or spaces |
| $\cong$: | isomorphism |
| $\approx$: | homeomorphism |
| iff: | if and only if |
| $X \subset Y$ or $Y \supset X$: | set-theoretic containment (not necessarily proper). |

## 1.11   Exercises

1. Given any $x \in \mathbf{R}$, let $S_x = \{y \in \mathbf{Q} : y < x\} \subset \mathbf{Q}$. Show that lub $S_x = x$.
2. Let $X$ and $Y$ be nonempty sets, and let $f : X \to Y$ and $g : Y \to X$ two mappings. Show that

   (i) If $g \circ f$ is the identity mapping on $X$, then $f$ is an injection and $g$ is a surjection.

   (ii) If $g \circ f$ is the identity mapping on $X$ and $f \circ g$ is the identity mapping on $Y$, then both $f$ and $g$ are bijections and $g = f^{-1}$.

3. Let $X$ be a nonempty set, and let $f \in X^X$. Prove that;

   (i) if $f \circ g = f \circ h$, then $g = h$ for all $g, h \in X^X$ iff $f$ is injective;

(ii) if $g \circ f = h \cdot f$, then $g = h$ for all $g, h \in X^X$ iff $f$ is a surjection.

4. Let $f : X \to Y$ be a surjection and $B \subset Y$.
   Prove the following statements:

   (i) $f^{-1}(\emptyset) = \emptyset$, and $f^{-1}(Y) = X$.
   (ii) $f^{-1}(Y - B) = X - f^{-1}(B)$, where $B \subset Y$.
   (iii) $B \subset A \subset Y$ implies $f^{-1}(B \subset (A) \subset X$.
   (iv) For any collection of subsets $\{B_\alpha\}$ of $Y$,

   $$f^{-1}(\bigcup\{B_\alpha\}) = \bigcup\{f^{-1}(B_\alpha)\} \text{ and } f^{-1}(\bigcap\{B_\alpha\}) = \bigcap\{f^{-1}(B_\alpha)\}.$$

   (v) $B \cap A = \emptyset$ in $Y$ implies $f^{-1}(B) \cap f^{-1}(A) = \emptyset$ in $X$.
   (vi) $D \subset f^{-1}(B)$ implies $f(D) \subset B$, where $D \subset X$ and $B \subset Y$.

5. Show that

   (i) The order-isomorphism of any well-ordered set onto itself is only its identity map.
   (ii) Any two order-isomorphic well-ordered set can be mapped order-isomorphically onto each other in a unique way.

6. (**Cayley transformation**) Let $\mathbf{R}$ be the real line and $S^1 = \{z \in \mathbf{C} : |z| = 1\}$ be the unit circle in the complex plane.
   Show that the map

   $$C_T : \mathbf{R} \to S^1 - \{1\}, t \mapsto \frac{t - i}{t + i}$$

   is a bijection.

7. Show that the Cantor set $C$ is equivalent to the set $\mathbf{I} = [0, 1]$.

8. Given two maps $f : X \to Y, g : Y \to Z$, with their inverse maps $f^{-1} : Y \to X, g^{-1} : Z \to Y$, show that the inverse of the composite map $g \circ f : X \to Z$ is the composite map $f^{-1} \circ g^{-1} : Z \to X$.

9. Show that every infinite set has an enumerable subset.

10. (**Fundamental theorem of arithmetic**) Show that every integer $n > 1$ is uniquely factorizable (up to order).

11. Show that

    (i) Every subset of a countable set is countable.
    (ii) The Cartesian product of a finite collection of countable sets is countable.
    (iii) Countable union of countable sets is countable.
    (iv) Every open interval $(a, b)$ has cardinality $c$.
    (v) Cantor set has cardinality $c$.

12. Show that card $(\mathbf{I}) = $ card $(\mathbf{I}^n)$.
    [Hint: Use that $\mathbf{I} \sim \mathbf{I}^n$.]

13. Show that card $(\mathbf{I}) = $ card $(\mathbf{R})$.

14. Let $G$ and $H$ be two groups with with $1_G$ and $1_H$ identity automorphisms of $G$ and $H$ respectively.

   Let $f : G \to H, g : H \to G$ be two homomorphisms. Show that

   (i) If $g \circ f : G \to G$ is $1_G$, then $f$ is a monomorphism.

   (ii) If $f \circ g : H \to G$ is $1_H$, then $f$ is an epimorphism.

   (iii) If $g \circ f$ is $1_G$ and $f \circ g$ is $1_H$, then $f$ is an isomorphism with $g$ its inverse.

15. Let $\mathcal{C}([0, 1])$ be the ring of real-valued continuous functions on $[0, 1]$. Show that $M_x = \{f \in \mathcal{C}([0, 1]) : f(x) = 0\}$ is a maximal ideal of $\mathcal{C}([0, 1])$.

16. Show the there exists a bijective correspondence between the set of all maximal ideals of the ring $\mathcal{C}([0, 1])$ and the points of $[0, 1]$.

# References

Adhikari, M.R.: Basic Algebraic Topology and Its Applications. Springer, India (2016)

Adhikari, M.R.: Basic Topology. Volume 3: Algebraic Topology and Topology of Fiber Bundles. Springer, India (2022)

Adhikari, M.R., Adhikari, A.: Groups. Rings and Modules with Applications. Universities Press, Hyderabad (2003)

Adhikari, M.R., Adhikari, A.: Textbook of Linear Algebra: An Introduction to Modern Algebra. Allied Publishers, New Delhi (2006)

Adhikari, M.R., Adhikari, A.: Basic Modern Algebra with Applications. Springer, New Delhi, New York, Heidelberg (2014)

Adhikari, A., Adhikari, M.R.: Basic Topology. Volume 2: Topological Groups, Topology of Manifolds and Lie Groups. Springer, India (2022)

Alexandrov, P.S.: Introduction to Set Theory and General Topology. Moscow (1979)

Chatterjee, B.C., Ganguly, S., Adhikari, M.R.: Introduction to Topology. Asian Books Pvt. Ltd., New Delhi (2002)

Conway, J.B.: A Course in Point Set Topology. Springer, Switzerland (1914)

Hu, S.T.: Introduction to General Topology. Holden-Day, San Francisco (1966)

Kelly, J.L.: General Topology. Van Nostrand, New York (1955); Springer, New York (1975)

Simmons, G.: Introduction to Topology and Modern Analysis. McGraw Hill, New York (1963)

# Chapter 2
# Metric Spaces and Normed Linear Spaces

This chapter starts a journey in metric spaces describing the concept of metrics, which is an abstraction of distance in the Euclidean space and conveys an axiomatic framework for this abstraction with a systemic study of elementary basic properties of **metric spaces**. It also discusses **normed linear spaces** which form a versatile class of metric spaces. This discussion includes a brief study of **Banach and Hilbert spaces**.

Metric spaces give the simplest setting for the study of certain problems arising in analysis and provide a rich supply of continuous functions as well as topological spaces. Most of the applications of topology to analysis arise through the metric spaces. In many areas of mathematics such as in geometry, analysis and algebra (specially, in matrix algebra), the concept of distance is generalized in an abstract setting by introducing the concept of metric spaces, which facilitates a study of continuous functions defined on abstract sets as well as convergent sequences on these sets.

The basic aim of this chapter is to address an introduction to metric spaces with an eye to make a preparatory grounding for a general topology course which officially begins in Chap. 3. The motivation of this approach is that metric spaces are immediate generalizations of real and complex number systems and lie between them and topological spaces. Thus metric spaces are less general than topological spaces. Metric spaces are discussed as our first step, because of their simplicity and wide usefulness in modern mathematics. Moreover, metric spaces have more structure than topological spaces, and they provide stepping stones to a variety of important topics in topology. Various interesting applications of metric spaces are available in Sect. 2.16. Throughout the book, $\mathbf{R}^n$ represents the Euclidean $n$-space with usual distance function $d$ formulated in Definition 2.3.10 unless specified otherwise.

**General topology (also called point set topology)** is developed in Euclidean $n$-space $\mathbf{R}^n$ as well as in the setting of more general metric spaces. The central concept in analysis and topology is the continuity of functions. This chapter motivates to study spaces such as metric spaces on which a workable definition of continuity of functions can be given, where the concept of continuity can be completely expressed in terms of open sets (see Theorem 2.12.6 for metric spaces and Chap. 3 for topological spaces). Continuity of a function between metric spaces can be characterized in the language of sequences (see Theorems 2.12.3 and 2.12.4).

**Urysohn lemma** for metric spaces provides a vast supply of real-valued continuous functions on the metric spaces. Another important theorem of this chapter is the **Banach contraction theorem** on complete metric spaces, which is applied to prove **Picard's theorem** on the existence of solution of a differential equation. In general, a metric space does not carry any algebraic structure. But there exist metric spaces which are also linear (vector) spaces, and their metrics are induced by certain norm functions defined on these linear spaces, which are known as **normed linear spaces**.

**Historically**, the concept of metric spaces was introduced by the French mathematician Maurice Fréchet (1878–1973) in 1906 in his Ph.D. thesis by generalizing the concept of Euclidean spaces. This concept is an abstraction of distance in the Euclidean space born through the well-known properties of the Euclidean distance in an abstract setting, and it provides a rich supply of continuous functions. The special structure of a metric space induces a topology (see Chap. 3) having many applications of topology in modern analysis. The term **Topology** was coined by J. B. Listing (1808–1882) in 1847, but Felix Hausdorff (1868–1942) popularized the term topology in 1914 and developed this subject in his book **Grundzüge der Mengenlehre** of 1914, which stemmed from analysis. His land-marking work sets out the systematic journey of general topology.

For this chapter, the books Bredon (1993), Simmons (1963), Stephen (1970), Patterson (1959) and some other books are referred in Bibliography.

## 2.1   Results of Analysis Leading to the Concept of Metric Spaces

Before presenting the formal definition of a metric space, this section conveys some fundamental results of classical analysis, which motivated the concept of metric spaces as well as its systematic study in an abstract setting. Most of the concepts related to metric spaces were born through the geometric ideas of the set **R** of real numbers. So, it has become necessary to study open sets, closed sets and continuity of functions in **R**, which play a key role in mathematics, particularly, in algebra, geometry and analysis. Throughout this chapter, **R** represents the real line endowed with Euclidean metric $d$, sometimes, called it the **Euclidean line**, formally formulated in Definition 2.1.1.

### 2.1.1 Open Sets in R

This subsection studies open sets in $\mathbf{R}$. Let "$\leq$" be the usual (natural) ordering on $\mathbf{R}$, and as usual, $(a, b)$ denotes an open interval, and $[a, b]$ denotes a closed interval in $\mathbf{R}$ for $a, b \in \mathbf{R}$, with $a < b$.

**Definition 2.1.1** The Euclidean distance $d(x, y)$ in $\mathbf{R}$ is defined by

$$d(x, y) = |x - y|, \quad \forall x, y \in \mathbf{R},$$

where $|x - y|$ denotes the absolute value of the real number $x - y$. The map

$$d : \mathbf{R} \times \mathbf{R} \to \mathbf{R}, (x, y) \mapsto |x - y|$$

is called the **Euclidean metric or usual metric** on $\mathbf{R}$, the set $\mathbf{R}$ equipped with the Euclidean metric is called the **Euclidean line**, and it is also denoted, sometimes by $\mathbf{R}^1$.

**Definition 2.1.2** A subset $X$ of $\mathbf{R}$ is said to be **open** if for each point $x \in X$, there exist points $a, b \in \mathbf{R}$, with $a < b$, such that $x \in (a, b) \subset X$.

**Example 2.1.3** Every open interval $(a, b)$ is an open set in $\mathbf{R}$, because, for any $x \in (a, b)$, there is an open interval $(x - \epsilon, x + \epsilon) \subset (a, b)$ obtained by taking $\epsilon = \min\{x - a, b - x\}$, which is positive. On the other hand, its converse is not necessarily true, since all open sets in $\mathbf{R}$ are not intervals. For example, the set $(2, 6) \cup (8, 9)$ is an open set in $\mathbf{R}$ by Corollary 2.6.13 but it is not an open interval. In general, the set $X = \bigcup_{n=1}^{\infty} (2n, 2n + 1)$ is an open set in $\mathbf{R}$, but it is not an open interval. The closed interval $[a, b]$ is not open set in $\mathbf{R}$.

**Remark 2.1.4** The set of all open sets in a nonempty set defined in Definition 2.1.2 satisfies the four axioms of open sets for a topology (see Theorem 2.6.15), called the **Euclidean topology** on $\mathbf{R}$. This leads to the concept of topology on an abstract setting (see Chap. 3) by defining open sets facilitating to define continuity of functions, which is one of the basic concepts in topology.

**Definition 2.1.5** Let $X$ be a subset of $\mathbf{R}$. A point $x \in X$ is said to be an **interior point** of $X$ if the point $x$ is in some open interval $O_x$ which is contained in $X$, i.e., $x \in O_x \subset X$. A subset $X$ of $\mathbf{R}$ is said to be **open** if every point of $X$ is an interior point of $X$. On the other hand, a subset $X$ of $\mathbf{R}$ is not open iff there exists a point $x \in X$ such that $x$ is not an interior point of $X$.

**Example 2.1.6** Every point of an open interval $(a, b)$ in $\mathbf{R}$ is an interior of $(a, b)$.

**Example 2.1.7** (i) The empty set $\emptyset$ in $\mathbf{R}$ is trivially an open set because there exists no point in $\emptyset$.

(ii) The real line $\mathbf{R}$ is itself an open set, because every point in $\mathbf{R}$ is an interior point of $\mathbf{R}$.

(iii)  Any open interval $(a, b)$ in **R** is an open set.
(iv)  The closed interval $[a, b]$ in **R** is not an open set, because the end points $a$ and $b$ are not interior points of $[a, b]$.
(v)  The infinite open intervals $(a, \infty) = \{x \in \mathbf{R} : a < x < \infty\}$ and $(-\infty, a) = \{x \in \mathbf{R} : -\infty < x < a\}$ in **R** are open sets, but infinite half-closed intervals $[a, \infty) = \{x \in \mathbf{R} : a \leq x < \infty\}$ and $(-\infty, a] = \{x \in \mathbf{R} : -\infty < x \leq a\}$ are not open in **R**, because the point $a \in \mathbf{R}$ is not an interior point of either of these two infinite half-closed intervals $[a, \infty)$ and $(-\infty, a]$.

Theorem 2.1.8 is a fundamental theorem on open sets of **R**.

**Theorem 2.1.8**    (i)  *The union of any number of open sets in* **R** *is an open set.*
(ii)  *The intersection of a finite number of open sets in* **R** *is an open set.*

**Proof**  Left as an exercise; otherwise, see Corollary 2.6.13.                      □

**Definition 2.1.9**  Let $X$ be a subset of **R**. A point $p$ in **R** is said to be a **limit point, cluster point or an accumulation point** of $X$ if every open set $U$ containing the point $p$ contains a point of $X$ other than the point $p$.

**Example 2.1.10**    (i)  Every point of **R** is a limit point of the set **Q** of rational numbers.
(ii)  For the subset $X = \{\frac{1}{n} : n \in \mathbf{N}\}$ of **R**, the point 0 is the only limit point of $X$.
(iii)  Every point in the interval $X = [a, b) \subset \mathbf{R}$ is a limit point of $X$. Even, the point $b \notin X$ is also a limit point of $X$.
(iv)  The subset $X = \{1, 1/2, 1/3, 1/4, \ldots\}$ of **R** has limit point 0, since any open interval $U$ (such as $(-a, b)$ with $-a < 0 < b$) containing the point 0 contains infinitely many points other than 0.

**Definition 2.1.11**  Let $X$ be a subset of **R**. The set of all limit points of $X$, abbreviated $X'$, is called the **derived set** of $X$.

**Example 2.1.12**    (i)  The point 0 is the only limit point of the subset $X = \{1, 1/2, 1/3, 1/4, \ldots\}$ of **R** and hence its derived set $X' = \{0\}$.
(ii)  The derived set of the set of integers **Z** is the empty set $\emptyset$.
(iii)  For $X = [a, b) \subset \mathbf{R}$, its derived set $X' = [a, b]$.

### 2.1.2  Closed Sets in **R**

This subsection studies closed sets in **R** which are complements of open sets in **R**.

**Definition 2.1.13**  A subset $F$ of **R** is said to be a **closed set** in **R** if its complement $F^c = \mathbf{R} - F$ is an open set in **R**.

**Example 2.1.14**    (i)  The complement of an open set in **R** is a closed set, and dually, the complement of a closed set in **R** is an open set in **R**.

(ii) The closed interval $[a, b] \subset \mathbf{R}$ is a closed set, because its complement in $\mathbf{R}$ is the open set $(-\infty, a) \cup (b, \infty)$, which is the union of two open sets in $\mathbf{R}$.

(iii) The subset $\mathbf{Z}$ of $\mathbf{R}$ is a closed set, because the complement of $\mathbf{Z}$ is the union of $\bigcup_{-\infty}^{\infty} (n, n + 1)$ of open intervals $(n, n + 1)$ of $\mathbf{R}$, is an open set and hence $\mathbf{Z}$ is a closed set in $\mathbf{R}$.

(iv) $\mathbf{R}$ and $\emptyset$ are closed sets, because their complements are open sets in $\mathbf{R}$.

A closed set in $\mathbf{R}$ is characterized in Theorem 2.1.15 by its limit points.

**Theorem 2.1.15** *A subset $F \subset \mathbf{R}$ is closed iff $F$ contains all of its limit points.*

**Proof** Left as an exercise. $\qquad \square$

**Example 2.1.16** The set $F = \{1, 1/2, 1/3, \ldots\} \subset \mathbf{R}$ is not closed in $\mathbf{R}$, because its limit point 0 is not in $F$. On the other hand, $[0, 1]$ is closed in $\mathbf{R}$.

### 2.1.3 Two Classical Theorems: Bolzano–Weierstrass Theorem and Heine–Borel Theorem

This subsection conveys Bolzano–Weierstrass Theorem and Heine–Borel Theorem, which are two classical results in real analysis. These theorems are generalized in Chap. 5 in the context of compactness in a topological setting. Bernard Bolzano (1781–1848) proved Theorem 2.1.19, in the 1817. This theorem was also proved by Karl Weierstrass (1815–1897) independently. On the other hand, Eduard Heine (1821–1881) used finite subcovers in a paper published in 1872 on his study of uniformly continuous functions. Emile Borel (1871–1956) proved that a countable open covers of closed intervals have finite subcovers in a paper published in 1894, which led to the concept of compactness studied in Chap. 5 of the present volume of the book series.

**Definition 2.1.17** Let $X$ be a subset of $\mathbf{R}$. A family $\mathcal{C} = \{X_i\}$ of open sets of $\mathbf{R}$ is said to be an **open covering** of $X$, if $X \subset \bigcup_i X_i$. A subset $X$ of $\mathbf{R}$ is said to be **compact** if every open covering of $X$ has a finite subcovering.

**Remark 2.1.18** An infinite set may not have a limit point. So the problem of existence of limit points is interesting. Bolzano and Weierstrass solved this problem in the classical Bolzano–Weierstrass Theorem 2.1.19, named after them.

**Theorem 2.1.19** *(Bolzano–Weierstrass theorem in $\mathbf{R}$) Let $X$ be a bounded infinite subset of $\mathbf{R}$. Then $X$ has at least one limit point.*

**Proof** Left as an exercise. $\qquad \square$

**Remark 2.1.20** The property of $\mathbf{R}$ embodied in Theorem 2.1.19 is said to be **Bolzano–Weierstrass (B-W) property of $\mathbf{R}$**. Its generalization in metric spaces asserts that a metric space $M$ is said to have the Bolzano–Weierstrass property if every infinite subset of $M$ has a limit point. A further generalization of this property in the topological language asserts that a metric space is compact iff its every infinite subset has a limit point (see Chap. 5).

**Corollary 2.1.21** *Every bounded sequence in* **R** *with the Euclidean metric has a convergent subsequence.*

**Remark 2.1.22** Another important property of **R** is the classical Heine–Borel Theorem 2.1.23, named after E. Heine and E. Borel, which asserts that every open covering $\mathcal{C}$ of a closed and bounded interval $X$ in **R** has a finite subcovering $\mathcal{S}$ in the sense that $\mathcal{S}$ is a finite subset of $\mathcal{C}$ such that $X$ is contained in a finite union of members of $\mathcal{S}$. This leads to the concept of compactness in the topological language which is studied in Chap. 5.

**Theorem 2.1.23** (*Heine–Borel theorem in* **R**) *Every closed bounded subset of* **R** *is compact in the sense that its every open covering has a finite subcovering.*

**Proof** Left as an exercise or see Chap. 5.                                                                 □

**Remark 2.1.24** The property of **R** embodied in Theorem 2.1.23 is said to be the **Heine–Borel property** of **R**, named after Heine and E. Borel. A generalization of this property in the topological language asserts that every closed and bounded subset of **R** is compact (in the usual metric topology) (see Chap. 5).

### 2.1.4   Continuity of Functions on **R**

This subsection addresses some basic properties of real-valued continuous functions defined on **R**. The concept of continuity of a function plays a key role in mathematics. It is defined at school level by utilizing the usual distance between two points in **R**, called "$\epsilon$-$\delta$" definition.

**Definition 2.1.25** A function $f : \mathbf{R} \to \mathbf{R}$ is said to be continuous at a point $a \in \mathbf{R}$, if for every real number $\epsilon > 0$, there exists a real number $\delta > 0$ such that $|f(x) - f(a)| < \epsilon$, whenever $|x - a| < \delta$, i.e., for every open set $O_{f(x)}$ containing the point $f(x)$, there exists an open set $O_x$ containing the point $x$ such that $f(O_x) \subset O_{f(x)}$. The function $f$ is said to be continuous if it is continuous at every point in **R**.

**Intuitively**, if a function $f : \mathbf{R} \to \mathbf{R}$ is continuous at a point $a$, then whenever $x$ is close to $a$, its image $f(x)$ becomes close to $f(a)$, and here closeness is measured by the Euclidean distance $d(x, a) = |x - a|$ on **R**.

**Definition 2.1.26** A function $f : \mathbf{R} \to \mathbf{R}$ is said to be continuous on a subset $S \subset \mathbf{R}$ if $f$ is continuous at each point in $S$.

**Remark 2.1.27** The continuity of a function $f : \mathbf{R} \to \mathbf{R}$ can be characterized in terms of open sets as shown in Corollary 2.12.7, which asserts that a function $f : \mathbf{R} \to \mathbf{R}$ is continuous iff the inverse image of every open set is open. Thus, a function $f : \mathbf{R} \to \mathbf{R}$ is not continuous iff there exists an open set $U \subset \mathbf{R}$ such that $f^{-1}(U)$ is not open.

**Example 2.1.28** The function

$$f : \mathbf{R} \to \mathbf{R}, \ x \mapsto \begin{cases} x - 1, & \text{if } x \leq 3 \\ \frac{x+5}{2}, & \text{if } x > 3 \end{cases}.$$

is not continuous, since its inverse image of every open interval is not open.

**Proposition 2.1.29** *Let a function $f : \mathbf{R} \to \mathbf{R}$ be continuous at every point $x \in [a, b]$ and satisfy the property that $f(a) < 0 < f(b)$. Then there exists a point $c \in [a, b]$ such that $f(c) = 0$.*

**Proof** Left as an exercise. □

Weierstrass Intermediate Value Theorem 2.1.30 proves the connectedness property of $\mathbf{R}$. Its generalization in a topological setting is available in Chap. 5.

**Theorem 2.1.30** *(**Weierstrass intermediate value theorem**) If $f : \mathbf{R} \to \mathbf{R}$ is a continuous function on the closed interval $[a, b] \subset \mathbf{R}$, then $f$ assumes every value between $f(a)$ and $f(b)$.*

**Proof** Without any loss of generality, let $f(a) < f(b)$ and $r$ be a real number such that $f(a) < r < f(b)$. Consider the continuous function

$$\psi : \mathbf{R} \to \mathbf{R}, x \mapsto f(x) - r.$$

Then $\psi(a) < 0 < \psi(b)$. Hence it follows by Proposition 2.1.29 that there exists a point $c \in [a, b]$ such that $\psi(c) = f(c) - r = 0$, which asserts that $f(c) = r$. □

## 2.2 Sequence of Real Numbers and Cauchy Sequence

This section discusses the concepts of sequences of real numbers and specially Cauchy sequences used in analysis to facilitate their generalizations in metric spaces (see Sect. 2.10).

### 2.2.1 Sequence of Real Numbers

This subsection studies sequences of real numbers, which play a key role in topology and analysis.

**Definition 2.2.1** A **sequence of real numbers** is a function (map) $f : \mathbf{N} \to \mathbf{R}$, and the image $f(n) = x_n \in \mathbf{R}$ is called the $n$th term of the sequence. It is said to be **bounded** if its image set Im $f = \{x_n : n \in \mathbf{N}\}$ is bounded in $\mathbf{R}$.

**Example 2.2.2**     (i) The sequence $\{1/2, 1/4, 1/8, \ldots\}$ defined by

$$f : \mathbf{N} \to \mathbf{R}, \, n \mapsto 1/2^n$$

is bounded in $\mathbf{R}$.

(ii) The sequence $\{2, 4, 6, 8, \ldots\}$ defined by

$$f : \mathbf{N} \to \mathbf{R}, n \mapsto 2n$$

is not bounded in $\mathbf{R}$.

**Definition 2.2.3** A sequence $\{x_n\} = \{x_1, x_2, \ldots, x_n, \ldots\}$ of real numbers is said to **converge** to a point $x \in \mathbf{R}$, if for every real number $\epsilon > 0$, there exists a positive integer $n_0$ such that $|x_n - x| < \epsilon$ for every integer $n \geq n_0$, i.e., every open set containing $x$ contains almost all, except possibly, a finite number of terms of the sequence.

**Example 2.2.4**     (i) The constant sequence $\{5, 5, \ldots, 5, \ldots\}$ converges to 5, because every open interval containing 5 contains every term of the sequence.

(ii) The sequence $\{1, 1/2, 1/3, 1/4, \ldots\}$ converges to 0, because every open interval containing 0 contains almost all terms of the sequence.

**Definition 2.2.5** In the sequence $\{x_n\} = \{x_1, x_2, \ldots, x_n, \ldots\}$ of real numbers, if $\{j_n\}$ is a sequence of positive integers such that $j_1 < j_2 < j_3 < \cdots$, then $\{x_{j_1}, x_{j_2}, x_{j_3}, \ldots\}$ is called a **subsequence** of $\{x_n\}$.

Bolzano–Weierstrass Theorem 2.1.19 is also expressed in terms of real sequences in Theorem 2.2.6.

**Theorem 2.2.6** *(Bolzano–Weierstrass theorem) Every bounded sequence of real numbers has a convergent subsequence.*

**Proof** It follows from Theorem 2.1.19 .                                                    □

### 2.2.2   Cauchy Sequence

This subsection studies a special class of sequences of real numbers, called Cauchy sequences, and characterizes convergence of sequences of real numbers with the help of Cauchy sequences in Theorem 2.2.13. This sequence is named after the French mathematician Augustin-Louis Cauchy (1789–1857), who is considered as a cofounder of mathematical analysis.

**Definition 2.2.7** A sequence $\{x_n\} = \{x_1, x_2, \ldots, x_n, \ldots\}$ of real numbers is said to be a **Cauchy sequence** if for every positive real number $\epsilon$, there exists a positive integer $m$ such that $|x_i - x_j| < \epsilon$ for all $i, j \geq m$, i.e., a sequence $\{x_n\}$ of real numbers is a Cauchy sequence iff the terms of the sequence $\{x_n\}$ come arbitrarily close to each other as $n$ becomes sufficiently large.

**Definition 2.2.8** A subset $X$ of real numbers is said to be **complete** if every Cauchy sequence $\{x_n\}$ of points in $X$ converges to a point in $X$.

**Example 2.2.9** Every convergent sequence of real numbers is a Cauchy sequence, but its converse is not true. For example, consider the metric space $X = (0, 1)$ with standard metric

$$d : X \times X, \ (x, y) \mapsto |x - y|.$$

Then the sequence $\{x_n = 1/n\}$ is a Cauchy sequence in $X$, but this sequence does not converge in $X$.

**Example 2.2.10** The set $\mathbf{Q}$ of rational numbers endowed with the usual metric

$$\mathbf{Q} \times \mathbf{Q}, \ (x, y) \mapsto |x - y|$$

is not complete, because there exist Cauchy sequences in $\mathbf{Q}$ which converge to an irrational number such as the sequence

$$1.4, 1.41, 1.4142, 1.41421, \ldots$$

converges to $\sqrt{2} \in \mathbf{R}$, is a Cauchy sequence in $\mathbf{Q}$, but it does not converge in $\mathbf{Q}$.

**Example 2.2.11** The set $\mathbf{R}$ of real numbers is complete by Corollary 2.2.14.

**Proposition 2.2.12** *Every Cauchy sequence of real numbers is bounded.*

**Proof** Let $\{x_n\} = \{x_1, x_2, \ldots, x_n, \ldots\}$ be a Cauchy sequence of real numbers. Then given an $\epsilon = 1$, there exists a positive integer $n_0$ such that $|x_i - x_j| < \epsilon = 1$ for all $i, j \geq n_0$. This shows that

$$|x_n| = |x_n - x_{n_0} + x_{n_0}| \leq |x_n - x_{n_0}| + |x_{n_0}| < 1 + |x_{n_0}|, \quad \forall n \geq n_0,$$

This asserts that the sequence $\{x_n\}$ is bounded.

$\square$

Theorem 2.2.13 characterizes convergence of real sequences by Cauchy sequences.

**Theorem 2.2.13** *(Cauchy) A sequence $\{x_n\} = \{x_1, x_2, \ldots, x_n, \ldots\}$ of real numbers is convergent iff it is a Cauchy sequence.*

**Proof** Left as an exercise.

$\square$

**Corollary 2.2.14** *The set* **R** *of real numbers is complete.*

**Proof** Since every Cauchy sequence in **R** converges to a point in **R** by Theorem 2.2.13, the Corollary follows. ☐

**Remark 2.2.15** Corollary 2.2.14 gives one of the fundamental properties of **R** that **R** is complete. On the other hand, the set **Q** of rational numbers is not complete by using the supporting Example 2.2.10.

**Example 2.2.16** Every Cauchy sequence in **R** is convergent, but it is not valid on a proper subset of **R**. For example, in the open interval $(0, 1)$ endowed with the Euclidean metric, the sequence

$$0.1, 0.01, 0.001, 0.0001, \ldots$$

is a Cauchy sequence, but it fails to converge to any point in $(0, 1)$.

**Proposition 2.2.17** *Every sequence* $\{x_n\}$ *in* **R** *has a monotonic subsequence.*

**Proof** Left as an exercise. ☐

**Proposition 2.2.18** *Let* $\{x_n\}$ *be a monotonic sequence in* **R**. *Then the sequence* $\{x_n\}$ *converges to a point in* **R** *iff the sequence* $\{x_n\}$ *is bounded.*

**Proof** Left as an exercise. ☐

## 2.3 Concept of Distance in Euclidean Spaces $\mathbf{R}^n$

We are familiar with the concept of distance in Euclidean spaces from school level. This concept permits the definitions of open sets, closed sets and continuity of functions from one Euclidean space to another. The most of the spaces of our interest are subsets of Euclidean spaces. This section generalizes the concepts of distance and continuity of functions defined in **R** for the Euclidean $n$-space $\mathbf{R}^n$. Since $\mathbf{R}^n$ represents the Euclidean $n$-space with usual distance function $d$ formulated in Definitions 2.3.10, 2.1.9 of limit points for a subset in **R** can be extended in a similar way for a subset in $\mathbf{R}^n$.

### 2.3.1 Open Sets of Euclidean Plane $\mathbf{R}^2$

This subsection studies open sets in the Euclidean plane $\mathbf{R}^2$ by using the usual distance in the Euclidean plane. Throughout the book, $\mathbf{R}^2$ stands for the Euclidean plane with usual metric $d$, given in Definition 2.3.1, unless specified otherwise.

**Definition 2.3.1** Let $x = (x_1, x_2)$, $y = (y_1, y_2) \in \mathbf{R}^2$ be two points. Then their usual distance $d(x, y)$ defined by

$$d(x, y) = [(x_1 - y_1)^2 + (x_2 - y_2)^2]^{1/2},$$

is called the **Euclidean distance** between the points $x$ and $y$ and the map

$$d : \mathbf{R}^2 \times \mathbf{R}^2 \to \mathbf{R}, \ (x, y) \mapsto [(x_1 - y_1)^2 + (x_2 - y_2)^2]^{1/2}, \ \text{for } x = (x_1, x_2),$$
$$y = (y_1, y_2) \in \mathbf{R}^2$$

is called the **Euclidean metric or usual metric** on $\mathbf{R}^2$. The set $\mathbf{R}^2$ endowed with the Euclidean metric is called the **Euclidean plane**.

**Definition 2.3.2** An **open disk or an open ball** in $\mathbf{R}^2$ centered at a point $x \in \mathbf{R}^2$ and radius $r > 0$ symbolized $B_x(r)$ is defined by

$$B_x(r) = \{y \in \mathbf{R}^2 : d(x, y) < r\}.$$

*Example 2.3.3* The open unit disk centered at a point $x \in \mathbf{R}^2$ and radius 1 is given by
$$B_x(1) = \{y \in \mathbf{R}^2 : d(x, y) < 1\}.$$

**Definition 2.3.4** Given a subset $X \subset \mathbf{R}^2$, a point $x \in X$ is said to be an **interior point of** $X$ if $x$ is an element of some open disk $B_x(r)$, equivalently, $x \in B_x(r) \subset X$ for some open disk $B_x(r)$. The set $X$ is said to be open if every point of $X$ is an interior point.

*Example 2.3.5*    (i) The Euclidean plane $\mathbf{R}^2$ is an open set.
  (ii) Any open disk $B_x(r)$ in $\mathbf{R}^2$ is an open set.
  (iii) The empty set $\emptyset$ is an open set.

*Remark 2.3.6* Definition 2.1.9 of limit points for a subset in $\mathbf{R}$ can be extended in a similar way for $\mathbf{R}^2$ and in general, for $\mathbf{R}^n$.

**Definition 2.3.7** Let $X$ be a subset of $\mathbf{R}^2$. A point $x \in \mathbf{R}^2$ is said to be an **accumulation point or limit point** of $X$ if every open set $U$ containing the point $x$ contains a point of $X$ which is different from $x$.

**Definition 2.3.8** Let $X$ be a subset of $\mathbf{R}^2$. The set of all limit points of $X$ abbreviated $X'$ is called the **derived set** of $X$.

*Example 2.3.9* In the Euclidean plane, every point of the set $A = \{(x, y) \in \mathbf{R}^2 : x = 0, |y| \leq 1\}$, i.e., every point on the $y$-axis lying between the points $(0, -1)$ and $(0, 1)$ is a limit point of the set

$$X = \left\{(x, y) \in \mathbf{R}^2 : y = \sin \frac{1}{x}, \ x > 0\right\}.$$

### 2.3.2 Distance Function in $\mathbf{R}^n$

The Euclidean $n$-space $\mathbf{R}^n$ for any finite dimension $n$ is defined as a natural general-ization of 2-dimensional Euclidean plane $\mathbf{R}^2$ and 3-dimensional Euclidean space $\mathbf{R}^3$. A point $x$ in the Euclidean $n$-space $\mathbf{R}^n$ can be represented by an ordered set of $n$-real numbers, $x = (x_1, x_2, \ldots, x_n)$. For convenience, a point $x \in \mathbf{R}^n$ is sometimes writ-ten as $(x) \in \mathbf{R}^n$. For infinite dimensional Euclidean space $\mathbf{R}^\infty$, see Example 2.14.12. The concept of distance between any two points in Euclidean line $\mathbf{R}$, Euclidean plane $\mathbf{R}^2$ and Euclidean 3-space $\mathbf{R}^3$ is well known to the students from their school level. This notion of distance in Euclidean $n$-space $\mathbf{R}^n$ is analogous.

**Definition 2.3.10** Given two points $x = (x_1, x_2, \ldots, x_n)$ and $y = (y_1, y_2, \ldots, y_n) \in \mathbf{R}^n$, their **Euclidean distance** $d(x, y)$ is defined by

$$d(x, y) = \left[ \sum_{i=1}^{n} (x_i - y_i)^2 \right]^{\frac{1}{2}},$$

and the function

$$d : \mathbf{R}^n \times \mathbf{R}^n \to \mathbf{R}, \ (x, y) \mapsto d(x, y)$$

is called the **Euclidean distance function on $\mathbf{R}^n$**.

This distance function formulated in Definition 2.3.11 facilitates to formulate an open ball in $\mathbf{R}^n$.

**Definition 2.3.11** The subset $B_x(\epsilon) = \{y \in \mathbf{R}^n : d(x, y) < \epsilon\} \subset \mathbf{R}^n$ is called an **open ball** of radius $\epsilon \ (> 0)$, centered at the point $x \in \mathbf{R}^n$.

### 2.3.3 Continuity of Functions $f : \mathbf{R}^n \to \mathbf{R}^m$

Continuity is a basic concept on which topology is founded. This subsection studies continuity of functions $f : \mathbf{R}^n \to \mathbf{R}^m$ by using the concept of distance function. Our study relies on the standard setting in $\mathbf{R}^n$ with an eye to generalize this study in arbitrary metric spaces in Sect. 2.12 and in topological spaces in Chap. 3 (where the concepts of distance, angle and derivatives are lacking). The continuity of a function in Euclidean spaces defined by generalizing "$\epsilon$-$\delta$" definition of continuity given in calculus or real analysis is equally well defined by its equivalent definition in terms of open balls in $\mathbf{R}^n$.

**Definition 2.3.12** A function $f : \mathbf{R}^n \to \mathbf{R}^m$ is said to be **continuous** at a point $x \in \mathbf{R}^n$ if for every $\epsilon > 0$, there exists a real number $\delta > 0$ such that if $d(x, y) < \delta$, then $d(f(x), f(y)) < \epsilon$, where $d$ is the Euclidean distance function given by $d(x, y) = |x - y|$ between any two points $x, y \in \mathbf{R}^n$ (accordingly, $f(x), f(y) \in$

$\mathbf{R}^m$). A function $f : \mathbf{R}^n \to \mathbf{R}^m$ is said to be continuous if it is continuous at every point $x \in \mathbf{R}^n$.

**Remark 2.3.13** Definition 2.3.12 is analogous to the "$\epsilon$-$\delta$" definition of continuity of a function given in calculus or real analysis. It is frequently used in the study of manifolds given in **Basic Topology, Volume 2** of the present series of books. **Geometrically**, this definition asserts that a function $f : \mathbf{R}^n \to \mathbf{R}^m$ is continuous if the points $f(x)$ and $f(y)$ in $\mathbf{R}^m$ can be made arbitrarily close by choosing the points $x$ and $y$ in $\mathbf{R}^n$ close enough according to the need. Moreover, it follows that a function $f : \mathbf{R}^n \to \mathbf{R}^m$ given by

$$f(x) = (f_1(x), f_2(x), \ldots, f_m(x))$$

is continuous at a point $a \in \mathbf{R}^n$ if and only if each of the component functions $f_i : \mathbf{R}^n \to \mathbf{R}$ is continuous at the point $a$, for $i = 1, 2, \ldots, n$. We formulate a definition of continuity of a function in Definition 2.3.14 in terms of open balls which is equivalent to the Definition 2.3.12.

**Definition 2.3.14** A function $f : \mathbf{R}^n \to \mathbf{R}^m$ is said to be continuous at a point $x \in \mathbf{R}^n$ if for every real number $\epsilon > 0$, there exists a real number $\delta > 0$ such that $y \in B_x(\delta)$ implies $f(y) \in B_{f(x)}(\epsilon)$ (i.e., $f(B_x(\delta)) \subset B_{f(x)}(\epsilon)$). The function $f : \mathbf{R}^n \to \mathbf{R}^m$ is said to be **continuous** if it is continuous at every point $x \in \mathbf{R}^n$.

**Remark 2.3.15 Equivalence of two definitions** of continuous functions given in Definitions 2.3.12 and 2.3.14, i.e., Definition 2.3.12 $\Leftrightarrow$ Definition 2.3.14, is established in Proposition 2.3.16.

**Proposition 2.3.16** *A function $f : \mathbf{R}^n \to \mathbf{R}^m$ is continuous if $f$ satisfies either of the conditions prescribed*

  (i) *in Definition 2.3.12 or*
  (ii) *in Definition 2.3.14.*

**Proof** Let $f : \mathbf{R}^n \to \mathbf{R}^m$ be a given function. First suppose that $f$ is continuous in terms of Definition 2.3.12. Let $y \in B_x(\delta)$ be an arbitrary point. Then by hypothesis, $d(f(x), f(y)) < \epsilon$. This shows that

$$y \in B_x(\delta) \implies f(y) \in B_{f(x)}(\epsilon).$$

Hence $f$ is also continuous in terms of Definition 2.3.14. Conversely, suppose that $f$ is continuous in terms of Definition 2.3.14. Let $y \in B_x(\delta)$ be an arbitrary point. Then the hypothesis $f(y) \in B_{f(x)}(\epsilon)$ shows that

$$d(x, y) < \delta \implies d(f(x), f(y)) < \epsilon.$$

Hence $f$ is also continuous in terms of Definition 2.3.12.

$\square$

**Definition 2.3.17** A subset $U$ of $\mathbf{R}^n$ is said to be **open** if for every element $x \in U$, there exists a real number $\delta > 0$ such that the open ball $B_x(\delta) \subset U$.

Theorem 2.3.18 characterizes continuity of a function in terms of open sets.

**Theorem 2.3.18** *A function $f : \mathbf{R}^n \to \mathbf{R}^m$ is continuous iff $f^{-1}(U)$ is an open set in $\mathbf{R}^n$ for every open set $U$ in $\mathbf{R}^m$.*

**Proof** Let $f : \mathbf{R}^n \to \mathbf{R}^m$ be a continuous function, $U$ be an open set in $\mathbf{R}^m$ and $x \in f^{-1}(U)$. As $U$ is open and $f(x) \in U$, there exists a real number $\epsilon > 0$ such that the open ball $B_{f(x)}(\epsilon) \subset U$. Again, since $f$ is continuous by hypothesis, there exists a real number $\delta > 0$ such that $f(B_x(\delta)) \subset B_{f(x)}(\epsilon)$. This asserts that

$$B_x(\delta) \subset f^{-1}(B_{f(x)}(\epsilon)) \subset f^{-1}(U).$$

This shows that $f^{-1}(U)$ is open in $\mathbf{R}^n$.

Conversely, suppose that $f^{-1}(U)$ is open in $\mathbf{R}^n$ for every open set $U$ in $\mathbf{R}^m$. Let $\epsilon > 0$ be given and $x \in \mathbf{R}^n$ be an arbitrary point. Since the open ball $B_{f(x)}(\epsilon)$ is open in $\mathbf{R}^m$, the set $f^{-1}(B_{f(x)}(\epsilon))$ is open in $\mathbf{R}^n$ and $x \in f^{-1}(B_{f(x)}(\epsilon))$. Consequently, there exists a real number $\delta > 0$ such that

$$B_x(\delta) \subset f^{-1}(B_{f(x)}(\epsilon)).$$

This implies that $f(B_x(\delta)) \subset B_{f(x)}(\epsilon)$ and hence $f$ is continuous by Definition 2.3.14.                                                                                           $\square$

**Remark 2.3.19** Some examples of continuous functions from the geometrical viewpoint are given in Sect. 2.16.4.

## 2.4  Metric Spaces: Introductory Concepts

This section begins with the introductory concept of metric spaces by defining several metrics on abstract settings and studies different properties of metric spaces. A metric space $X$ is a set of points admitting a quantitative measure of the degree of nearness between pair of points in $X$. The concept of "nearness" in a metric space through its metric generalizes "$\epsilon$-$\delta$" definition of continuity of a function in $\mathbf{R}$. The study of convergence of sequences of points in metric spaces is almost similar to the study of sequences of real numbers. For example, the concept of Cauchy sequence in a metric space is similar to the concept of Cauchy sequence of real numbers in analysis.

The continuity of functions can be studied with the help of convergence of sequences in metric spaces. The concept of distance available in many branches of mathematics can be extended to abstract sets by defining metrics. For example, given two real numbers $x, y \in \mathbf{R}$, there exists a nonnegative real number $d(x, y)$, called the distance between the points $x$ and $y$. This defines a function

$$d : \mathbf{R} \times \mathbf{R} \to \mathbf{R}$$

satisfying the properties (i)–(iii) prescribed in Definition 2.4.1. In the Euclidean $n$-space $\mathbf{R}^n$ a distance function can be defined in a similar way. These properties of $\mathbf{R}$ (or $\mathbf{R}^n$) are sufficient to study continuity of a function and motivate to extend a study of continuity in an abstract set $X$ admitting a distance function $d : X \times X \to \mathbf{R}$ given in Definition 2.4.1.

**Definition 2.4.1** A nonempty abstract set $X$ is said to have a **metric or a distance function**

$$d : X \times X \to \mathbf{R},$$

if for every pair of elements $x, y$ in $X$

  (i) (**positivity**) $d(x, y) \geq 0$, equality holds iff $x = y$;
  (ii) (**symmetry**) $d(x, y) = d(y, x)$;
  (iii) (**triangle inequality**) $d(x, y) + d(y, z) \geq d(x, z)$ for all $z \in X$.

$d(x, y)$ is called the distance between $x$ and $y$ and the pair $(X, d)$ is called a **metric space** or $X$ is said to be metricized by $d$.

**Remark 2.4.2** In a metric space $(X, d)$, the distance function $d$ conveys a quantitative measure of the **degree of closeness** of two points in $X$, and the triangle inequality in Definition 2.4.1 asserts the **transitivity of closeness** in the sense that given three points $x, y, z \in X$, if $x$ is close to $y$ and $y$ is close to $z$, then $x$ is close to $z$.

## Pseudo-metric Spaces

There are many spaces studied in modern analysis which are not metric spaces, but they behave almost like metric spaces. Such spaces, called pseudo-metric spaces, are formally defined in Definition 2.4.3.

**Definition 2.4.3** A nonempty set $X$ is said to have a pseudo-metric

$$d : X \times X \to \mathbf{R},$$

if it satisfies the following axioms:

  (i) $d(x, x) = 0$, $\forall x \in X$;
  (ii) (triangle inequality) $d(x, z) + d(y, z) \geq d(x, y)$, $\forall x, y, z \in X$.

Then the pair $(X, d)$ is called a **pseudo-metric space** with $d$ as its pseudo-metric.

***Example 2.4.4***   (i) Every metric space is a pseudo-metric space.

(ii) The converse of (i) is not true. For example let $X$ be set such that $\text{card}(X) > 1$. Consider the function

$$d : X \times X \to \mathbf{R}, (x, y) \mapsto 0.$$

Then $d$ is a pseudo-metric, but it is not a metric.

**Proposition 2.4.5** *If a set $X$ has a pseudo-metric $d : X \times X \to \mathbf{R}$, then*

(i) *(positivity)* $d(x, y) \geq 0, \ \forall x, y \in X$;
(ii) *(symmetry)* $d(x, y) = d(y, x), \ \forall x, y \in X$.

**Proof** Let $(X, d)$ be a pseudo-metric space.

(i) Apply the axioms of Definition 2.4.3 to the elements $x, x, y \in X$ to show that

$$2d(x, y) = d(x, y) + d(x, y) \geq d(x, x) = 0.$$

This implies that $d(x, y) \geq 0, \ \forall x, y \in X$.
(ii) Apply the axiom (ii) of Definition 2.4.3 to the elements $x, y, x \in X$ to show that

$$d(x, x) + d(y, x) \geq d(x, y).$$

This implies that $d(y, x) \geq d(x, y)$, since $d(x, x) = 0$. Similarly, it follows that $d(x, y) \geq d(y, x)$ and hence $d(x, y) = d(y, x)$.

□

**Remark 2.4.6** A metric on a nonempty set $X$ can be redefined as a pseudo-metric

$$d : X \times X \to \mathbf{R}$$

satisfying the following additional axiom:
   (**M**) for any two points $x, y \in X$, if $d(x, y) = 0$, then $x = y$.


## 2.5  Examples of Metrics Arising from Mathematical Analysis

This section presents some examples of useful metrics arising from mathematical analysis.

*Example 2.5.1*    (i) The function

$$d : \mathbf{R} \times \mathbf{R} \to \mathbf{R}, \ (x, y) \mapsto |x - y|,$$

defines a metric on $\mathbf{R}$, called the Euclidean metric or usual metric on $\mathbf{R}$.

(ii)  The function

$$d : \mathbf{R} \times \mathbf{R} \to \mathbf{R}, \ (x, y) \mapsto \begin{cases} |x - y|, & \text{if } xy \leq 0 \\ |x| + |y|, & \text{otherwise.} \end{cases}$$

also defines a metric on $\mathbf{R}$.

**Remark 2.5.2**  Example 2.5.1 shows that there may exist different nontrivial metrics on the same set.

**Example 2.5.3**  The function

$$d : \mathbf{C} \times \mathbf{C} \to \mathbf{C}, \ (z_1, z_2) \mapsto |z_1 - z_2|,$$

(absolute value of the complex number $z_1 - z_2$) is called the usual metric on $\mathbf{C}$. The complex plane is an ideal model of a metric space.

### 2.5.1  Norm Function

Norm functions given in Definition 2.5.4 on vector spaces provide a rich supply of metric spaces.

**Definition 2.5.4**  Let $V$ be a real vector space with 0 its zero vector. A real-valued function

$$\| \ \| : V \to \mathbf{R}$$

is called a **norm function** on $V$ if it satisfies the following axioms for all $x, y \in V$ and for all $r \in \mathbf{R}$:

**N(1)**  $\|x\| \geq 0$ and $\|x\| = 0$ iff $x = 0$;
**N(2)**  $\|x + y\| \leq \|x\| + \|y\|$;
**N(3)**  $\|rx\| = |r| \, \|x\|$.

**Example 2.5.5**  Let $\mathcal{C} = \mathcal{C}([0, 1])$ be the set of all real-valued continuous functions on $[0, 1]$. Then every function $f \in \mathcal{C}$ is also bounded, since $[0, 1]$ is compact (see Chap. 5). $\mathcal{C}$ is a real vector space under usual compositions of addition and scalar multiplication of functions. Using the different norm functions on it, some important metrics are defined on $\mathcal{C} = \mathcal{C}([0, 1])$. Such metric spaces are also born through the study of the problems in analysis.

(i)  ($l_1$-**metric on** $\mathcal{C}$) If a norm of $f \in \mathcal{C}$ is defined by

$$\|f\| = \int_0^1 |f(x)| \, \mathrm{d}x,$$

where the integral is the Riemann integral, then the function

$$d : C \times C \to \mathbf{R}, (f, g) \mapsto \|f - g\| = \int_0^1 |f(x) - g(x)| dx$$

is a metric on $C$. By a property of Riemann integral, $\int_0^1 |f(x) - g(x)| dx \geq 0$
implies that $d$ is well defined. Again, for $f \in C$, the integral $\int_0^1 f(x) dx = 0$
iff $f$ vanishes identically on $[0, 1]$. It shows that $d(f, g) = 0$ iff $f = g$. This
metric $d$ is called $l_1$-metric on $C$.

(ii) If a norm of $f \in C$ is defined by

$$\|f\| = \sup\{|f(x)| : x \in [0, 1]\} = \sup_{x \in [0,1]} \{|f(x)|\},$$

then the function $\rho$ defined by

$$\rho(f, g) = \|f - g\| = \sup_{x \in [0,1]} \{|f(x) - g(x)|\}$$

is a metric on $C$.
On the other hand,

$$\sigma : C \times C \to \mathbf{R}, (f, g) \mapsto \inf_{x \in [0,1]} \{|f(x) - g(x)|\}$$

is not a metric on $C$. Because for $f, g \in C$, defined by $f(x) = 0$, $g(x) = x$, $\forall x \in [0, 1]$, $\rho(f, g) = 0$ but $f \neq g$, since $f(1) = 0$, $g(1) = 1$.

**Geometrically**, this $d(f, g)$ (defined in Example 2.5.5 (i)) represents the area of the shaded region lying between the graphs of the functions $f, g \in C$ and the lines $x = 0$ and $x = 1$, as shown in Fig. 2.1.

**Geometrically**, this $\rho(f, g)$ (defined in Example 2.5.5 (ii)) represents the largest vertical gap $h$ between the graphs of the functions $f, g \in C$, as shown in Fig. 2.2.

**Example 2.5.6**   (i) Let $\mathcal{B}(X)$ be the set of all bounded real-valued functions defined on a given set $X$. Then the function $d : \mathcal{B}(X) \times \mathcal{B}(X) \to \mathbf{R}$ defined by

$$d(f, g) = \sup_{x \in X} |f(x) - g(x)|$$

is a metric on $\mathcal{B}(X)$, and hence $(\mathcal{B}(X), d)$ is a metric space.

(ii) Let $M(n, \mathbf{R})$ be the set of all $n \times n$ matrices over $\mathbf{R}$. Then identifying $M(n, \mathbf{R})$ with the Euclidean space $\mathbf{R}^{n^2}$, the function

**Fig. 2.1** Geometrical
representation of $d(f, g)$

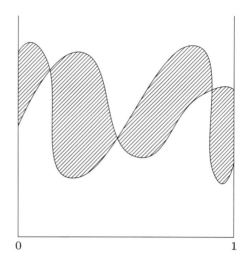

0                                                                                  1

**Fig. 2.2** Geometrical
representation of $\rho(f, g)$

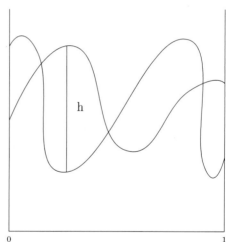

h

0                                                                                  1

$$d : M(n, \mathbf{R}) \times M(n, \mathbf{R}) \to \mathbf{R}, (A, B) \mapsto \sum_{i,j=1}^{n} |a_{i,j} - b_{i,j}|,$$

$$\forall\, A = (a_{i,j}),\, B = (b_{i,j}) \in M(n, \mathbf{R}),$$

defines a metric on $M(n, \mathbf{R})$ and hence $(M(n, \mathbf{R}), d)$ is a metric space.

(iii)  (**Discrete metric**): Given a nonempty set $X$, the function $d : X \times X \to \mathbf{R}$,
defined by

$$d(x, y) = \begin{cases} 0, & \text{if } x = y \\ 1, & \text{if } x \neq y \end{cases}$$

gives a metric, called the discrete metric on $X$, and $(X, d)$ is called a discrete metric space.

## 2.5.2  Euclidean Metric, $l_p$-Metric and $l_\infty$-Metric on $\mathbf{R}^n$

This subsection defines several commonly used metrics such as Euclidean metric, $l_p$-metric and $l_\infty$-metric on $\mathbf{R}^n$.

**Definition 2.5.7**    (i)  The metric

$$d : \mathbf{R}^n \times \mathbf{R}^n \to \mathbf{R}, \ ((x_1, x_2, \ldots, x_n), \ (y_1, y_2, \ldots, y_n)) \mapsto$$
$$[(x_1 - y_1)^2 + \cdots + (x_n - y_n)^2]^{\frac{1}{2}}$$

is called the **usual metric or Euclidean metric** on $\mathbf{R}^n$.

(ii)  For any integer $p \geq 1$, the metric

$$d_p : \mathbf{R}^n \times \mathbf{R}^n \to \mathbf{R}, \ ((x_1, x_2, \ldots, x_n), \ (y_1, y_2, \ldots, y_n)) \mapsto$$
$$[|x_1 - y_1|^p + \cdots + |x_n - y_n|^p]^{\frac{1}{p}}$$

is called the $l_p$-**metric** on $\mathbf{R}^n$.

(iii)  The metric

$$d_\infty : \mathbf{R}^n \times \mathbf{R}^n \to \mathbf{R}, \ ((x_1, x_2, \ldots, x_n), \ (y_1, y_2, \ldots, y_n)) \mapsto$$
$$\max\{|x_1 - y_1| + \cdots + |x_n - y_n|\}$$

is called the $l_\infty$-**metric** on $\mathbf{R}^n$.

## 2.5.3  $l_p$-Metric and $l_\infty$-Metric on $\mathcal{C}([0, 1])$

This subsection defines two special metrics, called $l_p$-metric and $l_\infty$-metric on the set of all real-valued continuous functions on $[0, 1]$.

**Definition 2.5.8** ($l_p$-**metric**). Let $\mathcal{C} = \mathcal{C}([0, 1])$ be the set of all real-valued continuous functions on $[0, 1]$ and $p > 0$ be an integer.

(i)

$$d_p : \mathcal{C} \times \mathcal{C} \to \mathbf{R}, \ (f, g) \mapsto \left[ \int_0^1 |f(x) - g(x)|^p dx \right]^{\frac{1}{p}}$$

is a metric on $\mathcal{C}$, called $l_p$-**metric on** $\mathcal{C}$. In particular, for $p = 2$, the metric $d_2$ is called $l_2$-metric on $\mathcal{C}$.

(ii)

$$d_\infty : \mathcal{C} \times \mathcal{C} \to \mathbf{R}, \ (f, g) \mapsto \max_{x \in [0,1]} |f(x) - g(x)|$$

is a metric on $\mathcal{C}$, called $l_\infty$-**metric on** $\mathcal{C}$.

Clearly, $d_\infty(x, y) = \lim_{p \to \infty} d_p(x, y)$.

### 2.5.4 $l_p$-Metric and $l_\infty$-Metrics on Sequences of Real Numbers

This subsection introduces the concepts of $l_p$- space, Hilbert space and $l_\infty$-space by defining the corresponding metrics on the set of all sequences of real numbers.

**Definition 2.5.9** ($l_p$-**metric**) Let $X$ be the set of all sequences $\{x_n\}$ over $\mathbf{R}$ with the property that for $p > 0$, $\sum_{n=1}^\infty |x_n|^p < \infty$. Then the function

$$d : X \times X \to \mathbf{R}, (f, g) \mapsto \left[ \sum_{n=1}^\infty |x_n - y_n|^p \right]^{\frac{1}{p}}, \ \forall f = \{x_n\}, \ g = \{y_n\} \in X,$$

is a metric on $X$, called the $l_p$-**metric on** $X$. The metric space $(X, \ l_p)$ is called $l_p$-metric space. Let $\mathbf{R}^\infty$ denote the family of real infinite sequences $\{x_n\}$ such that $\sum_{n=1}^\infty |x_n|^2 < \infty$, i.e., the family of sequences $\{x_n\}$ such that the series $\sum_{n=1}^\infty |x_n|^2$ is convergent. Then in particular, for $p = 2$, the metric $l_2$ is called the $l_2$-metric on $\mathbf{R}^\infty$, and the metric space $(\mathbf{R}^\infty, \ l_2)$ is called a **Hilbert space or** $l_2$-**space**.

**Definition 2.5.10** ($l_\infty$-**metric**) Let $X$ be the set of all bounded sequences $\{x_n\}$ over $\mathbf{R}$. Then the function

$$d : X \times X \to \mathbf{R}, (f, g) \mapsto \sup\{|x_n - y_n| : n \in \mathbf{N}\}, \ \forall f = \{x_n\}, g = \{y_n\} \in X$$

is well defined (since $f, g$ are bounded real sequences), and it is a metric on $X$, called the $l_\infty$-**metric on** $X$. The metric space $(X, \ l_\infty)$ is called $l_\infty$-**space**.

**Remark 2.5.11** For more study of Hilbert space or $l_\infty$-space, see Sect. 2.14.

## 2.5.5   p-Adic Metric on Q

This subsection introduces an interesting metric on $\mathbf{Q}$, known as $p$-adic metric by defining $p$-adic valuation of an integer, by using fundamental theorem of arithmetic.

**Definition 2.5.12** (*p*-**Adic metric**) Let $p$ be a given prime integer and $x \in \mathbf{Q}$ be nonzero. Then $x$ is represented as $x = p^r \frac{m}{n}$, where $r, m, n$ are integers such that $p$ divides neither $m$ nor $n$. Then the function $v_p$ defined by $v_p(x) = r$ is called the $p$-adic valuation of $x$.

**Example 2.5.13**

$$d_p : \mathbf{Q} \times \mathbf{Q} \to \mathbf{R}, \ (x, y) \mapsto \begin{cases} 0, & \text{if } x = y \\ p^{-v_p(x-y)}, & \text{if } x \neq y \end{cases}$$

defines a metric on $\mathbf{Q}$, called the $p$-adic metric or $p$-adic distance function on $\mathbf{Q}$, and $(\mathbf{Q}, d_p)$ is called a $p$-adic metric space.

**Proposition 2.5.14** *For $m, n \in \mathbf{Q}$,*

(i) $v_p(m + n) \geq \min\{v_p(m), v_p(n)\}$, *equality holds iff* $v_p(m) \neq v_p(n)$.
(ii) $v_p(mn) = v_p(m) + v_p(n)$.

**Proof** Left as an exercise.                                                              □

**Definition 2.5.15** Given a prime integer $p$, the function

$$\| \ \| : \mathbf{Q} \to \mathbf{R}, \ x \mapsto \begin{cases} 0, & \text{if } x = 0 \\ p^{-v_p(x)}, & \text{if } x \neq 0 \end{cases}$$

is said to be **the *p*-adic norm on Q**, abbreviated by $\| \ \|_p$.

**Proposition 2.5.16** *The $p$-adic norm function $\| \ \|_p$ has the following properties:*

(i) $\|n\|_p = 0$ *iff* $n = 0$;
(ii) $\|mn\|_p = \|m\|_p + \|n\|_p$, $\forall m, n$ *in* $\mathbf{Q}$;
(iii) $\|m + n\|_p \leq \max\{\|m\|_p, \|n\|_p\}$, $\forall m, n$ *in* $\mathbf{Q}\}$, *the equality holds when* $\|m\|_p \neq \|n\|_p$.

**Proof** Left as an exercise.                                                              □

**Definition 2.5.17** The $p$-adic norm function $\| \ \|_p$ is said to be nonArchimedean on $\mathbf{Q}$, if

$$\|m - n\|_p \leq \max\{\|m\|_p, \|n\|_p\}, \ \forall m, n \text{ in } \mathbf{Q}.$$

**Corollary 2.5.18** *The norm $\| \ \|_p$ is non Archimedean on $\mathbf{Q}$.*

Given a fixed prime integer $p$, the difference $x - y$ of any two $x, y \in \mathbf{Q}$ can be expressed uniquely as

$$x - y = \frac{r}{t} p^s \text{ such that } : r, s, t \in \mathbf{Z}, \text{ and } r, t \text{ are not divisible by } p.$$

Then the $p$-adic distance function $d_p$ can be given equally well by

$$d_p : \mathbf{Q} \times \mathbf{Q} \to \mathbf{R}, (x, y) \mapsto \begin{cases} 0, & \text{if } x = y \\ p^{-s}, & \text{if } x \neq y \end{cases}$$

***Example 2.5.19*** For $x = \frac{9}{5}$ and $y = \frac{3}{2}$, the $p$-adic distances $d_p(x, y)$ *for* $p = 2, 3, 5$, are $d_2(x, y) = 2$, $d_3(x, y) = \frac{1}{3}$, $d_5(x, y) = 5$, because, $x - y = 3 \cdot 2^{-1}$. $5^{-1}$ and hence $d_2(x, y) = 2^{-(-1)} = 2$, $d_3(x, y) = 3^{(-1)} = \frac{1}{3}$, $d_5(x, y) = 5^{-(-1)} = 5$.

## 2.6 Open Balls and Open Sets in Metric Spaces

This section addresses the concepts of open balls and open sets in metric spaces for our subsequent study by taking open balls in a metric space as open sets. Open balls in a metric space $X$ are generalizations of open intervals in $\mathbf{R}$, and closed balls in $X$ are generalizations of closed intervals in $\mathbf{R}$.

**Definition 2.6.1** Let $(X, d)$ be a metric space and $Y \subset X$ be a nonempty subset. If $d' : Y \times Y \to \mathbf{R}$ is the restriction of $d : X \times X \to \mathbf{R}$ to $Y \times Y$, then $d'$ is said to be a metric on $Y$ induced by the metric $d$, and $(Y, d')$ or simply $Y$ is said to be a **subspace of the metric space** $X$.

***Example 2.6.2*** Definition 2.6.1 asserts that every nonempty subset of a metric space inherits a metric structure from its mother metric space.This shows that every nonempty subset of the Euclidean space $\mathbf{R}^n$ is a metric space.

**Definition 2.6.3** Let $X$ be a metric space with metric $d$. The set

$$B_x(\epsilon) = \{y \in X : d(x, y) < \epsilon\},$$

for $x \in X$ and $\epsilon > 0$, is said to be the **open ball of radius** $\epsilon$, **centered at** $x$, also known as the "$\epsilon$-ball" about the point $x$. The set

$$B_x(\epsilon) = \{y \in X : d(x, y) \leq \epsilon\},$$

for $x \in X$ and $\epsilon > 0$ is said be the **closed ball of radius** $\epsilon$, **centered at** $x$.

**Definition 2.6.4** Let $(X, d)$ be a metric space with metric $d$. A subset $U$ of $X$ is said to be **open** if given $x \in U$, there is a positive real number $\epsilon$ such that the open ball $B_x(\epsilon) \subset U$.

**Example 2.6.5**    (i)  No one-pointic set $\{x\}$ on the real line **R** is open, because for each bounded open interval centered on the point $x$ contains points not in the set $\{x\}$.

(ii)  The subset $(0, 1)$ of **R** is an open set (see Corollary 2.6.9).

(iii)  The subset $[0, 1)$ of the real line **R** is not open.

**Theorem 2.6.6** *Let $(X, d)$ be a metric space. Then the emptyset $\emptyset$ and the whole set $X$ are open sets.*

**Proof** Let $x \in X$ be an arbitrary point. Then every open ball $B_x(\epsilon)$ centered at $x$ and radius $\epsilon > 0$, however small it may be, is contained in $X$. This shows that every open ball in $X$ centered on its every point is contained in $X$ and hence the whole set $X$ is open. Again, as there are no points in $\emptyset$, the empty set $\emptyset$ is automatically an open set.                                                                                 □

**Example 2.6.7** $[0, 1)$ is not an open subset of the real line **R** but $[0, 1)$ is an open set considered as a metric space $X$, i.e., $[0, 1) = X$, the whole metric space is open by Theorem 2.6.6. This shows that given set $B$, whether it is open or not, it depends on the metric space containing $B$, which we are considering.

**Theorem 2.6.8** *Let $(X, d)$ be a metric space with metric $d$. Then every open ball $B_{x_o}(\epsilon)$ in $X$ is an open set.*

**Proof** Let $x \in B_{x_o}(\epsilon)$. Then $d(x, x_0) < \epsilon$ implies that $r = \epsilon - d(x, x_0)$ is a positive real number. This implies that $B_x(r) \subset B_{x_o}(\epsilon)$. Because, for any point $z \in B_x(r)$, its distance $d(x, z) < r$ from $x$. Using the triangle inequality, it follows that

$$d(x_0, z) \leq d(x_0, x) + d(x, z)$$

This shows that

$$d(x_0, z) < \epsilon - r + r = \epsilon \implies z \in B_{x_0}(\epsilon) \implies B_x(r) \subset B_{x_0}(\epsilon)$$

This asserts by Definition 2.6.4 that every open ball in $X$ is an open set.

□

**Corollary 2.6.9** *Any open interval $(a, b)$ of **R** is an open set.*

**Proof** It follows from Theorem 2.6.8.                                                           □

Theorem 2.6.10 characterizes open sets in a metric space with the help of open balls.

**Theorem 2.6.10** *Let $(X, d)$ be a metric space. A subset $B$ of $X$ is open iff $B$ is a union of open balls in $(X, d)$.*

**Proof** Let $(X, d)$ be a metric space. First suppose that the subset $B$ of $X$ is open. We claim that $B$ is a union of open balls in $(X, d)$. If $B = \emptyset$, then $B$ is is the union of empty class of open balls. On the other hand, if $B \neq \emptyset$, then as $B$ is open in $(X, d)$, every point of $B$ is the center of an open ball contained in it, and $B$ is the union of all the open balls contained in it.

Conversely, if $B$ is the union of a class $\mathcal{B}$ of open balls in $(X, d)$, then we claim that $B$ is open in $(X, d)$. If $\mathcal{B} = \emptyset$, then $B$ is open by Theorem 2.6.6. On the other hand, if $\mathcal{B} \neq \emptyset$, then $B \neq \emptyset$. If $x \in B$, then $B$ being the union of the open balls in $\mathcal{B}$, $x \in B_{x_o}(\epsilon)$ in $\mathcal{B}$. This shows by Theorem 2.6.8 that $x$ is the center of an open ball $B_x(r) \subset B_{x_o}(\epsilon)$. Again as $B_{x_o}(\epsilon) \subset B$, it follows that there is an open ball centered at x and contained in $B$. This implies that $B$ is open in $(X, d)$. $\qquad \square$

**Remark 2.6.11** Given a metric space $X$, the empty set $\emptyset$ and the whole (universal) set $X$ are open sets by Theorem 2.6.6. We now claim that the union of any number of open sets in $X$ is an open set and the intersection of a finite number of open sets in $X$ is an open set.

**Theorem 2.6.12** *Let $(X, d)$ be a metric space. Then*

(i) *the union of any number of open sets in $(X, d)$ is an open set;*
(ii) *the intersection of a finite number of open sets in $(X, d)$ is an open set.*

**Proof** (i) Let $\{F_i\}$ be an arbitrary family (may be finite or infinite) of open sets in $(X, d)$. Claim that $F = \bigcup_i F_i$ is open in $(X, d)$. If each member in $\{F_i\}$ is empty, then $F = \emptyset$ is open by Theorem 2.6.6. On the other hand, if $\{F_i\} \neq \emptyset$, each $F_i$, being an open set by hypothesis, it is a union of open balls in $(X, d)$ by Theorem 2.6.10. Again applying Theorem 2.6.10 it follows that $F$ is an open set in $(X, d)$.

(ii) Let $\{F_i\}$ be a finite family of open sets in $(X, d)$. Claim that $F = \bigcap_i F_i$ is open in $(X, d)$. To prove this, we consider the following cases:
Case 1: If $\{F_i\} = \emptyset$, then $F = \bigcap_i F_i = \emptyset$ is open by Theorem 2.6.6.
Case 2: If $\{F_i\} \neq \emptyset$, and $\{F_i\} = \{F_1, F_2, \ldots, F_n\}$ for some $n \in \mathbf{N}$, then $F = \bigcap_i F_i$ is also open in $(X, d)$. Because, in this case, there is a point $x$ in each $F_i$, which is an open set. Hence, there exists a real number $\epsilon_i > 0$ such that $B_x(\epsilon_i) \subset F_i$ for every $i = 1, 2, \ldots, n$. Let $\epsilon$ be the smallest real number of the set $\{\epsilon_1, \epsilon_2, \ldots, \epsilon_n\}$ of real numbers. Then $\epsilon > 0$ and

$$B_x(\epsilon) \subset B_x(\epsilon_i) \subset F_i, \ \forall i = 1, 2, \ldots, n.$$

This asserts that $B_x(\epsilon) \subset F$. This implies that $F$ is an open set in $(X, d)$. $\qquad \square$

**Corollary 2.6.13** (i) *The union of any number of open sets in $\mathbf{R}$ is an open set.*
(ii) *The intersection of a finite number of open sets in $\mathbf{R}$ is an open set.*

**Proof** It follows from Theorem 2.6.12 as a particular case. $\qquad \square$

***Remark 2.6.14*** Given a metric space $(X, d)$, a subset $U \subset X$ is open if for every point $x \in U$, there exists an $\epsilon > 0$ such that the open ball $B_x(\epsilon) \subset U$. The discussion in Theorems 2.6.6 and 2.6.12 (taken together) is summarized in the following basic and important result embodied in Theorem 2.6.15 that leads to the axiomatic framework for open sets defining a topology on an abstract set (see Chap. 3).

**Theorem 2.6.15** *Let $(X, d)$ be a metric space. A subset $U \subset X$ is defined to be open in $X$, if for every point $x \in U$, there exists an $\epsilon > 0$ such that the open ball $B_x(\epsilon) \subset U$. Let $\tau$ be the set of all open sets in $X$. Then*

**OS(1)** *the empty set $\emptyset$ is an open set;*
**OS(2)** *the union of any number of open sets is an open set;*
**OS(3)** *the intersection of a finite number of open sets is an open set;*
**OS(4)** *the whole (universal) set $X$ is itself an open set.*

**Definition 2.6.16** The family $\tau$ of all open sets in $X$ given in Theorem 2.6.15 is said to form a topology on $X$, called the topology determined by the metric $d$, and the the pair $(X, \tau)$ is called the corresponding topological space.

***Example 2.6.17*** Theorem 2.6.12 shows that the collection of all open sets in a metric space is closed under arbitrary union and finite intersection. The restriction of finite intersections in this theorem is necessary. For example, consider the metric space **R** with usual metric and the sequence

$$\{(-1, 1), (-1/2, 1/2), (-1/3, 1/3), \dots, \},$$

which is the sequence of open intervals $\{X_n = (-\frac{1}{n}, \frac{1}{n})\}$. Then for each $n \in \mathbf{N}$, the open interval $X_n$ is an open set in **R**, but their infinite intersection $\bigcap_1^\infty X_n = \{0\}$ is not an open set in **R**.

**Theorem 2.6.18** *Every nonempty open set in the metric space $(\mathbf{R}, d)$ with usual metric $d$ on $\mathbf{R}$ is the union of a countable family of disjoint open intervals.*

***Proof*** Let $U \subset \mathbf{R}$ be a nonempty open subset. Then for each $x \in U$, there is a bounded open interval contained in $U$, since $U$ is open and hence the point $x$ is the center of a bounded open interval $\mathbf{I}_x$ contained in $U$. Let $U_x$ denote the union of all open intervals $\mathbf{I}_x$ such that $x \in \mathbf{I}_x \subset U$ and hence

$$U_x = \bigcup \{\mathbf{I}_x : x \in \mathbf{I}_x \subset U\}$$

is an open interval and contains every interval which contains x and is contained in $U$ with the property that if $y \in U_x$, then $U_x = U_y$. Again it follows that if $x, y \in U$ and $x \neq y$, then either $U_x = U_y$ or they are distinct, because if $z \in U_x \cap U_y$, then $U_x = U_z$, and $U_y = U_z$ imply that $U_x = U_y$. Let $\Omega$ be the set of all distinct subsets of $\mathbf{R}$ of the form $U_x$ for $x \in U$. Then $\Omega$ is a family of disjoint open intervals, and $U$ is their union. Let $U_r$ be the set of all rational points in $U$. Then $U_r \neq \emptyset$. Define a map

$$\psi : U_r \to \Omega, r \mapsto \mathbf{I}_r$$

where $\mathbf{I}_r$ is the unique open interval in $\Omega$ which contains the point $r$. Then $\psi$ is onto and $U_r$ is countable. This asserts that the set $\Omega$ is countable.                                □

**Definition 2.6.19** Let $(X, d)$ be a metric space and $A \subset X$ be a given subset. Then a point $x \in A$ is said to be an interior point of $A$ with respect to metric d if $x \in B_x(\epsilon) \subset A$ for some $\epsilon > 0$, i.e., $x$ is at the center of some open ball in $A$. The set of all interior points of $A$ is called **Interior of** $A$, denoted by $\text{Int}(A)$.

**Example 2.6.20** For the closed interval $A = [2, 3] \subset \mathbf{R}$, $\text{Int}(A) = (2, 3)$.

**Example 2.6.21** For a nonenumerable subset $A$ of a metric space $(X, d)$, the set $\text{Int}(A)$ may be $\emptyset$. For example, metric space $(\mathbf{R}, d)$ with usual metric $d$, let $A$ be the set of all irrational points. Then for any irrational point $x$, and any $\epsilon > 0$, the open ball $B_x(\epsilon)$ contains rational points. Hence $x$ cannot be an interior point of $A$. This implies that $\text{Int}(A) = \emptyset$.

**Proposition 2.6.22** *Let* $(X, d)$ *be a metric space and* $A \subset X$ *be any subset. Then*

(i) $\text{Int}(A)$ *is an open set;*
(ii) $\text{Int}(A)$ *is the largest open set contained in* $A$*;*
(iii) $A$ *is open iff* $\text{Int}(A) = A$.

**Proof** It follows from the Definition 2.6.19 of $\text{Int}(A)$.                                □

## 2.7 Neighborhoods in Metric Spaces

This section starts with the concept of neighborhoods (nbd) in the real line space $\mathbf{R}$ and generalizes this idea in metric spaces. Neighborhoods in the real line $\mathbf{R}$ play a key role in real analysis and are closely related to its open sets. So its generalization for an arbitrary metric space has become necessary for the study of topological spaces, in particular, manifolds. The entire (whole) metric space needs not be taken as a nbd of any point of the space.

### 2.7.1 Neighborhoods in R

This subsection considers neighborhoods in the metric space $\mathbf{R}$ with usual metric, which is the beginning of the framework for neighborhoods in metric spaces. A nbd of a point in a metric space is not necessarily an open set (see Example 2.7.2 (iv)).

**Definition 2.7.1** Let $x \in \mathbf{R}$. Then a subset $N_x \subset \mathbf{R}$ is said to be a **neighborhood in brief (nbd)** of $x$ if $x \in N_x$, and there is a real number $\delta > 0$ such that the open interval $(x - \delta, \ x + \delta) \subset N_x$.

***Example 2.7.2***   (i) $\mathbf{R}$ is itself a nbd of every point $x \in \mathbf{R}$.
  (ii) $\mathbf{Q}$ is not a nbd of any point $x \in \mathbf{Q}$, because for every $\delta > 0$, the open interval $(x - \delta, \ x + \delta)$ must contain irrational and rational points and hence the open interval $(x - \delta, \ x + \delta)$ is not contained in $\mathbf{Q}$.
  (iii) $[0, 2]$ is a nbd of 1, but it is not so for $x = 0$ or $x = 2$.
  (iv) Nbd of a point in $\mathbf{R}$ is not necessarily an open set. Because, for any $x \in \mathbf{R}$, each closed interval $[x - \delta, \ x + \delta]$ ( $\delta > 0$) is a nbd of $x$, since $x \in (x - \delta, \ x + \delta) \subset [x - \delta, \ x + \delta]$. This example shows that a closed interval in $\mathbf{R}$ may be a nbd of a point but it is not an open set. Thus a nbd of a point in $\mathbf{R}$ is not necessarily an open set.

## 2.7.2   Neighborhoods and Open Sets in Metric Spaces

This subsection considers neighborhood system in an arbitrary metric space, which is a generalization of neighborhood system in the metric space $\mathbf{R}$ with usual metric.

**Definition 2.7.3** Let $(X, d)$ be a metric space. Given $x \in X$ and $\epsilon > 0$, the subset $\{y \in X : d(x, y) < \epsilon\} \subset X$, denoted by $N_x(\epsilon)$, is called a **neighborhood** (in brief nbd) of $x$.

**Definition 2.7.4** Let $(X, d)$ be a metric space. Given $x \in X$, a subset $A \subset X$ is said to be a nbd of $x$ with respect to "$d$" if there exists an $\epsilon > 0$ such that $N_x(\epsilon) \subset A$. Equivalently, $A$ is said to be a nbd of $x$ with respect to "$d$" if there exists an $\epsilon > 0$ such that $x \in B_x(\epsilon) \subset U$. In particular, $N_x(\epsilon)$ is itself a nbd of $x$, called an $\epsilon$-nbd of $x$. The set of all nbds of $x$ abbreviated by $\mathcal{N}_x$ is called the **nbd system** of $x$.

***Remark 2.7.5*** For a metric space $X$, a nbd system $\mathcal{N}_x$ at the $x \in X$, is a family of nbds of $x$, such that for any nbd $U_x$ of $x$, there is a member $N_x \in \mathcal{N}_x$ with the property that $x \in N_x \subset U_x$.

***Example 2.7.6*** For any point $x \in \mathbf{R}^2$, each closed ball

$$B_x(\delta) = \{y \in \mathbf{R}^2 : d(x, y) \leq \delta\} \ (\delta > 0)$$

in $\mathbf{R}^2$ is a nbd of $x$.

**Definition 2.7.7** Let $(X, d)$ be a metric space. A subset $U$ of $X$ is said to be open in $(X, d)$, i.e., open with respect to the metric $d$, if for every $x \in U$, there is an $\epsilon > 0$ (depending on $x$) such that

$$N_x(\epsilon) \subset U.$$

*Example 2.7.8* Let $(X, d)$ be a metric space. Then

(i) every open ball $B_x(\epsilon)$ is an open set in $(X, d)$ for every choice of $x \in X$ and $\epsilon > 0$;

(ii) The whole set $X$ and $\emptyset$ are open sets in $(X, d)$.

## 2.8 Limit Points, Closed and Dense Sets in Metric Spaces

This section introduces the concepts of limit points of subsets, closed sets and dense sets in the setting of metric spaces and characterizes closed sets with the help of open sets in metric spaces. Open and closed sets are considered as the basic sets for the study of topological spaces (see Chap. 3).

**Definition 2.8.1** Let $(X, d)$ be a metric space and $A$ be a subset of $X$. Then a point $x \in X$ is said to be a **limit point of** $A$ if every open $\epsilon$-ball $B_x(\epsilon)$ centered at $x$ contains at least one point of $A$ different from the point $x$. The limit points of $A$ form a subset of $X$, called the **derived set of** $A$, and are written as $A'$.

**Remark 2.8.2** Definition 2.8.1 asserts that for a metric space $X$ with a metric $d$ and a subset $A$ of $X$, a point $x \in X$ is a limit point of $A$ if for every $\epsilon > 0$, there exists a point $y \in A, (x \neq y)$ such that $d(x, y) < \epsilon$. The limit point or points of $A$ (if exist) may or may not lie in $A$. **Geometrically**, it means that if $x$ is a limit point of $A$, there exist points $y \in A$ different from $x$ which are arbitrarily close to the point $x$.

**Definition 2.8.3** Let $(X, d)$ be a metric space and $A$ be a subset of $X$. Then the **closure of** $A$ denoted by $\overline{A}$ is defined by

$$\overline{A} = A \cup A'.$$

**Remark 2.8.4** Let $(X, d)$ be a metric space and $A$ be a subset of $X$. Then a point $x \in X$ is said to be a point of closure $\overline{A}$ of $A$ if either $x \in A$ or $x$ is a limit point of $A$. It follows that $\overline{A}$ is the **smallest closed set in** $(X, d)$ containing $A$ and if $A$ is itself closed in $(X, d)$, then $\overline{A} = A$.

**Remark 2.8.5** The concepts of Interior and Closure operations on subsets of a metric space are **dual** in the sense that if $A$ is any subset of a metric space $(X, d)$, then

$$\overline{X - A} = X - \text{Int}(A).$$

**Definition 2.8.6** Let $(X, d)$ be a metric space and $A$ be a subset of $X$. Then $A$ is said to be **dense or everywhere dense** in $X$, if $\overline{A} = X$.

If a subset $A$ of a metric space $(X, d)$ is dense in $X$, then every point $x \in X$ is a limit point of $A$.

***Example 2.8.7***    (i)  0 is the only limit point of the subset $X = \{1, \frac{1}{2}, \frac{1}{3}, \ldots\} \subset \mathbf{R}$
in the Euclidean line $\mathbf{R} = \mathbf{R}^1$.

(ii)  Every real number in the Euclidean line $\mathbf{R}$ is a limit point of the set of all
rationals $\mathbf{Q}$ in $\mathbf{R}$. The set $\mathbf{Q}$ is dense in $\mathbf{R}$, since $\overline{\mathbf{Q}} = \mathbf{R}$.

(iii)  The set of integral points $\mathbf{Z}$ on the Euclidean line $\mathbf{R}$ has no limit points in $\mathbf{R}$.

(iv)  Let $\mathbf{D}^2 = \{(x, y) \in \mathbf{R}^2 : x^2 + y^2 < 1\}$ be the open disk in the Euclidean plane
$\mathbf{R}^2$. Then the limit points of $\mathbf{D}^2$ are either the points of $\mathbf{D}^2$ or the points on the
unit circle $x^2 + y^2 = 1$. These are also the points of the closure of $\mathbf{D}^2$, and hence
the closure of $\mathbf{D}^2$ is the closed disk defined by $\{(x, y) \in \mathbf{R}^2 : x^2 + y^2 \leq 1\}$.

**Definition 2.8.8** Let $X$ be a metric space. A subset $A$ of $X$ is said to be **closed** if it
contains each of its limit points.

***Remark 2.8.9*** Theorem 2.8.10 characterizes closed sets in terms of open sets, asserting that these two are dual concepts.

**Theorem 2.8.10** *Let $(X, d)$ be a metric space. Then a subset $A$ of $X$ is closed iff its
complement $A^c = X - A$ is an open set.*

***Proof*** Let $A$ be closed in the metric space $X$. Claim that $A^c$ is an open in $X$. If
$A^c = \emptyset$, then it is automatically open. If $A^c \neq \emptyset$, take an arbitrary point $x \in A^c$.
Then the point $x$ is not in the closed set $A$, and hence the point $x$ cannot be a limit
point of $A$. This asserts that there exists an open ball $B_x(r)$ centered at the point $x$ and
disjoint from $A$. Since $x \in A^c$ is an arbitrary point, it follows from $x \in B_x(r) \subset A^c$
that

$$A^c \subset \bigcup B_x(r) \subset A^c$$

and hence $A^c = \bigcup B_x(r)$ is an open set in $X$.

Conversely, let $A^c$ be open in $(X, d)$. The set $A$ will not be closed set if it has a
limit point in $A^c$. But this not possible, because $A^c$ is open in $(X, d)$ and every point
of $A^c$ is the center of an open ball, which is disjoint from $A$, and hence no such point
can be a limit point of $A$.

$\square$

**Theorem 2.8.11** *Let $(X, d)$ be a metric space. Then each closed ball $B_x(\epsilon) = \{y \in
X : d(y, x) \leq \epsilon\}$ in $X$ is a closed set in $(X, d)$.*

***Proof*** Let $B_{x_0}(\epsilon)$ be a closed ball in the metric space $(X, d)$. To prove the theorem, it
is sufficient by using Theorem 2.8.10 to show that its complement $B_{x_0}(\epsilon)^c$ in $X$ is an
open set in $(X, d)$. If $B_{x_0}(\epsilon)^c = \emptyset$, the proof follows automatically. So, we assume that
$B_{x_0}(\epsilon)^c \neq \emptyset$. Take an arbitrary point $x \in B_{x_0}(\epsilon)^c$. Then $d(x, x_0) > \epsilon, \forall x \in B_{x_0}(\epsilon)^c$.
If $\epsilon_1 = d(x, x_0) - \epsilon$, then $\epsilon_1 > 0$. To show that $B_{x_0}(\epsilon)^c$ is open, we prove that
$B_x(\epsilon_1) \subset B_{x_0}(\epsilon)^c$. The latter inclusion is true, because for any $y \in B_x(\epsilon_1)$, the dis-
tance $d(y, x) < \epsilon_1$ and $d(x_0, x) \leq d(x_0, y) + d(y, x)$ imply that

$$d(x_0, y) \geq d(x_0, x) - d(y, x) > d(x_0, x) - \epsilon_1 = \epsilon.$$

This shows that $y \in B_{x_0}(\epsilon)^c$. Since $y \in B_x(\epsilon_1)$ is an arbitrary point, it follows that $B_x(\epsilon_1) \subset B_{x_0}(\epsilon)^c$. Hence the theorem follows from Theorem 2.8.10.

$\square$

**Remark 2.8.12**  The concepts of open balls and closed balls in a metric space are just higher dimensional analogues of open intervals and closed intervals in $\mathbf{R}$. By using **de Morgan rule**, the discussions in Theorems 2.6.6, 2.6.12, 2.6.15 and 2.8.10 are summarized in the basic and important result embodied in Theorem 2.8.13, which leads to the axiomatic framework for closed sets defining a topology ( see Chap. 3).

**Theorem 2.8.13**  *Let $X$ be a metric space with a metric $d$. Then*

**CS(1)**  *the whole (universal) set $X$ is itself a closed set;*
**CS(2)**  *the intersection of any number of closed sets is a closed set;*
**CS(3)**  *the union of a finite number of closed sets is a closed set;*
**CS(4)**  *the empty set $\emptyset$ is a closed set.*

**Definition 2.8.14**  Let $\mathcal{C}$ be a family of some subsets of a nonempty set $X$ satisfying the conditions $CS(1) - CS(4)$ of Theorem 2.8.13. Then there exists a unique topology $\tau$ on $X$ such that the closed sets in $X$ defined by $\tau$ are precisely the same as the given family $\mathcal{C}$.

Corollary 2.8.15 asserts that the Cantor set $C = \bigcap_{n=1}^{\infty} \mathbf{I}_n$ constructed in Chap. 1 is a closed set.

**Corollary 2.8.15**  *Cantor set $C = \bigcap_{n=1}^{\infty} \mathbf{I}_n$ is a closed set in $\mathbf{R}$.*

**Proof**  Since each closed interval is a closed set and each $\mathbf{I}_n$ is a finite union of closed intervals in $\mathbf{R}$, it follows by **CS(3)** that each $\mathbf{I}_n$ is a closed set for $n = 1, 2, 3, \ldots$. Again since Cantor set $C = \bigcap_{n=1}^{\infty} \mathbf{I}_n$, it follows by **CS(2)** that $C$ is a closed set in $\mathbf{R}$.

$\square$

## 2.9  Diameter of Sets and Continuity of Distance Functions on Metric Spaces

This section generalizes the concept of distance between a pair of points in a metric space for a point from a set and also between a pair of its subsets in a metric space. Moreover, the concept of diameter of a subset of metric space is introduced, and continuity of the function $f : X \to \mathbf{R}$, $x \mapsto d(x, A)$ for any nonempty subset $A$ of a metric space $(X, d)$ is proved in this section.

### 2.9.1 Distance of a Point from a Set in Metric Spaces

This subsection studies the distance function of a point from a set in a metric space.

**Definition 2.9.1** Let $(X, d)$ be metric space and $A \subset X$ be nonempty. Then the **distance** $d(x, A)$ from a point $x \in X$ to $A$ is defined by

$$d(x, A) = \inf \{d(x, a) : a \in A\},$$

which is the glb of the distances from the point $x$ to the points in $A$. In particular, if $x \in A$, then $d(x, A) = 0$.

***Example 2.9.2*** Let $X$ be a nonempty set with discrete metric $d$. If $A$ be a nonempty subset of $X$, then

(i) $d(x, A) = 0$, if $x \in A$;
(ii) $d(x, A) = 1$ if $x \notin A$.

**Proposition 2.9.3** *Let $(X, d)$ be a metric space and $A \subset X$ be nonempty. Then for all $x, y \in X$,*

$$|d(x, A) - d(y, A)| \leq d(x, y).$$

***Proof*** Given any two points $x, y \in X$,

$$d(x, a) \leq d(x, y) + d(y, a), \ \forall a \in A.$$

This shows that

$$d(x, A) = \inf \{d(x, a)) : a \in A\} \leq d(x, y) + \inf \{d(y, a) : a \in A\}$$
$$= d(x, y) + d(y, A).$$

Interchanging the role of $x$ and $y$, we obtain

$$d(y, A) \leq d(x, y) + d(x, A).$$

Hence it follows that

$$|d(x, A) - d(y, A)| \leq d(x, y), \ \forall x, y \in X.$$

$\square$

### 2.9.2 Distance Between Two Sets in Metric Spaces

The concept of distance of a point from a set defined in Definition 2.9.1 is now extended for a pair of subsets of a metric space.

**Definition 2.9.4** Let $(X, d)$ be a metric space and $A$, $B$ be a pair of nonempty subsets of $X$. Then their **distance** $d(A, B)$ is defined by

$$d(A, B) = \inf \{d(a, b) : a \in A, b \in b\}.$$

It is the glb of the distances of the points in $A$ from the points in $B$, which exists, since the set is nonempty and bounded below in **R**.

**Remark 2.9.5** $\inf \{d(a, b) : a \in A, b \in b\}$ in Definition 2.9.4 cannot be replaced by $min \{d(a, b) : a \in A, b \in b\}$, because, there may not exist any $a \in A$ and $b \in B$ such that $d(a, b) = d(A, B)$, which is necessary in the second case. For example, for $A = (0, 2)$ and $B = (3, 4)$, by Definition 2.9.4, their distance $d(A, B) = 1$, but there exist no points $a \in A, b \in B$ such that $d(a, b) = 1$, because, $d(a, b) > 1$, $\forall a \in A, b \in B$.

**Proposition 2.9.6** *Let $(X, d)$ be a metric space and $A$, $B$ be two subsets of $X$. If $d(A, B) > 0$, then there exist two open sets $U$ and $V$ in $X$ such that*

$$A \subset U, \ B \subset V \text{ and } U \cap V = \emptyset.$$

**Proof** Suppose $d(A, B) = r > 0$. Then for any point $a \in A$ and any point $b \in V$, $d(a, b) \geq r$. Take two open sets $U$ and $V$ as open balls defined by

$$U = B_a \left(\frac{r}{2}\right), \text{ and } V = B_b \left(\frac{r}{2}\right).$$

Then

$$A \subset U, \ B \subset V \text{ and } U \cap V = \emptyset.$$

$\square$

**Example 2.9.7** $d(A, B)$ given in Definition 2.9.4 does depend on the metric on $X$. For example, let $X = \mathbf{R}$, $A = [0, 1)$ and $B = (1, 2]$. Then

(i) under usual metric $d$ on **R**, the distance $d(A, B) = 0$;
(ii) under discrete metric $d$ on **R**, the distance $d(A, B) = 1$, because $A$ and $B$ are disjoint subsets of **R**. If $C = [2, 3)$, then $d(B, C) = 0$, since $B \cap C \neq \emptyset$.

**Example 2.9.8** Under Euclidean metric, for two sets subsets $A$ and $B$ in $\mathbf{R}^2$ defined by

$$A = \{(x, y) \in \mathbf{R}^2 : y = 0\}, \text{ and } B = \{(x, y) \in \mathbf{R}^2 : xy = 1\},$$

$d(A, B) = 0$ because the graph of $B$ will never intersect that of $A$ but $y \to 0$, when $x \to \infty$.

## 2.9.3  Diameter of a Set

This section introduces the concept of diameter of a nonempty subset of a metric space. This concept plays a key role in Cantor's Intersection Theorem 2.11.8. Bounded sets are defined in metric spices to facilitate the concept of diameter of a set.

**Definition 2.9.9**  Let $(X, d)$ be a metric space and $A \subset X$ be nonempty. Then $A$ is said to be **bounded**, if there exists a positive real number $K$ such that

$$d(x, y) \leq K, \quad \forall x, y \in A.$$

Otherwise, $A$ is said to be unbounded.

**Remark 2.9.10**  Boundedness in a metric space depends on the particular choice of its metric . More precisely, corresponding to every metric, there exists the standard bounded metric constructed in Definition 2.9.11.

**Definition 2.9.11**  Let $(X, d)$ be a metric space. Then the function

$$\rho_X : X \times X \to \mathbf{R}, \ (x, y) \mapsto \ \min \ \{d(x, y), \ 1\}$$

defines a metric on $X$ such that every subset of the metric space $(X, \rho_X)$ is bounded. The metric $\rho_X$ is called the **standard bounded metric** corresponding to the metric $d$ on $X$.

**Definition 2.9.12**  Let $(X, d)$ be a metric space and $A$ be a nonempty subset of $X$. Then the **diameter of** $A$, denoted by diam $A$, is defined by

$$\operatorname{diam} A = \sup\{d(a, y) : a, y \in A\},$$

which is the lub of the distances between the points in $A$, provided the set $\{d(a, y) : a, y \in A\}$ is bounded above in $\mathbf{R}$, written as diam $A < \infty$.

**Remark 2.9.13**  $\sup\{d(a, y) : a, y \in A\}$ in Definition 2.9.12 cannot be replaced by $\max \ \{d(a, y) : a, y \in A\}$, because, there may not exist any $a, b \in A$ such that $d(a, b) = $ diam $A$, which is necessary in the second case. For example, for $A = (0, 1)$, by Definition 2.9.12, diam $A = 1$, but there exist no points $a, b \in A$ such that $d(a, b) = 1$, because, $d(a, b) < 1, \ \forall a, b \in A$.

**Example 2.9.14**  Let $X = \mathbf{R}$ and $A = (0, 4)$. Then under usual metric $d$ on $\mathbf{R}$, diam $A = 4$.

**Proposition 2.9.15**  *Let $(X, d)$ be a metric space and $A$ be a nonempty subset of $X$. Then*

  (i) *if* diam $A < \infty$, *then the set $A$ is bounded;*
  (ii) diam $A = $ diam $\overline{A}$, *where $\overline{A}$ is the smallest closed set in $X$, which contains $A$.*

**Proof**  It follows from Definition of diam $A$.                                        □

### 2.9.4 Continuity of Distance Function on a Metric Space

This subsection proves the continuity of the real-valued function

$$f : X \to \mathbf{R}, \ x \mapsto d(x, A)$$

for any nonempty subset $A$ of a metric space $(X, d)$ in Proposition 2.9.16.

**Proposition 2.9.16** *Let $(X, d)$ be a metric space and $A$ be a nonempty subset of $X$. Then the map*

$$f : X \to \mathbf{R}, \ x \mapsto d(x, A)$$

*is continuous.*

**Proof** Let $\epsilon$ be an arbitrary small positive real number. Then there exists an open ball $B_x(\epsilon)$ such that whenever $y \in B_x(\epsilon)$, then $d(x, y) < \epsilon$. This asserts by Proposition 2.9.3 that whenever, $d(x, y) < \epsilon$, then

$$|d(x, A) - d(y, A)| < \epsilon.$$

This implies that

$$f : X \to \mathbf{R}, \ x \mapsto d(x, A)$$

is continuous.

$\square$

**Corollary 2.9.17** *Let $(X, d)$ be a metric space and $A = \{a\}$ be a one-pointic subset of $X$. Then the map*

$$f : X \to \mathbf{R}, \ x \mapsto d(x, a)$$

*is continuous.*

**Proof** It follows from Proposition 2.9.16. $\square$

## 2.10 Sequences, Convergence of Sequences and Cauchy Sequences in Metric Spaces

This section generalizes the concepts of convergence of sequences of real numbers and Cauchy sequences of analysis in metric spaces and studies them. The concept of limit of a convergent sequence $\{x_n\}$ and that of the limit point of the set of points of the sequence $\{x_n\}$ are in general different, but these two concepts coincide in a metric space under suitable condition laid down in Theorem 2.10.6.

## *2.10.1  Convergence of Sequences in Metric Spaces*

This subsection generalizes the convergence of sequences of real numbers for sequences in metric spaces. A sequence in a metric space $X$ is a way of enumerating some points in the space $X$. Its formal definition is now given in Definition 2.10.1.

**Definition 2.10.1** Let $(X, d)$ be a metric space. A sequence of points $x \in X$ is a function $f : \mathbf{N} \to X$, symbolized

$$\{x_n\} = \{x_1, x_2, \ldots, x_n, \ldots\},$$

where $f(n) = x_n$.

**Definition 2.10.2** A sequence $\{x_n\} = \{x_1, x_2, \ldots, x_n, \ldots\}$ in a metric space $(X, d)$ is said to **converge to a point** $x \in X$ if given a real number $\epsilon > 0$, there exists a natural number $n_0$ such that $d(x_n, x) < \epsilon$ for all $n \geq n_0$. Equivalently, for each open ball $B_x(\epsilon)$ in $X$, there exists a natural number $n_0$ such that $x_n \in B_x(\epsilon)$ for all $n \geq n_0$. This $x$ (if it exists) is unique and is called the limit point of the sequence $\{x_n\}$, and it is often symbolized as $x_n \to x$ or $\lim_{n\to\infty} x_n = x$, which asserts that $d(x_n, x) \to 0$ as $n \to \infty$. Otherwise, the sequence $\{x_n\}$ is said to be divergent in $X$.

**Remark 2.10.3** Definition 2.10.2 asserts that if a sequence $\{x_n\} = \{x_1, x_2, \ldots, x_n, \ldots\}$ in a metric space $(X, d)$ converges to a point $x \in X$, then for any nbd $U$ of $x$ in $X$, there exists a natural number $n_0$ such that $x_n \in U$ for all $n \geq n_0$. If $X = \mathbf{R}$, (endowed with Euclidean metric), then the concept of convergence of a sequence in Definition 2.10.2 coincides with its classical concept used in analysis. Let $x$ be the limit of the sequence $\{x_n\}$. Then the points in $\{x_n\}$ remain nearer and nearer to $x$ from certain term and onward of the sequence $\{x_n\}$ (i.e., for every $n \geq n_0$), and hence $x$ is the limit of the set $S = \{x_1, x_2, \ldots, x_n, \ldots\}$.

**Example 2.10.4** Let $(X, d)$ be a discrete metric space. A sequence $\{x_n\}$ in $X$ converges to $x$ iff there is an integer $n_0$ such that $x_n = x$ for every $n \geq n_0$.

**Example 2.10.5** The concepts of limit point of the set of points of a convergent sequence and limit of the sequence are different. For example, the sequence $\{5, 5, 5, \ldots\}$ is convergent and converges to the point 5. On the other hand, the set of points of this sequence has no limit point. But these two concepts coincide under certain conditions prescribed in Theorem 2.10.6.

**Theorem 2.10.6** *If $\{x_n\}$ is a convergent sequence in a metric space $(X, d)$ has infinitely many distinct points, then its limit is a limit point of the set of points of the sequence $\{x_n\}$.*

**Proof** Let $(X, d)$ be a metric space and $\{x_n\}$ be a convergent sequence in $(X, d)$ having infinitely many distinct points. Suppose $x$ is the limit of the convergent sequence $\{x_n\}$. If $x$ is not a limit point of the set of points $A = \{x_1, x_2, \ldots\}$, then the sequence

$\{x_n\}$ has only finitely many distinct points. Otherwise, there exists an open ball $B_x(\epsilon)$, which contains no point of the sequence $\{x_n\}$, different from $x$. Then all the points $x_n$'s, after a certain place, will lie in the ball $B_x(\epsilon)$, since $x$ is the limit of the sequence $\{x_n\}$. This implies that there exists an integer $n_0$ such that $x_n = x$, $\forall n \geq n_0$. This shows that the sequence $\{x_n\}$ has only a finitely many distinct points. This contradiction proves the theorem.

□

## 2.10.2 Cauchy Sequences in Metric Spaces

This subsection studies Cauchy sequences in metric spaces. **Cauchy's criterion of the convergence** of a sequence of real numbers in the space **R** is generalized in the context of an arbitrary metric space in this subsection. The Cauchy's criterion of convergence of real numbers in analysis asserts that a sequence $\{x_n\}$ of real numbers converges to a point $x$, if corresponding to an $\epsilon > 0$, there exists positive integer $n_0$ such that

$$d(x_n, x_m) < \epsilon, \ \forall m, n \geq n_0.$$

Equivalently,

$$|x_{n+k} - x_n| < \epsilon \ \forall n \geq n_0, \ k \geq 1.$$

This implies that for a Cauchy sequence $\{x_n\}$, its terms get arbitrarily closer to each other in **R** as $n$ becomes sufficiently large.

The concept Cauchy sequences $\{x_n\}$ in **R** is now generalized in Definition 2.10.7 for metric spaces.

**Definition 2.10.7** Let $(X, d)$ be a metric space. A sequence $\{x_n\} = \{x_1, x_2, \ldots, x_n, \ldots\}$ in $X$ is said to be a **Cauchy sequence** if for every positive real number $\epsilon$, there exists a positive integer $n_0$ such that

$$d(x_n, x_m) < \epsilon, \ \forall m, n \geq n_0.$$

**Example 2.10.8** Every convergent sequence in the metric space $(\mathbf{R}, d)$ with usual metric $d$ is a Cauchy sequence, but its converse is not true by Example 2.10.10.

**Proposition 2.10.9**  *(i) Every convergent sequence in a metric space has a unique limit.*

*(ii) Every convergent sequence in a metric space is a Cauchy sequence.*

**Proof** Let $(X, d)$ be a metric space and $\{x_n\}$ be a convergent sequence in $(X, d)$.

(i) To prove the uniqueness of limit of the convergent sequence $\{x_n\}$ in $(X, d)$, use the triangle inequality for the metric space $(X, d)$.

(ii) For the second part, as $\{x_n\}$ is a convergent sequence in $(X, d)$, if $x_n \to x \in X$, then given an $\epsilon > 0$, there exists a positive integer $n_0$ such that $d(x_n, x) < \epsilon/2$, $\forall n \geq n_0$. Now,

$$\forall m, n \geq n_0, \ d(x_n, x_m) \leq d(x_n, x) + d(x_m, x) < \epsilon,$$

which implies that $\{x_n\}$ is a Cauchy sequence.

$\square$

**Example 2.10.10**  The converse of Proposition 2.10.9 (ii) is not necessarily true in an arbitrary metric space, but it is valid under a situation specified in Proposition 2.10.12. For example, consider the sequence $\{\frac{1}{n}\}$ in the metric space $(X, d)$, where $X = (0, 1]$ and $d$ is the Euclidean metric on $X$ defined by $d(x, y) = |x - y|$. Since $x_n \to 0$ in **R**, it follows that $\{\frac{1}{n}\}$ is a Cauchy sequence in $(0, 1]$, but it is not convergent.

**Remark 2.10.11**  The converse of Proposition 2.10.9 (ii) holds under certain conditions prescribed in Proposition 2.10.12.

**Proposition 2.10.12**  *A Cauchy sequence in a metric space is convergent if it has a convergent subsequence.*

**Proof**  Let $(X, d)$ be a metric space and $\{x_n\}$ be a Cauchy sequence in $X$ such that it has a convergent subsequence $\{x_{n_k}\}$. If $x_{n_k} \to x \in X$, to prove this proposition, it is sufficient to prove that $x_n \to x$, *i.e.*, given an $\epsilon > 0$, there exists a positive integer $n_0$ such that $d(x_n, x) < \epsilon$, $\forall n \geq n_0$. By hypothesis, $\{x_n\}$ is a Cauchy sequence in $X$. Hence it follows that corresponding to the given $\epsilon > 0$, there exists a positive integer $n_0$ such that

$$d(x_n, x_m) < \epsilon/2, \ \forall m, n \geq n_0.$$

Since by hypothesis, $x_{n_k} \to x \in X$, it follows that for any integer $n \geq n_0$, there exists some $k \in \mathbf{N}$ such that

$$d(x_{n_k}, x) < \epsilon/2, \ \forall n_k \geq n_0.$$

Hence by triangle inequality,

$$d(x_n, x) \leq d(x_n, x_{n_k}) + d(x_{n_k}, x) < \epsilon/2 + \epsilon/2 = \epsilon, \ \forall n, n_k \geq n_0$$

implies that the Cauchy sequence $\{x_n\}$ is convergent.

$\square$

# 2.11 Complete Metric Spaces and Cantor's Intersection Theorem

This section studies complete metric spaces, which is an important concept and proves Cantor's intersection theorem, a basic theorem in metric spaces which characterizes a complete metric space in Theorem 2.11.8 by a decreasing sequence of nonempty subsets of the metric space.

## 2.11.1 Complete Metric Spaces

This subsection conveys the concept of a **complete metric space** in which every Cauchy sequence is convergent and characterizes the completeness of its proper subsets in Theorem 2.11.3.

**Definition 2.11.1** A metric space $(X, d)$ is said to be **complete** if every Cauchy sequence in $X$ converges in $X$.

**Example 2.11.2** (i) The Euclidean line $\mathbf{R}$ is a complete metric space, which follows directly from the completeness property of the real line $\mathbf{R}$ (see Corollary 2.11.10).

(ii) The complex plane $\mathbf{C}$ is a complete metric space.

(iii) $\mathbf{Q}$ is not a complete metric space by Corollary 2.11.12. Alternatively, as there exist sequences of rational numbers which converge to an irrational number (see Example 2.2.9), i.e., it has no limit in $\mathbf{Q}$. Hence it follows that $\mathbf{Q}$ is not a complete metric space.

(iv) $[0, 1]$ is a complete metric space, but $(0, 1]$ is not so by Theorem 2.11.3, since the Euclidean line $\mathbf{R}$ is a complete metric space.

Theorem 2.11.3 characterizes completeness property of a subset of a metric space by its closeness property.

**Theorem 2.11.3** *Let $(X, d)$ be a complete metric space and $A$ be a proper subset of $X$. Then $A$ is complete iff $A$ is closed in $X$.*

**Proof** Let $(X, d)$ be a complete metric space and $A$ be a proper subset of $X$. First suppose that $A$ is closed in $X$. To prove that $A$ is complete, let $\{a_n\}$ be a Cauchy sequence in $A$. Then $\{a_n\}$ is also a Cauchy sequence in $X$, and hence the sequence $\{a_n\}$ converges to a point $a \in X$. It is sufficient to show that $a \in A$. Suppose $\{a_n\}$ has only finitely many distinct points, then the point $a$ in the set of points of this sequence is to be repeated infinitely many times and hence $a \in A$. Again, if $\{a_n\}$ has infinitely many distinct points, then $a$ is a limit point of the set of points of the sequence $\{a_1, a_2, \ldots, a_n, \ldots\}$ by Theorem 2.10.6, and hence $a$ is also a limit point of $A$. By hypothesis $A$ is closed and hence $a \in A$. ·

Conversely, let the subset $A \subset X$ be complete in $(X, d)$. To show that $A$ is closed, let $a$ be a limit point of $A$. Now, for each $n \in \mathbf{N}$, the open ball $B_a(1/n)$ contains a point $a_n$ in $A$, and hence the sequence $\{a_n\}$ converges to the point $a \in X$. Since by hypothesis, $A$ is complete, it follows that the sequence $\{a_n\}$ is a Cauchy sequence in $A$ and $a \in A$. This asserts that $A$ is closed.

$\square$

### 2.11.2　Completion of a Noncomplete Metric Space

There are metric spaces which are not complete. This subsection provides a particular construction method for completion of a noncomplete metric space.

**Definition 2.11.4** Let $(X, d)$ be a metric space. A metric space $(\tilde{X}, \tilde{d})$ is said to be a **completion** of $(X, d)$ if

(i) the metric space $(\tilde{X}, \tilde{d})$ is complete and
(ii) there exists a map $f : X \to \tilde{X}$ such that $f$ is an isometry of $X$ into $\tilde{X}$ and $f(X)$ is dense in $\tilde{X}$.

*Example 2.11.5* Under Euclidean metric, $\mathbf{R}$ is a metric completion of the metric space $\mathbf{Q}$, because $\mathbf{R}$ is complete and $\mathbf{Q}$ is dense in $\mathbf{R}$.

**Theorem 2.11.6** *(Completion of a metric space) Let $(X, d)$ be a metric space. Then there exists a unique completion ( upto isometry) of $(X, d)$.*

**Proof I:** Let $(X, d)$ be an arbitrary metric space and $\mathcal{B}(X)$ be the set of all bounded real-valued functions defined on $X$. Then the function $d_\infty : \mathcal{B}(X) \times \mathcal{B}(X) \to \mathbf{R}$ defined by

$$(f, g) \mapsto \sup_{x \in X} |f(x) - g(x)|$$

is a metric. Hence $(\mathcal{B}(X), d_\infty)$ is a metric space. Given a fixed $x_0 \in X$, for every $x \in X$, define a map

$$f_x : X \to \mathbf{R}, y \mapsto d(y, x) - d(y, x_0).$$

Hence it follows that $f_x \in \mathcal{B}(X)$ and the map

$$\psi : X \to \mathcal{B}(X), x \mapsto f_x$$

is an isometry. Let $\tilde{X} \subset (X)$ be the closure $\overline{\psi(X)}$ and endowed with metric $\tilde{d}$ induced on $\tilde{X}$ induced from $\mathcal{B}(X)$. Since $\tilde{X}$ is a closed subset of the complete metric space $\mathcal{B}(X)$, the metric space $(\tilde{X}, \tilde{d})$ is complete. Finally, by construction $\psi(X)$ is dense in $\overline{X}$. This completes the proof of the theorem.

**Proof II**: Let $(X, d)$ be an arbitrary metric space and $Cs(X)$ be the set of all Cauchy sequences in $X$. Define an equivalence relation $\sim$ on $Cs(X)$ by the rule

$$\{x_n\} \sim \{y_n\} \text{ iff } \lim_{n \to \infty} d(x_n, y_n) = 0.$$

Then "$\sim$" identifies the Cauchy sequences in $X$ having the same limit. Let $\tilde{X} = Cs(X)/\sim$ be the quotient set. Define

$$\bar{d} : \tilde{X} \times \tilde{X} \to \mathbf{R}, \ (\{x_n\}, \{y_n\}) \mapsto \lim_{n \to \infty} d(x_n, y_n).$$

Then $\tilde{d}$ is well-defined, and it defines a metric on $\tilde{X}$. This metric space $(\tilde{X}, \tilde{d})$ is a complete metric space, which is a completion of the metric space $(X, d)$. Moreover, if $(\tilde{Y}, \rho)$ is any other completion of $(X, d)$ then $(\tilde{Y}, \rho)$ is isometric to $(\tilde{X}, \tilde{d})$. $\qquad\square$

### 2.11.3 Cantor's Intersection Theorem for Metric Spaces

This subsection proves Cantor's Intersection Theorem 2.11.8 for metric spaces, which characterizes a complete metric space by a decreasing sequence of its nonempty subsets satisfying certain conditions.

This theorem is named after Georg Cantor (1845–1918) who gave first the formal definition of the set of real numbers in 1871.

**Definition 2.11.7** Let $(X, d)$ be a metric space. A sequence $\{A_n\}$ of subsets of $X$ is said to be **decreasing** if the following set-theoretic relation holds

$$A_1 \supset A_2 \supset \cdots \supset A_n \supset \cdots.$$

**Theorem 2.11.8** *(Cantor's intersection theorem) Let $(X, d)$ be a metric space. Then it is complete iff whenever $\{A_n\}$ is a sequence of nonempty subsets of $X$ satisfy the conditions*

*(i) every $A_n$ is a closed set in $(X, d)$;*
*(ii) the sequence $\{A_n\}$ is decreasing;*
*(iii) diam $A_n \to 0$ as $n \to \infty$,*

*then $\bigcap_{n=1}^{\infty} A_n$ contains exactly one point.*

**Proof** Let $(X, d)$ be a metric space and $\{x_n\}$ be a Cauchy sequence in $X$. Let $A_n$ be the set defined by

$$A_n = \text{closure}\{x_n, x_{n+1}, \ldots\},$$

i.e., it is the smallest closed set containing the set of points of $\{x_n, x_{n+1}, \ldots\}$. First suppose that $\bigcap_{n=1}^{\infty} A_n$ contains exactly one point. Then the conditions (i) and (ii) are satisfied. For the condition (iii), let given $\epsilon > 0$, there is an integer $n_0$ such that $d(x_n, x_m) < \epsilon$ for $m, n \geq n_0$. But for $k \geq n_0$, the diameter diam $A_k = \sup\{d(x_n, x_m) : n, m \geq k\} \leq \epsilon$. This asserts that the sequence $\{A_n\}$ satisfies all the above three above conditions (i)–(iii). Hence by hypothesis $\bigcap_{n=1}^{\infty} A_n = \{x\}$ for some point $x \in X$. Again for any $n \geq 1$, $d(x, x_n) \leq$ diam $A_n \implies d(x, x_n) \to 0$ since diam $A_n \to 0$ as $n \to \infty$. This implies that $x_n \to x$ and hence the metric space $(X, d)$ is complete. Conversely, let the metric space $(X, d)$ be complete and $\{A_n\}$ be a sequence of nonempty subsets of $X$ satisfying all the 3 conditions of the theorem such that for each $n$, $x_n \in A_n$. We claim that $\{x_n\}$ is a Cauchy sequence. Given an $\epsilon > 0$, let $n_0$ be an integer such that diam $A_n < \epsilon$ for $n \geq n_0$. Hence, if $m, n \geq n_0$, then the condition $(ii)$ asserts that $x_n, x_m \in A_{n_0}$ and hence $d(x_n, x_m) \leq$ diam $A_{n_0} < \epsilon$. This implies that $\{x_n\}$ is a Cauchy sequence. As $(X, d)$ is a complete metric space, there is a point $x \in X$ such that $x_n \to x$. Again $x \in \bigcap_{n=1}^{\infty} A_n$, since $A_n$ is closed for each $n$. If $y \in \bigcap_{n=1}^{\infty} A_n$, then $d(x, y) \leq$ diam $A_n$ for every $n \geq 1$. Then the condition (iii) implies that $x = y$.                                                                                    □

**Another form of Cantor's Intersection Theorem is given in Corollary 2.11.9.**

**Corollary 2.11.9** *(Cantor's intersection theorem) Let $(X, d)$ be a complete metric space and $\{A_n\}$ be a decreasing sequence (i.e., $A_1 \supset A_2 \supset \cdots$) of nonempty closed sets in $X$ such that diam $A_n \to 0$ as $n \to \infty$. Then $\bigcap_{n=1}^{\infty} A_n$ contains exactly one point.*

**Proof** It follows from the proof of the second part of the Theorem 2.11.8.        □

**Corollary 2.11.10** *The Euclidean $\mathbf{R}$ is a complete metric space.*

**Proof** To prove the corollary, we use the following properties of $\mathbf{R}$.

 (i) If a nonempty subset $A$ of $\mathbf{R}$ has an upper bound, then $A$ has a supremum. By applying this property to the set $-A$, it follows that if a nonempty subset $B$ of $\mathbf{R}$ has a lower bound, then $B$ has a infimum.
 (ii) If a nonempty subset $A$ of $\mathbf{R}$ is bounded above and $x = \sup A$, then there is an increasing sequence $\{x_n\}$ in $A$ such that $x_n \to x$.
 (iii) If a nonempty subset $A$ of $\mathbf{R}$ is bounded below and $y = \inf A$, then there is an decreasing sequence $\{y_n\}$ in $A$ such that $y_n \to y$.
 (iv) If a bounded sequence in $\mathbf{R}$ is either increasing or decreasing, then it converges.

To prove the corollary, let $\{A_n\}$ be a sequence of nonempty subsets of $\mathbf{R}$ satisfying all the conditions of Cantor's Theorem 2.11.8. Then each diam $A_n$ is finite and hence it is bounded. By using the property (i) of $\mathbf{R}$, it follows that inf $A_n = x_n$ and sup $A_n = y_n$ exist. Since each $A_n$ is closed by hypothesis, it follows that $x_n, y_n \in A_n$ and $0 \leq y_n - x_n \leq$ diam $A_n \to 0$ as $n \to \infty$. Again, by hypothesis, the sequence $\{A_n\}$ is deceasing and hence $A_{n+1} \subset A_n$. This shows that $x_n \leq x_{n+1} \leq y_{n+1} \leq y_n$. This asserts that the increasing sequence $\{x_n\}$ converges to a point and the decreasing sequence $\{y_n\}$ converges to a point. Hence there exist points $x, y \in A_n$ for every integer $n$ such that $x_n \to x$ and $y_n \to y$. Moreover, $|y - x| \leq$ diam $A \to 0$ and hence $x = y$. It asserts that $\bigcap_{n=1}^{\infty} A_n = \{x\}$ and hence $\mathbf{R}$ is complete by Cantor's Theorem 2.11.8.                                                                                $\square$

**Corollary 2.11.11** *Euclidean m-space $\mathbf{R}^m$ is a complete metric space with usual metric for every integer $m \geq 1$.*

**Proof** $\mathbf{R}^m$ is a metric space with Euclidean metric $d$ for every integer $m \geq 1$. For $x_n = (x_{n_1}, x_{n_2}, \ldots, x_{n_m}) \in \mathbf{R}^m$, let $\{x_n\}$ be a Cauchy sequence in $\mathbf{R}^m$. Then $|x_{n_p} - x_{q_p}| \leq d(x_n, x_q)$ for $1 \leq p \leq m$. This asserts that each sequence $\{x_{np}\}$ is a Cauchy sequence in $\mathbf{R}$. Hence it follows by Corollary 2.11.10 that $x_{np} \to x_p$ for some $x_p \in \mathbf{R}$ This proves that $x_n \to x = (x_1, x_2, \ldots, x_m) \in \mathbf{R}^m$.                      $\square$

**Corollary 2.11.12** $\mathbf{Q}$ *with Euclidean metric is not a complete metric space.*

**Proof** Consider the subspace $(\mathbf{Q}, d)$ of the metric space $(\mathbf{R}, d)$ with Euclidean metric $d$. Applying Theorem 2.11.3, it follows that $(\mathbf{Q}, d)$ is not complete, since $\mathbf{Q}$ is not closed in $\mathbf{R}$.                                                                            $\square$

**Example 2.11.13** The condition that $\{A_n\}$ is to be a decreasing sequence (i.e., $A_1 \supset A_2 \supset \cdots$) of nonempty closed sets in $X$ in Cantor's Intersection Theorem 2.11.9 is necessary. In support, consider the sequence $\{A_n = (0, 1/n]\}$ of subsets in $\mathbf{R}$. Then $\{A_n\}$ forms a decreasing sequence such that diam $A_n \to 0$ as $n \to \infty$ but this sequence is not a sequence of closed sets in $\mathbf{R}$. The metric space $\mathbf{R}$ is complete but $\bigcap_{n=1}^{\infty} A_n = \emptyset$.

## 2.12  Continuity and Uniform Continuity in Metric Spaces

This section generalizes "$\epsilon$-$\delta$" definition of continuity of functions $f : \mathbf{R}^n \to \mathbf{R}^m$ of analysis (see Sect. 2.3.3) for arbitrary metric spaces by utilizing the concept of distance function. The motivation of this generalization was born through the concept of usual distance in Euclidean $n$-space $\mathbf{R}^n$, which permits the notion of continuity of functions from $\mathbf{R}^n$ to $\mathbf{R}^m$ by the usual "$\epsilon$-$\delta$" method. In analysis and geometry, the notion of usual Euclidean metric is available to develop the subjects. But as all spaces used in topology are not subsets of Euclidean spaces, it becomes necessary to extend the definition of continuity for a certain class of spaces where a suitable concept of distance is available such as in metric spaces. This section also studies uniform continuity in metric spaces with the help of Lipschitz functions.

### 2.12.1  Continuity of Functions and Convergence of Sequences in Metric Spaces

This subsection continues a study of continuity of functions and proves its equivalence with convergence of sequences in metric spaces, which implies that continuity of functions in metric spaces preserves convergence.

**Definition 2.12.1** Given metric spaces $(X, d)$ and $(Y, \rho)$, a function $f : X \to Y$ is said to be **continuous at a point** $a \in X$ if either of the following equivalent criteria holds for every real number $\epsilon > 0$ :

(i) there exists a number $\delta > 0$ such that $d(x, a) < \delta$ implies $\rho(f(x), f(a)) < \epsilon$; or

(ii) there exists a number $\delta > 0$ such that for each open ball $B_{f(a)}(\epsilon)$, there exists an open ball $B_a(\delta)$ in $X$ such that

$$f(B_a(\delta)) \subset B_{f(a)}(\epsilon).$$

**Remark 2.12.2** The condition (i) generalizes the concept of continuity used in calculus and the condition (ii) conveys (i) in the language of open balls. In the topological setting the concept of continuity of a function is given in a similar way in Chap. 3 by using the concept of open sets (or closed sets).

Theorems 2.12.3 and 2.12.4 characterize continuity of a function between metric spaces in the language of sequences.

**Theorem 2.12.3** *Let $(X, d)$ and $(Y, \rho)$ be two metric spaces. Then a function $f :$ $X \to Y$ is continuous iff for any sequence $\{x_n\}$ in $X$ converging to a point $a \in X$, the sequence $\{f(x_n)\}$ in $Y$ converges to the point $f(a) \in Y$.*

**Proof** Let $(X, d)$ and $(Y, \rho)$ be two metric spaces and $f : X \to Y$ be a given function. First suppose that a sequence $\{x_n\}$ in $X$ converges to a point $a \in X$ and $f$ is continuous at $a$. Then $x_n \to a$. Claim that $f(x_n) \to f(a)$. Since $f$ is continuous at $a \in X$, corresponding to the open ball $B_{f(a)}(\epsilon)$ in $Y$, there exists an open ball $B_a(\delta)$ such that

$$f(B_a(\delta)) \subset B_{f(a)}(\epsilon).$$

As $x_n \to a$, there exists an integer $n_0 \in \mathbf{N}$ such that

$$x_n \in B_a(\delta), \ \forall n \geq n_0.$$

Hence it follows that $f(x_n) \in B_{f(a)}(\epsilon)$, $\forall n \geq n_0$. This asserts that $f(x_n) \to f(a)$. Conversely, if $f$ is not continuous at the point $a \in X$, claim that $x_n \to a$ does not assert that $f(x_n) \to f(a)$. By hypothesis, there exists an open ball $B_{f(a)}(\epsilon)$ such that $f(B_a(\delta))$ is not contained in $B_{f(a)}(\epsilon)$. Construct the sequence of open balls $\{B_a(1/n) : n = 1, 2, \ldots\}$ in $X$ and consider a sequence points

$$\{x_n : x_n \in B_a(1/n) : n = 1, 2, \ldots, \text{ and } f(x_n) \notin B_{f(a)}(\epsilon)\}.$$

This shows that the sequence $\{x_n\}$ converges to the point $a$ but the sequence $\{f(x_n)\}$ does not converge to $f(a)$.

$\square$

A mapping $f : X \to Y$ between metric spaces is continuous if it is continuous at each point $x \in X$ and Theorem 2.12.3 conveys the concept of continuity in the language of sequences.

We summarize the above discussion in a basic and important result given in Theorem 2.12.4.

**Theorem 2.12.4** *Let $(X, d)$ and $(Y, \rho)$ be two metric spaces. Then a function $f : X \to Y$ is continuous iff for any sequence $\{x_n\}$ in $X$ converging to $x$, the sequence $\{f(x_n)\}$ in $Y$ converges to $f(x)$.*

**Remark 2.12.5** Theorem 2.12.4 shows that continuous maps $f : X \to Y$ between metric spaces are precisely those maps $f$ which send convergent sequences in $X$ to the convergent sequences in $Y$. The continuity of functions in metric spaces is also completely characterized in Theorem 2.12.6 by open sets by utilizing the Theorem 2.6.8 saying that every open ball $B_x(\epsilon)$ in a metric space $X$ is an open set. This idea is also employed for continuity of a function in a topological setting in Chap. 3 in an analogous way.

**Theorem 2.12.6** *Let $(X, d)$ and $(Y, \rho)$ be metric spaces. Then a function*

$$f : X \to Y$$

*is continuous iff $f^{-1}(U)$ is an open subset in $X$ for every open subset $U$ of $Y$.*

**Proof** Let $(X, d)$ and $(Y, \rho)$ be two metric spaces and $f : X \to Y$ be a given function. First suppose that $f : X \to Y$ is continuous, $U \subset Y$ is an open set and $f(x) \in U$. Then there exists a real number $\epsilon > 0$ such that the open ball $B_{f(x)}(\epsilon) \subset U$. As $f$ is continuous by hypothesis, there is a real number $\delta > 0$ such that $f$ maps the open ball $B_x(\delta)$ in $X$ centered at $x$ into the open ball $B_{f(x)}(\epsilon)$. This implies that $B_x(\delta) \subset f^{-1}(U)$. This asserts that $f^{-1}(U)$ is open. Next suppose that given $\epsilon > 0$, there is an open set $f^{-1}(B_{f(x)}(\epsilon))$ in $X$, containing the point $x \in X$. Then there exists a $\delta > 0$ such that $B_x(\delta) \subset f^{-1}(B_{f(x)}(\epsilon))$. Hence by using the distance functions of metric spaces it follows that $f$ is continuous, because, if $d(x, x') < \delta$, then $f(x') \in B_{f(x)}(\epsilon))$ and hence $\rho(f(x), f(x')) < \epsilon$, which shows the continuity of $f$ by $\epsilon$-$\delta$ definition. $\qquad\square$

**Equivalence of conditions for continuity**: The criterion of continuity of a function $f : \mathbf{R} \to \mathbf{R}$ given in Corollary 2.12.7 coincides with the $\epsilon$-$\delta$ definition of its continuity.

**Corollary 2.12.7** *Let $\mathbf{R}$ be the Euclidean line. Then a function $f : \mathbf{R} \to \mathbf{R}$ is continuous iff the inverse image of every open subset in $\mathbf{R}$ is also open in $\mathbf{R}$.*

**Proof** Since $\mathbf{R}$ is a metric space, the Corollary follows from Theorem 2.12.6. $\qquad\square$

**Corollary 2.12.8** *Let $(X, d)$ be a metric space and $I_d : X \to X$ be the identity function. Then $I_d$ is continuous.*

*Proof* It follows from Theorem 2.12.6.                                                    □

**Proposition 2.12.9** *Let $d_1$ and $d_2$ be two different metrics on the same set $X$ such that for every point $x \in X$ and an $\epsilon > 0$, there is a $\delta > 0$ with the properties*

(i) $d_1(x, y) < \delta$, whenever $d_2(x, y) < \epsilon$;
(ii) $d_2(x, y) < \epsilon$, whenever $d_1(x, y) < \delta$.
   *Then the metrics $d_1$ and $d_2$ determine the same open sets in $X$.*

*Proof* Left as an exercise.                                                    □

## 2.12.2 Uniform Continuity and Lipschitz Functions in Metric Spaces

This subsection conveys the concept of uniform continuity in metric spaces and relates this concept to Lipschitz functions given in Definition 2.12.16. By uniform continuity, we mean continuity together with an additional condition: For every $\epsilon > 0$, there exists a $\delta > 0$ such that it works uniformly over the whole space in the sense of Definition 2.12.10. Continuity of a function at a point is a local property of the function. On the other hand, uniform continuity of a function is its global property.

**Definition 2.12.10** Let $(X, d)$ and $(Y, \rho)$ be two metric spaces. A function $f : X \to Y$ is said to be **continuous** if for any $x_0 \in X$ and any $\epsilon > 0$ , there is a $\delta > 0$ (depending on $\epsilon$ and $x_0$) such that $d(x, x_0) < \delta$ implies $\rho(f(x), f(x_0)) < \epsilon$. If $\delta$ does not depend on $x_0$ in the sense that given any $\epsilon > 0$, there is a $\delta > 0$ such that every pair of points $x_1$ and $x_2$ in $X$ has the property:

$$d(x_1, x_2) < \delta \implies \rho(f(x_1), f(x_2)) < \epsilon.$$

Then the function $f$ is said to be **uniformly continuous** on $X$. Equivalently, $f$ is uniformly continuous on $X$ if every pair of sequences $\{x_n\}$ and $\{y_n\}$ in $X$ has the property :

$$d(x_n, y_n) \to 0 \implies \rho(f(x_n), f(y_n)) \to 0.$$

*Example 2.12.11* Consider the function $f : [1, \infty) \to \mathbf{R}, x \mapsto 1/x$. Then

$$f(x_1) - f(x_2) = \frac{|x_2 - x_1|}{x_1 x_2} \leq |x_1 - x_2| \text{ for every pair of points } x_1, x_2 \in [1, \infty).$$

This shows that given an $\epsilon > 0$, we may take $\delta = \epsilon$. Hence it follows the $f$ is uniformly continuous on $[1, \infty)$. But the function $f : (0, 1) \to \mathbf{R}, x \mapsto 1/x$ is not uniformly continuous.

***Remark 2.12.12*** Let $(X, d)$ and $(Y, \rho)$ be two metric spaces. If a function $f : X \to Y$ uniformly continuous, then for every $\epsilon > 0$, there is a $\delta > 0$ such that $\rho(f(x), f(y)) < \epsilon$, whenever $d(x, y) < \delta$. For continuity of a function, $\delta$ depends on both $x$ and $\epsilon$, but for uniform continuity, $\delta$ depends only on $\epsilon$. This asserts that every uniformly continuous function is continuous but its converse is not necessarily true (see Example 2.12.13).

***Example 2.12.13*** The function $f : (0, 1] \to \mathbf{R}$, $x \mapsto \sin(x^{-1})$ is continuous but not uniformly continuous. The continuity of $f$ follows from the fact that it is the composite of two continuous functions $x \mapsto x^{-1}$ and the sine function. It is not uniformly continuous, because, given $\epsilon = 1$ there exists $\delta > 0$, such that there exist points $x, y \in (0, \delta)$ for which $f(x) = 1$, $f(y) = -1$ and hence

$$|x - y| < \delta \text{ but } |f(x) - f(y)| = 2 > \epsilon = 1.$$

***Example 2.12.14*** The function $f : S^2 \to S^2$, $\alpha \mapsto -\alpha$ is continuous, where $S^2 = \{(x, y, z) \in \mathbf{R}^3 : x^2 + y^2 + z^2 = 1\}$ is the unit sphere in $\mathbf{R}^3$ and for $\alpha = (x, y, z) \in S^2$, the element $-\alpha = (-x, -y, -z) \in S^2$.

***Example 2.12.15*** The continuous function $f : \mathbf{R} \to \mathbf{R}, x \to x^2$ is not uniformly continuous, because, given any $\delta > 0$, for $x = n, y = n + \delta$, where $n$ is a positive integer, $|f(x) - f(y)| > 2n\delta$, which can assume a large value as we like, even for a small value of $\delta$.

**Definition 2.12.16** Let $(X, d)$ and $(Y, \rho)$ be two metric spaces. A function $f : X \to Y$ is said to be a **Lipschitz function** if there exists a positive constant $K$ such that

$$\rho(f(x), f(z)) \leq K d(x, z), \ \forall x, z \in X.$$

The positive constant $K$ is called a **Lipschitz constant**.

***Example 2.12.17*** Let $f : [a, b] \to \mathbf{R}$ be a continuously differential function with the property that $|f(x)| \leq K > 0$ for all $x \in [a, b]$. Then under the Euclidean metric on $\mathbf{R}$

$$|f(x) - f(z)| = \left| \int_z^x f'(t)dt \right| \leq \int_z^x |f'(t)|dt \leq K|x - z|,$$

where the integral is the **Riemann integral**, shows that $f$ is a Lipschitz function.

**Proposition 2.12.18** *Let $(X, d)$ and $(Y, \rho)$ be two metric spaces and $f : X \to Y$ be a Lipschitz function. Then $f$ is uniformly continuous.*

**Proof** By hypothesis, $f : X \to Y$ is a Lipschitz function. Then there exists a positive constant $K$ such that $\rho(f(x), f(z)) \leq K d(x, z)$ for all $x, z \in X$. Now, for every $\epsilon > 0$, there is a $\delta = \epsilon/K > 0$ such that $\rho(f(x), f(z)) < \epsilon$, whenever $d(x, z) < \delta$. Hence, it follows that $f$ is uniformly continuous. $\qquad \square$

*Example 2.12.19* Let $f : X \to Y$ be a map between two metric spaces $X$ and $Y$ and $\{x_n\}$ be a Cauchy sequence in $X$.

  (i) If $f$ is uniformly continuous, then image of the Cauchy sequence $\{x_n\}$ in $X$ is also a Cauchy sequence in $Y$. Because, every Cauchy sequence in $X$ is mapped to a Cauchy sequence in $Y$ by every uniformly continuous map.

 (ii) On the other hand, if $X$ is complete and $f$ is continuous, then image of the Cauchy sequence $\{x_n\}$ in $X$ is also a Cauchy sequence in $Y$. Because, as $X$ is complete, every Cauchy sequence $\{x_n\}$ in $X$ has a limit $x \in X$, and hence by continuity of $f$, the sequence $\{f(x_n)\}$ converges to $f(x)$, and every convergent sequence is a Cauchy sequence.

## 2.13   Homeomorphism and Isometry in Metric Spaces

This section addresses the concept of homeomorphism in metric spaces. Homeomorphisms form an important class of continuous functions to study metric spaces. It is a natural question: when two metric structures are equivalent? In group theory, two groups are equivalent if there exists an isomorphism between them. Similarly, two metric spaces are equivalent if there exists a homeomorphism between them. An isometry in metric spaces is a particular homeomorphism between them by Proposition 2.13.10. Finally, this section defines the concept of equivalent metrics and relates it to a homeomorphism in Sect. 2.13.3.

### 2.13.1   Homeomorphisms in Metric Spaces

This subsection introduces the concept of homeomorphisms for metric spaces, which plays a key role in classification of metric spaces.

**Definition 2.13.1** A bijective mapping $f : X \to Y$ between metric spaces $X$ and $Y$ is said to be a **homeomorphism** if both $f$ and $f^{-1}$ are continuous functions. If there exists a homeomorphism between the metric spaces $X$ and $Y$, then they are called **homeomorphic metric spaces**; they are also said to be **equivalent as metric spaces**.

*Example 2.13.2* The open intervals $(0, 1)$ and $(a, b)$ in **R** with usual metric are homeomorphic.

*Example 2.13.3* The metric spaces $S^2 - N$ and $\mathbf{R}^2$ (where $N$ is the north pole of the sphere $S^2$) are equivalent by stereographic projection (see Chap. 3).

**Definition 2.13.4** A property of a metric space $X$ which is preserved by a homeomorphism is said to a **topological property of** $X$.

**Example 2.13.5** The property of being open or closed sets in a metric space is a topological property.

**Example 2.13.6** Completeness in a metric space is not a topological property. Consider $\mathbf{R}$ with usual metric and its open interval $(-1, 1)$. The sequence $\{x_n : x_n = 1 - 1/n\}$ is a Cauchy sequence in $(-1, 1)$, but it does not converge in $(-1, 1)$, which shows that open interval $(-1, 1)$ is not complete. On the other hand, $\mathbf{R}$ is complete, and it is homeomorphic to $(-1, 1)$. On the other hand, compactness and connectedness in a metric space (also in a topological space) are both topological properties (see Sects. 2.17.1 and 2.17.5).

**Example 2.13.7** (**Nonhomeomorphic spaces**) While classifying metric spaces ( or topological spaces) up to equivalence, either we have to give an explicit expression for a homeomorphism between the given spaces or we have to show that no such homeomorphism exists. For example, a circle minus a point is not homeomorphic to a closed line segment. For topological spaces, we utilize certain topological property ( see Chap. 5) in this volume but for topological invariant (see **Volume 3** of the present series of books). For example, the space $X = [0, 2]$ is not homeomorphic to the subspace $[0, 1] \cup [2, 3]$ of $\mathbf{R}$, because $X$ is connected but $Y$ is not so (connectedness is a topological property which is proved in Chap. 5).

## 2.13.2 Isometry in Metric Spaces

This subsection introduces the concept of isometry in metric spaces and establishes its relation to a homeomorphism in Proposition 2.13.10.

**Definition 2.13.8** Let $(X, d)$ and $(Y, \rho)$ be two metric spaces. Then a bijective map $f : X \to Y$ is said to be $(d, \rho)$-isometry (or simply an **isometric mapping**) if

$$d(x, x') = \rho((f(x), f(x'))), \ \forall x, x' \in X.$$

Two metric spaces $(X, d)$ and $(Y, \rho)$ are said to be **isometric** if there exists a $(d, \rho)$-isometry between them.

**Remark 2.13.9** (**Importance of isometry**) Let two metric spaces $(X, d)$ and $(Y, \rho)$ be isometric. Then it asserts that the points in $X$ and $Y$ are in bijective correspondence by an isometric mapping $f : X \to Y$ in such a way that the distance between every pair of points $x, x' \in X$ and the distance between their image points $f(x), f(x')$ in $Y$ are the same. This implies that an isometry identifies the metric structures of two metric spaces and also identifies their closed sets and open sets.

Proposition 2.13.10 asserts that every isometry is a homeomorphism.

**Proposition 2.13.10** *Let* $f : X \to Y$ *be an isometry between metric spaces* $(X, d)$ *and* $(Y, \rho)$. *Then* $f$ *is a homeomorphism.*

**Proof** It follows from the defining condition of the isometry $f$ that $f$ is a bijection such that both $f$ and $f^{-1}$ are continuous.

$\square$

**Remark 2.13.11** Proposition 2.13.12 proves that completeness property of a metric space is preserved by an isometry and hence by a homeomorphism. An isometry preserves metric structures. On the other hand, a homeomorphism preserves their open sets, i.e., their topological structures (see Chap. 3).

**Proposition 2.13.12** *Let $f : X \to Y$ be an isometry between metric spaces $(X, d)$ and $(Y, \rho)$. Then $(X, d)$ is complete iff $(Y, \rho)$ is complete.*

**Proof** Let $\{x_n\}$ be a Cauchy sequence in $X$. Then $\{f(x_n)\}$ is a Cauchy sequence in $Y$. If $(X, d)$ is complete, then $\{x_n\} \to x$ for some $x \in X$. This implies that $\{f(x_n)\} \to f(x)$. Its converse part is similar.

$\square$

**Example 2.13.13** Every isometry is a homeomorphism by Proposition 2.13.10, but its converse is not necessarily true. In support, consider the metric spaces $(X, d)$ and $(X, \rho)$ with different metrics $d$ and $\rho$ on the same set $X$. Then the identity map $1_d : X \to X$ is an isometry iff the metrics $d$ and $\rho$ are equivalent in the sense of Definition 2.13.14.

### 2.13.3  Equivalent Metrics

This subsection introduces the concept of **equivalent metrics**, which is an important concept, because they induce the same topology (see Chap. 3) and the identity map on metric spaces admitting equivalent metrics is a homeomorphism.

**Definition 2.13.14** Let $X$ be a nonempty set and $d$, $\rho$ be two metrics on $X$. Then they are said to **equivalent metrics** if they define the same convergent sequences. Alternatively, the two metrics $d$, $\rho$ on $X$ are said to be equivalent if given any point $x \in X$, any open ball $B_x^d(r)$ in $(X, d)$ contains an open ball $B_x^\rho(r')$ in $(X, \rho)$ for some $r' > 0$ and any open ball $B_x^\rho(t)$ in $(X, \rho)$ contains an open ball $B_x^d(t')$ in $(X, d)$ for some $t' > 0$. Equivalently, the identity map $1_d : (X, d) \to (X, \rho)$ is a homeomorphism.

**Example 2.13.15** Let $(X, d)$ be a metric space. Define a metric

$$\rho : X \times X \to \mathbf{R}, (x, y) \mapsto \frac{d(x, y)}{1 + d(x, y)}.$$

This shows that $\rho(x, y) < 1$, $\forall x, y \in X$ and hence

$$\rho(x_n, x)[1 + d(x_n, x)] = d(x_n, x) \implies d(x_n, x) = \frac{\rho(x_n, x)}{1 - \rho(x_n, x)}.$$

Then $d$ and $\rho$ are equivalent metrics on $X$. Because, if $d(x_n, x) \to 0$, then $\rho(x_n, x) = \frac{d(x_n, x)}{1 + d(x_n, x)} \to 0$. Again, if $\rho(x_n, x) \to 0$, then $d(x_n, x) \to 0$.

## 2.14 Normed Linear Spaces: Banach Spaces, Hilbert Spaces and Hahn–Banach Theorem

This section studies a special type of metric spaces having an additional structure such as normed linear structure with a brief study of Banach and Hilbert spaces, which form important classes of normed linear spaces.

**Definition 2.14.1** A **normed linear space** is a linear space (vector space) $X$ over $\mathbf{R}$ or $\mathbf{C}$ together with a real-valued function $|| \, ||: X \to \mathbf{R}$, called a norm function on $X$ such that it satisfies the following conditions for all $x, y \in X$ and all $\alpha \in \mathbf{R}$ or $\mathbf{C}$:

**N(1)** $||x|| \geq 0$ and $||x|| = 0$ iff $x = 0$;
**N(2)** $||x + y|| \leq ||x|| + ||y||$;
**N(3)** $||\alpha x|| = |\alpha| \, ||x||$.

It is sometimes written as the pair $(X, || \, ||)$. **Geometrically**, the nonnegative real number $||x||$ called the norm of $x$ represents the length of the vector $x$ from the origin of the linear space.

**Example 2.14.2** Let $\mathcal{C} = \mathcal{C}([0, 1])$ be the set of all real-valued continuous functions on $[0, 1]$. Then the norm function defined in Example 2.5.5 makes the linear space $\mathcal{C}$ a normed linear space.

**Remark 2.14.3** Every normed linear space is a metric space with respect to the metric defined by $d(x, y) = ||x - y||$. This result facilitates to study normed linear spaces with the help of the corresponding metrics. For example, it defines the equivalent norms in terms of equivalent metrics.

**Definition 2.14.4** Two norms $|| \, ||_1$ and $|| \, ||_2$ on the same linear space $X$ are said to be **equivalent** if the corresponding metrics are equivalent in the sense of Definition 2.13.14.

**Example 2.14.5** The norms $|| \, ||_1$ and $|| \, ||_2$ on $\mathbf{R}^n$ defined by

(i) $|| \, ||_1 : \mathbf{R}^n \to \mathbf{R}, \ x \mapsto (\sum_{i=1}^n x_i^2)^{1/2}$ (Euclidean norm) and
(ii) $|| \, ||_2 : \mathbf{R}^n \to \mathbf{R}, \ x \mapsto (\sum_{i=1}^n |x_i|)$

are equivalent.

## 2.14.1  Pseudo-normed Linear Spaces

This subsection defines pseudo-normed linear spaces by defining pseudo-norm function, which is more general than a norm function in the sense that every norm function is a pseudo-norm function but its converse is not necessarily true (see Example 2.14.10). Moreover, $l_p$-metric defined in Sect. 2.5.3 is generalized in this section by defining pseudo-metric $\mathcal{L}_p$.

**Definition 2.14.6** A **pseudo-normed linear space** is a linear space $X$ over **R** or **C** such that there is a real-valued function $|| \ || : X \to \mathbf{R}$, called a **pseudo-norm function** if it satisfies the following conditions for all $x, y \in X$ and all $\alpha \in \mathbf{R}$ or **C** :

**PN(1)**  $||x + y|| \leq ||x|| + ||y||$;
**PN(2)**  $||\alpha x|| = |\alpha| \ ||x||$.

**Proposition 2.14.7** *Let $X$ be a pseudo-normed linear space. Then*

*PN(3)*  $||0|| = 0$;
*PN(4)*  $||x|| \geq 0, \ \forall x \in X$.

**Proof** $||0|| = 0$ follows from **PN(2)**, because $||0|| = ||00|| = |0|||0|| = 0$. Again, **PN(4)** follows from **PN(1)** $-$ **PN(3)**, because for any $x \in X$,

$$0 = ||0|| = ||x + (-x)|| \leq ||x|| + |(-1)|||x|| = 2||x||.$$

This shows that $||x|| \geq 0, \ \forall x \in X$.

$\square$

**Remark 2.14.8** A normed linear space is a pseudo-normed linear space $X$ such that

$$\mathbf{PN(5)} : for \ any \ x \in X, ||x|| = 0 \implies x = 0.$$

**Definition 2.14.9** On a pseudo-normed linear space $X$, the function

$$d : X \times X \to \mathbf{R}, \ (x, y) \mapsto ||x - y||$$

is called a pseudo-metric on $X$.

We now use the Minkowski's inequality which asserts that for any two points $x = (x_1, x_2, \ldots, x_n)$ and $y = (y_1, y_2, \ldots, y_n) \in \mathbf{R}^n$ or $\mathbf{C}^n$,

$$\left( \sum_{k=1}^{n} |x_k + y_k|^2 \right)^{\frac{1}{2}} \leq \left( \sum_{k=1}^{n} |x_k|^2 \right)^{\frac{1}{2}} + \left( \sum_{k=1}^{n} |y_k|^2 \right)^{\frac{1}{2}},$$

equivalently, $||x + y|| \leq ||x|| + ||y||$.

**Example 2.14.10** ($\mathcal{L}_p$-**space**) For any integer $p > 0$, the set $\mathcal{L}_p$ consists of all functions $f : [0, 1] \to \mathbf{R}$ such that the Lebesgue integral

$$\int_0^1 |f(x)|^p dx$$

exists. Thus

$$\mathcal{L}_p = \left\{ f \in \mathcal{C}([0, 1]) : \int_0^1 |f(x)|^p dx \text{ exists} \right\}$$

The set $\mathcal{L}_p$ forms a real linear space under usual addition and scalar multiplication of functions, called $\mathcal{L}_p$- space. Then the function

$$\| \ \| : \mathcal{L}_p \to \mathbf{R}, f \mapsto \left[ \int_0^1 |f(x)|^p dx \right]^{\frac{1}{p}},$$

exists and $\mathcal{L}_p$ becomes a pseudo-normed linear space by using the second Minkowski's inequality. But it is not a normed linear space, because the equality $\|f\| = 0$ holds for each function $f : [0, 1] \to \mathbf{R}$, which vanishes except on a subset $S$ of $[0, 1]$ having **Lebesgue measure** $0$.

## 2.14.2   Banach Spaces and Hahn–Banach Theorem

This subsection presents a special type of normed linear spaces, called Banach spaces named after S. Banach (1892–1945), which are complete as metric spaces. Banach is considered as one of the founders of functional analysis. Banach spaces provide a link between algebraic and metric structures.

This subsection proves Hahn–Banach theorem on Banach (or normed linear) spaces which guarantees a rich supply of functionals. This theorem is named after H. Hahn (1879–1934) and S. Banach .

**Definition 2.14.11** A **Banach space** $X$ is a normed linear space which is complete with respect to the metric defined by its norm, i.e., a Banach space is a normed linear space $X$ in which every Cauchy sequence is convergent.

**Example 2.14.12** Examples of Banach spaces are enormous.

(i)  For the real field $\mathbf{R}$, the Euclidean $n$-space $\mathbf{R}^n$ is a real Banach space, since $\mathbf{R}^n$ is a complete metric space with respect to usual Euclidean metric induced by the norm

$$\|x\| = \left( \sum_{i=1}^{n} |x_i|^2 \right)^{\frac{1}{2}}, \ \forall \ x = (x_1, x_2, \ldots, x_n) \in \mathbf{R}^n.$$

(ii) For the complex field $\mathbf{C}$, the complex linear space $\mathbf{C}^n$ is a complex Banach space with respect to the metric induced by the norm

$$\|z\| = \left( \sum_{i=1}^{n} |z_i|^2 \right)^{\frac{1}{2}}, \ \forall \ z = (z_1, z_2, \ldots, z_n) \in \mathbf{C}^n.$$

(iii) **(Infinite-dimensional Euclidean space $\mathbf{R}^\infty$)** The set of all infinite real sequences $\{x = (x_1, x_2, \ldots) : \sum_{i=1}^{\infty} |x_i|^2 < \infty\}$, denoted by $\mathbf{R}^\infty$, is a real linear space, which admits a norm function

$$\|x\| = \left( \sum_{i=1}^{\infty} |x_i|^2 \right)^{\frac{1}{2}}$$

makes $\mathbf{R}^\infty$ a real Banach space, called the infinite-dimensional Euclidean space. Clearly, the point $x = (2, 2, 2, \ldots) \notin \mathbf{R}^\infty$. On the other hand, the point $y = (1, \frac{1}{2}, \frac{1}{2^2}, \frac{1}{2^3}, \ldots) \in \mathbf{R}^\infty$.

(iv) **(Infinite-dimensional unitary space $\mathbf{C}^\infty$)** The set of all infinite complex sequences $\{z = (z_1, z_2, \ldots) : \sum_{i=1}^{\infty} |z_i|^2 < \infty\}$, denoted by $\mathbf{C}^\infty$, is a complex linear space, which admits a norm function

$$\|z\| = \left( \sum_{i=1}^{\infty} |z_i|^2 \right)^{\frac{1}{2}}$$

makes $\mathbf{C}^\infty$ a complex Banach space, called the infinite-dimensional unitary space.

**Proposition 2.14.13** *Let $X$ and $Y$ be two normed linear spaces over the same scalar field $\mathbf{F} = \mathbf{R}$ or $\mathbf{C}$ and $\mathcal{C}(X, Y)$ be the set of all continuous linear transformations from $X$ to $Y$. Then*

(i) *$\mathcal{C}(X, Y)$ is also a normed linear space under pointwise linear operations and the norm function defined by*

$$\|T\| = \sup\{\|T(x)\| : \|x\| \leq 1\}.$$

(ii) *moreover, if $Y$ is a Banach space, then $\mathcal{C}(X, Y)$ is also a Banach space.*

**Proof** By hypothesis, $\mathcal{C}(X, Y)$ is the set of all continuous linear transformations from $X$ to $Y$.

(i) $\mathcal{C}(X, Y)$ is clearly a normed linear space under the given linear operations and norm.

(ii) Let $Y$ be a Banach space. We claim that the normed linear space $\mathcal{C}(X, Y)$ is also a Banach space. Given a Cauchy sequence $\{T_n\}$ in $\mathcal{C}(X, Y)$, let $x \in X$ be an arbitrary point. Then

$$||T_k(x) - T_n(x)|| = ||(T_k - T_n)(x)|| \leq ||T_k - T_n|| ||x||$$

implies that $\{|T_n(x)|\}$ is also a Cauchy sequence in $Y$. By hypothesis $Y$ is a Banach space and hence the sequence $\{|T_n(x)|\}$ is complete. Then there exists a point (vector ) $T(x) \in Y$ (say) such that $T_n(x) \to T(x)$. This assignment defines a linear map $T : X \to Y$ by using the continuity of both addition and scalar multiplication.

$T$ **is continuous**: $T$ is bounded, because

$$||T(x)|| = || \lim T_n(x)|| = \lim ||T_n(x)|| \leq \sup(||T_n|| ||x||)$$
$$= ( \sup ||T_n||) ||x||, \ \forall x \in X.$$

This implies that $T$ is continuous.

Next, to show that $||T_n - T|| \to 0$, take an $\epsilon > 0$ and an $k_0 \in \mathbf{N}$ such that

$$||T_k(x) - T_n(x)|| < \epsilon \ \forall k, n \geq k_0.$$

If $||x|| \leq 1$, then for $k, n \geq k_0$

$$||T_k(x) - T_n(x)|| = ||(T_k - T_n)(x)|| \leq ||T_k - T_n|| \, ||x|| \leq ||T_k - T_n|| < \epsilon.$$

Keeping $k$ fixed and allowing $n \to \infty$, it follows that

$$||T_k(x) - T_n(x)|| \to ||T_k(x) - T(x)||.$$

This implies that for all $x$ satisfying $||x|| \leq 1$.

$$||T_k(x) - T(x)|| < \epsilon \ \forall k \geq k_0.$$

It proves that
$$||T_k - T|| \leq \epsilon \ \forall k \geq k_0.$$

This proves that $||T_n - T|| \to 0$.

$\square$

Hahn–Banach Theorem 2.14.14 is a basic theorem in linear space theory. It provides a rich supply of functionals. It is proved by using Zorn's lemma.

**Theorem 2.14.14** *(**Hahn–Banach theorem**) Let $X$ be a normed linear space and $Y$ be a linear subspace of $X$. If $T$ is a functional on $Y$, then it can be extended to a functional $\tilde{T}_0$ over $X$ with the property that $||\tilde{T}_0|| = ||T||$.*

***Proof*** Let $\mathcal{F}$ be the set of all extensions of $T$ to functionals $\tilde{T}$ having the same norm on subspaces containing $Y$. Then $\mathcal{F} \neq \emptyset$. If $\text{dom}\,\tilde{T}$ represents the domain of $\tilde{T}$, then this domain set is partially ordered by the relation : $\tilde{T} \leq \tilde{T}'$ if $\text{dom}\,\tilde{T} \subset \text{dom}\,\tilde{T}'$ and $\tilde{T}(x) = \tilde{T}'(x)$ for all $x \in \text{dom}\,\tilde{T}$. Consider a chain $\{C_i : i \in \mathbf{A}\}$ in $\mathcal{F}$ and the union of any chain of members of the family $\mathcal{F}$. Then this union is an upper bound for the chain. Now use Zorn's lemma to show that there exists a maximal extension $\tilde{T}_0$. Again $\text{dom}\,\tilde{T} = X$, because, otherwise, $\tilde{T}_0$ can be further extended, which will contradict the maximality of $\tilde{T}_0$.                                                                             □

**Corollary 2.14.15** *Let $X$ be a real normed linear space and $x_0 \in X$ be a nonzero vector. Then there exists a functional $T_0 : X \to \mathbf{R}$ such that*

$$T_0(x_0) = ||x|| \text{ and } ||T_0|| = 1.$$

***Proof*** By hypothesis, $x_0 \in X$ be a nonzero vector. Then $Y = \{rx_0 : r \in \mathbf{R}\}$ is a linear subspace of $X$, called the subspace of $X$ spanned by $x_0$. Define

$$T : Y \to \mathbf{R}, \ rx_0 \mapsto r||x||.$$

Then $T$ is a functional on $Y$ such that

$$T(x_0) = ||x_0|| \text{ and } ||T|| = 1.$$

Then by Hahn– Banach Theorem 2.14.14, there exists an extension of $T$ to a functional

$$T_0 : X \to \mathbf{R}$$

such that

$$T_0(x_0) = ||x|| \text{ and } ||T_0|| = 1.$$

□

### 2.14.3   $l_p$-Space

This subsection gives examples of $l_p$-spaces which are used in subsequent study.

***Example 2.14.16***   (i) Let $p$ be a real number such that $1 \leq p < \infty$. Then the linear space $l_p^n$ of all $n$-tuples $\{x = (x_1, x_2, \ldots, x_n)\}$ over $\mathbf{R}$ or $\mathbf{C}$ is a Banach space with respect to the norm function defined by

$$\|x\|_p = \left(\sum_{i=1}^{n} |x_i|^p\right)^{\frac{1}{p}}$$

(ii) Let $p$ be a real number such that $1 \le p < \infty$. Then the linear space $l_p$ of all sequences $\{x = (x_1, x_2, \ldots, x_n \ldots, )\}$ over $\mathbf{R}$ or $\mathbf{C}$ is a Banach space with respect to the norm function defined by

$$\|x\|_p = \left(\sum_{i=1}^{\infty} |x_i|^p\right)^{\frac{1}{p}}.$$

***Example 2.14.17***    (i) The **real** $l_2$-**space** is precisely the infinite-dimensional Euclidean space $\mathbf{R}^\infty$ with metric

$$d : \mathbf{R}^\infty \times \mathbf{R}^\infty \to \mathbf{R}, \ (x, y) \mapsto \|x - y\| = \left(\sum_{i=1}^{\infty} |x_i - y_i|^2\right)^{\frac{1}{2}},$$

called the $l_2$-metric on $\mathbf{R}^\infty$. This metric space is called real $l_2$-space with the $l_2$-metric.

(ii) The **complex** $l_2$-**space** (defined in an analogous way) is precisely the infinite-dimensional unitary space $\mathbf{C}^\infty$.

## *2.14.4  Hilbert Spaces and Examples*

This subsection presents a special type of complex Banach spaces, called Hilbert spaces, whose norm is defined by an inner product. Banach spaces fail to provide the angle between two vectors; on the other hand, Hilbert spaces provide the concept of orthogonality of two nonzero vectors. Hilbert space is named after D. Hilbert ( 1862–1943), and modern development of such spaces is stimulated by the operator theory.

An inner product space $X$ is said to be a Hilbert space iff $X$ is complete with respect to the inner product norm. For example, a complete Hilbert space is formulated in Definition 2.14.18.

**Definition 2.14.18**  A complex Hilbert space is a complex Banach space $X$ with a function

$$\langle \, , \, \rangle : X \times X \to \mathbf{C}$$

such that for all $x, y, z \in X$ and $\alpha, \beta \in \mathbf{C}$,

**H(1)** $\langle \alpha x + \beta y, z \rangle = \alpha \langle x, z \rangle + \beta \langle y, z \rangle$;

**H(2)** $\overline{\langle x, y \rangle} = \langle y, x \rangle$, where $\overline{\langle x, y \rangle}$ is the complex conjugate of $\langle x, y \rangle$;

**H(3)** $\langle x, x \rangle = \|x\|^2$.

A **real Hilbert space** is defined in an analogous way.

***Example 2.14.19***  Examples of Hilbert spaces are plenty. For example,

(i)  the space $l_2^n$ with respect to inner product of two vectors defined by

$$\langle x, y \rangle = \sum_{i=1}^{n} x_i \overline{y}_i,$$

where $x = (x_1, x_2, \ldots, x_n) \in l_2^n$ is a Hilbert space.

(ii)  the space $l_2$ with respect to inner product of two vectors defined by

$$\langle x, y \rangle = \sum_{i=1}^{\infty} x_i \overline{y}_i,$$

where $x = (x_1, x_2, \ldots, x_n, \ldots) \in l_2$, is a Hilbert space, because the above series is convergent and converges to a complex number for every $x, y \in l_2$ by Cauchy's inequality.

## 2.15  Continuity of Functions on Normed Linear Spaces

This section studies continuity of linear transformations and linear functionals on normed linear spaces. An analogous study from the viewpoint of topological vector spaces is available in Basic Topology, Volume 3 of the present series of books.

### 2.15.1  Continuous Linear Transformations

This subsection defines continuous linear transformations between normed linear spaces which are continuous linear transformations between metric spaces. Since every normed linear space is a metric space, the properties which hold for continuous linear transformations between metric spaces are also satisfied by continuous linear transformations between normed linear spaces.

**Definition 2.15.1** Let $X$ and $Y$ be normed linear spaces over the same field **R** and $T : X \to Y$ be a linear transformation. Then $T$ is said to be **continuous** if it is continuous as a mapping from the metric space $X$ to the metric space $Y$ and $T$ said to be **bounded** if there exists a real number $K > 0$ such that

$$\|T(x)\| \le K \|x\|, \ \forall x \in X.$$

A continuous linear transformation $T : X \to Y$ between normed linear spaces $X$ and $Y$ asserts that whenever $x_n \to x$ in $X$, then $T(x_n) \to T(x)$ in $Y$.

**Definition 2.15.2** Let $X$ and $Y$ be normed linear spaces over **R** and $T : X \to Y$ be a linear transformation. Then $T$ is said to be an **isometrically isomorphism**, if $T$ is a bijective map such that

$$\|T(x)\| = \|x\|, \ \forall x \in X.$$

The normed linear spaces $X$ and $Y$ are said to be isometrically isomorphic if there exists an isometric isomorphism between them.

**Remark 2.15.3** Two isometrically isomorphic normed linear spaces are essentially the same from the viewpoint of their normed linear structures like two isomorphic group structures.

**Theorem 2.15.4** *Let $X$ and $Y$ be normed linear spaces over the same field **R** and $T : X \to Y$ be a linear transformation. Then the following statements on $T$ are equivalent:*

*(i)   $T$ is continuous at every point $x \in X$;*
*(ii)  for a sequence $\{x_n\}$ in $X$, the term $x_n \to 0$ implies that $T(x_n) \to 0$, i.e., $T$ is continuous at the origin $0$;*
*(iii) $T$ is bounded in the sense that there exists a real number $K > 0$ such that*

$$\|T(x)\| \leq K\|x\|, \ \forall x \in X;$$

*(iv) if $B = \{x \in X : \|x\| \leq 1\}$ is the closed unit ball in $X$, then $T(B)$ is bounded in $Y$.*

*Proof* (i) $\Leftrightarrow$ (ii). Suppose $T$ is continuous. Then $T(0) = 0$ shows that $T$ is continuous at the origin $0$. Conversely, let $T$ be continuous at $0$. Then $x_n \to x \Leftrightarrow x_n - x \to 0$. This implies that $T(x_n - x) \to 0 \Leftrightarrow T(x_n) - T(x) \to 0 \Leftrightarrow T(x_n) \to T(x) \Rightarrow T$ is continuous at every point $x \in X$.

(ii) $\Leftrightarrow$ (iii) Let there exist a real number $K > 0$ such that $\|T(x)\| \leq K\|x\| \ \forall x \in X$. Then $x_n \to 0$ implies $T(x_n) \to 0$. Conversely, let $T$ be continuous at $0$ and (ii) hold. If possible, there exists no such $K > 0$ such that $\|T(x)\| \leq K\|x\|, \forall x \in X$. Then for each $n \in \mathbf{N}$, there exists a vector $x_n$ such that $||T(x_n)|| > n\,||x_n||$. This shows that $\frac{\|T(x_n)\|}{n\|x_n\|\|} > 1$. Take $y_n = \frac{x_n}{n\|x_n\|}$. Then $y_n \to 0$ but $T(y_n)$ does not tend to $0$. This failure shows that $T$ is not continuous at $0$. This is a contradiction.

(iii) $\Leftrightarrow$ (iv) A nonempty subset of a normed linear space is bounded iff it is contained in closed ball $B_0(1)$ with center at $O$ and radius $1$. This shows that (iii) implies (iv), because if $\|x\| \leq 1$, then $\|T(x)\| \leq K$. Conversely, let $T(B)$ be contained in a closed ball $B_0(K)$ with center at $O$ and of radius $K$. For $x = 0$, $T(x) = 0$ implies $\|T(x)\| \leq K\|x\|$. For $x \neq 0$, $x/\|x\| \in B_0(K)$ and hence $\frac{\|T(x)\|}{\|x\|} \leq K$. This asserts $\|T(x)\| \leq K\|x\|, \ \forall x \in X \Rightarrow (iii)$.

(iv) $\Leftrightarrow$ (i): Left as an exercise.

$\square$

**Definition 2.15.5**    The **norm** $\|T\|$ of a continuous linear transformation $T : X \to Y$ is defined by

$$\|T\| = \sup\{\|T(x)\| : \|x\| \leq 1\}.$$

**Proposition 2.15.6**    *Let X and Y be two normed linear spaces over the same scalar field* $\mathbf{F} = \mathbf{R}$ *or* $\mathbf{C}$ *and* $\mathcal{C}(X, Y)$ *be the set of all continuous linear transformations from X to Y. Then* $\mathcal{C}(X, Y)$ *is also a normed linear space under pointwise linear operations and the norm function defined by*

$$\|T\| = \sup\{\|T(x)\| : \|x\| \leq 1\}.$$

*Moreover, if Y is a Banach space, then* $\mathcal{C}(X, Y)$ *is also a Banach space.*

**Proof**    Left as an exercise.                                                                                          □

## 2.15.2    Continuous Linear Functionals

This section continues the study of normed linear spaces over $\mathbf{R}$ by introducing the concept of linear functional and provides the requirement for continuity of linear transformations in some equivalent forms, which are convenient for our further study.

**Definition 2.15.7**    Let $X$ be a normed linear space over $\mathbf{R}$. Then a continuous function $f : X \to \mathbf{R}$ is said to be a **continuous linear functional** if

$$f(rx + sy) = rf(x) + sf(y),$$

for all $x, y \in X$ and all $r, s \in \mathbf{R}$.

**Example 2.15.8**    Given a point $r = (r_1, r_2, \ldots, r_n) \in \mathbf{R}^n$, the function

$$f : \mathbf{R}^n \to \mathbf{R}, x \to \sum_{i=1}^{n} r_i x_i$$

is a continuous linear functional.

**Example 2.15.9**    The elements of the set $\mathcal{C}(X, \mathbf{R})$ or $\mathcal{C}(X, \mathbf{C})$ defined in Proposition 2.15.6 are continuous linear functionals or simply called functionals, and this set of functionals of $X$ is abbreviated as $X^*$.

## 2.16 Applications

This section conveys some applications related to metric spaces such as Banach contraction theorem on complete metric spaces, Urysohn's lemma on metric spaces and some geometrical applications.

### *2.16.1 Banach Contraction Principle*

This subsection proves Banach contraction theorem on complete metric spaces named after the Polish mathematician Stefan Banach (1892–1945) and applies it to prove Picard's theorem on the existence of solutions of a differential equation (see Theorem 2.16.7 and Exercise 28 of Sect. 2.18), named after the French mathematician Charles Emile Picard (1856–1941).

Fixed-point theorems play a key role in the study of solution of certain system of equations:
Let

$$f_j(x_1, x_2, \ldots, x_n) = 0, \ j = 1, 2, \ldots, n,$$

be a given system of $n$-equations in $n$-unknowns, where $f_j$ are continuous real-valued functions of the $n$ real variables $x_i : i = 1, 2, \ldots, n$. Suppose

$$g_j(x_1, x_2, \ldots, x_n) = f_j(x_1, x_2, \ldots, x_n) + x_j,$$

for any point $x = (x_1, x_2, \ldots, x_n)$. If the function

$$g(x) = (g_1(x), g_2(x), \ldots, g_n(x))$$

has a fixed point $x_0 \in \mathbf{R}^n$, then $x_0$ is a solution of the above system of equations.

**Definition 2.16.1** $(X, d)$ be a metric space. Then a map $T : X \rightarrow X$ is said to satisfy **Lipschitz condition** if there exists a real number $r \geq 0$ such that

$$d(T(x_1), T(x_2)) \leq rd(x_1, x_2), \ \forall x_1, x_2 \in X.$$

The smallest $r$ satisfying the above inequality is called the **Lipschitz constant** for the map $T$. If $r \leq 1$, then the map $T$ is said to be nonexpansive, and it is said to be a contraction if $0 < r < 1$.

**Definition 2.16.2** Let $(X, d)$ be a metric space. A mapping $T : X \rightarrow X$ is said to be a **contraction** if there exists positive real number $r < 1$ such that

$$d(T(x), T(y)) \leq rd(x, y) < d(x, y), \ \forall x, y \in X.$$

**Remark 2.16.3** For a contraction mapping $T$ on a metric space $(X, d)$, the distance between the images $T(x)$ and $T(y)$ of any two points $x, y \in X$ under $T$ is less than the distance between the points $x$ and $y$. Such a mapping is (uniformly) continuous, since it is continuous at every point $x \in X$. Because given an $\epsilon > 0$, if $d(x, y) < \epsilon$, then by defining criterion of a contraction mapping, it follows that

$$d(T(x), T(y)) \le rd(x, y) < d(x, y) < \epsilon \ \forall x, y \in X.$$

**Theorem 2.16.4** *(Banach contraction theorem) Let $(X, d)$ be a complete metric space and $T : X \to X$ be a contraction mapping. Then $T$ has a unique fixed point, i.e., there exists a unique point $x \in X$ such that $T(x) = x$.*

**Proof** Let $x_0 \in X$ be an arbitrary point and $T : X \to X$ be a contraction mapping. Then there exists a positive real number $r < 1$ such that

$$d(T(x), T(y)) \le rd(x, y) < d(x, y), \ \forall x, y \in X.$$

($r$ may be taken the Lipschitz constant $(0 < r < 1)$ for the contraction mapping $T$, given in Definition 2.16.1). Denote

$$x_1 = T(x_0), x_2 = T(x_1) = T^2(x_0), \ldots, x_n = T(x_{n-1}) = T^n(x_0).$$

Suppose $k < n$. Then

$$d(x_k, x_n) = d(T^k(x_0), T^n(x_0)) = d(T^k(x_0), T^k \circ T^{n-k}(x_0)) \le r^k d(x_0, T^{n-k}(x_0))$$
$$= r^k d(x_0, x_{n-k})$$

$\le r^k[d(x_0, x_1) + d(x_1, x_2) + \cdots + d(x_{n-k-1}, x_{n-k})] \le r^k d(x_0, x_1)[1 + r + \cdots + r^{n-k-1}] < r^k d((x_0, x_1)/((1 - r))$. As $r < 1$, the sequence $\{x_n\}$ is a Cauchy sequence. Hence by completeness of $X$, it follows that there exists a point $x \in X$ with the property that $x_n \to x$. By continuity of $T$ (see Remark 2.16.3), it follows that

$$T(x) = T(\lim x_n) = \lim (T(x_n)) = \lim x_{n+1} = x,$$

which shows that $x$ is a fixed point of $T$. To prove the uniqueness of fixed point of $T$, let $y \in X$ be also a fixed point of $T$. Then $T(x) = x$ and $T(y) = y$ assert that

$$d(x, y) = d(T(x), T(y)) \le r \, d(x, y)$$

which shows that $d(x, y) = 0$ *or* $y = x$, since $r < 1$. It asserts that the fixed point of $T$ is unique. $\square$

**Example 2.16.5** Theorem 2.16.4 known as **Banach contraction principle or simply contraction principle** may not be true for spaces which are not complete metric.

For example, consider the metric space $X = (0, 1]$. It is not complete. The map $T : X \to X, x \mapsto x/3$ has no fixed point.

***Example 2.16.6*** There are subjective maps on metric spaces, which are not continuous, but they have unique fixed points. For example, consider the map

$$T : \mathbf{I} \to \mathbf{I}, \ x \mapsto \begin{cases} 1/2 + 2x, & \text{if } x \in [0, 1/4] \\ 1/2, & \text{if } x \in (1/4, 1] \end{cases}$$

The map $T$ has a unique fixed point which is $x = 1/2$. The map $T$ is onto but not continuous.

## 2.16.2  Further Application to Analysis: Picards's Theorem

This subsection proves Picards's Theorem 2.16.7 on existence of solution of differential equations.

**Theorem 2.16.7** (***Picards's theorem***) *Let the functions $f(x, y)$ and $\frac{\partial f}{\partial y}$ be both continuous in a closed rectangle $R = \{(x, y) \in \mathbf{R}^2 : a \leq x \leq b \text{ and } c \leq y \leq d\}$. If $(x_0, y_0)$ is an interior point of the rectangle $R$, then the differential equation*

$$\frac{dy}{dx} = f(x, y)$$

*has a unique solution $y = h(x)$ with $h(x_0) = y_0$.*

***Proof*** By hypothesis of continuity of both the functions $f(x, y)$ and $\frac{\partial f}{\partial y}$, it follows that there exist positive constants $A$ and $B$ such that

$$|f(x, y)| \leq A \text{ and } |\frac{\partial f(x, y)}{\partial y}| \leq B, \ \forall (x, y) \in \mathbf{R}^2.$$

Take a number $k > 0$ such that $Bk < 1$ and a closed rectangle $R^*$ determined by $|x - x_0| \leq k$ and $|y - y_0| \leq Ak$. Let $X$ be the set of all continuous real-valued functions $y = h(x)$ defined on the closed interval $|x - x_0| \leq k$ with $|h(x) - y_0| \leq Ak$. Let $\mathbf{I}_0 = [x_0 - a, x_0 + a]$ and $\mathcal{C}(\mathbf{I}_0)$ be the complete metric space of all real-valued continuous functions on $I_0$. Then $X$ is a closed subspace of $\mathcal{C}(I_0)$, and hence $X$ is a complete metric space. Consider the map

$$\phi : X \to X, \ g \mapsto w, \text{ where } w(x) = y_0 + \int_{x_0}^{x} f(t, g(t)) dt.$$

Then $\phi$ is a contraction mapping on the complete metric space $X$. Finally, using Banach contraction Theorem 2.16.4 on $\phi$, it follows that $\phi$ has the unique fixed point $h$ and hence $\phi(h) = h$. □

**Remark 2.16.8** An alternative form of Picard's Theorem 2.16.7 is given in Exercise 28 of Sect. 2.18.

### 2.16.3 Urysohn Function and Urysohn Lemma for Metric Spaces

This subsection proves Urysohn's Lemma 2.16.10 for metric spaces, which is a very significant result in metric spaces with wide applications. It provides a rich supply of continuous functions. This result was given by P. S. Urysohn (1898–1924). We utilize the Corollary 2.9.16 to prove Urysohn's lemma on metric spaces. For its more general result, see Chap. 6.

**Definition 2.16.9** Two disjoint subsets $A$ and $B$ of a metric space $(X, d)$ are said to be **separated by real-valued continuous functions x**, $f \in \mathcal{C}(X, \mathbf{R})$ with the property:

$$f(x) = \begin{cases} 0, & \text{for all} x \in A \\ 1, & \text{for all} x \in B \end{cases}$$

and

$$0 \le f(x) \le 1 \text{ for all } x \in X.$$

Such a function $f$ is called **Urysohn function** corresponding to the pair of disjoint subsets $A$ and $B$.

Urysohn's Lemma 2.16.10 guarantees the existence of Urysohn functions.

**Lemma 2.16.10** *(Urysohn lemma) Let $(X, d)$ be a metric space and $A$, $B$ be two disjoint closed subsets of $X$. Then there is a continuous function $f : X \to \mathbf{R}$ such that*

  *(i) $f(x) \in [0, 1]$ for all $x \in X$;*
  *(ii) $f(x) = 0$ for all $x \in A$;*
  *(iii) $f(x) = 1$ for all $x \in B$.*

**Proof** By hypothesis, $A$ and $B$ are closed sets in $X$. Since $x$ is a limit point of $A$ iff $d(x, A) = 0$ and $x$ is a limit point of $B$ iff $d(x, B) = 0$. This implies that $d(x, A) = 0$ iff $x \in A$ and $d(x, B) = 0$ iff $x \in B$. Since $A$ and $B$ are disjoint closed sets, $d(x, A)$ and $d(x, B)$ cannot vanish simultaneously. Hence it follows that $d(x, A) + d(x, B)$ never vanishes. Define a function

$$f : X \to \mathbf{R}, \ x \mapsto \frac{d(x, A)}{d(x, A) + d(x, B)}.$$

Then $f$ is a well-defined continuous function such that

$$f(x) = \begin{cases} 0, & \text{for all } x \in A \\ 1, & \text{for all } x \in B \end{cases}$$

and

$$0 \leq f(x) \leq 1 \text{ for all } x \in X.$$

$\square$

**Corollary 2.16.11** *Let $(X, d)$ be a metric space and $U$ be an open set containing a closed subset $G$ of $X$. Then there is a continuous function $f : X \to \mathbf{R}$ such that*

*(i) $f(x) \in [0, 1]$ for all $x \in X$;*
*(ii) $f(x) = 1$ for all $x \in G$;*
*(iii) $f(x) = 0$ for any $x \in X$ but not lying in $U$.*

*Proof* The corollary follows from Urysohn Lemma 2.16.10 by taking in particular $A = X - U$ and $B = G$, since here $A$ and $B$ are disjoint closed sets in $X$. $\square$

*Remark 2.16.12* **Urysohn's lemma** 2.16.10 given by P. S. Urysohn (1898–1924) prescribes a method of construction of a real-valued continuous function on a metric space. Its generalization for specific topological spaces is given in Chap. 6 of this book. This is an outstanding result connecting a particular class of topological spaces to the real number space which characterizes this type of topological spaces (normal spaces). The tenure of his mathematical work was only three years.

## *2.16.4 Geometrical Applications*

This subsection presents some geometrical examples of continuous functions arising through Euclidean spaces of different dimensions, which will be used throughout the book. We are interested in geometrical configurations of certain metric spaces, which are subsets of Euclidean line $\mathbf{R}$ (or $\mathbf{R}^1$), Euclidean plane $\mathbf{R}^2$ or in general, Euclidean $n$-space $\mathbf{R}^n$, where there are Cartesian coordinates having each point expressed uniquely by coordinates of an ordered $n$-tuple of real numbers $(x_1, x_2, \ldots, x_n)$. In particular, it represents a point in the Euclidean line $\mathbf{R}$ (or $\mathbf{R}^1$) for $n = 1$, a point in the Euclidean plane $\mathbf{R}^2$ for $n = 2$ or in general, a point in Euclidean $n$-space $\mathbf{R}^n$.

***Example 2.16.13***     (i)   (**Translation**) A translation $T_a$ by $a \in \mathbf{R}$ on the Euclidean line $\mathbf{R}$ is a continuous function

$$T_a : \mathbf{R} \to \mathbf{R}, \; x \mapsto x + a.$$

This translation carries a point $x \in \mathbf{R}$ to the point $x + a \in \mathbf{R}$. Similarly, given $a, b \in \mathbf{R}$, a translation

$$T_{(a,b)} : \mathbf{R}^2 \to \mathbf{R}^2, \; (x, y) \mapsto (x + a, y + b)$$

is continuous function which carries the point $(x, y) \in \mathbf{R}^2$ to the point $(x + a, y + b) \in \mathbf{R}^2$.

(ii)   (**Rotation**) A rotation of the plane $\mathbf{R}^2$ is described by specifying its center and the angle of rotation. For example, the continuous function

$$R_\theta : \mathbf{R}^2 \to \mathbf{R}^2, \; (x, y) \mapsto (x \cos \theta - y \sin \theta, \; x \sin \theta + y \cos \theta)$$

represents a rotation of the plane $\mathbf{R}^2$ about its origin through an angle $\theta$.

(iii)   (**Reflection**) Reflection on a line l in $\mathbf{R}^2$ is a continuous function $R_l : \mathbf{R}^2 \to \mathbf{R}^2$, which keeps every point of the line l fixed and takes the mirror image of points of the two sides of  l, i.e., interchanges the two sides of the line  l.

***Remark 2.16.14*** Any translation, rotation, reflection or any reflection followed by a translation described in Example 2.16.13 keep the shape and size of the configurations in the plane unchanged, but their positions may be altered. Such functions are the congruences of elementary geometry.

**Definition 2.16.15   A rigid function (motion)** on $\mathbf{R}^n$ with Euclidean distance function $d$ is a continuous function $f : \mathbf{R}^n \to \mathbf{R}^n$ such the distance between every pair of points is preserved under $f$ in the sense that $d(f(x), f(y)) = d(x, y), \; \forall x, y \in \mathbf{R}^n$.

***Example 2.16.16*** Every translation, rotation, reflection or any reflection followed by a translation described in Example 2.16.13 is a rigid motion.

**Proposition 2.16.17** *Every rigid motion $f$ on $\mathbf{R}^n$ with Euclidean distance function $d$ is a continuous function $f : \mathbf{R}^n \to \mathbf{R}^n$ .*

***Proof*** Let $f : \mathbf{R}^n \to \mathbf{R}^n$ be a rigid motion. Then $d(f(x), f(y)) = d(x, y), \; \forall x, y \in \mathbf{R}^n$. For any $x \in X$ and for every $\epsilon > 0$, take $\delta = \epsilon$. If $y \in B_x(\delta)$, then

$$d(x, y) < \delta = \epsilon \implies d(f(x), f(y)) = d(x, y) < \epsilon \implies f(y) \in B_{f(x)}(\epsilon),$$
$$\forall \, y \in B_x(\delta)$$

This asserts that

$$f(B_x(\delta)) \subset B_{f(x)}(\epsilon).$$

Since $x \in X$ is an arbitrary point of $X$, it follows by Definition 2.3.14 that $f$ is continuous. □

**Corollary 2.16.18** *(i) Every rigid motion of the plane is continuous.*
*(ii) Every translation, rotation, reflection or any reflection followed by a translation described in Example 2.16.13 are continuous.*

**Proof** (i) It follows from Proposition 2.16.17 for $n = 2$.
(ii) It follows from Example 2.16.16 by using Proposition 2.16.17.

□

## *2.16.5 Separable Metric Spaces*

This subsection studies **separable metric spaces** introduced by M. Fréchet, which form an important class of metric spaces. Separable spaces in the topological setting are discussed in Chaps. 3 and 7.

**Definition 2.16.19** Let $(X, d)$ be a metric space and $A$ be a nonempty subset of $X$. Then $A$ is said to be **dense** in $X$ if

$$B_x(\epsilon) \cap A \neq \emptyset, \text{ for every } x \in X \text{ and for all } \epsilon > 0.$$

Thus if $A$ is dense in $X$, then every open ball centered at any point in X contains a point of $A$.

**Definition 2.16.20** A metric space $(X, d)$ is said to **separable** if there exists a countable dense subset $A$ of $X$.

*Example 2.16.21* The Euclidean line **R** is a separable metric space, since its subset **Q** is countable, and **Q** is dense in **R**, because every open interval in **R** contains rational numbers. On the other hand, **R** equipped with the discrete metric

$$d : \mathbf{R} \times \mathbf{R} \to \mathbf{R}, \ (x, y) \mapsto \begin{cases} 0, & \text{if } x = y \\ 1, & \text{if } x \neq y, \end{cases}$$

is not separable.

Example 2.16.22 gives a generalization of Example 2.16.21.

*Example 2.16.22* The Euclidean $n$-space $\mathbf{R}^n$ is a separable metric space. Because, its subset

$$\mathbf{Q}^n = \{(x_1, x_2, \ldots, x_n) : x_1, x_2, \ldots, x_n \in \mathbf{Q}\} \subset \mathbf{R}^n$$

is countable and it is dense in the metric space $\mathbf{R}^n$.

Theorem 2.16.23 characterizes separable metric spaces $(X, d)$ with the help of its distance function $d$.

**Theorem 2.16.23** *A metric space $(X, d)$ is separable iff there exists a countable subset $A$ of $X$ such that for every $\epsilon > 0$ and $x \in X$, there exists $y \in A$ satisfying the condition $d(x, y) < \epsilon$.*

**Proof** Let $A$ be a countable subset of the metric space $X$ satisfying the given hypothesis. Then every point $x \in X$ is either a point of $A$ or is a limit point of $A$. This asserts that $\overline{A} = X$ and hence $X$ is separable. Conversely, suppose that the metric space is separable. Then there exists a countable dense subset $A$ of $X$, which implies that $A$ is countable and $\overline{A} = X$. Then given $\epsilon > 0$ and $x \in X$, it follows that $x \in \overline{A}$ and the open ball $B_x(\epsilon)$ contains a point $y \in A$. This asserts that $d(x, y) < \epsilon$.

□

## 2.17 Compact Subsets of Metric Spaces

This section addresses the concept of compact subsets in a metric space, which is an abstraction of the concept of compactness derived from the Heine–Borel property of the real line **R** expressed in Theorem 2.1.23. On the other hand, a study of compactness for a topological space generalizing Heine–Borel Theorem 2.1.23 is given in Chap. 5.

**Definition 2.17.1** Let $(X, d)$ be a metric space. $X$ is said to have an **open covering** if there exists a collection of open sets $\mathcal{U} = \{U_a : a \in \mathbf{A}\}$ such that $X = \bigcup\{U_a : a \in \mathbf{A}\}$, i.e., every point of $X$ belongs to at least one $U_a$. A subfamily of an open covering $\mathcal{U}$ is said to be subcovering of $X$ if this subfamily forms itself an open covering of $X$. A metric space $(X, d)$ is said to be **compact** if every open covering of $X$ has a finite subcovering. In the language of sequences, $(X, d)$ is said to be compact (or sequentially compact) if every sequence in $X$ has a convergent subsequence.

**Remark 2.17.2** Definition 2.17.1 asserts that every compact metric space is complete. But its converse is not true. For example, the the Euclidean line **R** is a complete metric space, but it is not compact.

**Example 2.17.3** Consider the Euclidean line **R** and its open interval $(0, 1)$ with metric induced from the standard metric of **R**. Then $\mathcal{U} = \{(-n, n) : n \in \mathbf{N}\}$ forms an open covering of **R**. There exists different coverings of the same subset of **R**. For example, $\mathcal{U}_1 = \{(1/n, 1) : n > 1 \text{ and } n \in \mathbf{N}\}$ forms an open covering of $(0, 1)$, and $\mathcal{U}_2 = \{(\frac{1}{n}, \frac{n}{n+1}) : n > 1 \text{ and } n \in \mathbf{N}\}$ forms another open covering of $(0, 1)$.

**Example 2.17.4** Let $(X, d)$ be a metric space. Then given $r > 0$, the family $\mathcal{B} = \{B_x(r) : x \in X\}$ of all open balls in $X$ of radius $r$ centered at $x$ forms an open covering of $X$. This gives a nontrivial open covering of any metric space $(X, d)$.

**Example 2.17.5** Let $(X, d)$ be a metric space and $Y \subset X$ be finite. If $d_Y$ is the metric on $Y$ induced by the metric $d$ of $X$, then $(Y, d_Y)$ is compact.

**Example 2.17.6** Under usual metric on $\mathbf{R}$, the closed and bounded interval $[a, b]$ is compact, but the open interval $(a, b)$ is not compact. Because **Heine–Borel Theorem** 2.1.23 in $\mathbf{R}$ asserts that every closed bounded subset of $\mathbf{R}$ is compact in the sense that its every open covering has a finite subcovering. On the other hand, $(0, 1)$ is not compact, because $\mathcal{U}_1 = \{(1/n, 1) : n > 1 \text{ and } n \in \mathbf{N}\}$ forms an open covering of $(0, 1)$ but it does not admit any subcovering.

Theorem 2.17.7 determines precisely the compact subsets of the Euclidean space $\mathbf{R}^n$, which is a generalization of the classical Heine–Borel theorem in analysis. Its partial generalization in an arbitrary metric space is given in Exercise 46 of Sect. 2.18.

**Theorem 2.17.7** (*Generalization of Heine–Borel theorem in $\mathbf{R}^n$*) *A subset $A$ in the Euclidean plane $\mathbf{R}^n$ is compact iff it is closed and bounded.*

**Proof** Let $A$ be a compact subset of $\mathbf{R}^n$. Then the family of open balls $\mathcal{B} = \{B_0(n) : n \in \mathbf{N}\}$ with center at origin $0 \in \mathbf{R}^n$ forms an open covering of $A$ admitting a finite covering of $A$ by compactness property of $A$. Then

$$A \subset \bigcup B_0(n) \text{ for finite values of } n.$$

Consequently, it follows that $A$ is closed and bounded. For its converse part see Chap. 5. $\qquad\square$

## 2.17.1 Compactness of Metric Spaces is a Topological Property

This subsection shows that the compactness of metric spaces is a topological property in the sense that if two metric spaces $X$ and $Y$ are homeomorphic and if $X$ is compact, then $Y$ is also compact and conversely, if $Y$ is compact, then $X$ is also. More precisely, Theorem 2.17.8 proves that compactness in metric spaces is preserved by a continuous map and hence it follows that compactness in a metric space is a topological property (see Corollary 2.17.10). In topological setting, it is also proved in Chap. 5 that compactness is a topological property.

**Theorem 2.17.8** *Let $(X, d)$ and $(Y, \rho)$ be two metric spaces and $f : X \to Y$ be a continuous onto map. If $X$ is compact, then $Y$ is also compact.*

**Proof** Let $(X, d)$ be a compact metric space. We claim that $f(X) = Y$ is also a compact metric space. To prove it, take any open covering $\mathcal{U} = \{U_i : i \in \mathbf{A}\}$ of $Y$. Then $\mathcal{V} = \{V_i : i \in \mathbf{A}\}$ forms an open covering of $X$, where $V_i = f^{-1}(U_i)$. By hypotheses, $X$ is compact. Hence there exists a finite subcovering $\{V_{i_k} : 1 \leq k \leq n\}$ of $\mathcal{V}$. This asserts that the finite subfamily of open sets $\{U_{i_k} : 1 \leq k \leq n\}$ of $\mathcal{U}$ is an open covering of $Y$. This proves that $Y$ is also compact. $\qquad\square$

**Corollary 2.17.9** *Let $(X, d)$ and $(Y, \rho)$ be two metric spaces and $f : X \to Y$ be a continuous map. If $X$ is compact, then $f(X)$ is also so.*

**Proof** It follows from Theorem 2.17.8.                                                           □

**Corollary 2.17.10** *Let $(X, d)$ and $(Y, \rho)$ be two homeomorphic metric spaces. Then $X$ is compact iff $Y$ is also so.*

**Proof** It follows from Theorem 2.17.8.                                                           □

**Remark 2.17.11** More study on **compactness** in topological setting is available in Chap. 5.

### 2.17.2  Lebesgue Lemma and Lebesgue Number

The notion of a Lebesgue number stems from Lebesgue's work on measure theory, starting with his dissertation in 1902. This subsection studies Lebesgue lemma and Lebesgue number named after H. Lebesgue (1875–1941) for an open covering of a compact metric space establishing a relation between such a space and Lebesgue number and proves Lebesgue lemma providing a technical result on open covering of a compact metric space.

**Definition 2.17.12** Given an open covering $\mathcal{F} = \{U_\alpha : \alpha \in \mathbf{A}\}$ of a compact metric space $X$, there exists a real number $\delta > 0$ (called **Lebesgue number** of $\mathcal{F} = \{U_\alpha\}$) such that every open ball $B_x(\epsilon)$ in $X$ for some $\epsilon > 0$, is contained in at least one open set $\{U_\alpha\} \in \mathcal{F}$.

The existence of Lebesgue number is proved in Lemma 2.17.13.

**Lemma 2.17.13** *(Lebesgue covering lemma) Let $X$ be a compact metric space. Given an open covering $\mathcal{F} = \{U_\alpha : \alpha \in \mathbf{A}\}$ of $X$, there exists a Lebesgue number $\delta > 0$ of the covering $\mathcal{F} = \{U_\alpha\}$ such that whenever $Y \subset X$ and $\mathrm{diam}(Y) < \delta$, then $Y \subset U_\alpha$ for some $\alpha \in \mathbf{A}$.*

**Proof** Let $X$ be a compact metric space with metric $d$. Then for an arbitrary point $x \in X$, there is an $\epsilon(x) > 0$ depending on $x$ such that the open ball $B_x(2\epsilon(x)) \subset U_\alpha$ for some $\alpha \in \mathbf{A}$. Since $X$ is compact, there is a finite number of the open balls $B_x(2\epsilon(x))$, suppose for $x = x_1, x_2, \ldots, x_m$. Let $\delta = \min\{\epsilon(x_j) : j = 1, 2, \ldots, m\}$. If $\dim(Y) < \delta$ and $y_0 \in Y$, there exists an index $j$ such that $1 \leq j \leq m$ with the property that $d(y_0, x_j) < \epsilon(x_j)$. Again, for $y \in Y, d(y, y_0) < \delta \leq \epsilon(x_j)$. Hence by triangle inequality for the metric space $X$, it follows that

$$d(y, x_j) \leq d(y, y_0) + d(y_0, x_j) < 2\epsilon(x_j).$$

This asserts that $Y \subset B_x(2\epsilon(x)) \subset U_\alpha$ for some $\alpha \in \mathbf{A}$.

□

***Example 2.17.14*** In the Euclidean line **R**,

(i) consider an open covering of $\mathcal{U}$ of $X = [0, 1]$ given by

$$\mathcal{U} = (-1/10, 1/10) \cup \{U_t = (t/2, 2) : 0 < t \le 1\}.$$

It has a Lebesgue number $\delta = 1/10$. Because if $Y \subset X$ and $\dim(Y) < 1/10$, then $t = \mathrm{Inf}\ Y$ is either positive or 0. For the first case, $Y \subset U_t$ and for the second case, $Y \subset (-1/10, 1/10)$.

(ii) consider an open covering of $\mathcal{U}$ of $X = (-1, 0) \cup (0, 1)$ given by

$$\mathcal{U} = \{(-1, 0), (0, 1)\}.$$

Then $\mathcal{U}$ has no Lebesgue number.

***Remark 2.17.15*** More study on **Lebesgue Covering Lemma** 2.17.13 and **Lebesgue number** is available in Chap. 5.

### 2.17.3 A Characterization of Totally Bounded Complete Metric Spaces

Every compact metric space is complete, but its converse is not true (see Remark 2.17.2). So, it is a natural question: does there exist any additional condition on a complete metric space to make it a compact space? The search of such a condition leads to the concept of totally boundedness in a metric space. This concept gives a satisfactory answer to this question saying that a metric space $(X, d)$ is compact iff it is complete and totally bounded (see Exercise 49 of Sect. 2.18). This subsection characterizes a totally bounded complete metric space $(X, d)$ in Theorem 2.17.19 with the help of a sequence in $X$ having a convergent subsequence, which is equivalent to the compactness of $(X, d)$ by Exercise 49 of Sect. 2.18).

**Definition 2.17.16** A metric space $(X, d)$ is said to be **totally bounded** if, for every $\epsilon > 0$, the space $X$ is covered by a finite number of $\epsilon$-balls in the sense that $X = \bigcup B_x(\epsilon)$ (union of finite number of $\epsilon$-balls).

**Proposition 2.17.17** *Every totally bounded metric space is bounded.*

***Proof*** Let $(X, d)$ be a totally bounded metric space. Then for every $\epsilon > 0$, the space $X$ is covered by a finite number of $\epsilon$-balls. For $\epsilon = 1/2$, let

$$X \subset B_{x_1}(1/2) \cup B_{x_2}(1/2) \cup \cdots \cup B_{x_n}(1/2)$$

This asserts that

$$\mathrm{diam}\ X \le 1 + \max\ \{d(x_i, x_k)\}.$$

It proves that $X$ is bounded.

$\square$

***Example 2.17.18*** Total boundedness implies boundedness in a metric space by Proposition 2.17.17. Its converse is not true in general. For example,

(i) Let $(\mathbf{R}, d)$ be the metric space with metric

$$d : \mathbf{R} \times \mathbf{R} \to \mathbf{R}, (x, y) \mapsto \min \{1, |x - y|\}.$$

This asserts that $(\mathbf{R}, d)$ is bounded. But $(\mathbf{R}, d)$ is not totally bounded.
(ii) Converse of Proposition 2.17.17 is true for finite-dimensional Euclidean space $\mathbf{R}^n$ but it is not true for the infinite-dimensional Euclidean space $\mathbf{R}^\infty$ (see Example 2.14.12). Because, the closed infinite-dimensional unit sphere $S^\infty$ in $\mathbf{R}^\infty$ defined by

$$S^\infty = \{x \in \mathbf{R}^\infty : ||x|| \leq 1\} \subset \mathbf{R}^\infty$$

is not totally bounded but it is bounded.

Theorem 2.17.19 characterizes a totally bounded complete metric space by its sequences.

**Theorem 2.17.19** *Let* $(X, d)$ *be a metric space. Then it is totally bounded and complete iff every sequence in* $X$ *has a convergent subsequence.*

***Proof*** Let $(X, d)$ be a totally bounded complete metric space and $\{x_n\}$ be any sequence in $X$. As the metric space $X$ is totally bounded by hypothesis, $X$ can be covered by only a finite number of 1-balls of the form $B_x(1)$. Then there is some 1-ball $B_1$, say, out of these finite number of 1-balls, such that $B_1$ contains $x_n$ for an infinite number of $n$. Since $B_1 \subset X$ and $X$ is totally bounded, $B_1 \subset X$ can also be covered by a finite number of $\frac{1}{2}$- balls of the form $B_x(\frac{1}{2})$. Then there is some $\frac{1}{2}$-ball, $B_2$, say, out of these finite number of $\frac{1}{2}$-balls such that $B_2$ contains $x_n$ for infinitely many $n$. Hence $B_1 \cap B_2$ must contain $x_n$ for an infinite number of $n$. Repeating this process for $n = 1, 2, 3, \ldots$, it follows that there is some $\frac{1}{n}$-ball, say $B_n$, of the form $B_x(\frac{1}{n})$, such that $B_1 \cap B_2 \cap \cdots \cap B_n$ contain $x_k$ for an infinite number of $k$. So there exists a subsequence $\{x_{n_k}\}$ of the given sequence $\{x_n\}$ such that

$$x_{n_k} \in B_1 \cap B_2 \cap \cdots \cap B_n, \forall k.$$

If $k < m$, then both the elements $x_{n_k}$ and $x_{n_m}$ are in $B_k$ and hence $d(x_{n_k}, x_{n_m}) < \frac{1}{k}$. This asserts that the subsequence $\{x_{n_k}\}$ is a Cauchy sequence and hence it is convergent by Proposition 2.10.12.

Conversely, let every sequence in $X$ have a convergent subsequence and $\{x_n\}$ be a Cauchy sequence in $X$. Then by hypothesis, there is a convergent subsequence $\{x_{n_k}\}$ of the sequence $\{x_n\}$ and hence $x_{n_k} \to x$ for some $x \in X$. Applying triangle inequality, it follows that $x_n \to x \in X$. This shows that $X$ is a complete metric space. To show that $X$ is totally bounded, assume to the contrary that $X$ is not covered by a finite number of open $\epsilon$-balls in $X$. Then there exist points $x_1, x_2, \ldots, x_m, \ldots, x_k, \ldots$ in $X$ such that

$$d(x_k, x_m) > \epsilon, \ \forall k > m.$$

This implies that such a sequence $\{x_n\}$ cannot have a convergent subsequence, which contradicts the hypothesis that every sequence in $X$ has a convergent subsequence. This shows that $X$ is totally bounded.

$\square$

## 2.17.4 Connectedness in Metric Spaces and Connected Subsets of **R**

This subsection introduces the concepts of connectedness in metric spaces and studies connected subsets of the Euclidean line **R**. The intermediate value theorem in calculus says that if $f : [a, b] \to \mathbf{R}$ is continuous and $r \in \mathbf{R}$ lies between $f(a)$ and $f(b)$, then there exists a point $\alpha \in [a, b]$ such that $f(\alpha) = r$. There is a natural question: does there exist a generalization of this theorem? This theorem depends not only on continuity of $f$ but also on a special property of $[a, b]$, now called connectedness.

**Definition 2.17.20** Let $(X, d)$ be a metric space. A **separation** of $X$ by a pair $\{U, V\}$ of open sets in $X$ means

$$X = U \cup V \text{ and } U \cap V = \emptyset.$$

For this separation, $U$ and $V$ are both open and closed in $X$. This separation is said to be trivial if either $U$ or $V$ is $\emptyset$. Otherwise, it is said to be nontrivial.

**Definition 2.17.21** Let $(X, d)$ be a metric space. It is said to be **connected** if it has no nontrivial separation. Otherwise, $X$ is said to disconnected. Alternatively, $X$ is connected if $X$ has no nonempty proper subset that is both open and closed in $X$.

Theorem 2.17.22 characterizes connectedness in terms of closed-open sets and hence it gives another formulation of Definition 2.17.21, which can be used as an equivalent definition of connectedness.

**Theorem 2.17.22** *Let $(X, d)$ be a metric space. It is connected iff the only subsets of $X$ which are both open and closed in $X$ are the set $X$ itself and the emptyset $\emptyset$.*

**Proof** Suppose that $X$ is connected. If $Y \subset X$ be a proper subset of $X$, which is both open and closed in $X$, then choose $U = Y$ and $V = X - Y$. Since $U \cap V = \emptyset$, the pair $\{U, V\}$ forms a separation of $X$. Since by hypothesis, $X$ is connected, no such proper subset of $X$ exists. In other words, the only subsets of $X$ which are both open and closed in $X$ are the set $X$ itself and $\emptyset$. Conversely, suppose that $\{U, V\}$forms a separation of $X$. Then $U \neq \emptyset$ and $U \neq X$ and it is both open and closed.    $\square$

**Definition 2.17.23** Let $(X, d)$ be a metric space and $(Y, d_Y)$ be a subspace of $(X, d)$ with metric $d_Y$ induced from metric $d$ of $X$. Then a pair $\{U, V\}$ of disjoint open sets is said to form a separation of $Y$, if $Y = U \cup V$ and neither of them contains a limit point of the other.

**Proposition 2.17.24** *Let $(X, d)$ be a metric space and $(Y, d_Y)$ be a subspace of $(X, d)$ with metric $d_Y$ induced from metric $d$ of $X$. Then $Y$ is connected iff there is no separation of $Y$.*

*Proof* If possible, let $\{U, V\}$ form a separation of $Y$. Then $U$ and $V$ are both open and closed in $Y$. If $\overline{U}$ denotes the closure of $U$ and $\overline{V}$ denotes the closure of $V$, then $\overline{U} = \overline{U} \cap Y$, which implies that $U = \overline{U} \cap Y$, since $U$ is closed in $Y$ and hence $\overline{U} \cap V = \emptyset$, because $\overline{U} = U \cup U'$ (derived set of $U$) and $V$ contains no point of $U'$. Similarly, it follows that $U \cap \overline{V} = \emptyset$. This asserts that $\{U, V\}$ cannot form a separation of $Y$. This contradiction shows that $Y$ has no separation and hence $Y$ is connected. Conversely, suppose that $U$ and $V$ are two disjoint subsets of $Y$ such that $Y = U \cup V$ and $U \cap \overline{V} = \emptyset = \overline{U} \cap V$. Then $U \cap \overline{V} = V$ and $\overline{U} \cap V = U$. Hence it follows that both $U$ and $V$ are closed sets in $Y$ and they are also open on $Y$, since $U = Y - V$ and $V = Y - U$.    $\square$

In view of Proposition 2.17.24, connectedness of a metric space is redefined in Definition 2.17.25.

**Definition 2.17.25** Let $(X, d)$ be a metric space. Then it is connected if whenever $X$ is decomposed as $X = U \cup V$ of two nonempty subsets, then either $U \cap \overline{V} \neq \emptyset$ or $\overline{U} \cap V \neq \emptyset$.

Theorem 2.17.26 proves the connectedness of the Euclidean line **R** , which is a motivating example of connectedness.

**Theorem 2.17.26** *The Euclidean line **R** is connected.*

*Proof* Suppose $\mathbf{R} = A \cup B$ for some disjoint nonempty subsets $A$ and $B$ of **R**. To prove the theorem, it is sufficient to show that there is a point of $B$, which is a limit point of $A$ or there is a point of $A$, which is a limit point of $B$, which asserts that either $A$ intersects $\overline{B}$ or $\overline{A}$ intersects $B$. To show it, let $a \in A, b \in B$, with $a < b$. Define the set $X = \{x \in A : x < b\} \subset A$. Since $a \in X$, the set $X \neq \emptyset$. Suppose $\sup X = m$. Then $m \in A$ *or* $m \in \overline{A}$ (by definition of supremum). If $m \in A$, then $m < b$ and all the points $x$ such that $m < x < b$ are in $B$, since $m$ is an upper bound for $X$. This asserts that $m$ is a limit point of $B$. Again, if $m \notin A$, then $m \in B$, since $\mathbf{R} = A \cup B$. Hence, in this case, $m$ is a limit point of $A$. This implies that either $A \cap \overline{B} \neq \emptyset$ or $\overline{A} \cap B \neq \emptyset$ and hence **R** is connected.

$\square$

**Definition 2.17.27** Let $(X, d)$ be a metric space and $Y$ be a discrete space. Then a continuous map $f : X \to Y$ is called a **discrete-valued map**.

A connected space is characterized in Theorem 2.17.28 with the help of a discrete-valued map on it. This theorem is used to show that in Theorem 2.17.33 that every continuous image of a connected space is also connected.

**Theorem 2.17.28** *Let $(X, d)$ be a metric space. Then $X$ is connected iff every discrete-valued map on $X$ is a constant map.*

*Proof* First suppose that every discrete-valued map on $X$ is constant but $X$ is not connected. Then there exist disjoint clopen sets $A$ and $B$ in $X$ such that $X = A \cup B$. Define the map

$$f : X \to \{0, 1\}, \; x \mapsto \begin{cases} 0, \text{ for all } x \in A \\ 1, \text{ for all } x \in B \end{cases}$$

But this discrete-valued map $f$ on $X$ is not constant. This contradiction implies that $X$ is connected. Conversely, let $X$ be connected and $f : X \to Y$ be a discrete-valued map. If $y \in Y$ is such that $y \in \text{Im}(f)$, then the set $\{f^{-1}(y)\}$ is nonempty and clopen in $X$, and hence this set is the same as $X$. This implies that $f(x) = y$ for all $x \in X$. This proves that $f$ is a constant map. $\qquad\square$

Theorem 2.17.29 determines the connected subsets of the Euclidean line.

**Theorem 2.17.29** *Let $A$ be nonempty subset of the Euclidean line $\mathbf{R}$. If $A$ consists of at least two distinct points, then $A$ is connected iff $A$ is an interval.*

*Proof* First suppose that $A$ is connected. If $A$ is not an interval, there exist points $a, b \in A$ and a point $c \in \mathbf{R} - A$ such that $a < c < b$. Consider the sets

$$X = \{x \in A : x < c\} \subset A$$

and

$$Y = A - X.$$

Let $\overline{X}$ be the closure of $X$ in $A$ and $\overline{Y}$ be the closure of $Y$ in $A$. As the point $c$ is not in $A$, every point $\alpha$ of $\overline{X}$ in $A$ is such that $\alpha < c$, and every point $\beta$ of $\overline{Y}$ in $X$ is such that $\beta > c$. This implies that $X \cap \overline{Y}$ and $\overline{X} \cap Y$ are both $\emptyset$. This shows that $A$ is not connected. This contradiction asserts that $A$ is an interval. For its converse, consider the interval $X = [a, b]$ as a subspace in the Euclidean line space $\mathbf{R}$. We prove that $X$ is connected. If it is not so, then there exist nonempty disjoint open sets $U, V$ in $\mathbf{R}$ such that $(X \cap U) \cup (X \cap V) = X$. Consider the map

$$f : X \to \mathbf{R}, \; x \mapsto \begin{cases} 0, \text{ for all } x \in X \cap U \\ 1, \text{ for all } x \in X \cap V. \end{cases}$$

Since the inverse image of $f$ of any open set in $\mathbf{R}$ is either $X \cap U$, $X \cap V$, $\emptyset$ or $X$, each of which is an open set in $X$, it follows that $f$ is continuous. By definition of $f$, for any point $a \in X \cap U$, $f(a) = 0$ and for any point $b \in X \cap V$, $f(b) = 1$. Then by intermediate value theorem, there exists a point $c \in X$ such that $f(c) = 1/2$, which is different from 0 or 1. But this not possible. This asserts that $X$ is connected. □

Theorem 2.17.26 proves that intervals of the Euclidean line $\mathbf{R}$ are its only connected subsets, which may be open, closed, half-open or it can be stretched to infinity in either direction and all other subsets of $\mathbf{R}$ have gaps and hence consist of several distinct pieces which are described in Corollary 2.17.30.

**Corollary 2.17.30** *All the intervals:* $[a, b]$, $(a, b)$, $[a, b)$, $(a, b]$, $(-\infty, a]$, $[a, +\infty)$ *of the Euclidean line* $\mathbf{R}$ *are connected subsets of* $\mathbf{R}$.

**Proof** It follows from Theorem 2.17.26. □

**Example 2.17.31** Consider the subspace $X = (-1, 7) \cup (7, 10)$ of the Euclidean line $\mathbf{R}$. Then $X$ is not connected. On the other hand the interval $(-1, 10)$ is connected. The punctured Euclidean line $\mathbf{R}^* = \mathbf{R} - \{0\}$ is not connected.

**Theorem 2.17.32** *(Intermediate value theorem) Let* $f : X = [a, b] \rightarrow \mathbf{R}$ *be a continuous function such that* $f(a) \neq f(b)$. *Then for each real number* $r$ *with* $f(a) < r < f(b)$, *there is a real number* $c \in [a, b]$ *such that* $f(c) = r$.

**Proof** As $X = [a, b]$ is connected and $f$ is continuous, it follows that $f(X) \subset \mathbf{R}$ is an interval. Then $f(a) \neq f(b) \in f(X)$ and $r \in f(X)$, since $f(X)$ is an interval. By hypothesis $r$ is a real number between $f(a)$ and $f(b)$. This asserts that there is a point $c \in X$ such $f(c) = r$. □

### 2.17.5  Connectedness of Metric Spaces is a Topological Property

This subsection shows that the connectedness of metric spaces is a topological property. In topological setting, it is also proved in Chap. 5 that the connectedness is a topological property. More precisely, Theorem 2.17.33 proves that connectedness in metric spaces is preserved by a continuous map and hence it follows that connectedness in a metric space is a topological property (see Corollary 2.17.35).

**Theorem 2.17.33** *Let* $(X, d)$ *and* $(Y, \rho)$ *be two metric spaces and* $f : X \rightarrow Y$ *be a continuous onto map. If* $X$ *is connected, then* $Y$ *is also connected.*

**Proof** Let $X$ be connected and $U$ and $V$ be two disjoint open sets in $Y = f(X)$ such that $U \cup V = Y$. Then $f^{-1}(U)$ and $f^{-1}(V)$ are open sets in $X$ such that $f^{-1}(U) \cup f^{-1}(V) = X$. As the space $X$ is connected, one of $f^{-1}(U)$ and $f^{-1}(V)$ must be an empty set $\emptyset$. Again since $f$ is onto, this asserts that either $U$ or $V$ must be $\emptyset$. This shows that the space $Y$ is connected. □

**Corollary 2.17.34** *Let $(X, d)$ and $(Y, \rho)$ be two metric spaces and $f : X \to Y$ be a continuous map. If $X$ is connected, then $f(X)$ is also so.*

**Proof** It follows from Theorem 2.17.33. □

**Corollary 2.17.35** *Let $(X, d)$ and $(Y, \rho)$ be two homeomorphic metric spaces. Then $X$ is connected iff $Y$ is also so.*

**Proof** It follows from Theorem 2.17.33. □

**Remark 2.17.36** More study on **connectedness** in topological setting is available in Chap. 5.

## 2.17.6 Other Applications

This subsection presents more applications related to metric spaces.

**Proposition 2.17.37** *Let $X \subset \mathbf{R}^n$ be an uncountable subset. Then there is sequence of distinct points in $X$ converging to a point in $X$.*

**Proof** If possible, $X$ contains no limit point. Then for each $x \in X$, there is a $\delta_x > 0$ (depending on x) such that $B_x(\delta_x) \cap X = \{x\}$. If $\epsilon_x = \delta_x/2$, then the balls $B_x(\epsilon_x)$ are disjoint, and hence there exist distinct points taken one from each ball with rational coordinates. As the set of points in $\mathbf{R}^n$ with rational coordinates is countable, the set $X$ is countable, which contradicts the fact that $X$ is uncountable. □

**Proposition 2.17.38** *Let $(X, d)$ be a complete metric space. Let $f : X \to X$ be a map such that $f^2 = f \circ f$ is a strict contraction map in the sense that there is a positive real number $r < 1$ such that for all $x, y \in X, d(f^2(x), f^2(y)) < rd(x, y)$. Then $f$ has a unique fixed point in $X$.*

**Proof** By hypothesis, $(X, d)$ is a complete metric space such that the map $f^2 : X \to X$ is a strict contraction. Then by Banach Contraction Theorem 2.16.4, there is a unique point $x \in X$ such that $f^2(x) = x$. If $f(x) = y$, then $f^2(y) = f^3(x) = f(x) = y$ shows that $y$ is a fixed point of $f^2$. It implies that $x = y$. It asserts that any fixed point of $f$ is also a fixed point of $f^2$. It concludes that $f$ has a unique fixed point. □

**Example 2.17.39** Let $f : [0, 1] \to [0, 1]$ be a map such that

$$|f(x) - f(y)| \leq \frac{1}{2}|x - y|, \ \forall \ x, y \in [0, 1].$$

Then $f$ is a contraction mapping from the complete metric space $[0, 1]$ to itself, since, $[0, 1]$ is closed in $\mathbf{R}$ and hence it is complete. Hence it has a unique fixed point.

**Proposition 2.17.40** *Let* $(X, d)$ *be a complete metric space and* $X_1, X_2, \cdots$ *be nonempty closed subsets of* $X$ *with* $X_{n+1} \subset X_n$, $\forall n \geq 1$. *If* $\lim_{n \to \infty} \operatorname{diam}(X_n) = 0$, *then* $\bigcap_{n=1}^{\infty} X_n \neq \emptyset$, *where the diameter* $\operatorname{diam}(X_n)$ *is defined by*

$$\operatorname{diam}(X_n) = \sup\{d(x, y) : x, y \in X_n\}.$$

*Proof* It follows from Cantor's Intersection Theorem 2.11.8. However, for an independent proof, construct a sequence

$$\{x_n : x_n \in X_n, \ \forall n \geq 1\}.$$

This construction is possible, as every $X_n$ is nonempty. For a given $\epsilon > 0$, there is a positive integer $n_0$ (depending on $\epsilon$) such that $\operatorname{diam} X_n < \epsilon$, $\forall n > n_0$. Hence it follows that $d(x_n, x_m) < \epsilon$, $\forall n, m > n_0$. By hypothesis, as $X$ is a complete metric space, then this Cauchy sequence converges to a point $x \in X$. Again, by hypothesis, since each $X_n$ is closed and it contains the sequence $\{x_n\}$, it follows that $x \in X_n$ for all $n \geq 1$. This asserts that $x \in \bigcap_{n=1}^{\infty} X_n$ and hence $\bigcap_{n=1}^{\infty} X_n \neq \emptyset$.                     □

## 2.18 Exercises

(Assume that **R** is endowed with the Euclidean metric unless stated otherwise)

1.  (i) Let $X$ be a nonempty set. If $x = \{x_n\}$ and $y = \{y_n\}$ are two sequences in $X$, show that the function

$$d : X \times X \to \mathbf{R}, \ (x, y) \mapsto \sum_{n=1}^{\infty} \frac{1}{2^n} \cdot \frac{|x_n - y_n|}{1 + |x_n - y_n|}$$

defines a metric on $X$.

(ii) Let $X$ be the set of all mappings $f : \mathbf{N} \to \mathbf{R}$, i.e., $X$ be the set of all sequences over **R**. If $f = \{x_n\}$ and $g = \{y_n\}$ are in $X$, show that the function

$$d : X \times X \to \mathbf{R}, \ (f, g) \mapsto \sum_{n=1}^{\infty} \frac{1}{2^n} \cdot \frac{|x_n - y_n|}{1 + |x_n - y_n|}$$

defines a metric on $X$.

2. Show that every open set in **R** is a union of disjoint open intervals.

3. Show that the Cantor set $C = \{x \in \mathbf{R} : x = \sum_{k=1}^{\infty} a_n/3^n, a_n = 0, 2\}$ (see Chap. 1)

(i) is contained in $[0, 1]$;

(ii) does not meet the open interval $(1/3, 2/3)$;

(iii) is closed in **R**.

4. Let $(X, d)$ be a metric space. Show that

(i) given a point $a \in X$, the function $f : X \to \mathbf{R}$, $x \mapsto d(x, a)$ is continuous;

(ii) given a subset $A \subset X$, the function $f : X \to \mathbf{R}$, $x \mapsto d(x, A) = \inf\{d(x, a) : a \in A\}$ is continuous.

5. Let $(X, d)$ be a metric space and $A, B$ be two nonempty subsets of $X$. If $diam\, A$ denotes the diameter of $A$ and $d(A, B)$ denotes the distance between $A$ and $B$ defined by

$$d(A, B) = \inf\{d(x, y) : x \in A, y \in B\},$$

show that

$$d(A \cup B) \le \operatorname{diam} A + \operatorname{diam} B + d(A, B).$$

6. Show that

(i) given a continuous function $f : \mathbf{R} \to \mathbf{R}$, the set $A = \{x \in \mathbf{R} : f(x) = x\}$, i.e., the set of points of $\mathbf{R}$ which are kept fixed by $f$ is a closed subset of $\mathbf{R}$;

(ii) the function $g : \mathbf{R} \to (0, 1)$, $x \mapsto e^x/(1 + e^x)$ is a homeomorphism.

7. Let $(X, d)$ be a metric space. Declaring a subset $A$ of $X$ to be open if its complement $A^c = X - A$ is closed, show that the set $A$ is open iff for every point $x \in A$, there exists an open ball $B_x(\epsilon)$ such that $B_x(\epsilon) \subset A$.

8. Let $(X, d)$ be a discrete metric space. Show that every nonempty subset in $X$ is closed.

9. Let $(X, d)$ and $(Y, d')$ be two metric spaces and $f; X \to Y$ be a function. Show that $f$ is continuous at a point $x \in X$ iff whenever $\{x_n\}$ is a sequence with $x_n \to x$ in $X$, then $f(x_n) \to f(x)$ in $Y$.

Further show that the following statements are equivalent for $f$:

(i) $f$ is continuous on $X$.

(ii) For every open subset $U$ of $Y$, $f^{-1}(U)$ is an open subset of $X$.

(iii) For every closed subset $U$ of $Y$, $f^{-1}(U)$ is a closed subset of $X$.

10. Let $(X, d)$ be a metric space and $A \subset X$. Show that $A$ is bounded iff for every point $x \in X$, there is an $\epsilon > 0$ such that $A \subset B_x(\epsilon)$.

11. Let $(X, d)$ be a metric space. Prove the following statements:

(i) any finite union of bounded sets in $X$ is a bounded set;

(ii) any Cauchy sequence in $X$ is a bounded set.

12. Let $X$ and $Y$ be two metric spaces. If a function

$$f : X \to Y$$

is uniformly continuous and $\{x_n\}$ is a Cauchy sequence in $X$, show that

(i) $\{f(x_n)\}$ is a Cauchy sequence in $Y$;

(ii) moreover, if the metric space $Y$ is complete, then given a subset $A \subset X$ with its closure $\overline{A}$, every uniformly continuous function

$$f : X \to Y$$

can be extended to a uniformly continuous function

$$\psi : \overline{A} \to Y.$$

In particular, if $A$ is dense in $X$, then the extension $\psi : X \to Y$ is unique.

13. **(Equivalent definitions of continuity)** Let $(X, d)$ and $(Y, \rho)$ be two metric spaces. If

$$f : X \to Y$$

is continuous, show that the following statements are equivalent:

   (i) $f$ is continuous at a point $x \in X$.
   (ii) Corresponding to any $\epsilon > 0$, there exists a $\delta > 0$ such that if $d(x, x') < \delta$, then $\rho(f(x), f(x')) < \epsilon$.
   (iii) Corresponding to an open set $U$ containing $f(x)$ in $Y$, there exists an open set $V$ containing $x$ such that $f(V) \subset U$.

14. Let $X$ be a normed linear space over $\mathbf{R}$ and $f : X \to \mathbf{R}$ be an arbitrary linear functional. Show that the following statements are equivalent:

   (i) $f$ is uniformly continuous.
   (ii) $f$ is continuous.
   (iii) for every $\epsilon > 0$, there exists a $\delta > 0$ with the property that $|f(x)| < \epsilon$ for every $x \in X$ satisfying $\|x\| < \delta$.
   (iv) $f$ is bounded.

15. Let $X$ be a normed linear space over $\mathbf{R}$. Show that its dual space $X^*$ is a Banach space.

16. Let $X$ and $Y$ be normed linear spaces and $T : X \to Y$ be a linear transformation. Show that $T$ is continuous iff $T$ is continuous at the origin (i.e., iff $x_n \to 0$ asserts that $T(x_n) \to 0$).

17. Show that

$$d : \mathbf{N} \times \mathbf{N} \to \mathbf{R}, (m, n) \mapsto \left| \frac{1}{m} - \frac{1}{n} \right|$$

defines a metric on $\mathbf{N}$ such that each singleton set $\{x\}$ in $\mathbf{N}$ is open.

18. Let $X$ be a metric space with metric $d$. Show that

   (i) the function $f : X \times X \to \mathbf{R}$, $(x, y) \to d(x, y)/1 + d(x, y)$ is a metric;
   (ii) the function $g : X \times X \to \mathbf{R}$, $(x, y) \to \min\{d(x, y), 1\}$ is a metric;
   (iii) the metrics $d$ and $\rho = d/(1 + d)$ are equivalent on $X$.

19. Let $d$ and $\rho$ be two equivalent metrics on $X$. Show that the metric space $(X, d)$ is complete iff $(Y, \rho)$ is also so.

20. (**Characterization of complete metric spaces**) Let $(X, d)$ be a metric space. Show that the following statements are equivalent:

   (i) The metric space $(X, d)$ is complete;
   (ii) Each sequence $\{x_n\}$ in $X$ having the property $\sum_{n=1}^{\infty} d(x_{n+1}, x_n) < \infty$ is convergent.
   (iii) Each Cauchy sequence $\{x_n\}$ in $X$ has a convergent subsequence.

21. Let $h : \mathbf{R} \to \mathbf{R}$ be a function such that

   (i) $h(0) = 0$.
   (ii) $h(x + y) \leq h(x) + h(y)$ for all $x, y \in \mathbf{R}$.

   If $h$ is monotonic increasing and $d$ is a metric on $\mathbf{R}$, show that the function

   $$f : \mathbf{R} \times \mathbf{R} \to \mathbf{R}, \quad (x, y) \to h(d(x, y))$$

   is also a metric.

22. Show that the composite of continuous functions in metric spaces is also continuous.

23. Let $I_d : \mathbf{R}^n \to \mathbf{R}^n$ be the identity function. If $d, d'$ are two metrics on $\mathbf{R}^n$, show that the identity functions

   $$I_d : (\mathbf{R}^n, d') \to (\mathbf{R}^n, d)$$

   and

   $$I_d : (\mathbf{R}^n, d) \to (\mathbf{R}^n, d')$$

   are both continuous.

24. Let $f : \mathbf{R} \to \mathbf{R}$ be continuous at a point $a \in \mathbf{R}$. If $f(a) > 0$, show that there exists a number $\delta > 0$ such that whenever $|x - a| < \delta$, then $f(x) > f(a)/2$.

25. Show that in a metric space $X$, if $\lim_{n \to \infty} x_n = x$ and $\lim_{n \to \infty} x_n = y$, then $x = y$.

26. Show that every subspace of a complete metric space is complete iff it is closed.

27. Show that the space $\mathcal{C}[0, 1]$ of real-valued continuous maps with the metric $d_1$ defined by

   $$d_1(f, g) = \max_{0 \leq x \leq 1} |f(x) - g(x)|$$

   is complete, but it is not so with the metric $d_2$ defined by

   $$d_2(f, g) = \left[ \int_0^1 |f(x) - g(x)|^2 dx \right]^{\frac{1}{2}}$$

28. (**An Alternative form of Picard's theorem**) Given a differential equation

$$\frac{dy}{dx} = f(x, y),$$

let there be a nbd $U$ of the $(x_0, y_0)$ in which $f(x, y)$ is continuous and a constant $K > 0$ such that

$$|f(x, y_1) - f(x, y_2)| \le K|y_1 - y_2|, \quad \forall (x, y_1), (x, y_2) \in U \text{ (Lipschitz condition)}$$

Show that the given differential equation $\frac{dy}{dx} = f(x, y)$ has a unique solution $y = h(x)$ such that $h(x_0) = y_0$.

29. Show that the sequence

$$\left\{ 1, \ 1 + \frac{1}{1!}, \ 1 + \frac{1}{1!} + \frac{1}{2!}, \cdots \right\}$$

is a Cauchy sequence in the metric space $(\mathbf{Q}, d)$, where $d$ is the usual metric. [Hint: Use the result that every convergent sequence in a metric space is a Cauchy sequence .]

30. Let $(X, d)$ be a complete metric space and $\{x_{ij}, i, j \in \mathbf{N}\}$ be a doubly indexed set (double sequence) in $X$ such that

$$d(x_{ij}, x_{m,n}) \le \min\{\max\{1/i, 1/m\}, \max\{1/j, 1/n\}\}.$$

Show that the repeated limits $\lim_{i \to \infty} \lim_{j \to \infty} x_{ij}$ and limits $\lim_{j \to \infty} \lim_{i \to \infty} x_{ij}$ exist and they are equal.

31. Let $(X, d)$ and $(Y, d')$ be two metric spaces. If $f : X \to Y$, $g : Y \to X$ are two inverse functions in the sense that $f \circ g = 1_Y$ and $g \circ f = 1_X$. Show that the following statements are equivalent:

   (i) The functions $f$ and $g$ are continuous;
   (ii) A subset $U \subset X$ is open iff $f(U) \subset Y$ is open;
   (iii) A subset $A \subset X$ is closed iff $f(A) \subset Y$ is closed ;
   (iv) Given a point $a \in X$ and a subset $B \subset X$, the subset $B$ is a nbd of $a$ iff $f(B)$ is a nbd of the point $f(a)$.

32. (**Bolzano–Weierstrass Theorem**) Prove that every bounded sequence in $\mathbf{R}$ with the Euclidean metric has a convergent subsequence.

33. Let $X = \mathcal{C}([0, 1])$ be the set of all continuous real-valued functions on $\mathbf{I} = [0, 1]$. Define for $f \in X$, its norm

$$\|f\| = \max_{t \in \mathbf{I}} |f(t)|$$

and define a metric

$$d : X \times X \to \mathbf{R}, \ (f, g) \mapsto \|f - g\|.$$

Show that

(i) $(X, d)$ is a Banach space;
(ii) $(X, d)$ is a complete metric space.
[Hint: Use the result that every uniformly convergent sequence of continuous functions converges to a continuous function.]

34. Let $(M, d)$ be a metric space and $X \subset M$. Show that

(i) if $X$ is complete, then $X$ is closed in $M$;
(ii) if $M$ is complete and $X$ is closed in $M$, then $X$ is complete.

35. Let $p > 1$ be a prime integer. Define a function

$$d_p : \mathbf{Z} \times \mathbf{Z} \to \mathbf{R}, \ (m, n) \mapsto \begin{cases} 0, & \text{if } m = n \\ 1/p^r, & \text{if } m \neq n \end{cases},$$

where $p^r$ is the nonnegative power of $p$ that divides $m - n$. Show that $d_p$ is a metric on $\mathbf{Z}$.

36. Given normed linear spaces $X$ and $Y$ over $\mathbf{R}$, show that the set $\mathcal{C}(X, Y)$ of all continuous linear transformations $T : X \to Y$ forms a normed linear space under pointwise linear operations with norm function $\| \ \|$ defined by

$$\|T\| = \sup\{\|T(x)\| : \|x\| \leq 1\}$$

such that if $Y$ is a Banach space, then this $\mathcal{C}(X, Y)$ is also so.

37. (**Canonical embedding theorem**) Let $X$ be a given normed linear space with $X^*$ and $X^{**}$ as its dual and bidual spaces. Show that the function

$$\psi : X \to X^{**}, x \mapsto \psi_x, \text{ where } \psi_x(f) = f(x), \ \forall x \in X \text{ and } \forall f \in X^*,$$

equivalently, the function

$$\psi : X \to X^{**}, \ (f, x) \mapsto f(x), \ \forall x \in X \text{ and } \forall f \in X^*$$

is an isometric linear operator such that $\|\psi\| = 1$. ($\psi$ is called the canonical embedding of $X$ into $X^{**}$).

38. A normed linear space is said to be **reflexive** if the canonical embedding $\psi$ defined in Exercise 37 is surjective. Show that any reflexive normed linear space is complete.
[Hint: Use that the map $\psi$ is bijective and the space $X^{**}$ is complete.]

39. Let M be a metric space and $f : M \to M$ be a contracting mapping. Show that the map $f$ is continuous.

40. Let M be a complete metric space and $f : M \to M$ be a contracting mapping. Show that the map $f$ has a unique fixed point.

41. Let $(X, d)$ be a metric space and $\{U_n\}$ be a countable family of open sets each of which is dense in $X$. Show that $\bigcap_{n=1}^{\infty} U_n \neq \emptyset$.

42. Given a metric space $(X, d)$ and a nonempty closed subset $A$ of $X$, show that $d(x, A) = 0$ iff $x \in A$.

43. Show that
    (i) every convex subset of $\mathbf{R}^n (n \geq 1)$ is connected;
    (ii) the Euclidean $n$-space $\mathbf{R}^n (n \geq 1)$ is connected.
    [Hint: First part: Let $X$ be a convex subset of $\mathbf{R}^n$. Suppose that $X = A \cup B$ for two nonempty separated open sets $A$ and $B$. Take a point $a \in A$ and a point $b \in B$. Then the line segment $Y = [a, b]$ has the property

$$Y \cap A = A_Y(\text{say}) \neq \emptyset, Y \cap B = B_Y(\text{say}) \neq \emptyset \text{ and } Y = A_Y \cup B_Y.$$

It contradicts the connectedness of the line segment $[a, b]$. Second part follows from First part.]

44. (**Existence of $n$-th roots**) For any integer $n \in \mathbf{N}$ and any point $a \in [0.\infty) \subset \mathbf{R}$, show that there a unique point $x \in [0.\infty)$ such that $x^n = a$.
    [Use Archimedean property of $\mathbf{R}$ and apply intermediate value theorem.]

45. Let $X$ and $Y$ be normed linear spaces over $\mathbf{R}$ and $T : X \to Y$ be a linear transformation. Show that $T$ is continuous iff $T$ is a Lifschitz function.

46. Show that every compact subset of any metric space is closed and bounded.

47. Show that the real Hilbert space $\mathbf{R}^\infty$ is isometric to a proper subspace of $\mathbf{R}^\infty$.
    [Hint: Consider the proper subspace of $\mathbf{R}^\infty$ consisting of all points in $\mathbf{R}^\infty$ having first coordinate zero.]

48. Let $(X, d)$ be a compact metric space and $(Y, \rho)$ be a metric space. Show that every continuous map $f : X \to Y$ is uniformly continuous.

49. (**Characterization of compact metric spaces** ) Let $(X, d)$ be a metric space. Show that for $(X, d)$, the following statements are equivalent:

    (i)   $X$ is compact in the sense that every open covering of $X$ has a finite sub-covering;
    (ii)  $X$ is complete and totally bounded;
    (iii) Every infinite subset of $X$ has an accumulation point;
    (iv)  Every sequence $\{x_n\}$ in $X$ has a convergent subsequence.

## *Multiple Choice Exercises*

Identify the correct alternative(s) (there may be more than one) from the following list of exercises:

1. Let $d_n : X_n \times X_n \to \mathbf{R}$, $n = 1, 2, 3, 4$ be the four metrics defined as follows:

   (i)  For $X_1 = (0, \pi/2) \subset \mathbf{R}$, the metric $d_1 : X_1 \times X_2 \to \mathbf{R}$, $(x, y) \mapsto |\tan x - \tan y|$ is complete.
   (ii)  For $X_2 = [0, 1] \subset \mathbf{R}$, the metric $d_2 : X_2 \times X_2, \to \mathbf{R}$, $(x, y) \mapsto \frac{|x-y|}{1+|x-y|}$ is complete.
   (iii)  For $X_3 = \mathbf{Q}$, the metric $d_3 : X_3 \times X_3 \to \mathbf{R}$, $(x, y) \mapsto 1$, $if$ $x \neq y$ and 0, otherwise, is complete.
   (iv)  For $X_4 = \mathbf{R}$, the metric $d_4 : X_4 \times X_4 \to \mathbf{R}$, $(x, y) \mapsto |e^x - e^y|$ is complete.

2. Let $(X, d)$ be a metric space and $d(A, B)$ be the usual distance between two nonempty subsets $A$ and $B$ of $X$.

   (i)  If $A$ and $B$ are disjoint and closed, then $d(A, B) > 0$.
   (ii)  If $A$ and $B$ are disjoint and compact, then $d(A, B) > 0$.
   (iii)  If $A$ and $B$ are disjoint and compact, then there exist points $a \in A$ and $b \in B$ such that $d(A, B) = d(a, b)$.

3. Let $d_n : \mathbf{R} \to \mathbf{R} \to \mathbf{R}$, $n = 1, 2, 3$ be three functions defined as follows:

   (i)  If $d_1 : \mathbf{R} \times \mathbf{R} \to \mathbf{R}$, $(x, y) \mapsto \frac{|\,|x|-|y|\,|}{1+|x|\,|y|}$, then $d_1$ is a metric on $\mathbf{R}$.
   (ii)  If $d_2 : \mathbf{R} \times \mathbf{R} \to \mathbf{R}$, $(x, y) \mapsto [|x - y|]^{1/2}$, then $d_2$ is a metric on $\mathbf{R}$.
   (iii)  If $f : \mathbf{R} \to \mathbf{R}$ is a strictly monotonic increasing function, then

   $$d_3 : \mathbf{R} \times \mathbf{R} \to \mathbf{R}, \ (x, y) \mapsto |f(x) - f(y)|$$

   is a metric on $\mathbf{R}$.

4. Let $X$ be a complete normed linear space and $\mathbf{B}$ be a basis for $X$ as a vector space.

   (i)  The set $\mathbf{B}$ is a finite.
   (ii)  The set $\mathbf{B}$ is countably infinite.
   (iii)  If the set $\mathbf{B}$ is infinite, then it is an uncountable set.

5. Let $(X, d)$ be a metric space, $A$ and $B$ be two nonempty subsets of $X$ and $d(A, B)$ be the usual distance between $A$ and $B$.

   (i)  The function $f : X \to \mathbf{R}$, $x \mapsto d(x, A)$ is continuous.
   (ii)  $d(A, B) = 0 \implies A \cap B = \emptyset$.
   (iii)  $d(x, B) = 0$ iff $x \in B$.

# References

Adhikari, M.R.: Basic Algebraic Topolgy and Its Applications. Springer, India (2016)

Adhikari, M.R.: Basic Topology. Volume 3: Algebraic Topology and Topology of Fiber Bundles. Springer, India (2022)

Adhikari, M.R., Adhikari, A.: Basic Modern Algebra with Applications. Springer, New Delhi, New York, Heidelberg (2014)

Adhikari, A., Adhikari, M.R.: Basic Topology. Volume 2: Topological Groups, Topology of Manifolds and Lie Groups. Springer, India (2022)

Borisovich, Y.C.U., Blznyakov, N., Formenko, T.: Introduction to Topology (Translated from the Russia by Oleg Efimov). Mir Publishers, Moscow (1985)

Bredon, G.E.: Topology and Geometry. Springer, New York (1993)

Chatterjee, B.C., Ganguly, S., Adhikari, M.R.: Introduction to Topology. Asian Books Pvt. Ltd., New Delhi (2002)

Conway, J.B.: A Course in Point Set Topology. Springer, Switzerland (1914)

Hu, S.T.: Introduction to General Topology. Holden-Day, San Francisco (1966)

Kelly, J.L.: General Topology. Van Nostrand, New York, (1955); Springer, New York (1975)

Kumaresan, S.: Topology of Metric Spaces. Narosa, New Delhi (2019)

Mendelson, B.: Introduction to Topology. College Mathematical Series. Allyn and Bacon, Boston (1962)

Munkres, J.R.: Topology. Prentice-Hall, New Jersey (2000)

Patterson, E.M.: Topology. Oliver and Boyd, Edinburgh (1959)

Simmons, G.: Introduction to Topology and Modern Analysis. McGraw Hill, New York (1963)

Stephen, W.: General Topology. Addison-Wesley, Boston (1970)

# Chapter 3
# Topological Spaces and Continuous Maps

The subject **Topology** sets out its official journey in this chapter through the address of the concepts of topological spaces and their continuous maps, which are the basic objects and the basic functions in topology. They are presented in an axiomatic framework by introducing the concept of open or closed sets, where a notion of nearness is defined without any distance function or a metric. This approach is given in a convenient language to study the situation when different points in a space come near to each other. Classical analysis traditionally studies real-valued functions in the Euclidean space $\mathbf{R}^n$. To extend the study of continuous functions between abstract sets, it is necessary to endow them with topological structures, which are then called topological spaces. A topology is defined on an abstract nonempty set by using the concepts of open sets or closed sets, and continuity of functions is also introduced in the language of open sets or closed sets. This leads to the concepts of **topological spaces and their continuous maps**. Though the term **Topology** was coined by J. B. Listing (1808–1882) in 1830s, this term failed to attract the mathematicians till Felix Hausdorff (1869–1942) popularized this term in 1914 and developed this subject in his book **Grundzüge der Mengenlehre** of 1914, which stemmed from analysis. His land-marking work sets out the journey of **general topology,** which is also called **set topology**.

Present day, the subject **Topology** has become very powerful and beautiful, as it provides various key tools to solve problems in almost all areas of mathematics such as in algebra, theory of numbers, analysis, geometry, knot and graph theories, differential equations in many other areas of mathematics and even outside mathematics. As this subject studies in a general sense, a precise concept of the intuitive ideas of nearness and continuity of functions, it plays an important role in science and technology. Various interesting applications of topological spaces are also available in Sect. 3.18, and a historical note on beginning of topology through the work of Euler is available in Sect. 3.19.

The concept of topological spaces is motivated by the concept of metric spaces. But the concept of metric spaces studied in Chap. 2, generalizing the Euclidean spaces

A. Adhikari and M. R. Adhikari, *Basic Topology 1*,
https://doi.org/10.1007/978-981-16-6509-7_3
123

fails to solve many important mathematical problems, specially, where a metric is not available. For example, the cylinder, Möbius band (Möbius strip), torus and the Klein bottle are constructed in Sect. 3.16.7 from the metric space $X = \mathbf{I} \times \mathbf{I}$ with the Euclidean metric $d$. As they are not subsets of the metric space $(X, d)$, the metric $d$ fails to contribute anything to their study. So, a generalization of metric spaces is needed. A general concept of spaces known as the **theory of topological spaces and their continuous maps was born** during the twentieth century for the development of mathematics. An open interval $(a, b)$ in the real line $\mathbf{R}$ is the prototype of the abstract concept of an open set in topology. In metric spaces, an open set is defined as a set which contains an open ball around each of its points. Every metric space defines a topology, called **a metric topology** (see Sect. 3.8). There are various choices of defining open sets on a nonempty set, and each choice of open sets defines a topology on the given set. The open sets and the corresponding topologies play a key role in general topology as well as different branches of topology. This chapter also discusses certain topologies such as **Euclidean topology, metric topology, Zariski topology, Kuratowski closure topology, Sierpinski topology** and some other topologies. The basic motivation of metric topology comes from the standard distance function in the Euclidean line and Euclidean space which induce Euclidean topology. A study of this topology leads to the concept of basis for a topology in a general setting. Like a vector space, every open set in a topological space can be expressed as a union of elements (open sets) of the basis. Though every metric space induces a topological space, it is very difficult to define a metric on some interesting sets. Instead of making them topological space via metric, they are made topological spaces directly by properly choosing open sets or closed sets. They are more general than the metric spaces in the sense that every metric space induces a topology on itself but its converse is not necessarily true. For example, the Zariski topology given in Definition 3.17.7 is a particular topology defined by Oscar Zariski (1899–1986) around 1950 in algebraic geometry that conveys the algebraic nature of varieties. This topology is not induced by any metric (see Chap. 4).

The subject Topology has now several branches. **General topology (or point-set topology), differential topology and algebraic topology**, which are its principal areas of study. This book in three volumes intends to study basic topology with applications in modern analysis, geometry and algebra in a series of three volumes. The present Volume (**Volume 1**) of the series of the books studies Metric Spaces and General topology. General topology conveys the basic set-theoretic definitions and constructions used in topology and formulates the basic concepts used in all other branches of topology such as the concepts of continuity, compactness, and connectedness, thereby it establishes the foundational aspects of topology and investigates properties of topological spaces and concepts inherent to topological spaces. On the other hand, **Volume 2** addresses topological groups, manifolds and differential topology, Lie groups and **Volume 3** addresses fundamentals of algebraic topology and topology of fiber bundles.

For this chapter, the books, Alexandrov (1979), Bredon (1983), Chatterjee et al. (2002), Dugundji (1966), Kelly (1955), Adhikari and Adhikari (2014, 2022), Adhikari (2016, 2022), Armstrong (1983), Borisovich et al. (1985), Hilbert and Cohn-Vossen (1952), Lipschutz (1988), Mendelson (1962), Munkres (2000),

Patterson (1959), Prasolov (1995), Singer and Thorpe (1967), Stephen (1970) and some others are referred in Bibliography.

## 3.1 Topological Spaces: Introductory Concepts

This section presents introductory concepts of topological spaces which are much general than metric spaces. For example, the continuity of a function in a metric space can be studied in terms of convergence of sequences. On the other hand, convergence of sequences in an arbitrary topological space is not adequate to study continuity. But the framework for topology started in Chap. 2 through a study of metric spaces and the continuity of functions in Euclidean spaces $\mathbf{R}^n$, facilitates to build up a general theory of topology, which provides powerful tools to invade many problems of mathematics. The axiomatic definition of topology was introduced by the Poland mathematician K. Kuratowski (1896–1980) in 1922. This topology is now known as Kuratowski closure topology in his honor. While studying metric spaces in Chap. 2, the concepts of open sets and continuity of functions with their basic interesting properties are explained using open sets. This suggests a possible way to define continuous functions in abstract sets, whenever an abstract concept of open sets or closed sets is available. The open sets in a metric space $(X, d)$ have the following four properties (see Chap. 2):

**OM**(1)     the empty set $\emptyset$ is an open set;
**OM**(2)     the union of any number of open sets is an open set;
**OM**(3)     the intersection of a finite number of open sets is an open set;
**OM**(4)     the whole (universal) set $X$ is itself an open set.

As the central concept in topology is the notion of continuity of a function, it is necessary to consider spaces on which a workable definition of continuity can be given, even in the absence of a metric structure (see Definition 3.6.1). This leads to the concept of topological spaces by specifying open sets in an axiomatic way in Definition 3.1.1. The geometric objects studied in topology are called topological spaces, whose definition is extremely general and hence all topological spaces are not geometric. For obtaining an abstract concept of open sets, the following postulates are taken formulated in Definition 3.1.1, which are analogous to the above properties **OM(1)–OM(4)** of metric spaces.

**Definition 3.1.1** Let $X$ be a nonempty set and $\tau$ be a family of some subsets of $X$. Then $\tau$ **is called a topology** of $X$ (or a topology on $X$), and subsets in $\tau$ are called **open sets (or open subsets)** of $X$ if the following axioms **OS(1)–OS(4)**, called axioms for open sets are satisfied:

**OS**(1)     the empty set $\emptyset$ is an open set, i.e., $\emptyset \in \tau$;
**OS**(2)     the union of any number of open sets is an open set;
**OS**(3)     the intersection of a finite number of open sets is an open set;
**OS**(4)     the whole (universal) set $X$ is itself an open set, i.e., $X \in \tau$.

The ordered pair $(X, \tau)$ is called a **topological space** and it is sometimes called $X$ a topological space with a **topology** $\tau$ on $X$.

**Remark 3.1.2** In order to define a topological space, a nonempty set is specified and certain subsets of the set are chosen satisfying the above axioms **OS(1)–OS(4)**, called the open sets. The merits of these axioms are justified by the growth of topology as a separate subject and its various applications in mathematics and other fields also. We write generally a topological space as $(X, \tau)$ to avoid any confusion regarding the topology $\tau$ on $X$. Sometimes, $(X, \tau)$ is written simply by $X$, where there is no confusion of its topology. The sets $U \in \tau$ are called the open sets of the topological space $X$, and $\tau$ is said to determine a topological structure on $X$. The elements of $X$ are called points of the topological space $X$. There exist different topologies on the same nonempty set. Their comparison is available in Sect. 3.1.3.

**Definition 3.1.3** (*Trivial or indiscrete topology*) Let $X$ be a nonempty set. The two subsets $\emptyset$ and $X$ constitute a topology of $X$, called the trivial topology or the indiscrete topology (sometimes, called the **chaotic topology**) of $X$.

**Definition 3.1.4** (*Discrete topology*) The family of all subsets of a nonempty set $X$ constitutes a topology of $X$, called the discrete topology of $X$.

**Remark 3.1.5** Trivial and discrete topologies are the two extreme topologies. If $X$ has more than one element, then the trivial topology and the discrete topology of $X$ are different. The trivial topology contains the smallest number of open sets, and the discrete topology contains the largest number of open sets. Usually, there exist other topologies of $X$, such as cofinite topology and topology of countable complements, which lie between these two extreme topologies and they are used throughout the book.

**Example 3.1.6** (*Cofinite topology*) Let $X$ be an infinite set, and $\tau$ consists of the null set $\emptyset$ and all those subsets of $X$ whose complements in $X$ are finite subsets. Then $\tau$ forms a topology on $X$, called the cofinite topology or the **topology of finite complements** on $X$.

**Example 3.1.7** (*Topology of countable complements*) Let $X$ be a noncountable set, and $\tau$ consist of the null set $\emptyset$ and those subsets of $X$ whose complements in $X$ are countable subsets. Then $\tau$ forms a topology of $X$, called the topology of countable complements.

### 3.1.1 Natural Topology on $\mathbf{R}$, $\mathbf{R}^2$ and $\mathbf{R}^n$

This subsection defines the natural topology on the Euclidean line $\mathbf{R}$, Euclidean plane $\mathbf{R}^2$ and Euclidean $n$-space $\mathbf{R}^n$, which are used throughout the present book series. They are also equally well-defined by open base. For more study of the natural topology on $\mathbf{R}^n$, see Sects. 3.3 and 3.8. If a ring is regarded as a generalization of the concept of real numbers $\mathbf{R}$, then a topological space may also be regarded as a generalization of $\mathbf{R}$. The natural operations of addition and multiplication in $\mathbf{R}$ are generalized for rings. On the other hand, the concept of limit point in $\mathbf{R}$ is generalized for topological spaces. This concept stays at the foundation of the structure of a topological space, which will be realized on subsequent development of topological structure.

**Definition 3.1.8** (*Real line space*) A topology $\sigma$ is defined on $\mathbf{R}$ by declaring precisely the following family of subsets of $\mathbf{R}$ as open sets:

(i) $\emptyset$ is an open set;
(ii) $\mathbf{R}$ is an open set;
(iii) all open intervals $(a_i, b_i)$ with $a_i, b_i \in \mathbf{R}$ and $a_i < b_i$, are open sets;
(iv) the unions $U = \bigcup_i (a_i, b_i)$ are open sets.

The set $\mathbf{R}$ together this family of open sets $\sigma$ in $\mathbf{R}$ is called the real line space, denoted by $(\mathbf{R}, \sigma)$ and this topology $\sigma$ called the **natural topology or usual topology on** $\mathbf{R}$. This topology is also induced by the usual metric

$$d: \mathbf{R} \times \mathbf{R} \to \mathbf{R}, (x, y) \mapsto \|x - y\|$$

on $\mathbf{R}$. (see Example 3.8.7).

**Remark 3.1.9** An alternative approach of defining natural topology on $\mathbf{R}$ is given in Corollary 3.2.6 by using the concept of open base on $\mathbf{R}$.

Definition 3.1.10 formulates the concepts of natural topology on $\mathbf{R}^2$ and $\mathbf{R}^n$ as a natural generalization of usual topology on $\mathbf{R}$.

**Definition 3.1.10** (*Natural topology* on $\mathbf{R}^2$ and $\mathbf{R}^n$) A topology $\sigma$ on the set $\mathbf{R}^2 = \mathbf{R} \times \mathbf{R}$ is defined by declaring precisely the following family of subsets of $\mathbf{R}^2$ as open sets with the help of Euclidean metric on $\mathbf{R}^2$ :

(i) $\emptyset$ is an open set;
(ii) $\mathbf{R}^2$ is an open set;
(iii) a subset $U$ of $\mathbf{R}^2$ is said to be open if for every element $x \in U$, there exists a real number $\delta > 0$ such that the open circle or open ball $B_x(\delta) \subset U$.

This family $\sigma$ of open sets in $\mathbf{R}^2$ forms a topology on $\mathbf{R}^2$ is called the **natural or usual topology on** $\mathbf{R}^2$. The definition of **natural topology** $\sigma$ on $\mathbf{R}^n$ is defined in an analogous way, and this topology is also induced by the usual metric

$$d: \mathbf{R}^n \times \mathbf{R}^n \to \mathbf{R}, (x, y) \mapsto \|x - y\|$$

on $\mathbf{R}^n$ (see Example 3.8.7).

### 3.1.2  Construction of Topologies on Some Finite Sets

This subsection constructs specific topologies on some finite sets, consisting of two, three or four distinct elements for clear understanding of the concepts of topology and topological spaces which are used in future study.

**Example 3.1.11** Let $X = \{x, y\}$ be the set consisting of **two distinct elements**. Then each $\tau_i$: $i = 1, 2, 3, 4$ defined below

(i)   $\tau_1 = \{\emptyset, \{x\}, \{y\}, \{x, y\}\}$ (discrete topology);
(ii)  $\tau_2 = \{\emptyset, \{x\}, \{x, y\}\}$ (Sierpinski topology given in Definition 3.10.5);
(iii) $\tau_3 = \{\emptyset, \{y\}, \{x, y\}\}$ (Sierpinski topology) and
(iv)  $\tau_4 = \{\emptyset, \{x, y\}\}$ (trivial topology).

constitutes a topology on $X = \{x, y\}$ of various types, showing that there exist different topologies on the same set and making $X$ different topological spaces.

**Example 3.1.12** Let $X = \{x_1, x_2, x_3\}$ be the set consisting of **three distinct elements**. Then

$$\tau = \{\emptyset, \{x_1\}, \{x_2\}, \{x_1, x_2\}, X\}$$

constitutes a topology on $X = \{x_1, x_2, x_3\}$. It is one of the topologies on $X$.

**Example 3.1.13** Let $X = \{x_1, x_2, x_3, x_4\}$ be the set consisting of four distinct elements. Then

$$\tau = \{\emptyset, \{x_2\}, \{x_4\}, \{x_1, x_2\}, \{x_2, x_3\}, \{x_2, x_4\}, \{x_1, x_2, x_4\}, \{x_2, x_3, x_4\}, \{x_1, x_2, x_3\}, X\}$$

constitutes a topology on $X = \{x_1, x_2, x_3, x_4\}$. It is one of the topologies on $X$.

### 3.1.3  Comparison of Topologies

This subsection gives a comparison among different topologies defined on the same set. Any topology $\tau$ of a given set $X$ is a subset of the power set $\mathcal{P}(X)$, i.e., $\tau \subset \mathcal{P}(X)$. Hence, it is possible to establish an **order relation** among the topologies that can be defined on $X$, in terms of the set-theoretic inclusion relation for subsets.

**Definition 3.1.14** Given two topologies $\tau_1$ and $\tau_2$, defined on a set $X$, the topology $\tau_1$ is said to be **coarser or weaker** than the topology $\tau_2$, if $\tau_1 \subset \tau_2$, and in that case $\tau_2$ is said to be **finer or larger or stronger** than $\tau_1$. If the inclusion relation is proper in the sense that $\tau_1 \subset \tau_2$, but $\tau_1 \neq \tau_2$, then $\tau_1$ is said to be **strictly weaker or coarser** than $\tau_2$, and $\tau_2$ is said to be **strictly stronger or finer** than $\tau_1$.

**Example 3.1.15**   (i)  Among all the topologies on a set $X$, the trivial topology is the weakest topology and the discrete topology is the strongest topology on $X$ (see Definitions 3.1.3 and 3.1.4).

(ii) For an infinite set $X$, the cofinite topology is strictly stronger than the trivial topology and is strictly weaker than the discrete topology;

(iii) For a noncountable set $X$, the topology of countable complements is **strictly stronger** than the cofinite topology and is **strictly weaker** than the discrete topology.

(iv) Each of the lower-limit, upper-limit and $\mathcal{K}$-topologies on $\mathbf{R}$ defined in Sect. 3.4.1 is strictly stronger than the natural topology on $\mathbf{R}$ by Proposition 3.4.11.

**Definition 3.1.16** Two topologies $\tau_1$ and $\tau_2$ defined on a set $X$ are said to be **noncomparable** if neither of them is weaker than the other.

**Example 3.1.17** Consider the topologies $\tau_1$, $\tau_2$ and $\tau_3$ on $X = \{x, y\}$ consisting of two distinct points $x$ and $y$ defined in Example 3.1.11. Then the topologies $\tau_2$ and $\tau_3$ are noncomparable. On the other hand, $\tau_2 \subset \tau_1$ and $\tau_3 \subset \tau_1$.

**Example 3.1.18** The lower-limit topology and upper-limit topology on $\mathbf{R}$ defined in Sect. 3.4.1 are noncomparable.

**Remark 3.1.19** The family $\Omega$ of all topologies on a nonempty set forms a lattice which is a **complete lattice** under the weaker or stronger relations on this family $\Omega$. (see Sect. 3.5).

## 3.1.4 Neighborhoods and Limit Points

This subsection conveys the concept of limit points of a given subset of a topological space generalizing the concept of limit points in analysis, which started in classical analysis with the study of limit process and continuity of functions. This concept is now studied by neighborhood systems which is utilized to study open sets containing a given point of the topological space.

**Definition 3.1.20** Let $(X, \tau)$ be a topological space. A subset $N_x$ of $X$ is said to be a **neighborhood (abbreviated as nbd)** of a point $x \in X$ if there exists an open set $U \in \tau$ such that $x \in U \subset N_x$.

**Example 3.1.21** A nbd of a point in a topological space is not necessarily an open set. For example, in the real line space $(\mathbf{R}, \sigma)$, the closed interval $[2, 5]$ is not an open set but it is a nbd of the point $3 \in \mathbf{R}$, because

$$3 \in (2, 5) \subset [2, 5].$$

An open set in a topological space is characterized in Proposition 3.1.22 with the help of nbds.

**Proposition 3.1.22** *Let $(X, \tau)$ be a topological space and $U \subset X$. Then $U$ is open in $X$ iff $U$ contains a nbd of each of its points.*

**Proof** If $U$ is an open set in $X$, it is a nbd of each of its points by definition of nbd. Conversely, let $U$ be subset in $X$ such that it contains a nbd of each of its points. Then given an arbitrary point $x$ in $U$, there exists a nbd $N_x$ of the point $x \in X$ and an open set $U_x$ such that

$$x \in U_x \subset N_x.$$

Then

$$U = \bigcup_{x \in U} \{x\} \subset \bigcup_{x \in U} U_x \subset U \implies \bigcup_{x \in U} U_x = U \implies U \text{ is open in } X.$$

$\square$

**Definition 3.1.23** Let $(X, \tau)$ be a topological space and $x \in X$ be an arbitrary point. Then the family $\mathcal{N}_x$ of all nbds of $x$ in $X$ is said to be a **nbd system** of the point $x$ in the given topological space.

**Proposition 3.1.24** *Let $(X, \tau)$ be a topological space and $x \in X$ be an arbitrary point. If $\mathcal{N}_x$ is a nbd system of the point $x$ in the given topological space $X$, then*

(i) *every finite intersection of members belonging to $\mathcal{N}_x$ also belongs to $\mathcal{N}_x$ and*
(ii) *every subset in $X$ which contains a member of $\mathcal{N}_x$ also belongs to $\mathcal{N}_x$.*

**Proof** It follows from Definition 3.1.20.                                                   $\square$

**Definition 3.1.25** Let $(X, \tau)$ be a topological space and $A$ be a subset of $X$. A point $x \in X$ is said to be a **limit point (or accumulation point or cluster point)** of the set $A$ if every nbd of $x$ intersects $A$ in at least one point other than the point $x$. The set formed by all the limit points of $A$ is called the **derived set** of $A$, denoted by $A'$.

**Example 3.1.26** In the real line space **R**,

(i) the set $X = \{1/n : n = 1, 2, \ldots, \}$ has only one limit point, which is 0;
(ii) the limit points of the set $X = (0, 1)$ fill the line segment $I = [0, 1]$;
(iii) the limit points of the sets $X = (0, 1]$ and $Y = [0, 1)$ also fill the line segment $I = [0, 1]$;
(iv) the set $X = \{0, 1\}$ has no limit point. In general, if $X$ is a finite subset of **R**, then $X$ has no limit point.
(v) every point $x \in \mathbf{R}$ is a limit point of the set **Q** of all rational numbers, since every open set contains rational points. Similarly, every point $x \in \mathbf{R}$ is a limit point of the set $\mathbf{R} - \mathbf{Q}$ of irrational numbers.

### 3.1.5   Closed Sets

This subsection studies closed sets of a topological space via its open sets, which are our primitive undefined terms. A topology on a set is usually defined by specifying which sets are open. There is another approach to describe a topology by declaring which sets are closed. This subsection first defines closed sets, which are sometimes convenient to define a topology on a set by the axioms of closed sets, which are duals of the axioms of open sets for a topology. The unique topology determined in Theorem 3.10.2, called **Kuratowski closure topology** satisfying the four axioms of closed sets **C(1)–C(4)** of Definition 3.1.28, plays an important role in topology.

**Definition 3.1.27** A subset $F \subset X$ of a topological space $(X, \tau)$ is said to be **closed** if its complement $F^c = X - F$ $(= X \setminus F)$ is open in $(X, \tau)$.

Instead of open sets, Definition 3.1.28 formulates a topological space in terms of closed sets.

**Definition 3.1.28** Let $X$ be a nonempty set. It is called a topological space if there is a family of subsets of $X$, called closed sets such that

**C(1)** the union of finitely many closed sets is a closed set;
**C(2)** the intersection of any number of closed sets is a closed set;
**C(3)** $X$ is a closed set;
**C(4)** the empty set $\emptyset$ is a closed set.

*Example 3.1.29* In the real line space $(\mathbf{R}, \sigma)$ (with usual topology $\sigma$),

(i) $\emptyset$ and $\mathbf{R}$ are closed sets, since their complements $\emptyset^c = \mathbf{R}$ and $\mathbf{R}^c = \emptyset$ in $\mathbf{R}$, are open sets in $(\mathbf{R}, \sigma)$.
(ii) every closed interval $[a, b] \subset \mathbf{R}$ is a closed set, because its complement $[a, b]^c = (-\infty, a) \cup (b, \infty)$ is an open set in $(\mathbf{R}, \sigma)$. On the other hand, the open-closed interval $(a, b]$ is neither closed nor open under the usual topology on $\mathbf{R}$.

*Example 3.1.30* Let $X$ be an infinite set and $\tau$ be the cofinite topology on $X$ given in Example 3.1.6. Then the closed sets in the topological space $(X, \tau)$ are precisely the finite subsets of $X$ together with the emptyset $\emptyset$ and the whole set $X$.

*Example 3.1.31* The union of infinitely many closed sets may not be a closed set. For example, consider the sequence of closed sets $A_n = [\frac{1}{n}, 1]$ in the real line space $(\mathbf{R}, \sigma)$. Their infinite union $\bigcup A_n = (0, 1]$ is not closed in $(\mathbf{R}, \sigma)$.

Proposition 3.1.32 asserts that any family of some subsets of a nonempty set satisfying the conditions **C(1)** –**C(4)** of Definition 3.1.28 completely determines the closed sets in any topological space.

**Proposition 3.1.32** *Let $\mathcal{C}$ be a family of some subsets of a nonempty set $X$ satisfying the conditions **C(1)–C(4)** of Definition 3.1.28. Then there exists **a unique topology** $\tau$ on $X$ such that the closed sets in $X$ defined by $\tau$ are precisely the same as the given family $\mathcal{C}$.*

**Proof** Let a subset $U$ of $X$ be called open, if its complement $X - U \in \mathcal{C}$ and $\tau$ be the family of all open sets defined in this way. By hypothesis, $\mathcal{C}$ satisfies the conditions **C(1)–C(4)** of Definition 3.1.28. Then by dualizing the properties **C(1)–C(4)** in $X$, it follows that the family $\tau$ satisfies the conditions **O(1)–O(4)** of Definition 3.1.1 for open sets. Hence, $(X, \tau)$ forms a topological spaces by using the concept of open sets. The closed sets in this topological space $(X, \tau)$ are precisely the complements of open sets, which are exactly the members of $\mathcal{C}$. The topology $\tau$ thus determined on $X$ is clearly unique.                                                           $\square$

*Example 3.1.33* Let $\mathcal{C}$ be a family of some subsets of an infinite set $X$, consisting of $X$ and all finite subsets of $X$. Then $\mathcal{C}$ satisfies all the conditions **C(1)–C(4)** of Proposition 3.1.32. The topology $\tau$ determined by $\mathcal{C}$ is the cofinite topology on $X$ as given in Example 3.1.6.

*Example 3.1.34* In a cofinite topological space $(X, \tau)$ given in Example 3.1.6, all finite subsets of $X$ are closed by Proposition 3.1.32 but all of its infinite subsets may not be open (see Example 3.1.35).

*Example 3.1.35* In the set **N** of all positive integers, its finite subsets such as $\{1\}$, $\{2, 3, 4\}$ are closed in the cofinite topology on **N**. Hence, their complements in **N**, which are $\{2, 3, 4, \ldots\}$ and $\{1, 5, 6, \ldots\}$ are open sets in the cofinite topology on **N**. On the other hand, the set of odd positive integers is not closed in cofinite topology, because the set of even positive integers which is its complement is infinite.

Proposition 3.1.36 characterizes closed sets of a topological space in terms of its limit points.

**Proposition 3.1.36** *Let* $(X, \tau)$ *be a topological space. Then a set* $A \subset X$ *is closed iff it contains all its limit points.*

**Proof** Left as an exercise.                                                               $\square$

*Example 3.1.37* In the real line space $(\mathbf{R}, \sigma)$ (with usual topology $\sigma$), the subset $A = \{1, \frac{1}{2}, \frac{1}{3}, \frac{1}{4}, \ldots\} \subset \mathbf{R}$ is not closed, because its limit point 0 is not in $A$.

**Corollary 3.1.38** *Let* $(X, \tau)$ *be a topological space. Then for any subset* $A \subset X$, *the set* $A \cup A'$ *is closed, where* $A'$ *is the derived set of* $A$.

**Proof** It follows by using Proposition 3.1.36.                                         $\square$

## 3.1.6  Closed and Open (Clopen) Sets

There may or may not exist nontrivial subsets of a topological space which are both open and closed sets. On the other hand, $\emptyset$ and the whole set are both open and closed sets of any topological space.

**Definition 3.1.39** Let $(X, \tau)$ be a topological space. If a subset of $X$ is both open and closed in $X$, then it is designated as a **clopen set**.

**Proposition 3.1.40** *In the real number space* $\mathbf{R}$, *the subset* $\mathbf{Q}$ *is neither closed nor open.*

**Proof** The set $\mathbf{Q}$ is to be open in $\mathbf{R}$, if for each $x \in \mathbf{Q}$, there exist $a, b \in \mathbf{R}$ with $a < b$ such that $x \in (a, b) \subset \mathbf{Q}$. Claim that $\mathbf{Q}$ does not contain any such open interval $(a, b)$, with $a < b$. Otherwise, since between any two distinct real numbers, there is an irrational number, there exists some $y \in (a, b)$ such that $y \notin \mathbf{Q}$. This contradicts the assumption that $(a, b) \subset \mathbf{Q}$. This shows that $\mathbf{Q}$ does not contain any interval $(a, b)$, and hence, $\mathbf{Q}$ is not an open in $\mathbf{R}$. Claim that $\mathbf{Q}$ is not also closed; otherwise, the set $\mathbf{R} - \mathbf{Q}$ must be an open set. Since between two distinct real numbers, there is a rational number, $\mathbf{R} - \mathbf{Q}$ does not contain any interval $(a, b)$. This implies that $\mathbf{R} - \mathbf{Q}$ is not open in $\mathbf{R}$, and hence, $\mathbf{Q}$ is not closed in $\mathbf{R}$. □

**Corollary 3.1.41** *The only closed and open (clopen) subsets of the real number space* $\mathbf{R}$ *are* $\mathbf{R}$ *and* $\emptyset$.

**Proof** **Proof I**: In the real number space $\mathbf{R}$, the set $\mathbf{Z}$ of all integers is a closed subset of $\mathbf{R}$, since the complement of $\mathbf{Z}$ in $\mathbf{R}$ is the union $\bigcup_{-\infty}^{\infty}(n, n + 1)$, which is an open set in $\mathbf{R}$. But $\mathbf{Z}$ is not open in $\mathbf{R}$. Again $\mathbf{Q}$ is neither open nor closed in $\mathbf{R}$ by Proposition 3.1.40. Hence, the set of irrational numbers $\mathbf{Q}^c = \mathbf{R} - \mathbf{Q}$ is neither open nor closed in $\mathbf{R}$. Again, the open intervals in $\mathbf{R}$ are not closed and closed intervals in $\mathbf{R}$ are not open in $\mathbf{R}$.

**Proof II**: Let $U \neq \emptyset$ be an open set in $\mathbf{R}$ such that $U$ is both open and closed. Since $U$ is open, by Definition 3.3.1, for $x \in U$, there exist some $\epsilon > 0$ such that $(x - \epsilon, x + \epsilon) \subset U$. Let $A = \sup \{y \in \mathbf{R} : (x - \epsilon, y) \subset U, y > x\}$. Then $A \neq \emptyset$. It is not bounded above; otherwise, the openness property of $U$ would produce a contradiction.

**Proof III**: Use the connectedness property of $\mathbf{R}$ (see Chap. 5) and use the result asserting that a topological $(X, \tau)$ is connected iff the only subsets of $X$ which are both open and closed in $(X, \tau)$ are $\emptyset$ and $X$ itself. □

**Example 3.1.42** In the topological space $\mathbf{R}$ endowed with discrete topology, every subset of $\mathbf{R}$ is clopen, since all subsets of $\mathbf{R}$ that constitute the discrete topology on $\mathbf{R}$ are both open and closed.

### 3.1.7 Closure of a Set

This subsection defines the concept of the closure of a nonempty set as the intersection of specified closed sets and illustrates this concept by examples.

**Definition 3.1.43** Let $(X, \tau)$ be a topological space and $A$ be a subset of $X$. The intersection of all closed sets of $(X, \tau)$ containing $A$ is called the **closure** of $A$, denoted by $\bar{A}$. It is the smallest closed set in $(X, \tau)$ containing $A$.

**Remark 3.1.44** The concepts of closure of a set and limit points of the set are closely related. Let $(X, \tau)$ be a topological space and $A \subset X$. Then the closure $\bar{A}$ is given by $\bar{A} = A \cup A'$, where $A'$ is the derived set of $A$. This asserts that a subset $A$ of $X$ is closed iff $A$ contains all of its limit points in $X$ (see Proposition 3.1.36).

***Example 3.1.45*** Let $(\mathbf{R}, \sigma)$ be the real line space with usual topology $\sigma$. Then in the space $(\mathbf{R}, \sigma)$,

(i) the closure of $\mathbf{Q}$ is $\mathbf{R}$. Since every real number $x \in \mathbf{R}$ is a limit point of $\mathbf{Q}$, it follows that $\overline{\mathbf{Q}} = \mathbf{R}$;

(ii) the closure of the open interval $(0, 1)$ is the closed interval $[0, 1]$;

(iii) the closure of the closed interval $[0, 1]$ is $[0, 1]$ itself.

### 3.1.8  Interior of a Set

This subsection conveys the concept of interior of a nonempty subset of a topological space, which is a dual concept of its closure with illustrative examples.

**Definition 3.1.46** Let $A$ be a subset of a topological space $(X, \tau)$. The union of all open sets of $(X, \tau)$ contained in $A$ is called the **interior of** $A$, denoted by Int($A$). It is sometimes denoted by $\mathring{A}$ or by simply Int$A$.

***Remark 3.1.47*** Since the empty set $\emptyset$ is contained in every subset $A$ of a topological space, Int($A$) of $A$ always exists and it consists precisely of all points $a \in A$ for which $A$ is a nbd of $a$ in $X$. Int($A$) is **the largest open set** in $X$ contained in $A$ and consists of all interior points of $A$ in $X$.

***Example 3.1.48*** Let $(\mathbf{R}, \sigma)$ be the real line space with usual topology $\sigma$. Then in the space $(\mathbf{R}, \sigma)$,

(i) Int($[0, 2)$) $= (0, 2)$, since $[0, 2)$ is not open in the real line space $(\mathbf{R}, \sigma)$ but the open interval $(0, 2)$ is the largest open set in $(\mathbf{R}, \sigma)$, contained in $[0, 2)$.

(ii) Int($\mathbf{Q}$) $= \emptyset$ in the real line space $(\mathbf{R}, \sigma)$, since any interval in $\mathbf{R}$ contains an irrational point, so $\mathbf{Q}$ does not contain a nonempty open set in $(\mathbf{R}, \sigma)$.

(iii) In the real line space $(\mathbf{R}, \sigma)$, for the set $A$ defined by

$$A = \{x \in \mathbf{Q} : 0 < x < 1\} \subset \mathbf{Q}$$

the interior $Int(A) = \emptyset$.

**Proposition 3.1.49** *Let $(X, \tau)$ be a topological space. Then for any subset $A \subset X$,*

(i) *a point $x$ is an interior point of $A$ iff there is an open set $U$ with the property that $x \in U \subset A$;*

(ii) *Int($A$) $\subset A$;*

(iii) *Int($A$) is largest open set contained in $A$;*

(iv) *$A$ is an open set iff Int($A$) $= A$;*

(v) *$A \subset B \subset X$ implies that Int($A$) $\subset$ Int($B$);*

(vi) *Int($A$) $= X - \overline{X - A}$.*

(vii) *$A$ is closed in $X$ iff $\overline{A} = A$.*

***Proof*** Left as an exercise. $\qquad\qquad\qquad\qquad\qquad\qquad\qquad\qquad\qquad\qquad\qquad\qquad$ □

### 3.1.9 Exterior of a Set

This subsection conveys the concept of exterior of a nonempty subset $A$ of a topological space $X$, which is the interior of $(X - A)$ and illustrates this concept by examples.

**Definition 3.1.50** Let $A$ be a subset of a topological space $(X, \tau)$. The set $\text{Int}(X - A)$ is called the **exterior of** $A$, **denoted by** $\text{Ext}(A)$ or simply by $\text{Ext}A$.

**Remark 3.1.51** Definition 3.1.50 asserts that $\text{Ext}(A)$ consists of all exterior points of $A$ in $(X, \tau)$ and $\text{Ext}(A) \cap A = \emptyset$. Moreover, $\overline{A} = X - \text{Ext}(A)$.

**Example 3.1.52** Let X be the real line space **R** and $A = (0, 1] \subset \mathbf{R}$. Then

$$\text{Ext}(A) = \text{Int}(X - A) = (-\infty, 0) \cup (1, \infty), \text{ since } X - A = (-\infty, 0] \cup (1, \infty).$$

Proposition 3.1.53 characterizes location of exterior points of a subset of a topological space with the help of its open sets.

**Proposition 3.1.53** *Let $X$ be a topological space and $A \subset X$. Then a point $x \in X$ is an exterior point of $A$ iff there is an open set $U$ such that $x \in U \subset X - A$.*

**Proof** It follows from Proposition 3.1.49, since an exterior point of $A$ is an interior point of $X - A$. $\qquad\qquad\square$

### 3.1.10 Boundary of a Set

This subsection conveys the concept of boundary of a nonempty subset $A$ of a topological space $(X, \tau)$, which formulates generalizing the intuitive idea of a separator between a region of Euclidean space and its exterior and illustrates this concept by examples.

**Definition 3.1.54** Let $A$ be a nonempty subset of a topological space $(X, \tau)$. Then the set

$$X - (\text{Int}(A) \cup \text{Ext}(A))$$

is called the **boundary of** $A$, denoted by $\partial A$. It consists of all boundary points of $A$ in the topological space $(X, \tau)$.

**Example 3.1.55** Let X be the real line space **R** and $A = (0, 1] \subset \mathbf{R}$. Then $\partial A = \{0, 1\}$.

Proposition 3.1.56 characterizes location of boundary point of a subset of a topological in terms of its open sets.

**Proposition 3.1.56** *Let* $(X, \tau)$ *be a topological space and* $A \subset X$. *Then a point* $x \in X$ *is a boundary point of* $A$ *iff every open set in* $(X, \tau)$ *containing the point* $x$ *intersects both the sets* $A$ *and* $X - A$.

**Proof** Left as an exercise.                                                                              □

## 3.1.11   Interrelations Among Closure, Interior, Exterior and Boundary Operators

This subsection conveys interrelationships among the concepts of different operators: Closure, Interior, Exterior and Boundary Operators defined in this section. Some properties of closure operator proved in Theorem 3.10.1 lead to define a topology in Definition 3.10.3, known as **Kuratowski closure topology.** For any nonempty subset $A$ of a topological space $X$, only three possible **locations of a point** $x \in X$ occur:

(i) $x \in \text{Int } A$ asserts that there is an open set $U$ in $X$ such that $x \in U \subset A$;

(ii) $x \in \text{Int } (X - A)$ asserts that there is an open set $U$ in $X$ such that $x \in U \subset X - A$;

(iii) $x \in \partial A$ asserts that there is an open set $U$ in $X$ such that $x \in U$, $U \cap A \neq \emptyset$ and $U \cap (X - A) \neq \emptyset$.

The points $x$ satisfying (i) are the points with the property that every open set $U$ containing the point $x$ intersects the set $A$, and hence, the limit points of $A$ and the set of these limit points are the set $\overline{A}$, the closure of $A$.

**Definition 3.1.57** Given a topological space $(X, \tau)$, the operators such as Closure, Interior, Exterior and Boundary on a subset $A \subset X$ can be considered as mappings on the power $\mathcal{P}(X)$ of $X$:

(i) **Closure operator** is a mapping $Cl: \mathcal{P}(X) \to \mathcal{P}(X)$, $A \mapsto Cl(A) = \overline{A}$.

(ii) **Interior operator** is a mapping $Int: \mathcal{P}(X) \to \mathcal{P}(X)$, $A \mapsto Int(A)$.

(iii) **Exterior operator** is a mapping $Ext: \mathcal{P}(X) \to \mathcal{P}(X)$, $A \mapsto Ext(A)$.

(iv) **Boundary operator** is a mapping $\partial: \mathcal{P}(X) \to \mathcal{P}(X)$, $A \mapsto \partial A$.

**Theorem 3.1.58** *Let* $(X, \tau)$ *be a topological space and* $A$ *be a subset of* $X$. *Then*

*(i)* $\partial A = \overline{A} \cap \overline{(X - A)}$;

*(ii)* $\partial A = \overline{A} - Int(A)$;

*(iii)* $\overline{A} = A \cup \partial A$;

*(iv)* $Int(A) = A - \partial A$;

*(v)* *A is closed in X iff* $\partial A \subset A$, *i.e., A is closed in X iff it consists of all points of its boundary* $\partial A$.

*(vi)* *A is open in X iff* $\partial A \cap A = \emptyset$, *i.e., A is open in X iff it contains no point of its boundary* $\partial A$.

**Proof** It follows from the respective definitions.                             □

**Example 3.1.59** For $(X, \tau) = (\mathbf{R}, \sigma)$ (real-line space) and $A = (0, 1]$, verify the validity of the properties (i)-(iv) of Theorem 3.1.58.

**Example 3.1.60** The composite of the interior and closure operators on a subset of an arbitrary topological space is not commutative. In support, consider the real line space $\mathbf{R}$ with its natural topology $\sigma$ and the subspace $\mathbf{Q}$ of rarional numbers. Then $\text{Int}(\mathbf{Q}) = \emptyset$ and $\overline{\mathbf{Q}} = \mathbf{R} \implies (\text{Int} \circ Cl)(\mathbf{Q}) = \mathbf{R}$ but $(Cl \circ \text{Int})(\mathbf{Q}) = \emptyset$.

## 3.1.12   Subspace Topology

This subsection defines a topology, called relative topology or subspace topology induced on a subset of a given topological space and illustrates this concept by examples. Given a nonempty subset $Y$ of a topological space $X$, the problem is: how to define a topology on $Y$ so that $Y$ becomes a topological space compatible with the topology of $X$? Definition 3.1.61 provides a solution of this problem.

**Definition 3.1.61** Let $X$ be a topological space and $Y$ be a subset of $X$. A set $U \subset Y$ is defined to be open in $Y$ if there exists an open set $V \subset X$ such that $V \cap Y = U$. This topology on $Y$ is called the **subspace or relative topology** induced by the topology of $X$ on $Y$. The resulting topological space is said to be a **subspace of** $X$.

**Example 3.1.62** Let $(X, \tau)$ be a topological space and $(Y, \tau_Y)$ be a subspace of $(X, \tau)$. Then the topology $\tau_Y$ induced by the canonical injection $i: Y \hookrightarrow X$ is given by

$$\tau_Y = \{i^{-1}(U): U \in \tau\} = \{U \cap Y: U \in \tau\}.$$

**Example 3.1.63**   (i)   The set $\mathbf{I} = [0, 1]$ endowed with relative topology inherited from the natural topology on $\mathbf{R}$ is called the space of unit interval, which is used throughout the book.

(ii)   The topology of a surface in $\mathbf{R}^3$ such as the 2-sphere $S^2 = \{(x, y, z) \in \mathbf{R}^3: x^2 + y^2 + z^2 = 1\} \subset \mathbf{R}^3$ is customarily taken to be the topology induced by the Euclidean topology on $\mathbf{R}^3$.

(iii)   Let $(M, d)$ be a metric space and $X$ be a nonempty subset of $X$. Then $X$ can be endowed with a metric $d_X$ induced by the metric $d$. The topology $\tau_{d_X}$ on $X$ induced by the metric $d_X$ (see Definition 3.8.5) is the topology on $X$ induced by the metric $d$.

## 3.1.13   Dense and Nowhere Dense Sets

This subsection continues a study of closed sets by defining dense and nowhere dense sets in topological spaces and illustrates these concepts by examples.

**Definition 3.1.64** Let $(X, \tau)$ be a topological space and $A \subset X$. Then

(i) $A$ is said to be a **dense (or everywhere dense) set** in $X$ if $\overline{A} = X$;
(ii) $A$ is said to be a **nowhere dense (or a nondense)** set if $\text{Int}(\overline{A}) = \emptyset$, where $\text{Int}(\overline{A})$ is the union of all open sets of $X$ contained in $\overline{A}$, called the interior of $\overline{A}$ in $X$.

**Remark 3.1.65** A subset A of a topological space $X$ is dense in $X$ if for any point $x \in X$, any nbd of $x$ contains at least one point from $A$ (i.e., $A$ has a nonempty intersection with every nonempty open subset of $X$). In other words, $A$ is dense in $X$ if the only closed subset of $X$ containing $A$ is $X$ itself.

**Example 3.1.66** The set of all rational numbers $\mathbf{Q}$ is dense in $\mathbf{R}$ with usual topology, since in this topology, every real number is a limit point of $\mathbf{Q}$ and hence $\overline{\mathbf{Q}} = \mathbf{R}$. On the other hand, every one-pointic set in $\mathbf{R}$ is nowhere dense.

**Example 3.1.67** The set of all irrational numbers is also dense in $\mathbf{R}$ with usual topology, since $\overline{(\mathbf{R} - \mathbf{Q})} = \mathbf{R}$.

**Example 3.1.68** In the real line space $\mathbf{R}$, the set $A$ defined by

$$A = \{x \in \mathbf{Q} : 0 < x < 1\},$$

is not nowhere dense in $\mathbf{R}$, because $\overline{A} = [0, 1]$ and hence $\text{Int}(\overline{A}) = (0, 1) \neq \emptyset$.

**Example 3.1.69** The real number space $\mathbf{R}$ with the natural topology has the rational numbers $\mathbf{Q}$ as a countable dense subset. This implies that the cardinality of a dense subset of a topological space may be strictly smaller than the cardinality of the space itself.

**Example 3.1.70** The subset $A = \{1, \frac{1}{2}, \frac{1}{3}, \frac{1}{4}, \ldots\}$ of the real number space $\mathbf{R}$ (with usual topology) is nowhere dense in $\mathbf{R}$, because $\overline{A} = \{0, 1, \frac{1}{2}, \frac{1}{3}, \frac{1}{4}, \ldots\}$ has no interior point.

**Example 3.1.71** Let $(X, \tau)$ be a topological space and $A$ be a subset of $X$.

(i) If $A$ is open in $(X, \tau)$, then $\partial A$ is nowhere dense in $(X, \tau)$.
(ii) If $A$ is closed in $(X, \tau)$, then $\partial A$ is nowhere dense in $(X, \tau)$.
(iii) If $A$ is a closed in $X$, then $A$ is nowhere dense iff its complement is everywhere dense.

Proposition 3.1.72 characterizes nowhere dense subsets of a metric space with the help of its open balls.

**Proposition 3.1.72** *Let $(X, d)$ be a metric space. A subset $A$ of $X$ is nowhere dense*

(i) *iff $\overline{A}$ does not contain any nonempty open ball; equivalently,*
(ii) *iff every nonempty open set has a nonempty open ball disjoint from $\overline{A}$.*

**Proof** It follows from Definition 3.1.64(ii).                                   □

**Example 3.1.73** Let $M(n, \mathbf{R})$ be the set of all $n \times n$ matrices over $\mathbf{R}$. The set $M(n, \mathbf{R})$ identified with the $n^2$-Euclidean space $\mathbf{R}^{n^2}$ and endowed with its usual product topology forms a topological space (see Sect. 3.18.1). This topological space has some interesting properties. For example,

(i) The general linear group $GL(n, \mathbf{R}) = \{A \in M(n, \mathbf{R}) : \det A \neq 0\}$ is the set of all nonsingular matrices of order $n$ over $\mathbf{R}$. It is an open and a dense subset but it is not closed in $M(n, \mathbf{R})$, (see Proposition 3.18.2).

(ii) Let $X = \{A \in M(n, \mathbf{R}) : A \text{ is singular}\}$. Then $X$ is nowhere dense in the topological space $M(n, \mathbf{R})$, (see Proposition 3.18.3).

## 3.2 Open Base and Subbase for a Topology

This section introduces the concepts of open base $\mathcal{B}$ and subbase for a topology $\tau$ on a set $X$ to determine completely the open sets in $\tau$ as a union of some members belonging to $\mathcal{B}$ for open base for the topology $\tau$ and as an intersection of a finite number of members belonging to $\mathcal{B}$ for a subbase for the topology $\tau$. In doing so, a subset $U \subset X$ is said to be open in X if for every point $x \in U$, there exists a member $B \in \mathcal{B}$ such that $x \in B \subset U$. The importance of an open base or a subbase for a topology lies in the result that the topology of a topological space is completely determined by an open base or a subbase. For example, the **concept of compact open topology** on a function space defined by a subbase is an important concept in topology (see Chap. 5) and the topology generated by a family of functions is available is Sect. 3.9.

### 3.2.1 Open Base

To study a topological space, it is not necessary to describe completely (i.e., all) its open sets. Rather, in doing so, this subsection studies the work of specifying a topology $\tau$ by taking only chosen open sets to generate all the open sets (similar to a basis of a vector space) by introducing the concept of open base for the topology $\tau$. Many topological properties can be proved by using the concept of an open base generating the relevant topology. In linear algebra, it is proved that every vector space has a basis and every vector in this vector space is a linear combination of the members of the basis. An analogue is true in topology, which asserts that every open set in a topological space $(X, \tau)$ can be expressed as a union of certain collection of open sets in $(X, \tau)$. The open sets in $\mathbf{R}$, $\mathbf{R}^2$ and $\mathbf{R}^n$ under metric topology are respectively defined in terms of open intervals, open disks and open balls and arbitrary open sets are expressed as unions of a certain collection open sets in terms of open intervals. open disks and open balls in the respective cases. Such a special collection of open sets determines completely the open sets. This idea leads to the concept of basis for a topology in Definition 3.2.1. Many important topologies such as order topology, Euclidean topology, metric topology and important topological spaces such as a second-countable space are defined by the concept of a base.

**Definition 3.2.1** Let $(X, \tau)$ be a given topological space. A collection of open sets $\mathcal{B}$ in $X$ is said to form an **open base (base or basis)** for the topology $\tau$, if every open set in $(X, \tau)$ is expressible as the union of some open sets belonging to $\mathcal{B}$. Equivalently, $\mathcal{B}$ forms an open base if for any point $x$ belonging to an open set $U$ in $(X, \tau)$, there exists a member $V \in \mathcal{B}$ such that $x \in V \subset U$ (see Theorem 3.2.4).

***Example 3.2.2*** (i) Any topology $\tau$ on any nonempty set always forms an open base of itself.

(ii) For the discrete topology on any nonempty set, $\emptyset$ and the singletons (i.e., subsets consisting of one point only) also form an open base. Some authors do not include $\emptyset$ for forming a base, because if $U = \emptyset$, it satisfies vacuously the defining criterion of an open set.

***Example 3.2.3*** Let $X$ be a given topological space with topology $\tau$. The unions of all subcollections of an open base $\mathcal{B}$ of the topology $\tau$ constitute the topology $\tau$. Thus, the topology $\tau$ is completely determined by any open base $\mathcal{B}$. For example, the natural topology $\sigma$ on **R** is completely determined by the family of open intervals in **R** in Corollary 3.2.6.

An open base of a topological space is characterized in Theorem 3.2.4.

**Theorem 3.2.4** *Let $(X, \tau)$ be a topological space. Then a given collection of open sets $\mathcal{B}$ forms an open base of the topology $\tau$ if and only if for any open set $U$ and any point $x \in U$, there exists a set $V \in \mathcal{B}$ such that $x \in V \subset U$.*

**Proof** Let the given condition be satisfied for $\mathcal{B}$, and let $U$ be any open set. Then for any point $x \in U$, there exists a set $V(x) \in \mathcal{B}$, such that $x \in V(x) \subset U$. Let $x$ run over $U$. Then we obtain

$$\bigcup_{x \in U} \{x\} \subset \bigcup_{x \in U} \{V(x)\} \subset U, \text{ i.e., } U \subset \bigcup_{x \in U} \{V(x)\} \subset U.$$

This shows that the open set $U = \bigcup_{x \in U} \{V(x)\}$ is a union of some members of $\mathcal{B}$.

Conversely, let $\mathcal{B}$ be an open base for the topology $\tau$ of $X$. Then any open set $U$ is a union of some sets lying in $\mathcal{B}$. Consequently, for any point $x \in U$, there exists a set $V \in \mathcal{B}$, such that $x \in V \subset U$. Hence, $\mathcal{B}$ forms an open base for $\tau$. $\square$

***Remark 3.2.5*** Theorem 3.2.4 asserts that any nonempty open set $U$ in $X$ is expressible as the union of open sets belonging in the open base $\mathcal{B}$. Conversely, if a nonemptyset $X$ is expressible as the union $X = \bigcup_{a \in \mathbf{A}} V_a$, then the family $\mathcal{B} = \{V_a : a \in \mathbf{A}\}$ forms an open base for a topology on $X$.

**Corollary 3.2.6** *The family of open intervals in* **R** *forms an open base for the natural topology* $\sigma$ *on* **R**.

*Proof* Let **R** be the set of real numbers with natural ordering. A subset $U \subset \mathbf{R}$ is said to be open if for every point $x \in U$, there exists an open interval $(a, b)$ such that $x \in (a, b) \subset U$. Then for any open set $U \subset \mathbf{R}$ and $x \in U$, by the openness of $U$, there exists an open interval $(a, b)$ such that

$$x \in (a, b) \subset U.$$

Hence, the corollary follows from Theorem 3.2.4. It is assumed that $(a, a) = \emptyset$, $\forall a \in$ **R**. On the other hand $(a, b) \neq \emptyset$ for any $a < b$ in **R**. $\qquad\square$

*Example 3.2.7* The null set $\emptyset$ and all open intervals $(a, b)$, where $a$ and $b$ are real numbers with $a < b$, form a base $\mathcal{B}$ for a topology of the set **R**. This topology is the **natural topology or usual topology** of **R** and the set **R** endowed with this topology is the real number space. Again the null set $\emptyset$ and all open intervals $(a, b)$, where $a$ and $b$ are rational numbers forms a countable a base $\mathcal{B}_1$ for the natural topology or usual topology of **R**. Sometimes, it is simply said that open intervals in **R** form a base for the natural topology or usual topology on **R** on the assumption that $(a, a) = \emptyset$,

*Remark 3.2.8* The open base for a topology is not unique in the sense that there may exist different open bases for any particular topology. In support, see Examples 3.2.2, 3.2.9 and 3.2.10.

*Example 3.2.9* Three different bases the usual topology on **R**. Consider three bases $\mathcal{F}$, $\mathcal{F}_r$ and $\mathcal{F}_i$ for generating the usual topology on **R**..

(i) The family $\mathcal{F}$ of all open intervals forms a base for the usual topology $\sigma$ on **R**.
(ii) The family $\mathcal{F}_r$ of open intervals with rational endpoints forms a countable base for the usual topology $\sigma$ on **R**.
(iii) The family $\mathcal{F}_i$ of open intervals with irrational endpoints forms a base (not countable) for the usual topology $\sigma$ on **R**.
    The basic sets $\mathcal{F}_r$ and $\mathcal{F}_i$ are disjoint but both of them are properly contained in $\mathcal{F}$.

*Example 3.2.10* The Euclidean plane $\mathbf{R}^2$ with natural topology $\tau$ has different open bases:

(i) The family $\mathcal{D}$ of open disks in $\mathbf{R}^2$ forms an open base for the topology $\tau$. Because if $U \in \tau$, and $x \in U$, then there exists an open disk $D_x$ in $\mathbf{R}^2$ with center $x$ such that $x \in D_x \subset U$.
(ii) The family $\mathcal{S}$ of open squares in $\mathbf{R}^2$ having horizontal and vertical sides forms an open base for the topology $\tau$.
(iii) The family $\mathcal{R} = \{(x, y) \in \mathbf{R}^2 : a \leq x < b, c \leq y < d\}$ of half-open rectangles in $\mathbf{R}^2$ forms an open base for the topology $\tau$.

A criterion for an aggregate of subsets of a given set $X$ to form an open base for a suitable topology of $X$ is given in Theorem 3.2.11.

**Theorem 3.2.11** *A collection of subsets $\mathcal{B}$ of a set $X$ forms an open base for a suitable topology of $X$ if, and only if,*

(i) *the null set $\emptyset \in \mathcal{B}$,*
(ii) *$X$ is the union of some sets lying in $\mathcal{B}$, and*
(iii) *the intersection of any two sets lying in $\mathcal{B}$ is the union of some sets lying in $\mathcal{B}$.*

**Proof** Let the conditions (i), (ii) and (iii) hold for a given collection of subsets $\mathcal{B}$ of $X$ and $\tau$ be the family of all those subsets of $X$, which are expressible as unions of some members of $\mathcal{B}$. Then $\tau$ forms a topology on $X$, of which $\mathcal{B}$ is an open base. Clearly, **OS**(1) and **OS**(4) hold for $\tau$, by the conditions (i) and (ii). Also, since every member of $\tau$ is a union of some members of $\mathcal{B}$, it follows that the union of any aggregate of members of $\tau$ is expressible as union of some members of $\mathcal{B}$ and is therefore a member of $\tau$. Consequently, **OS**(2) holds for $\tau$. Finally, the intersection of two members of $\tau$ is the intersection of two unions of members of $\mathcal{B}$, which is (by the distributive property for union and intersection of subsets) a union of intersections of pairs of members of $\mathcal{B}$; and this is a union of some members of $\mathcal{B}$, by (iii); i.e., this is a member of $\tau$. Hence, **OS**(3) is also satisfied for $\tau$.

Conversely, let $\mathcal{B}$ form an open base for a topology $\tau$ of $X$. Since the null set $\emptyset$ and the set $X$ are open sets, and $\emptyset$ cannot expressed as the union of nonempty sets, the conditions (i) and (ii) must hold. Again, since each member of $\mathcal{B}$ is an open set, the intersection of any two members of $\mathcal{B}$ is an open set and is therefore expressible as the union of some sets lying in $\mathcal{B}$. Thus (iii) also holds for $\mathcal{B}$.                                          $\square$

**Remark 3.2.12** The conditions (ii) and (iii) prescribed in Theorem 3.2.11 to form a base $\mathcal{B}$ for a topology on $X$ are sometimes replaced by their equivalent conditions (ii a) and (iii a):

(ii a): Every point of $X$ is contained in at least one member of $\mathcal{B}$;
(iii a): If $x \in U$ *and* $V$, where $U, V \in \mathcal{B}$, then there exists a member $W \in \mathcal{B}$ such that $x \in W \subset U \cap V$.

**Definition 3.2.13** The topology $\tau$ determined in Theorem 3.2.11 by a collection of subsets $\mathcal{B}$ of a set $X$ forms an **open base** for a suitable topology of $X$ is called the **topology generated or induced by the base** $\mathcal{B}$. A subset $U \subset X$ is called open in $X$, if $U \in \tau$, and for every $x \in U$, there exists a member $V \in \mathcal{B}$ such that $x \in V \subset U$.

**Example 3.2.14** The family of all open intervals in **R** forms an open base for a topology on **R**, since the intersection of any two open intervals is either $\emptyset$ or an open interval.

Theorem 3.2.15 establishes a necessary and sufficient condition for two bases to generate the same topology on a nonempty set.

**Theorem 3.2.15** *Two open bases $\mathcal{B}_1$ and $\mathcal{B}_2$ defined on a given nonempty set $X$ generate the same topology on $X$ iff the following two conditions are satisfied:*

*(i) given an $U_2$ in $\mathcal{B}_2$ and any point $x \in U_2$, there is an $U_1$ in $\mathcal{B}_1$ with the property that $x \in U_1 \subset U_2$;*

*(ii) given an $U_1$ in $\mathcal{B}_1$ and any point $y \in U_1$, there is an $U_2$ in $\mathcal{B}_2$ with the property that $y \in U_2 \subset U_1$.*

**Proof** Use Exercise 15 of Sect. 3.20 to prove the theorem. Because if the topology $\tau_1$ is generated on $X$ by the open base $\mathcal{B}_1$ and the topology $\tau_2$ is generated on $X$ by the open base $\mathcal{B}_2$, then $\tau_1 = \tau_2$ iff $\tau_1$ is finer than $\tau_2$ and $\tau_2$ is finer than $\tau_1$. □

**Example 3.2.16** The natural topology $\sigma$ on $\mathbf{R}$ is generated by the open intervals, and hence, every open set in $\sigma$ is a union of open intervals. There are also other topologies on $\mathbf{R}$. For example, let $\sigma$ be the natural topology on $\mathbf{R}$ and $\tau$ be the topology on $\mathbf{R}$ defined by $\tau = \{U \subset \mathbf{R}: U = \emptyset \text{ or } U \text{ is a union of open intervals}\}$. Then the topologies $\sigma$ and $\tau$ coincide. This gives another way of description of the natural topology on $\mathbf{R}$, which shows that any nonempty set in $\sigma$ is a union of open intervals and conversely every such union is an open set. Let $\sigma_{\mathbf{I}}$ be the subspace topology on $\mathbf{I} = [0, 1]$ induced by the natural topology $\sigma$ on $\mathbf{R}$. A nonempty proper subset $U \subset \mathbf{I}$ is open in $\sigma_{\mathbf{I}}$ iff $U$ is a union of open intervals of the form $(a, b)$ or of the form of an half-open interval of the form $(x, 1]$ or $[0, y)$, for $0 \le x < y \le 1$.

### 3.2.2 Local Base at a Point in a Topological Space

This subsection defines a local base or a neighborhood basis of a topological space at a point in the space, which is sometimes conveniently used to prove the continuity of a function. The concept of a neighborhood basis at a point gives the intuitive concept of the smallness of the neighborhood. The criterion of Theorem 3.2.4 motivates to define a local base at a point.

**Definition 3.2.17** Let $(X, \tau)$ be a topological space and $x \in X$ be an arbitrary point. Then a collection $\mathcal{B}_x$ of open sets in $(X, \tau)$ with each member containing the point $x$ is called an open base (or a **local base**) about the point $x$, if for every open set $U$ containing $x$, there exists a set $V_x \in \mathcal{B}_x$ such that $x \in V_x \subset U$.

There is another way of formulating a local base given in Definition 3.2.18 in the language of open nbds.

**Definition 3.2.18** Let $(X, \tau)$ be a topological space and $x \in X$ be an arbitrary point. Then a collection $\mathcal{B}_x$ of open sets in $(X, \tau)$ with each member containing the point $x$, is called **local base(or a nbd base)** at the point $x$, if for every open set $U$ containing $x$, there exists a set $V_x \in \mathcal{B}_x$ such that $x \in V_x \subset U$. Each member of $\mathbf{B}_x$ is called a nbd of $x$ in $X$. The members of nbd base $\mathbf{B}_x$ are called the **basic nbds** of $x$ in $(X, \tau)$.

**Theorem 3.2.19** *Let $(X, \tau)$ be a topological space. If $\mathcal{B}$ is an open base for the topology $\tau$, then the totality of all those members of $\mathcal{B}$, which contain a particular point $x$, forms a local base at the point $x$.*

**Proof** The proof follows from Theorem 3.2.4.                                                   □

**Example 3.2.20** The collection $\mathbf{B}_a = \{(a - \delta,\ a + \delta): \delta > 0\}$ in the real line space $\mathbf{R}$ for any point $a \in \mathbf{R}$. forms a local base at the point $a$. To show it consider for every $\delta > 0$, the $\delta$-**nbd** of the point $a$ in $\mathbf{R}$

$$N_a(\delta) = \{x \in \mathbf{R}: |x - a| < \delta\}$$

If

$$\{\delta_n\} = \{\delta_1, \delta_2, \ldots, \delta_n, \ldots\}$$

is any sequence of positive real numbers converging to the point $0 \in \mathbf{R}$, then the collection $\{N_a(\delta_n)\}$ forms a local base at the point $a$ in $\mathbf{R}$.

**Theorem 3.2.21** *Let $X$ be a nonempty set and $\mathbf{B}_x$ be a collection of subsets of $X$ containing the point $x$ such that every nbd of $x$ in $X$ contains some member of $\mathbf{B}_x$ and each member of $\mathbf{B}_x$ is a nbd of $x$ in $X$. Then there exists a unique topology $\tau$ on $X$ such that $\mathbf{B}_x$ forms a nbd basis of the point $x$, for every point $x \in X$.*

**Proof** Left as an exercise.                                                                   □

Proposition 3.2.22 gives a relation between a base for a topology and a local base at a point.

**Proposition 3.2.22** *Let $(X, \tau)$ be a topological space and $\mathbf{B}$ be a base for the $\tau$. Then for any point $x \in X$, the members of $\mathbf{B}$ that contain the point $x$ form a local base at the point $x$.*

**Proof** Left as an exercise.                                                                   □

**Example 3.2.23** For the usual topology $\sigma$ on $\mathbf{R}$, the collection of all open intervals $\{(a, b)\}$ in $\mathbf{R}$ forms a base for the topology $\sigma$; on the other hand, the collection $\mathbf{B}_a = \{(a - \delta,\ a + \delta): \delta > 0\}$ in $\mathbf{R}$ forms a local base at the point $a \in \mathbf{R}$.

**Definition 3.2.24** Let $X$ be a nonempty set and $\mathbf{B}_x$ be a collection of subsets of $X$ defined in Theorem 3.2.21. The family $\mathbf{B}_x$ is called a **nbd filter** of the point $x$.

### 3.2.3  Subbase for a Topology

This subsection studies the work of specifying a topology $\tau$ on a set $X$ by taking only a collection $\mathcal{S}$ of chosen open sets to generate all open sets in $\tau$ completely, i.e., to form an open base $\mathcal{B}$ for $\tau$ as an intersection of a finite subcollection of $\mathcal{S}$ of subsets of $X$. This solves a natural problem: given a family $\mathcal{F}$ of subsets of $X$, is it possible to find a subfamily of $\mathcal{S}$ consisting of all unions of finite intersections of members of $\mathcal{S}$ to form a topology on $X$? To solve this problem, the concept of subbase for a topology is introduced in Definition 3.2.25.

**Definition 3.2.25** Given a topological space $(X, \tau)$, a family of open subsets $\mathcal{S}$ in $(X, \tau)$ is said to be a **subbase or subbasis** for the topology $\tau$ if the subsets obtained by the intersection of all finite subcollections of $\mathcal{S}$ constitute a base of the topology $\tau$. The open sets in $\mathcal{S}$ are called **subbase open sets** of $X$. In other words, a subbasis $\mathcal{S}$ for a topology $\tau$ is a family of subsets of $X$ whose union is $X$ and the topology generated by $\mathcal{S}$ on $X$ is a family of subsets of $X$ whose union is the set $X$ and the topology $\tau$ is generated by $\mathcal{S}$ is defined to be the family $\tau$ of all unions of finite intersections of members belonging to $\mathcal{S}$.

*Example 3.2.26* Let $(\mathbf{R}, \sigma)$ be the real line space with natural topology $\sigma$. For each real number $r$, the sets

$$L_r = \{x \in \mathbf{R} : x \leq r\}$$

and

$$R_r = \{x \in \mathbf{R} : x \geq r\},$$

called the **open-half lines** determined by $r \in \mathbf{R}$ are such that the open interval $(r, s) = R_r \cap Ł_s$. The collection $\mathcal{S}$ of all such open-half lines constitutes a subbase for the natural topology on $\mathbf{R}$. This asserts that the family of all infinite open intervals in $\mathbf{R}$ is a subbase for the natural topology $\sigma$ on $\mathbf{R}$, because, every open interval $(r, s)$ in $\mathbf{R}$ is the intersection of two infinite open intervals $(r, s) = (r, \infty) \cap (-\infty, s)$. On the other hand, the family of all open intervals in $\mathbf{R}$ is a base for the topology $\sigma$ on $\mathbf{R}$ by Corollary 3.2.6.

Theorem 3.2.27 establishes a necessary and sufficient condition for a family of subsets to form a subbase for a topology.

**Theorem 3.2.27** *A family $\mathcal{F}$ of subsets of a given set $X$ constitutes a subbase for a suitable topology of $X$, iff*

(i) *either the emptyset $\emptyset \in \mathcal{F}$ or the intersection of a finite number of subsets belonging to $\mathcal{F}$ is also in $\mathcal{F}$;*

(ii) *$X$ is the union of the subsets belonging to $\mathcal{F}$.*

*Proof* Let $\mathcal{F}$ form a subbase for a topology $\sigma$ on $X$ and $\mathcal{B}$ be the base generated by $\mathcal{F}$. Since $\emptyset \in \mathcal{B}$, either $\emptyset \in \mathcal{F}$ or $\mathcal{F}$ contains a finite number of subsets, whose intersection is $\emptyset$. Since every point $x \in X$ is in at least one member of $\mathcal{B}$, the point $x$ is in at least one member belonging to $\mathcal{F}$. This shows that $X$ is the union of the subsets belonging to $\mathcal{F}$. This implies that the conditions (i) and (ii) hold for $\mathcal{F}$.

Conversely, let $\mathcal{F}$ satisfy the conditions (i) and (ii). If $\mathcal{B}$ be the set consisting of all finite intersections of the members of $\mathcal{F}$, then $\emptyset \in \mathcal{B}$ by (i). Again, since $\mathcal{F} \subset \mathcal{B}$, it follows that by condition (ii) that $X$ is the union of the subsets belonging to $\mathcal{B}$. Let $x \in U, x \in V$ for some $U, V \in \mathcal{B}$. Then $x \in U \cap V = W$ and

$$U = U_1 \cap U_2 \cap \cdots \cap U_m \text{ and } V = V_1 \cap V_2 \cap \cdots \cap V_n$$

for some members $U_1, U_2, \ldots, U_m$ and $V_1, V_2, \ldots, V_n \in \mathcal{F}$. Hence, it follows that

$$x \in W = U \cap V = U_1 \cap U_2 \cap \cdots \cap U_m \cap V_1 \cap V_2 \cap \cdots \cap V_n \in \mathcal{B}.$$

This asserts that $\mathcal{B}$ forms an open base generated by $\mathcal{F}$, which is a subbase for a suitable topology $\sigma$ on $X$.     □

**Remark 3.2.28** The topology of a topological space is completely determined by an open base or a subbase. Because a family $\mathcal{F}$ of open sets forms a subbase for the $\tau$ of a topological space $(X, \tau)$, if every open set $U \in \tau$ and every point $x \in X$, there are finite number of open sets in $\mathcal{F}$, say $V_1, V_2, \ldots, V_r$ such that

$$x \in V_1 \cap V_2 \cap \cdots \cap V_r \subset U.$$

**Theorem 3.2.29** *The weakest topology $\sigma$ on a set $X$ containing a given family $\mathcal{F}$ of subsets of $X$ is the topology generated by the subbase $\mathbf{B}$ formed by $\emptyset, X$ and all those subsets, which are in $\mathcal{F}$.*

**Proof** Let $\mathcal{A}$ be the collection of the subsets $\emptyset, X$ and all those subsets, which are in $\mathcal{F}$. Then by Theorem 3.2.27 $\mathcal{A}$ forms a subbase for a topology $\sigma$ on $X$. Let $\tau$ be a topology on $X$, such that $\mathcal{F} \subset \tau$. Then $\mathcal{A} \subset \tau$. Let $\mathcal{B}$ be the base generated by $\mathcal{A}$. Then $\mathcal{B} \subset \tau$. Since the topology $\sigma$ is obtained from $\mathcal{B}$ by taking unions of finite number of elements of all subcollections of $\mathcal{B}$, it follows that $\sigma \subset \tau$. This asserts that $\sigma$ is weakest topology on $X$ containing the given family $\mathcal{F}$ of subsets of $X$.     □

**Remark 3.2.30** For better understanding of the concept of a subbase, a geometrical illustration is given in Example 3.2.31.

**Example 3.2.31 (Geometrical example)** Let $\mathbf{R}^2$ be the Euclidean plane and $\mathbf{R}$ be the Euclidean line. Then $p_1: \mathbf{R}^2 \to \mathbf{R}, (x, y) \mapsto x$ and $p_2: \mathbf{R}^2 \to \mathbf{R}, (x, y) \mapsto y$ are projection maps such that the inverse images of an open interval $(a, b)$; i.e., $p_1^{-1}(a, b)$ and $p_2^{-1}(a, b)$ geometrically represent infinite open strips in $\mathbf{R}^2$. Such strips constitute a subbase for the natural (Euclidean) topology on $\mathbf{R}^2$.

**Example 3.2.32** In the real line space $\mathbf{R}$,

(i) a base for the subspace topology induced on the open interval $(2, 3)$ is the collection

$$\{(x, y) \cap (2, 3) : x, y \in \mathbf{R}, \text{ with } x < y\} = \{(x, y) : x, y \in \mathbf{R}, \ 2 \le x < y \le 3\};$$

(ii) a base for the subspace topology induced on the closed interval $[2, 3]$ is the collection

$$\{(x, y) \cap [2, 3]: x, y \in \mathbf{R}, \ x < y\},$$

which is the same as the collection

$$\{(x, y): 2 \leq x < y \leq 3\} \cup \{[2, y): 2 < y \leq 3\} \cup \{(x, 3): 2 \leq x < 3\} \cup \{[2, 3]\}$$

## 3.3 Euclidean Topology

This section studies Euclidean topology on $\mathbf{R}^n$ by using the concept of open base. This topology is the same as the metric (usual) topology induced by the Euclidean metric (usual metric) on $\mathbf{R}^n$. Its more study is available in Sect. 3.8. The concept of **locally Euclidean space** plays a key role in the study the topological manifolds. It is discussed in **Basic Topology, Volume 2** of the present book series.

### 3.3.1 Euclidean Topology on R

This subsection studies Euclidean topology on the Euclidean line $\mathbf{R}$. It uses the concept of open base.

**Definition 3.3.1** A subset $U \subset \mathbf{R}$ is said to be open for a topology $\sigma$ on $\mathbf{R}$ if for every point $x \in \mathbf{R}$, there are points $a, b \in \mathbf{R}$, with $a < b$ such that

$$x \in (a, b) \subset U.$$

Then the collection $\mathcal{B}$ of such sets $\{U\}$ forms an open base by Theorem 3.2.4 for the usual topology $\sigma$, which is also called the **Euclidean topology** on $\mathbf{R}$. This topology is the same as the metric topology induced by the standard Euclidean metric (usual metric)

$$d: \mathbf{R} \times \mathbf{R} \to \mathbf{R}, (x, y) \mapsto |x - y|$$

(see Example 3.8.6)

**Proposition 3.3.2** *Under the Euclidean topology on $\mathbf{R}$, all open intervals are open sets and all closed intervals are closed sets.*

**Proof** Left as an exercise. $\qquad\square$

**Remark 3.3.3** **There exist several important topologies on R:** Other than Euclidean topology, the set $\mathbf{R}$ has also several topologies such as lower-limit topology, upper-limit topology defined in Sect. 3.4.1 and metric topology defined in Sect. 3.8.1 and others.

### 3.3.2   Euclidean Topology on $\mathbf{R}^2$

This subsection introduces the concept of Euclidean topology on $\mathbf{R}^2$ by extending the concept of Euclidean topology on $\mathbf{R}$. It uses the concept of open base.

**Definition 3.3.4** Let $a, b, c, d \in \mathbf{R}$ be such that $a < b, c < d$, and $\mathcal{B}_2 = \{(x, y) \in \mathbf{R}^2 : a < x < b, \ c < y < d\} \subset \mathbf{R}^2$. Then $\mathcal{B}_2$ forms an open base for a topology on $\mathbf{R}^2$, called the **Euclidean topology** on $\mathbf{R}^2$.

**Geometrically**, an open base for the Euclidean topology on $\mathbf{R}^2$ given in Definition 3.3.4, is the set of open rectangles of the form $U_i \times V_j$, where $U_i$ and $V_j$ are open intervals. The collection $\mathcal{B}_2$ of all equilateral triangles with bases parallel to $x$-axis also forms an open base for the Euclidean topology on $\mathbf{R}^2$.

### 3.3.3   Euclidean Topology on $\mathbf{R}^n$

This subsection introduces the concept of Euclidean topology on $\mathbf{R}^n$ by extending the concept of Euclidean topology on $\mathbf{R}$. It uses the concept of open base.

**Definition 3.3.5** Let $x = (x_1, x_2, \ldots, x_n)$, $a = (a_1, a_2, \ldots, a_n)$, and $b = (b_1, b_2, \ldots, b_n)$ be arbitrary points in $\mathbf{R}^n$ with $a_i < b_i : i : i = 1, 2, \ldots, n$. If $U_{a,b} = \{x \in \mathbf{R}^n : a_i < x_i < b_i, \ \forall i : i = 1, 2, \ldots, n\}$, then the collection of sets

$$\mathcal{B} = \{U_{a,b}\}$$

forms a base for a topology on $\mathbf{R}^n$, called the **Euclidean topology on $\mathbf{R}^n$**. This topology is the same as the metric topology induced by the standard Euclidean metric (usual metric)

$$d : \mathbf{R} \times \mathbf{R} \to \mathbf{R}, (x, y) \mapsto \|x - y\|$$

(see Example 3.8.7).

**Geometrically**, the set $U_{a,b}$ given in Definition 3.3.5, represents a parallelopiped in $\mathbf{R}^n$ with sides parallel to axes and this collection forms an open base for the Euclidean topology on $\mathbf{R}^n$. It is natural to choose a base with the least possible number of countable number of elements for this topology. For example, there exists a countable base for $\mathbf{R}^n$, which are precisely the parallelopipeds with rational vertices such as

$$V_{r,t} = \{x \in \mathbf{R}^n : r_i < x_i < t_i, \ r = (r_1, r_2, \ldots, r_n),$$
$$t = (t_1, t_2, \ldots, t_n), r_i < t_i, \ \forall i : i = 1, 2, \ldots, n\}.$$

where $r_i$ and $t_i$ are rational points in $\mathbf{R}$.

**Definition 3.3.6** Let $X \subset \mathbf{R}^m$ and $Y \subset \mathbf{R}^n$. A function $f: X \to Y$ is said to be continuous at a point $x \in X$, if given a real number $\epsilon > 0$, there is a $\delta > 0$ such that if $x' \in X$ and $\|x - x'\| < \delta$, then

$$\|f(x) - f(x')\| < \epsilon.$$

If $f$ is continuous at every point $x \in \mathbf{X}$, then it is said to be continuous.

**Example 3.3.7**    (i)  In the Euclidean line $\mathbf{R}$, let $X = [0, 2\pi]$ and in the Euclidean plane $\mathbf{R}^2$, let $Y = S^1 = \{(x, y) \in \mathbf{R}^2 : x^2 + y^2 = 1\}$ with subspace topology. Then the function

$$f: X \to Y, x \mapsto (\cos x, \sin x)$$

is continuous.

(ii)  Let $M(n, \mathbf{R})$ be the set of all $n \times n$ matrices over $\mathbf{R}$ identified with $\mathbf{R}^{n^2}$ and endowed with its usual product topology (see Sect. 3.12.2) on $\mathbf{R}^{n^2}$ and $GL(n, \mathbf{R})$ be the set of all nonsingular real matrices with subspace topology of $M(n, \mathbf{R})$. The determinant function

$$\det: GL(n, \mathbf{R}) \to \mathbf{R}, M \mapsto \det M$$

is continuous (see Proposition 3.18.1).

(iii)  Let $GL(n, \mathbf{C})$ be the set of all nonsingular complex matrices. The determinant function

$$\det: GL(n, \mathbf{C}) \to \mathbf{C}, M \mapsto \det M$$

is continuous (see Proposition 3.18.4).

### 3.3.4  Special Examples of Open and Closed Sets in $\mathbf{R}^n$

This subsection presents a few special examples of open and closed subsets in $\mathbf{R}^n$ endowed with Euclidean topology which are other than the whole space $\mathbf{R}^n$ and the empty set $\emptyset$.

**Example 3.3.8** Given a set of continuous functions $f_i: \mathbf{R}^n \to \mathbf{R} : i = 1, 2, \ldots, n$, the set of solutions of the system of $n$-equations $f_i = 0$ is a closed set in $\mathbf{R}^n$, since the complement of the solution set is the union of the open sets determined by $\{f_i \neq 0, i = 1, 2, \ldots, n\}$. As an immediate applications of this result, it follows that the 2-sphere $S^2 = \{(x, y, z) \in \mathbf{R}^3 : x^2 + y^2 + z^2 = 1\}$ is a closed set in $\mathbf{R}^3$.

**Example 3.3.9** The open disk

$$\mathbf{D}^n = \{x \in \mathbf{R}^n : ||x|| < 1\}$$

is an open set in the Euclidean $n$-space $\mathbf{R}^n$. On the other hand,

(i) the closed disk

$$\overline{\mathbf{D}^n} = \{x \in \mathbf{R}^n : ||x|| \leq 1\}$$

(ii) and its boundary

$$S^{n-1} = \{x \in \mathbf{R}^n : ||x|| = 1\}, \text{ the } (n-1) \text{ - sphere } S^{n-1}$$

are both closed sets in $\mathbf{R}^n$.

## 3.4  Topology on Linearly Ordered Sets

This section studies topology on a linearly ordered set as a generalization of the usual topology on $\mathbf{R}$ with a particular study of order topology on $\mathbf{R}$, $\mathbf{Q}$ and $\mathbf{Z}$. In this section, $X$ denotes an arbitrary linearly ordered set with an order relation " $\leq$" such that for every pair of elements $x, y \in X$ either $x \leq y$ or $y \leq x$. Then $x < y$ means that $x \leq y$ but $x \neq y$. If $x < y$, then $x$ is said to be less (or smaller) than $y$ and $y$ is said to be greater than $x$.

**Definition 3.4.1** Let $X$ be a **linearly ordered set (or order set )** with an order relation " $\leq$". Given $a < b$ in $X$, the open interval $(a, b)$, closed interval $[a, b]$, the right-half open interval $[a, b)$, left-half open interval $(a, b]$ are defined by

   (i) $(a, b) = \{x \in X : a < x < b\}$;
  (ii) $[a, b] = \{x \in X : a \leq x \leq b\}$;
 (iii) $[a, b) = \{x \in X : a \leq x < b\}$;
 (iv) $(a, b] = \{x \in X : a < x \leq b\}$.

**Theorem 3.4.2** *Let $X$ be a linearly ordered set with an order relation " $\leq$". If $X$ has neither any greatest element nor any least element, then the family of all open intervals $(a, b)$ together with the empty set $\emptyset$ form an open base $\mathcal{B}$ for a topology $\tau$ on $X$, called the* **order topology or interval topology** *on $X$.*

***Proof*** Let $X$ be a linearly ordered set with an order relation " $\leq$" and $x \in X$. Then by hypothesis $X$ has neither a least element nor a greatest element. This shows that there exist elements $a, b \in X$ such that $a < b$, $a < x$ and $x < b$. Consequently, $x \in (a, b) \in \mathcal{B}$. Again, for $x \in (a, b)$ and $x \in (c, d)$, the element $x \in (z, t) \subset (a, b) \cap (c, d)$, where $z$ is greater than $a$, $c$ and $t$ is smaller than $b$, $d$. Hence, the proof follows from Remark 3.2.12.                                                                    □

**Theorem 3.4.3** *$X$ be a linearly ordered set with an order relation " $\leq$". If $X$ has no greatest element, then*

   (i) *the emptyset $\emptyset$ and the right half open intervals $[a, b) = \{x \in X : a \leq x < b\}$ constitute a base for a topology $\tau_l$ on $X$, called the lower-limit topology or right-half open interval topology on $X$;*

*(ii) the emptyset $\emptyset$ and all the improper intervals $(\infty, x)$ constitute a base for a topology on $X$, called the left-hand topology on $X$;*

**Proof** Proceed as in Theorem 3.4.2.                                                     □

**Definition 3.4.4** Let $X$ be a linearly ordered set with an order relation " $\leq$". If $X$ has no least element, then the **upper-limit or left-half open interval topology** defined on $X$ is generated by a base consisting of the emptyset $\emptyset$ and all the left-half open interval $(a, b] = \{x \in X : a < x \leq b\}$ and the **right-hand topology** on $X$ is generated by a base consisting of the emptyset $\emptyset$ and all the improper intervals $(x, \infty)$.

**Example 3.4.5** Order topologies on **R** by using usual order relation " $\leq$" on **R** are available in Sect. 3.4.1

### 3.4.1  Order Topologies on **R**

By using usual order relation " $\leq$" on **R**, this subsection defines the natural, lower-limit, upper-limit topologies and $K$-topology on **R** and compare them.

**Definition 3.4.6** (*Natural topology on* **R**) Let $\mathcal{B}$ be the family of open intervals in **R** given by

$$\mathcal{B} = \{(a, b) : a, b \in \mathbf{R}, \ a < b\}.$$

$\mathcal{B}$ forms a base for the order topology on **R**. Natural topology on **R** is the order topology derived from the natural ordering on **R**.

**Definition 3.4.7** (*Lower-limit topology on* **R**) Let $\mathcal{B}$ be the family of closed-open intervals in **R** given by

$$\mathcal{B} = \{[a, b) : a, b \in \mathbf{R}, \ a < b\}.$$

Then **R** is the union of the members of the family $\mathcal{B}$, since every real number $x$ is in some closed-open interval belonging to $\mathcal{B}$. Again for $[a, b)$, $[c, d) \in \mathcal{B}$, their intersection $[a, b) \cap [c, d)$ is either $\emptyset$ or in $\mathcal{B}$, Because, for

$$a < c < b < d \implies [a, b) \cap [c, d) = [c, b) \in \mathcal{B}.$$

This asserts that the family $\mathcal{R}_l$ consisting of unions of closed-open intervals in **R** form a topology for which $\mathcal{B}$ is a base. This topology $\mathcal{R}_l$ is called the lower-limit topology on **R**.

**Definition 3.4.8** (*Upper-limit topology on* **R**) Let $\mathcal{B}$ be the family of open-closed intervals in **R** given by

$$\mathcal{B} = \{(a, b] : a, b \in \mathbf{R}, \ a < b\}.$$

Then as before, $\mathcal{B}$ forms a base for a topology $\mathcal{R}_u$ on $\mathbf{R}$, called the upper-limit topology on $\mathbf{R}$.

**Example 3.4.9** Let $\mathcal{B}$ be the family of intervals in $\mathbf{R}$ given by

$$\mathcal{B} = \{(a, b): a, b \in \mathbf{R}, \ a < b \text{ and} a, b \text{ rational}\}.$$

Then $\mathcal{B}$ forms a base for a topology on $\mathbf{R}$, which is countable, different from both lower-limit topology and upper-limit topology on $\mathbf{R}$.

**Definition 3.4.10** ($\mathcal{K}$-topology on $\mathbf{R}$) Let $K = \{x_n = \frac{1}{n}\}_{n \in \mathbf{N}} \subset \mathbf{R}$ and $\mathcal{B}$ be the family of all open intervals $(a, b)$ in $\mathbf{R}$ together with the sets of the form $(a, b) - K$. Then the topology $\mathcal{R}_K$ generated by $\mathcal{B}$ as a base is called the $\mathcal{K}$-topology on $\mathbf{R}$.

**Proposition 3.4.11** *Each of the lower-limit topology $\mathcal{R}_l$, upper-limit topology $\mathcal{R}_u$ and the $\mathcal{K}$-topology $\mathcal{R}_K$ on $\mathbf{R}$ is strictly stronger than the natural (usual) topology on $\mathbf{R}$. Moreover, they are not comparable.*

**Proof** For any $x \in (a, b)$, the element $x \in (a, c] \subset (a, b)$ and for any $x \in [d, b) \subset (a, b)$, where $x \le c < b$ and $a < d \le x$. On the other hand, if $x \in (a, b]$, then there is no open interval $(z, t)$ such that $x \in (d, t) \subset (a, b]$, for $b = x$. This asserts that the upper-limit topology is strictly stronger than the usual topology on $\mathbf{R}$. Similarly, the lower limit topology is strictly stronger than the usual topology on $\mathbf{R}$. Finally, the $K$-topology is strictly stronger than the usual topology on $\mathbf{R}$, because, given $x \in (a, b)$, the same interval $(a, b)$ is a basic set for the $\mathcal{K}$-topology that contains the point $x$. But corresponding to the basic set $B = (-1.1) - K$ for the topology $\mathcal{R}_K$ and the point $0 \in B$, there exists no interval such that it contains 0 and wholly lies in $B$. The last part follows from the respective definitions.                                    □

**Example 3.4.12** The set $\mathbf{R}$ (with natural ordering) endowed with the natural or usual topology $\sigma$ is the real line space. On the other hand, the set $\mathbf{R}$ endowed with the lower-limit topology $\sigma_l$ (or $\mathcal{R}_l$) is called **the Sorgenfrey line space**. This space has some special properties discussed in subsequent chapters and is referred throughout the book. For **Sorgenfrey plane** $(\mathbf{R}^2, \sigma)$; see Example 3.12.9.

### 3.4.2 Order Topologies on Q and Z

This subsection studies the lower-limit topology and the upper-limit topology on $\mathbf{Q}$ and $\mathbf{Z}$ defined in a way analogous to these topologies defined on $\mathbf{R}$ (see Sect. 3.4.1).

**Example 3.4.13** In each of the ordered sets $\mathbf{R}$, $\mathbf{Q}$, and $\mathbf{Z}$ with natural ordering, there are neither a greatest element nor a least element. Hence for each of them, there exist

  (i)  the usual topology or the natural topology;
  (ii) the lower-limit topology;
  (iii) the upper-limit topology.

**Proposition 3.4.14** *(i) The usual topology, lower-limit topology and upper-limit topology are identical on* **Z**.

*(ii) The lower-limit topology and upper-limit topology on* **Q** *are strictly stronger than the usual topology on* **Q**,

**Proof** (i) Since

$$(x, y) = (x, y - 1] = [x + 1, y), \ \forall x, y \in \mathbf{Z},$$

it asserts (i).

(ii) Similar to Proposition 3.4.11. ☐

### 3.4.3 Ordinal Space

This subsection defines ordinal topology on the set of all ordinal numbers by using the concept of order topology. Let $(X, <)$ be a well-ordered set; i.e., every nonempty subset of $X$ has a first element. Its order type is called an **ordinal number** (see Chap. 1).

**Definition 3.4.15** Given an ordinal number $\alpha$, let $[0, \alpha]$ be the set of all ordinal numbers less than or equal to $\alpha$. The set $[0, \ \alpha]$ endowed with the order topology $\tau$ is called the closed **ordinal space**.

**Example 3.4.16** The order topology $\tau$ of the ordinal space $[0, \alpha]$ is generated by a base consisting of all sets

$$\mathcal{B} = (\beta, \delta + 1) = (\beta, \delta] = \{x : x \in (\beta, \delta]\} = \{x : \beta < x < \delta + 1\},$$

where $\beta, \delta + 1 \in [0, \alpha]$. It asserts that $[\beta, \delta)$ is an open set in $\tau$ iff either $\beta = 0$ *or* $\beta$ has an immediate predecessor.

## 3.5 Lattice of Topologies

This section is a continuation of Sect. 3.1.3 with a basic result that the family of all topologies defined on a nonempty set forms a complete lattice under the weaker or stronger relations between two topologies. The motivation of the lattice of topologies comes from the Propositions 3.5.1, 3.5.4 and 3.5.7. The main result of this section is the Theorem 3.5.8 which asserts that the family of all topologies defined on a nonempty set forms a **complete lattice**.

**Proposition 3.5.1** *Let $X$ be a nonempty set and $\Omega$ be the set of all topologies defined on the same set $X$.*

(i) *Let $\rho_w$ be the relation on $\Omega$ of being a topology $\tau$ to be weaker than a topology $\sigma$ on $X$. Then $(\Omega, \rho_w)$ is a partially ordered set.*

(ii) *Let $\rho_s$ be the relation on $\Omega$ of being a topology $\tau$ to be stronger than a topology $\sigma$ on $X$. Then $(\Omega, \rho_s)$ is also a partially ordered set.*

**Proof** Since the relation $\rho_w$ on $\Omega$ is reflexive, transitive and antisymmetric, it follows that $(\Omega, \rho_w)$ is a partially ordered set. Similarly, since the relation $\rho_s$ is reflexive, transitive and antisymmetric, it follows that $(\Omega, \rho_s)$ is also a partially ordered set.□

**Definition 3.5.2** A partially ordered set $X$ is said to be a

(i) lattice if every pair of elements in $X$ has both the lub and glb in X;
(ii) complete lattice if every nonempty collection of elements in $X$ has both the lub and glb in X.

**Example 3.5.3** Consider the set **N** partially ordered by divisibility relation. Under this relation, **N** is a partial ordered set for which the lub is the least common multiple (lcm) and the glb is the greatest common divisor (gcd).

**Proposition 3.5.4** *Let $\Omega$ be the set of all topologies defined on a given set $X$, partially ordered by $\rho_w$ and $\{\tau_i : i \in \mathbf{A}\}$ be any of its subfamily. Then their intersection also forms a topology on $X$ such that it is the glb of this family of topologies.*

**Proof** Let $X$ be a nonempty set and $\{\tau_i : i \in \mathbf{A}\}$ be a family of topologies on $X$ and $\tau = \bigcap\{\tau_i : i \in \mathbf{A}\}$. Since $\emptyset$ and $X$ are in each $\tau_i$, it follows that $\emptyset$ and $X$ are also in $\tau$. Let $U, V \in \tau$. Then they are in each $\tau_i$. Hence, it follows that their intersection $U \cap V \in \tau$. Again, let $\{V_i\}$ be any subfamily of $\tau$. Then $V_i \in \tau_i$ for each $i$, and hence their union $\bigcup\{V_i\}$ is in each $\tau_i$. It shows that it is also in $\tau$. Consequently, $\tau$ forms a topology on $X$. Finally, to show that $\tau$ is the glb of the family $\{\tau_i : i \in \mathbf{A}\}$, let $\sigma$ be a topology weaker than every topology $\tau_i$. Then $\sigma \subset \tau_i$, $\forall i \in \mathbf{A}$ and hence $\sigma \subset \tau$. This asserts that $\tau$ is the glb of the family $\{\tau_i : i \in \mathbf{A}\}$ of topologies. This shows that there exists glb of any given subfamily of topologies on $X$ partially ordered by $\rho_w$.                                                                    □

**Corollary 3.5.5** *Let $X$ be a nonempty set and $\Omega$ be the set of all topologies defined on $X$. Then there exists a glb of any subfamily of this family $\Omega$.*

**Proof** It follows from Proposition 3.5.4.                                        □

**Example 3.5.6** Let $\{\tau_i : i \in \mathbf{A}\}$ be a family of topologies on $X$. Then their union $\bigcup\{\tau_i : i \in \mathbf{A}\}$ may not be a topology on $X$. On the other hand, the family of all topologies defined on a nonempty set forms a complete lattice by Theorem 3.5.8.

**Proposition 3.5.7** *Let $\Omega$ be the set of all topologies defined on a given set $X$ and $\{\tau_i : i \in \mathbf{A}\}$ be any of its subfamily. Then there exists lub of this subfamily.*

**Proof** Let $\sigma$ be the topology on $X$ generated by the subbase **B**, formed by $\emptyset$, $X$ and all those subsets of $X$ which are contained in all the topologies belonging in the subfamily $\{\tau_i : i \in \mathbf{A}\}$. Then it follows by Theorem 3.2.29 that $\sigma$ is the lub of this subfamily $\{\tau_i : i \in \mathbf{A}\}$.                                              □

The above discussion is summarized in the following basic and important result.

**Theorem 3.5.8** *The family of all topologies defined on a nonempty set forms a* **complete lattice** *under the both weaker and stronger relations between topologies on this family.*

## 3.6 Continuous Maps

This section addresses the concept of continuous maps or functions between topological spaces. It is the central basic concept of topology, and its workable definition is given in Definition 3.6.1, even in the absence of a metric structure. For metric spaces, this concept is given with the help of distance functions but this notion is generalized for arbitrary topological spaces by using the concept of open sets in this section. Topological spaces and their continuous maps are the basic topics for the study of the subject "*Topology*." The concept of a homeomorphism, which is a special type of continuity, plays a key role for classification of topological spaces, is studied in Sect. 3.7.

### 3.6.1 Problems Leading to Continuous Functions

To extend the study of continuous functions in $\mathbf{R}^n$ for abstract sets , it is necessary to endow some topology on them making them topological spaces. This subsection discusses three types of problems on continuity of functions in topology, which we encounter frequently.

(i) Given a map $f: X \to Y$ between two topological spaces $X$ and $Y$, how can we examine whether $f$ is continuous or not?
(ii) Given a map $f: X \to Y$ from a topological space $X$ to a nonempty set $Y$, how can we endow $Y$ with a topology so that $f$ is continuous?
(iii) Given a map $f: X \to Y$ from a nonempty set $X$ to a topological space $Y$, how can we endow $X$ with a topology so that $f$ is continuous?

### 3.6.2 Continuous Functions: Introductory Concepts

This subsection studies continuity of maps or functions in topological settings. Although most of the spaces of our interest are metric spaces or can be endowed with metrics, where continuity is defined by metrics (see Chap. 2), but there are some important spaces where there is no concept of a metric. Continuity of functions on such spaces is formulated by using the concepts of open or closed sets in Definition 3.6.1 and in Theorem 3.6.7. The motivating example for the study of continuity is the continuity of a real function $f : \mathbf{R} \to \mathbf{R}$. In analysis, its continuity is studied by using "$\epsilon - \delta$" **method or in terms of limits**. But in topological setting, there is a third definition of continuity of a function in the language of open sets given in Definition 3.6.1 and all these three definitions are equivalent.

**Definition 3.6.1** Let $(X, \tau)$ and $(Y, \sigma)$ be two topological spaces. A function $f : X \to Y$ is said to be **continuous** if $f^{-1}(U) \subset X$ is an open set in $X$ for every open set $U$ in $Y$.

**Example 3.6.2** Let $(X, \tau)$ and $(Y, \sigma)$ be two topological spaces. Then

(i) the constant function $c : X \to Y$, $y \mapsto y_0 \in Y$ is continuous;
(ii) the identity function $1_X : X \to X$, $x \mapsto x$ is continuous; and
(iii) the map $f : (0, 2\pi) \to \mathbf{R}^2$, $x \mapsto (\sin x, \ \sin 2x)$ is continuous under usual topology for this particular choice of $(X, \tau)$ and $(Y, \sigma)$.

**Example 3.6.3** If the condition prescribed in Definition 3.6.1 fails for some function, then the function cannot be continuous. For example, consider any function $f : \mathbf{R} \to \mathbf{R}$ on the real-line space $\mathbf{R}$, which fails to satisfy the condition of Definition 3.6.1. Then there exists an open set $U$ in $\mathbf{R}$ such that $f^{-1}(U)$ in not an open set in $\mathbf{R}$. Then there exists a point $p \in f^{-1}(U)$ such that there is no interval $(a, b) \subset \mathbf{R}$ with the property that

$$p \in (a, b) \subset f^{-1}(U)$$

Then it follows that $f$ is is not continuous at $p$.

**Remark 3.6.4** Definition 3.6.1 gives a partial answer of the problem (i) raised in Sect. 3.6.1. To solve problem (ii) raised in the same subsection, it is necessary that given a topology $\tau = \{U\}$ on $X$, a topology $\sigma$ is to be endowed on $Y$, by declaring a subset $V$ of $Y$ to be open in $Y$, i.e., $V \in \sigma$ iff $f^{-1}(V) = U \in \tau$. Then the collection $\{V\}$ forms the topology $\sigma$ on $Y$. It is largest topology on $Y$ such that the map $f : (X, \tau) \to (Y, \sigma)$ is continuous by Proposition 3.6.5. The topology $\tau_f$ defined in Proposition 3.6.5 solves the problem (ii) posed in Sect. 3.6.1.

**Proposition 3.6.5** *Let $(X, \tau)$ be a topological space and $Y$ be a nonempty arbitrary set. Given a map $f : X \to Y$, the family*

$$\tau_f = \{V \subset Y : f^{-1}(V) \in \tau\}$$

*of subsets of $Y$ forms a topology on $Y$, which is the **strongest (largest) topology on** $Y$ such that the map $f : (X, \tau) \to (Y, \tau_f)$ is continuous.*

**Proof** Clearly, $\tau_f$ forms a topology on $Y$ such that the map

$$f : (X, \tau) \to (Y, \tau_f)$$

is continuous by Definition 3.6.1. To prove that $\tau_f$ is the strongest (largest) topology on $Y$ such that the map $f\colon (X, \tau) \to (Y, \tau_f)$ is continuous, let $\sigma$ be a topology on $Y$ such that the given map

$$f\colon (X, \tau) \to (Y, \sigma)$$

is continuous. Let $U \in \sigma$ be an arbitrary open set in $(Y, \sigma)$. Then by continuity of $f$, it follows that $f^{-1}(U) \in \tau$ and hence $U \in \tau_f$. This implies that $\sigma \subset \tau_f$.

$\square$

**Definition 3.6.6** The topology $\tau_f$ given in Proposition 3.6.5 is called the **topology induced** on $Y$ by the map $f\colon X \to Y$ from the topological space $(X, \tau)$ to a nonempty set $Y$.

The concept of continuity can also be equally well formulated by closed sets in topological settings in Definition 3.6.8 by utilizing Theorem 3.6.7 .

**Theorem 3.6.7** *Let $(X, \tau)$ and $(Y, \sigma)$ be two topological spaces. A map $f\colon X \to Y$ is continuous iff $f^{-1}(A) \subset X$ is a closed set in $X$ for every closed set $A$ in $Y$.*

**Proof** Let $f\colon X \to Y$ be continuous and $A$ be an arbitrary closed set in $Y$. Then $Y - A$ is open set in $Y$. Hence by continuity of $f$, its inverse $f^{-1}(Y - A)$ is an open set in $X$. Since $f^{-1}(Y - A) = X - f^{-1}(A)$, which is an open set, it follows that $f^{-1}(A)$ is a closed set in $X$. Conversely, let $f^{-1}(A)$ be a closed set in $X$ for every closed set $A$ in $Y$ and $V$ be an open set in $Y$, then $Y - U$ is closed in $Y$ and hence $f^{-1}(Y - V) = X - f^{-1}(V)$ is a closed set in $X$. This asserts that $f^{-1}(V)$ is an open set in $X$, and hence, $f$ is continuous.

$\square$

**Definition 3.6.8** Let $(X, \tau)$ and $(Y, \sigma)$ be two topological spaces. A map $f\colon X \to Y$ is said to be continuous if

$$f^{-1}(V) \subset X$$

is a closed set in $X$ for every closed set $V$ in $Y$.

**Remark 3.6.9** To solve problem (iii) raised in Sect. 3.6.1, it is necessary that given a map $f : X \to Y$ and a topology $\sigma = \{V\}$ on $Y$, a topology $\tau$ is to be endowed on $X$, by declaring a subset $U$ of $X$ to be open in $X$, if $U = f^{-1}(V)$ for some open set $V \in \sigma$. Then the collection $\{U\}$ forms a topology $\tau$ on $X$, which is the smallest topology on $X$ such that the map $(X, \tau) \to (Y, \sigma)$ is continuous.

The topology $\sigma_f$ defined in Proposition 3.6.10 solves the problem (iii) posed in Sect. 3.6.1. Its immediate applications are given in Sect. 3.16

**Proposition 3.6.10** *Let $X$ be a nonempty arbitrary set and $(Y, \sigma)$ be a topological space. Given a map $f\colon X \to Y$, the topology*

$$\sigma_f = \{U\} = \{f^{-1}(V), \ \forall V \in \sigma\}$$

*is the **weakest (smallest) topology** on $X$ such that the function $f: (X, \sigma_f) \to (Y, \sigma)$ is continuous.*

**Proof** Proceed as in Proposition 3.6.5 . □

**Definition 3.6.11** The topology $\sigma_f$ given in Proposition 3.6.10 is called the topology on $X$ induced by the function $f: X \to Y$ from the nonempty set $X$ to the topological space $(Y, \sigma)$.

**Remark 3.6.12** The classical "$\epsilon - \delta$" definition of continuity in analysis has no generalization in an arbitrary topological spaces unless there is a concept of a distance function, which a metric space carries. But its equivalent formulations are available in arbitrary topological spaces (see Exercise 12 of Sect. 3.20). Any one of them may be taken as a definition of continuity.

**Theorem 3.6.13** *The composite of two continuous maps is also continuous (in topological settings).*

**Proof** Let $(X, \tau), (Y, \sigma)$ and $(Z, \rho)$ be three topological spaces. If $f: (X, \tau) \to (Y, \sigma)$ and $g: (Y, \sigma) \to (Z, \rho)$ are two continuous maps, then their composite map

$$g \circ f: (X, \tau) \to (Z, \rho), \ x \mapsto g(f(x))$$

is also continuous. Because, given an open set $W$ in $(Z, \rho)$, its inverse image $(g \circ f)^{-1}(W) = f^{-1}(g^{-1}(W))$ is such that $g^{-1}(W))$ is open in $(Y, \sigma)$ by continuity of $g$. Hence, $(g \circ f)^{-1}(W)$ is open in $(X, \tau)$ by continuity of $f$. This asserts that $g \circ f$ is continuous.

□

### 3.6.3 Neighborhoods and Continuity at a Point

The intuitive idea of smallness of a nbd of a point in a topological space is reflected though the concept of local base at the point, which plays a convenient role in the study of continuity of functions (maps). A nbd is not necessarily an open set and the idea of smallness of a nbd dictates that the entire space need not be taken as a nbd of any of its points.

**Definition 3.6.14** Let $(X, \tau)$ and $(Y, \sigma)$ be two topological spaces. Then a function $f: X \to Y$ is said to be continuous at a point $x \in X$, if given any nbd $N_{f(x)}$ in $Y$, there is a nbd $N_x$ of $x$ in $X$ such that

$$f(N_x) \subset N_{f(x)}.$$

**Remark 3.6.15** As $f(f^{-1}(N_{f(x)})) \subset N_{f(x)}$, to prove the continuity of function $f: X \to Y$ between topological spaces at a point $x \in X$, it is sufficient to show that given any nbd $N_{f(x)}$ in $Y$ belonging to some local base at $f(x)$, there is a nbd $f^{-1}(N_{f(x)})$ of

$x$ in $X$. The equivalence of the two definitions of continuity of functions by open sets given in Definition 3.6.1 and nbds given in Definition 3.6.14 is established in Theorem 3.6.16.

**Theorem 3.6.16** *Let $(X, \tau)$ and $(Y, \sigma)$ be two topological spaces. Then a function $f : X \to Y$ is continuous iff it is continuous at every point $x \in X$.*

**Proof** Let $f : X \to Y$ be continuous by Definition 3.6.1. We claim that $f$ is continuous at every point $x \in X$. By continuity of $f$, it follows that $f^{-1}(V)$ is an open set in $X$ for every open set $V$ in $Y$. Let $N_{f(x)}$ be an arbitrary nbd of $f(x)$ in $Y$. Then by definition of nbd, there is an open set $V$ in $Y$ such that

$$f(x) \in V \subset N_{f(x)}.$$

This asserts that $x \in f^{-1}(V) \subset f^{-1}(N_{f(x)})$. Since $f^{-1}(V)$ is an open set in $X$, it follows that $f^{-1}(N_{f(x)})$ is a nbd of $x$ in $X$. This asserts by Definition 3.6.14 and Remark 3.6.15 that $f$ is continuous at the point $x$. Conversely, let $f : X \to Y$ be continuous at every point $x \in X$ and $V$ be an arbitrary open set in $Y$. Then for every point $x \in f^{-1}(V) \subset X$, the subset $f^{-1}(V)$ is a nbd of $x$ in $X$. This asserts that there is an open set $U_x$ in $X$ such that

$$x \in U_x \subset f^{-1}(V).$$

This shows that

$$\bigcup_{x \in f^{-1}(V)} \{x\} \subset \bigcup_{x \in X} U_x \subset f^{-1}(V)$$

and hence $f^{-1}(V) = \bigcup_{x \in X} U_x$ is an open set in $X$ implies that $f^{-1}(V)$ is an open set in $X$. Since $V$ is an arbitrary open set in $Y$, it follows that $f$ is continuous by Definition 3.6.1. $\qquad\square$

### 3.6.4 Pasting or Gluing Lemma

This section studies Pasting or Gluing Lemma, which is an important result used throughout the present book series in proving continuity of a certain class of maps. For example, this lemma is used to define product path of paths (see Definition 3.6.22).

**Lemma 3.6.17** (Pasting or Gluing Lemma) *Let $X$ be a topological space and $A$, $B$ be closed subsets in $X$ such that $X = A \cup B$. Given a topological space $Y$, if $f_1 : A \to Y$ and $f_2 : B \to Y$ are continuous maps such that $f_1(x) = f_2(x)$, $\forall\ x \in A \cap B$, then the map*

$$f : X \to Y, x \mapsto \begin{cases} f_1(x), \text{if } x \in A \\ f_2(x), \text{if } x \in B \end{cases}$$

*is continuous.*

**Proof** The map $f: X \to Y$ defined in this lemma is the well-defined unique map such that $f|_A = f_1$ and $f|_B = f_2$. We show that $f$ is continuous. Let $K$ be a closed set in $Y$. Then $f^{-1}(K) = (A \cup B) \cap f^{-1}(K) = (A \cap f^{-1}(K)) \cup (B \cap f^{-1}(K)) = f_1^{-1}(K) \cup f_2^{-1}(K)$. Since each of $f_1$ and $f_2$ is continuous, $f_1^{-1}(K)$ and $f_2^{-1}(K)$ are both closed in $X$. This implies that $f^{-1}(K)$ being the union of two closed set is closed in $X$. This asserts that $f$ is continuous. $\qquad\qquad\square$

**Remark 3.6.18** **Geometrically**, the map $f$ defined in Lemma 3.6.17 is obtained by "gluing" $f_1$ and $f_2$ together along their common domain $A \cap B$. For its immediate application see Sect. 3.6.5.

A generalization of the Pasting Lemma 3.6.17 is given in Proposition 3.6.19.

**Lemma 3.6.19** (Generalized Pasting or Gluing Lemma) *Let $X$ be topological space such that it is a finite union of closed subsets $X_i$, i.e., $X = \bigcup_{i=1}^{n} X_i$. For a given topological space $Y$, if there are continuous maps $f_i: X_i \to Y$ such that $f_i|_{X_i \cap X_j} = f_j|_{X_i \cap X_j}$, $\forall\ i, j$ (i.e., they agree on their common domain), then $\exists$ a unique continuous map $f: X \to Y$ with the property that $f|_{X_i} = f_i$, $\forall\ i$.*

**Proof** Proceed as in Lemma 3.6.17. $\qquad\qquad\square$

### 3.6.5 Path in a Topological Space

This subsection continues the study of continuous functions through a study of paths in an arbitrary topological space. For more study of paths and loops, **Basic Topology, Volume 3** of the present book series is referred.

**Definition 3.6.20** (*Path*) Let $\mathbf{I} = [0, 1]$ be the unit closed interval with subspace topology inherited from the usual topology on $\mathbf{R}$. Given a topological space $(X, \tau)$, a continuous map $\alpha : \mathbf{I} \to X$ is said to be a path in $X$. If $\alpha(0) = x_0$ and $\alpha(1) = x_1$, then it is said to be a path in $X$ from the point $x_0$ to the point $x_1$. The points $x_0, x_1 \in X$ are called the initial point and the terminal point of the path $\alpha$, respectively. It is said to be a **loop in $X$ based at the point** $x_0 \in X$ if $\alpha(0) = x_0 = \alpha(1)$.

**Example 3.6.21** Given two points $x_0, x_1 \in \mathbf{R}^2$, the map $\alpha: \mathbf{I} \to \mathbf{R}^2, t \mapsto (1-t)x_0 + tx_1$ is a path in $\mathbf{R}^2$ from $x_0$ to $x_1$. On the other hand, the map $\beta: \mathbf{I} \to \mathbf{R}^2, t \mapsto tx_0 + (1-t)x_1$ is a path in $\mathbf{R}^2$ from $x_1$ to $x_0$, called the **reverse path** of $\alpha$ in $\mathbf{R}^2$.

**Definition 3.6.22** (*Product-path*) Let $f_1$ and $f_2$ be two paths in a topological space $(X, \tau)$ such that the terminal point of $f_1$ is the same as the initial point of the path $f_2$, i.e., $f_1(1) = f_2(0)$. Then their product path $f_1 * f_2: \mathbf{I} \to X$ is defined by

$$(f_1 * f_2)(t) = \begin{cases} f_1(2t), & 0 \le t \le 1/2 \\ f_2(2t - 1), & 1/2 \le t \le 1 \end{cases} \tag{3.1}$$

The map $f_1 * f_2 \colon \mathbf{I} \to X$ is continuous, since it is well-defined by the defining condition that $f_1(1) = f_2(0)$ and is continuous by pasting Lemma 3.6.17. It is called the product of the paths $f_1$ and $f_2$ or their product path. Here, take $A = [0, \frac{1}{2}]$ and $B = [\frac{1}{2}, 1]$. Then $X = \mathbf{I} = A \cup B$ and $f_1, f_2$ agree with $A \cap B = \{t = \frac{1}{2}\}$. **Product of two loops** in $X$ at a point $x_0 \in X$ is defined in an analogous way.

### 3.6.6 Open and Closed Maps

This subsection conveys the concepts of open and closed maps for topological spaces, which are used throughout the book. For example, they are used in Proposition 3.7.4 and in Proposition 3.16.8 to ascertain a homeomorphism or an identification map. An open map is characterized by interior operator and a closed map is characterized by closure operator in Theorem 3.6.31.

**Definition 3.6.23** Let $(X, \tau)$ and $(Y, \sigma)$ be topological spaces. A map $f \colon (X, \tau) \to (Y, \sigma)$ is said to be

(i) **open** if $f$ sends every open set $U \in \tau$ to the open set $f(U) \in \sigma$;
(ii) **closed** if $f$ sends every closed set $K$ in $(X, \tau)$ to the closed set $f(K)$ in $(Y, \sigma)$.

*Remark 3.6.24* The concepts of continuity, closeness and openness of a map between topological spaces are independent of each other. In support see Examples 3.6.25–3.6.30.

*Example 3.6.25* A continuous map may be neither open nor closed. For example,

(i) let $(\mathbf{R}, \tau)$ be the space $\mathbf{R}$ with the discrete topology $\tau$ and $(\mathbf{R}, \sigma)$ be the real line space with usual topology $\sigma$. Then the identity map $f \colon (\mathbf{R}, \tau) \to (\mathbf{R}, \sigma)$, $x \mapsto x$ is continuous. But it is nether open nor nor closed. Because its inverse map $f^{-1} \colon (\mathbf{R}, \sigma) \to (\mathbf{R}, \tau)$, $x \mapsto x$ is not continuous, and hence, it is neither open nor nor closed by Proposition 3.7.4.
(ii) in the real line space $\mathbf{R}$, the map the map $f \colon \mathbf{R} \to \mathbf{R}$, $x \mapsto e^x \cos x$ is continuous, but $f$ is not open, because $f(-\infty, 0)$ is not open in $\mathbf{R}$. It is not closed, because, for $S = \{-t\pi : t \in \mathbf{R}\}$, its image $f(S) = \{e^x \cos x : x = -t\pi, t \in \mathbf{R}\}$ is not closed in $\mathbf{R}$.

*Example 3.6.26* An open and closed map may not be continuous. For example,

(i) let $(\mathbf{R}, \tau)$ be the space $\mathbf{R}$ with the discrete topology $\tau$ and $(\mathbf{R}, \sigma)$ be the real line space with usual topology $\sigma$. Then the identity map $f \colon (\mathbf{R}, \sigma) \to (\mathbf{R}, \tau)$, $x \mapsto x$ is not continuous. But it is both open and closed by Proposition 3.7.4, since $f^{-1}$ is continuous.

(ii) for the unit circle $S^1 = \{(x, y) \in \mathbf{R}^2 : x^2 + y^2 = 1\}$ with subspace topology inherited from the usual topology on $\mathbf{R}^2$ and the space $A = [0, 2\pi)$ with the topology inherited from the usual topology on $\mathbf{R}$, the map

$$f : S^1 \to A, \ (x, y) \mapsto \alpha, \text{ where } x = \cos \alpha, y = \sin \alpha, \text{ and } \alpha \in A$$

is both open and closed but it is not continuous at the point $(1, 0) \in S^1$.

***Example 3.6.27*** An open map may be neither continuous nor closed. For example, let $\tau$ be the topology on $\mathbf{R}^2$ consisting of the emptyset $\emptyset$ and the complements of countable sets in $\mathbf{R}^2$ and $\sigma$ be the topology on $\mathbf{R}$ consisting of the emptyset $\emptyset$ and the complements of finite sets in $\mathbf{R}$. Then the projection map $p_1 : (\mathbf{R}^2, \tau) \to (\mathbf{R}, \sigma), (x, y) \to x$ is open but it is neither closed nor continuous.

***Example 3.6.28*** A continuous and open map may not be closed. For example, let $(\mathbf{R}, \sigma)$ be real line space with usual topology $\sigma$ and $(\mathbf{R}^2, \sigma_2)$ be the Euclidean plane space with Euclidean topology $\sigma_2$. Then the projection lap $p_1 : \mathbf{R}^2 \to \mathbf{R}, (x, y) \to x$ is continuous and open but it is not closed.

***Example 3.6.29*** A continuous and closed map may not be open. For example, let $X = [0, 2] = Y$ be endowed with the subspace topology inherited from the usual topology on $\mathbf{R}$. Then the map

$$f : X \to Y, \ x \mapsto \begin{cases} 0, & \text{if } 0 \leq x \leq 1 \\ x - 1, & \text{if } 1 < x \leq 2 \end{cases}$$

is closed and continuous. Moreover, $X$ and $Y$ are compact metric spaces. But the map $f$ is not open, because, $f(0, 1)$, the image of the open set $(0, 1)$ in $X$ under $f$ is not open in $Y$.

***Example 3.6.30*** A closed map may be neither continuous nor open. For example, for the unit circle $S^1 = \{(x, y) \in \mathbf{R}^2 : x^2 + y^2 = 1\}$ with subspace topology inherited from the usual topology on $\mathbf{R}^2$ and the space $A = [0, 2\pi)$ with the topology inherited from the usual topology on $\mathbf{R}$. Then the map

$$f : S^1 \to A, \ (\cos \alpha, \sin \alpha) \mapsto \begin{cases} 0, & \text{if } \alpha \in [0, \pi] \\ \alpha - \pi, & \text{if } \alpha \in (0, 2\pi) \end{cases}$$

is closed but it is neither open nor continuous.

Theorem 3.6.31 characterizes open maps by interior operator and closed maps by closure operator.

**Theorem 3.6.31** *Let* $(X, \tau)$, $(Y, \sigma)$ *be two topological spaces and* $A \subset X$ *be an arbitrary subset. Then a map* $f : (X, \tau) \to (Y, \sigma)$ *is*

(i) *open iff*
$$f(Int(A)) \subset Int(f(A)) \text{ for all subsets } A \text{ of } X.$$

*(ii) closed iff*

$$\overline{f(A)} \subset f(\overline{A}) \text{for all subsets } A \text{ of } X.$$

**Proof** (i) Let $f$ satisfy the condition: $f(\text{Int}(A)) \subset \text{Int}(f(A))$, $\forall$ subsets $A \subset X$ and $U$ be an open set in $(X, \tau)$. Then

$$f(U) = f(\text{Int}(U)) \subset \text{Int}(f(U)) \text{by hypothesis} \implies f(U) = \text{Int}(f(U)).$$

This shows that $f(U)$ is an open set in $(Y, \sigma)$. It asserts that $f$ is an open map. Conversely, let $f$ be an open map and $A$ be an arbitrary subset of $X$. Since $\text{Int}(A)$ is an open set in $(X, \tau)$, $\text{Int}(A) \subset A$ and $f$ is open by hypothesis, it follows that

$$f(\text{Int}(A)) = \text{Int}(f(\text{Int}(f(A)))) \subset \text{Int}(f(A)), \text{ because}, f(\text{Int}(A)) \subset f(A).$$

(ii) Proceed as in (i).

□

### 3.6.7 Lebesgue Sets of a Continuous Function

This subsection defines Lebesque sets of a continuous function $f: \mathbf{R}^n \to \mathbf{R}$ and their basic properties. For more study on Lebesque sets of a continuous function, see Exercise 9 of Sect. 3.20 and for its generalization see Chap. 6.

**Definition 3.6.32** Let $f: \mathbf{R}^n \to \mathbf{R}$ be a continuous function and $r$ be any real number. Then the sets

$$X_r = \{x \in \mathbf{R}^n : f(x) < r\} \subset \mathbf{R}^n;$$

$$Y_r = \{x \in \mathbf{R}^n : f(x) \leq r\} \subset \mathbf{R}^n;$$

and

$$Z_r = \{x \in \mathbf{R}^n : f(x) = r\} \subset \mathbf{R}^n$$

are called **Lebesque sets** of the function $f$ corresponding to $r$.

**Proposition 3.6.33** *Let* $f: \mathbf{R}^n \to \mathbf{R}$ *be a continuous function and* $r$ *be any real number. Then the Lebesque sets have the following properties:*

  *(i)* $X_r$ *is open;*
  *(ii)* $Y_r$ *is closed;*
 *(iii)* $Z_r$ *is closed;*
 *(iv)* $\overline{X}_r \subset Y_r$.

**Proof** Left as an exercise.                                                  $\square$

## 3.7  Homeomorphism, Topological Embedding, Topological Property and Topological Invariant

This section continues the study of Sect. 3.6 and discusses the problems on homeomorphism, topological embedding, topological properties and topological invariants. The concept of homeomorphism is a basic tool used to determine the equivalence of topological spaces. One of the main problems of topology is the classification of topological spaces up to homeomorphism, like classification of groups in algebra up to isomorphism. For such classification of topological spaces, the concept of a homeomorphism is essential. A homeomorphism in topology is a bijective map that preserves topological structures involved and is analogous to the concept of an isomorphism between algebraic objects such as groups or rings, which is also a bijective map that preserves the algebraic structures involved.

### 3.7.1  Problems Leading to Homeomorphism

The main aim of topology is to ascertain whether two given topological spaces are identical or different from the topological viewpoint. So, it needs define equivalence of topological spaces like congruence of figures in school geometry. Two figures in topology are said to be equivalent if one figure is obtained from the other by a continuous deformation (see **Basic Topology, Volume 3**) of the present series of books, which presents the qualitative properties of geometric figures .To understand this meaning, we consider a sphere to be a rubber balloon. It can be stretched and shrunk without torning it or gluing any two distinct points together in any manner. Each of such transformations is called a homeomorphism, and the geometric objects obtained by homeomorphism are called homeomorphic to each other. The qualitative properties of geometric figures are conventionally called **topological properties** (see Sect. 3.7.4). On the other hand, characteristics shared by homeomorphic spaces are called **topological invariants** (see Sect. 3.7.5).

## 3.7.2 Homeomorphism

This subsection starts with a formal definition of a homeomorphism and illustrates this concept with several examples. The subject topology is also called a **qualitative geometry** in the sense that if one geometric object is continuously deformed into another, then the two objects are said to topologically same and each such transformations is called a **homeomorphism** and the geometric objects obtained by a homeomorphism are called homeomorphic to each other. For example, the qualitative property distinguishes the circle from the figure-eight curve (formed by two circles touching at a single point), because the number of connected pieces are different after deletion of any single point from the circle and deletion of the point of contact from the figure-eight curve, and hence, the circle and figure-eight curve can not be homeomorphic. Definition 3.7.1 formulates the concept of a homeomorphism.

**Definition 3.7.1** Let $(X, \tau)$ and $(Y, \sigma)$ be two topological spaces. Then a continuous map $f : X \to Y$ is said to be a **homeomorphism** if $f$ is bijective and $f^{-1} : Y \to X$ is also continuous. If $f$ is a homeomorphism, then $X$ and $Y$ are said to be **homeomorphic spaces abbreviated** $X \approx Y$. Its equivalent definitions are available in Proposition 3.7.4 and in Exercise 13 of Sect. 3.20.

**Example 3.7.2** (i) In $\mathbf{R}^2$ with usual topology, a circle and a square are homeomorphic.

(ii) The real line space $\mathbf{R}$ and the open interval $(0, 1)$ with subspace topology inherited from the usual topology on $\mathbf{R}$ are homeomorphic.

(iii) The open ball $B = \{x = (x_1, x_2) \in \mathbf{R}^2 : \|x\| < 1\}$ is homeomorphic to the whole plane $\mathbf{R}^2$.

(iv) For more examples, see Example 3.7.6.

**Example 3.7.3** The continuity of a bijective map can not guarantee the continuity of its inverse. To solve such a problem, a search is made to establish a necessary and sufficient condition under which a continuous bijective map will have its continuous inverse. More precisely, let $X$ and $Y$ be topological spaces. The inverse map $f^{-1}$ of a bijective map $f : X \to Y$ always exists but it may not be a homeomorphism. For example,

(i) let $\mathbf{R}$ be the set of real numbers with usual topology $\sigma$ and $\mathbf{R}$ be the same set with discrete topology $\tau$ or the lower-limit topology $\sigma_l$. Then the identity map $f : (\mathbf{R}, \sigma) \to (\mathbf{R}, \sigma_l)$, $x \mapsto x$ is a bijection but it is not a homeomorphism;

(ii) let $(\mathbf{R}, \sigma)$ be the topological space $\mathbf{R}$ endowed with usual topology $\tau$ and $(\mathbf{R}, \tau)$ be the topological space endowed with discrete topology $\tau$, then the identity map $f : (\mathbf{R}, \tau) \to (\mathbf{R}, \sigma)$, $x \mapsto x$ is a bijection and continuous but its inverse is not continuous.

Proposition 3.7.4 gives a necessary and sufficient condition under which a continuous bijective map has its continuous inverse, which solves the problem appearing in Example 3.7.3.

**Proposition 3.7.4** *Let* $(X, \tau)$ *and* $(Y, \sigma)$ *be two topological spaces. If* $f: (X, \tau) \to$ $(Y, \sigma)$ *is a continuous bijective map, then the following statements are equivalent:*

(i) $f^{-1}$ *is continuous;*

(ii) $f: X \to Y$ *is open;*

(iii) $f: X \to Y$ *is closed.*

**Proof** It follows from the definition of continuity of a map, definitions of open and closed maps. $\square$

**Corollary 3.7.5** *Let* $(X, \tau)$ *and* $(Y, \sigma)$ *be two topological spaces. A continuous bijective map* $f: (X, \tau) \to (Y, \sigma)$ *is a homeomorphism if it satisfies any one of the equivalent conditions prescribed in Proposition 3.7.4.*

**Example 3.7.6** (i) The map $f: (0, 2\pi) \to \mathbf{R}^2$, $x \mapsto (\sin x, \sin 2x)$ is continuous and bijective but $f^{-1}$ is not continuous.

(ii) A homeomorphism $f: (0, 1) \to \mathbf{R}$ cannot be extended over $\mathbf{I} = [0, 1]$.

(iii) The open square $A = \{(x, y) \in \mathbf{R}^2 : 0 < \langle x, y \rangle < 1\}$ is homeomorphic to the open ball $B = \{x = (x_1, x_2) \in \mathbf{R}^2 : \|x\| < 1\}$.

(iv) The cone $A = \{(x, y, z) \in \mathbf{R}^3 : x^2 + y^2 = z^2, z > 0\}$ is homeomorphic to the plane $\mathbf{R}^2$.

(v) Let $S^n$ be the n-sphere defined by $S^n = \{x \in \mathbf{R}^{n+1} : \|x\| = 1, n \geq 1\}$, $N = (0, 0, \ldots, 1) \in \mathbf{R}^{n+1}$ be the north pole of $S^n$ and $S = (0, 0, \ldots, -1) \in \mathbf{R}^{n+1}$ be the south pole of $S^n$. Then

(a) $S^n - S$ is homeomorphic to $S^n - N$;

(b) The standard homeomorphism

$$f: S^n - N \to \mathbf{R}^n$$

is called the stereographic projection. Its precise expression is the homeomorphism

$$f: S^n - N \to \mathbf{R}^n, x \mapsto \frac{1}{1 - x_{n+1}}(x_1, x_2, \ldots, x_n),$$
$$\forall x = (x_1, x_2, \ldots, x_{n+1}) \in S^n - N.$$

For every point $P \equiv (x_1, x_2, , x_{n+1}) \in S^n - N$, its image point $f(P)$ is the unique point $P' \equiv \frac{1}{1-x_{n+1}}(x_1, x_2, \ldots, x_n) \in \mathbf{R}^n$. Clearly, $f(x) = x$ iff $x \in S^{n-1}$ (equator). **The stereographic projection $f$ is geometrically represented in** Fig. 3.1

(vi) A circle minus a point is homeomorphic to a line segment, and a closed arc is homeomorphic to a closed line segment.

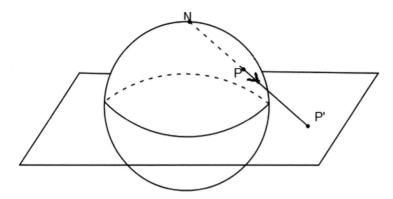

**Fig. 3.1** Stereographic projection from north point

*Example 3.7.7* **Analogue of Schroeder–Bernstein theorem does not exist in Topology.** For example, consider the subspaces $X = (0, 1)$ and $Y = [0, 1]$ in the real line space $(\mathbf{R}, \sigma)$ with usual topology $\sigma$. Then the subspaces $A = [1/4, 3/4]$ and $B = (1/2, 2/3)$ are such that $X$ is homeomorphic to $B$ and $Y$ is homeomorphic to $A$. But $X$ and $Y$ are not homeomorphic (see Chap. 5). This shows that an analogue of Schroeder–Bernstein theorem (see Chap. 1) for bijections does not hold for homeomorphisms.

*Remark 3.7.8* Just after the concept of homeomorphisms is clearly defined, the subject of topology begins to study those properties of geometric figures which are preserved by homeomorphisms with an eye to classify topological spaces up to homeomorphism, which stands the ultimate problem in topology, where a geometric figure is considered to be a point set in the Euclidean space $\mathbf{R}^n$. But this undertaking becomes hopeless, when there exists no homeomorphism between two given topological spaces.

(i) The concept of **topological property** (see Definition 3.7.15) such as compactness and connectedness introduced in general topology, solves this problem in a very few cases which is studied in **Basic Topology, Volume 1**. A study of the subspaces of the Euclidean plane $\mathbf{R}^2$ gives an obvious example.

(ii) On the other hand, the subject **Algebraic Topology** (studied in Basic Topology, Volumes III) was born to solve the problems of impossibility in many cases with a shift of the problem by associating **invariant objects** in the sense that homeomorphic spaces have the same object (up to equivalence). Initially, these objects were integers and subsequent research reveals that more fruitful and interesting results can be obtained from the algebraic invariant structures such as groups and rings. For example, homology and homotopy groups are very important algebraic invariants which provide strong tools to study the structure of topological spaces.

### 3.7.3   Embedding of Topological Spaces

This subsection conveys the concept of an embedding of topological spaces and illustrates this concept by examples. An embedding is a homeomorphism from a topological space onto its image space and is a very important concept in topology.

**Definition 3.7.9** A topological space $(X, \tau)$ is said to be **embedded in a topological space** $(Y, \sigma)$ if there exists a homeomorphism from $X$ onto a subspace of $Y$. An injective continuous map $f : X \to Y$ between topological spaces is said to be an **embedding**, if

$$f : X \to f(X) \subset Y$$

is a homeomorphism, i.e., if $f$ is a homeomorphism from a topological space onto its image space. By an embedding $f : X \to Y$, the space $X$ may be considered as a subspace of $Y$.

**Example 3.7.10** Let $X$ and $Y$ be topological space and $y_0 \in Y$ be a fixed point. Then the map $f : X \to X \times Y, x \mapsto (x, y_0)$ is an embedding.

**Example 3.7.11** Let $(\mathbf{R}, \sigma)$ be the real line space with usual topology $\sigma$. Since it is homeomorphic to any open interval $(a, b)$ with $a < b$ in $\mathbf{R}$, it follows that the topological space $(a, b)$ with relative topology induced on $(a, b)$ by $\sigma$ is embedded in $(\mathbf{R}, \sigma)$. The real line space $(\mathbf{R}, \sigma)$ is also embedded to its every open interval $(a, b)$ with relative topology induced by $\sigma$ on $(a, b)$.

**Definition 3.7.12** An embedding $f : S^1 \to \mathbf{R}^2$ of the circle $S^1$ in $\mathbf{R}^2$ is said to be a **Jordan curve**.

**Definition 3.7.13** An embedding $f : S^1 \to \mathbf{R}^3$ of the circle $S^1$ in $\mathbf{R}^3$ is said to be a **knot**.

**Remark 3.7.14** Jordan curve and knot are studied in **Basic Topology: Volume 3** of the present book series.

### 3.7.4   Topological Property

This subsection is a continuation of the study of homeomorphism and conveys the concept of topological property which is a property shared by homeomorphic spaces, i.e., a property which is preserved by every homeomorphism. To make it precise, let $(X, \tau)$ and $(Y, \sigma)$ be two topological spaces. For a homeomorphism $f : X \to Y$, both $f$ and $f^{-1}$ are continuous means that $f$ sends not only the points of $X$ to the points of $Y$ in a $(1{-}1)$ manner, but $f$ also sends the open sets of $X$ to the open sets of $Y$ in a $(1{-}1)$ manner. This asserts that two homeomorphic spaces differ only in the nature of their points but they are same from the topological viewpoint in the sense that if $X$ and $Y$

are homeomorphic spaces, then a topological property enjoyed by $X$ is also enjoyed by $Y$ and conversely. This implies that if any property of $X$ is expressed completely by the open sets of $X$, then $Y$ has also the corresponding property. This correspondence leads to the concept of a topological property in Definition 3.7.15. Thus, if $X \approx Y$, then any topological property enjoyed by $X$ is also enjoyed by $Y$ and conversely. For example, Proposition 3.8.26 asserts that metrizability is a topological property. For other powerful topological properties, consider a homeomorphism $f : X \to Y$. Then $X$ is compact (or connected) iff $Y$ is compact (or connected) (see Chap. 2 for metric spaces and Chap. 5 for topological spaces), which shows that compactness and connectedness are both topological properties. The concept of topological property is very useful to solve **classification problems** of topological spaces. **Historically**, C. F. Klein (1849–1925 defined in his Analysis Situs in 1872, now called topology as the geometry whose basic aim is to study topological properties. Topology presents the qualitative properties of geometric figures. The qualitative properties of geometric figures are conventionally called topological properties.

**Definition 3.7.15** A **topological property** of a topological space $X$ is a property of $X$ which when possessed by $X$ is also possessed by every topological space homeomorphic to $X$, i.e., a property of a topological space $X$ which is preserved by any homeomorphic image of $X$ is said to a topological property of $X$.

*Example 3.7.16* The properties of a set of being open or closed in a topological space are topological properties.

*Example 3.7.17* Metrizability of a topological space is a topological property (see Proposition 3.8.26)

*Example 3.7.18* Completeness of a metric space is not a topological property1. For example, consider the following examples.

(i) Consider **R** with usual metric and its open interval $(-1, 1)$. The sequence $\{x_n : x_n = 1 - 1/n\}$ is a Cauchy sequence in $(-1, 1)$ but it does not converge in $(-1, 1)$, which shows that open interval $(-1, 1)$ is not complete. On the other hand, **R** is complete and it is homeomorphic to $(-1, 1)$.

(ii) Consider the real line space **R** and its subspace $(a, b)$ with the usual topology. Both **R** and $(a, b)$ are metric spaces with usual metrics, and they are also homeomorphic. The space **R** is complete but $(a, b)$ is not so (see Chap. 2).

(iii) Let $X = (-\infty, +\infty)$ and $Y = (0, 1)$ with subspace topology induced by usual topology on **R**. The spaces $X$ and $Y$ are homeomorphic. $X$ is complete but $Y$ is not so.

(iv) Consider the metric spaces $X = (-\pi/2, \pi/2) \subset$ **R** and $Y = $ **R** with usual metric. Then the map $f : X \to Y, x \mapsto \tan x$ is a homeomorphism from the topological spaces $X$ to the topological space $Y$ having the topology induced by the their usual metric. As a metric space $Y$ is complete, but $X$ is not so.

*Example 3.7.19* Connectedness, compactness, regularity, normality and paracompactness (discussed in Chaps. 4 and 5) are topological properties. On the other hand, in Euclidean geometry, distance as well as angles are not topological properties, since they can be changed by suitable continuous deformations.

**Remark 3.7.20** **A common problem in topology** is to decide whether two topological spaces are homeomorphic or not. To prove that two spaces are not homeomorphic, it is sufficient to find a topological property which is not shared by them. This book studies several topological properties with their various applications. They include first countability, second countability, separability, connectedness, regularity, normality, compactness and paracompactness which are important topological properties.

## 3.7.5  Topological Invariant with a Historical Note

This subsection continues the study of homomorphism through the concept of topological invariant, which plays a key role in classification problem of topological spaces by converting a topological problem into an algebraic one for better chance for solution. Characteristic which is shared by homeomorphic spaces is called a topological invariant. The problem of attempting to show that given two topological spaces are not homeomorphic is of separate nature. It is not possible to examine every function between two spaces. The concept of topological invariants solves many such problems. For example, the **Euler characteristic** invented by L. Euler (1703–1783) in 1752 is an integral invariant, which distinguishes nonhomeomorphic spaces. The search of other invariants has established connections between topology and modern algebra in such a way that homeomorphic spaces have isomorphic algebraic structures (see **Basic Topology, Volume 3**) of the present series of books. **Historically**, the concept of fundamental group and homology groups invented by Henri Poincaré (1854–1912) in 1895 are the first powerful topological invariants in homotopy and homology theories which came from such a search. His work explained the difference between curves deformable to one another and curves bounding a larger space. The first one led to the concepts of homotopy and fundamental group. The fundamental group is one of the basic homotopy invariants. It is a very important invariant in algebraic topology and is the first of a series of algebraic invariants $\pi_n$ associated with a topological space with a base point. On the other hand, the second idea of Poincaré led to the concept of homology theory. Both the homotopy and homology theories are studied in **Basic Topology, Volume 3** of the present series of books.

**Definition 3.7.21** Characteristics of topological spaces which are shared by homeomorphic spaces are called **topological invariants** in the sense that a characteristic of a topological space is an invariant which is preserved by a homeomorphism.

**Remark 3.7.22** Several topological invariants such as **Euler characteristics, fundamental group, higher homotopy groups and homology groups, which are basic topological invariants** are studied in Basic Topology, Volume 3 of the present series of books.

# 3.8 Metric Topology and Metrizability of Topological Spaces

This section conveys the concepts of metric topology and metrizability of topological spaces born through a specified topology defined on metric spaces by using distance functions. Metrizable spaces provide a special class of topological spaces and carry more structure than most of the topological spaces. A metrizable space is a certain topological space such that it admits a metric which generates the original topology. Theorem 3.8.23 characterizes metrizability of a topological space in terms of a continuous function. Moreover, Urysohn metrizable theorem gives a sufficient condition of metrizability of a topological space (see Chap. 7).

## 3.8.1 Metric Topology

This subsection continues a study of metric spaces and relates this study with a special class of topological spaces by introducing a topology $\tau_d$ generated by the metric $d$ on the metric space $(X, d)$. This topology plays a key role in the study modern analysis. However, most of the applications of topology and analysis arise through metric spaces. Moreover, metric spaces provide a rich supply of topological spaces and their continuous functions. For example, the Euclidean topology on $\mathbf{R}^n$ defined in Sect. 3.3 plays a key role in analysis and topology, which can also be equally well-defined by the Euclidean metric on $\mathbf{R}^n$.

**Definition 3.8.1** Let $X$ be a metric space with metric $d$. The set $B_x(\epsilon) = \{y \in X : d(x, y) < \epsilon\}$, for $x \in X$ and $\epsilon > 0$ is said be the open ball of radius $\epsilon$, centered at $x$.

**Proposition 3.8.2** *Let $X$ be a metric space with metric $d$. A subset $U$ of $X$ is open if given $x \in U$, there is a positive real number $\epsilon$ such that the open ball $B_x(\epsilon) \subset U$.*

**Proof** Left as an exercise. □

Proposition 3.8.2 leads to define an open set in a metric space as given in Definition 3.8.3.

**Definition 3.8.3** Let $X$ be a metric space with metric $d$. Then a subset $U$ of $X$ is said to be open if given $x \in U$, there is a positive real number $\epsilon$ such that the open ball $B_x(\epsilon) \subset U$.

**Theorem 3.8.4** *Let $(X, d)$ be a metric space. Then the collection $\mathcal{B} = \{B_x(\epsilon) : x \in X, \epsilon > 0\}$ of open balls in $X$ together with $\emptyset$ form an open base for a topology of $X$.*

**Proof** Let $x \in X$ be an arbitrary point. Since $x \in B_x(\epsilon)$, it follows that every point of $X$ is contained in at least one member of $\mathcal{B}$. Claim that if $y \in B_x(\epsilon)$, there exists an open ball $B_y(\delta) \in \mathcal{B}$ such that $B_y(\delta) \subset B_x(\epsilon)$. To prove this, take $\delta = \epsilon - d(x, y) > 0$. Then

$$z \in B_y(\delta) \implies d(y, z) < \delta = \epsilon - d(x, y) \implies d(x, z) \leq d(x, y) + d(y, z) < \epsilon$$

This asserts that $B_y(\delta) \subset B_x(\epsilon)$. Finally, let $B_1^d$ and $B_2^d$ be any two members of $\mathcal{B}$ and $y \in B_1^d \cap B_2^d$. By above discussion, there exist $\delta_1 > 0$ and $\delta_2 > 0$ such that

$$B_y(\delta_1) \subset B_1^d \quad \text{and} \quad B_y(\delta_2) \subset B_2^d$$

Take $\delta = \min\{\delta_1, \delta_2\}$. Then $\delta > 0$ and $B_y(\delta) \subset B_1^d \cap B_2^d$. This proves that there exists a member $B_3^d = B_y(\delta)$ of $\mathcal{B}$ such that

$$y \in B_3^d \subset B_1^d \cap B_2^d.$$

This concludes by Theorem 3.2.11 that $\mathcal{B}$ forms an open base for a topology on $X$. $\square$

A metric space $X$ with a metric $d$ can be made into a topological space in a natural way by using Theorem 3.8.4.

**Definition 3.8.5** (*Metric Topology*) Let $(X, d)$ be a metric space. Then the collection $\mathcal{B} = \{B_x(\epsilon) : x \in X, \ \epsilon > 0\}$ of open balls in $X$ together with $\emptyset$ form an open base for a topology of $X$ by Theorem 3.8.4. This topology denoted by $\tau_d$ or $\tau(d)$ on $X$, called the metric topology on the metric space $(X, d)$ induced by the metric $d$.

**Example 3.8.6** Let $\mathbf{R}$ be endowed with the usual metric $d \colon \mathbf{R} \times \mathbf{R} \to \mathbf{R}$, $(x, y) \mapsto |x - y|$. Then the open ball $B_x(\epsilon)$ in $\mathbf{R}$ is formulated by

$$B_x(\epsilon) = (x - \epsilon, \ x + \epsilon) \subset \mathbf{R} \ (\text{open interval}).$$

However, $y \in B_x(\epsilon) \iff d(x, y) < \epsilon \iff |x - y| < \epsilon \iff y \in (x - \epsilon, \ x + \epsilon)$.

**Example 3.8.7** The usual topology and metric topology on $\mathbf{R}^n$ is the same. Because, the usual topology on $\mathbf{R}^n$ is induced by the usual metric on $\mathbf{R}^n$. More precisely, the usual topology on $\mathbf{R}$ is induced by the usual metric $d \colon \mathbf{R} \times \mathbf{R} \to \mathbf{R}$, $(x, y) \mapsto |x - y|$ because the metric topology, called the Euclidean topology on $\mathbf{R}$ induced by the usual metric $d$ and the usual topology on $\mathbf{R}$ are both generated by open intervals. Similarly, the usual metric $d \colon \mathbf{R}^n \times \mathbf{R}^n \to \mathbf{R}$, $(x, y) \mapsto \|x - y\|$ induces the usual topology on $\mathbf{R}^n$, because the open sets in $\mathbf{R}^n$ under usual topology on $\mathbf{R}^n$ are characterized as: a subset $U \subset \mathbf{R}^n$ is open iff given a point $x \in U$, there exist an $\epsilon > 0$ and an open ball $B_X(\epsilon)$ such that $x \in B_X(\epsilon) \subset U$. This asserts that the Euclidean topology on $\mathbf{R}^n$ induced by the usual metric $d$ and the usual topology on $\mathbf{R}^n$ are both generated by open balls.

**Definition 3.8.8** Let $(X_1, d_1)$ and $(X_2, d_2)$ be two metric spaces with corresponding metric topologies $\tau_{d_1}$ and $\tau_{d_2}$ respectively. A function $f \colon (X_1, \tau_{d_1}) \to (X_2, \tau_{d_2})$ is said to be continuous at a point $p \in X_1$ if corresponding to a given $\epsilon > 0$, there exists a real number $\delta > 0$ such that whenever $d_1(x, p) < \delta$, then $d_2(f(x), f(p)) < \epsilon$.

**Example 3.8.9** Different metrics on a set may or may not give identical topology. In support, consider the Euclidean $n$-space $\mathbf{R}^n$. For points $x = (x_1, x_2, \ldots, x_n)$ and $= (y_1, y_2, \cdots, y_n)$ in $\mathbf{R}^n$, the metrics defined by

(i) $d_1(x, y) = \max_{1 \le i \le n} |x_i - y_i|$;
(ii) $d_2(x, y) = |x_1 - y_1| + |x_2 - y_2| + \cdots + |x_n - y_n|$;
(iii) $d_3(x, y) = [(x_1 - y_1)^2 + (x_2 - y_2)^2 + \cdots + (x_n - y_n)^2]^{\frac{1}{2}}$

give rise to the same topology, which is the usual topology on $\mathbf{R}^n$. On the other hand, consider another Example 3.8.13 which shows that different metrics on a given nonempty set may not give rise the same topology.

Proposition 3.8.10 is an immediate application of metric topology.

**Proposition 3.8.10** *Let $(X, d)$ be a metric space and $A$ be nonempty subset of $X$. Then $\overline{A} = \{x \in X : d(x, A) = 0\}$.*

**Proof** To prove the proposition, it is sufficient to show that $d(x, y) = 0$ if $x \in \overline{A}$. Since the open balls $B_x(\epsilon)$ for $x \in X$ and $\epsilon > 0$, constitute an open base for the metric topology $\tau_d$ on $X$, it follows that $x \in \overline{A}$ iff $A \cap B_x(\epsilon)$ is nonempty for all open balls $B_x(\epsilon)$, i.e., iff for every $\epsilon > 0$, there exists a point $a_\epsilon \in A$ such that $d(x, a_\epsilon) < \epsilon$, i.e., iff $d(x, y) = 0$. This shows that

$$\overline{A} = \{x \in X : d(x, A) = 0\}.$$

$\square$

## 3.8.2 Equivalent Metrics from Viewpoint Topology

This subsection conveys the concept of equivalent metrics from the viewpoint topology. In Chap. 2, this concept is given from the viewpoint of convergent sequence.

**Definition 3.8.11** Let $d$ and $\rho$ be two metrics on the same set $X$. Then the metrics $d$ and $\rho$ are said to be **equivalent**, denoted by $d \sim \rho$, if both of them define the same topology on $X$.

**Example 3.8.12** The metrics $d_1, d_2$ and $d_3$ defined in Example 3.8.9 are equivalent metrics, as they induce the same topology on $\mathbf{R}^n$.

**Example 3.8.13** All metrics on a given nonempty set may not give the same topology. For example, consider the topologies $\tau_{d_1}$ and $\tau_{d_2}$ induced by the metrics $d_1$ and $d_2$ on $\mathbf{R}$, defined by

(i)

$$d_1: \mathbf{R} \times \mathbf{R} \to \mathbf{R}, (x, y) \mapsto |x - y|;$$

(ii)

$$d_2: \mathbf{R} \times \mathbf{R} \to \mathbf{R}, (x, y) \mapsto \begin{cases} 0, \text{if} x = y \\ 1, \text{if } x \neq y \end{cases}.$$

The open sets in the topology $\tau_{d_1}$ defined by $d_1$ are generated by the open intervals in $\mathbf{R}$ and hence the open sets in $\tau_{d_1}$ are the usual open sets in $\mathbf{R}$. On the other hand, the collection of all open sets in the topology $\tau_{d_2}$ defined by the metric $d_2$ is the set $2^{\mathbf{R}}$ which is the collection of all subsets of $\mathbf{R}$. However, the open balls $B_x(1/2)$ relative to the metric $d_2$ are given by $B_x(1/2) = \{x\}$, $\forall x \in \mathbf{R}$, and hence each point in $\mathbf{R}$ is an open set for the topology $\tau_{d_2}$ generated by the metric $d_2$. This asserts that each union (finite or infinite) of points $x \in R$ is an open set in $\tau_{d_2}$, and hence, the subsets of $\mathbf{R}$ are precisely the open sets in the topology $\tau_{d_2}$.

**Remark 3.8.14** Example 3.8.13 shows that there may exist different metrics on the same set inducing different topologies. On the other hand, Proposition 3.8.15 asserts that given a metric space $(X, d_1)$, there may exist another metric $d_2$ on $X$ such that the metrics $d_1$ and $d_2$ induce the same topology on $X$.

**Proposition 3.8.15** *Let $(X, d_1)$ be a metric space. Consider metric $d_2$*

$$d_2: X \times X \to \mathbf{R}: (x, y) \mapsto \begin{cases} 1, \text{ if } d_1(x, y) > 1 \\ d(x, y), \text{ if } d_1(x, y) \leq 1, \end{cases}.$$

*Then $d_1$ and $d_2$ induce the same topology on $X$.*

**Proof** Since the topology on $X$ depends only on the open balls $B_x(\epsilon)$ for small $\epsilon > 0$, it follows from the defining conditions of $d_1$ and $d_2$ that they induce the same topology on $X$. $\qquad\square$

**Corollary 3.8.16** *The metrics $d_1$ and $d_2$ defined in Proposition 3.8.15 on $X$ are equivalent.*

**Proof** It follows from Proposition 3.8.15. $\qquad\square$

**Example 3.8.17** The metric $d_2$ on $X$ defined in Proposition 3.8.15 is bounded.

**Proposition 3.8.18** *Let $(X, d)$ be a metric space. Then the metric*

$$\rho: X \times X \to \mathbf{R}, (x, y) \mapsto \frac{d(x, y)}{1 + d(x, y)}$$

*is equivalent to the metric $d$.*

**Proof** Let $B_x^d(\epsilon)$ be the open ball in $(X, d)$ with center $x$ and radius $\epsilon$ and $B_x^\rho(\epsilon')$ be the open ball in $(X, \rho)$ with center $x$ and radius $\epsilon'$. Then

$$B_x^d(\epsilon) = y \in X : d(x, y) < \epsilon\} = \{y \in X : \frac{d(x, y)}{1 + d(x, y)} < \frac{\epsilon}{1 + \epsilon}$$

$$= y \in X : \rho(x, y) < \frac{\epsilon}{1 + \epsilon}\} = B_x^\rho(\frac{\epsilon}{1 + \epsilon}).$$

Since the induced topology $\tau_d$ on a metric space $(X, d)$ depends on the open balls $B_x^d(\epsilon)$ in $X$, the equality $B_x^d(\epsilon) = B_x^\rho(\frac{\epsilon}{1+\epsilon})$ asserts that $\tau_d = \tau_\rho$. It proves that the metrics $d$ and $\rho$ are equivalent. $\qquad\square$

**Corollary 3.8.19** *The standard (Euclidean) metric $d$ on $\mathbf{R}$*

$$d : \mathbf{R} \times \mathbf{R} \to \mathbf{R}, (x, y) \mapsto |x - y|$$

*and the metric $\rho$*

$$\rho : \mathbf{R} \times \mathbf{R} \to \mathbf{R}, (x, y) \mapsto \frac{|x - y|}{1 + |x - y|}$$

*are equivalent.*

**Proof** It follows from Proposition 3.8.18 by taking $X = \mathbf{R}$ and the metric $d$ the standard (Euclidean) metric on $\mathbf{R}$. $\qquad\square$

### 3.8.3 Metrizable Spaces

Metrizable spaces form an important family of topogical spaces. This subsection introduces the concept of metrizability of a topological space $(X, \tau)$, having its topological structure $\tau$ identical with the topology $\tau_d$ induced by some metric structure $d$ (if it exists) on $X$. Thus, a metrizable space $X$ is a topological space such that while considering its open sets, it is essentially, the same open sets as induced by a metric on $X$. It gives a natural problem: When a given topological space can be equipped with a metric topology? One of its positive answers is given in Theorem 3.8.23. It is proved that every metrizable space can be metrized by a bounded metric in Proposition 3.8.22 and metrizability is a topological property in Proposition 3.8.26.

**Definition 3.8.20** A topological space $X$ with topology $\tau$ is said to be **metrizable** if at least one metric $d$ can be defined on $X$ such that the family of open sets $\tau_d$ defined by the metric $d$ coincides with the family of open sets in $\tau$, i.e., if $\tau = \tau_d$; i.e., a topological space $(X, \tau)$ is metrizable if there exists at least one metric on $X$ whose class of generated open sets is precisely the given topology $\tau$ on $X$.

***Example 3.8.21***    (i) Every finite space is metrizable iff it is discrete.

(ii) The real line space $(\mathbf{R}, \sigma)$ with usual topology $\sigma$ is metrizable. Because the usual metric on $\mathbf{R}$ induces the usual topology on $\mathbf{R}$. Hence, in general, the Euclidean $n$-space $(\mathbf{R}^n, \sigma)$ with usual metric on $\mathbf{R}^n$ is also metrizable by Theorem 3.13.1.

(iii) The closed unit interval $\mathbf{I} = [0.1]$ with natural topology is metrizable, and hence, the unit $n$-cube $\mathbf{I}^n$ with natural topology is metrizable by Theorem 3.13.1.

(iv) Let $X$ be a indiscrete space with indiscrete topology $\tau$ such that $|X| > 1$. It is not metrizable. However, $\tau = \{\emptyset, X\}$, and hence, $\emptyset$ and $X$ are the only closed sets in $(X, \tau)$.

**Proposition 3.8.22** *Every metrizable space can be metrized by a bounded metric.*

***Proof*** Let $(X, \tau)$ be a metrizable space. Then there exists at least one metric $d$ such that $\tau = \tau_d$. It follows from Proposition 3.8.18 that the metric $d$ is equivalent to a bounded metric $\rho$ defined by

$$\rho : X \times X \to \mathbf{R}, \ (x, y) \mapsto \frac{d(x, y)}{1 + d(x, y)}.$$

$\square$

Metrizability of a topological space is characterized in Theorem 3.8.23.

**Theorem 3.8.23** *Let $(X, \tau)$ be a topological space. Then it is metrizable iff there exist a metric space $(Y, d)$ and an embedding*

$$f : (X, \tau) \to (Y, \tau_d),$$

*where $\tau_d$ is the topology on $Y$ induced by the metric $d$.*

***Proof*** Let $(X, \tau)$ be a topological space. Suppose there exist a metric space $(Y, d)$ and an embedding

$$f : (X, \tau) \to (Y, \tau_d),$$

where $\tau_d$ is the topology on $Y$ induced by the metric $d$ on $Y$. Define a function

$$\psi : X \times X \to \mathbf{R}, \ (x, y) \mapsto d(f(x), f(y)).$$

Since $f$ is an embedding by hypothesis, it follows that $\psi$ is a metric and its induced topology on $X$ is the same as $\tau$. This asserts that $(X, \tau)$ is metrizable. Conversely, let the topological space $(X, \tau)$ be metrizable. Then there exists a metric $d$ on $X$ such that its induced topology $\tau_d$ is identical with $\tau$. Let $(Y, d)$ be this metric space, where $Y = X$. Then the identity map

$$1d : (X, \tau) \to (Y, \tau_d), x \mapsto x$$

is a homeomorphism, and hence, it is an embedding.

$\square$

**Corollary 3.8.24** *Every subspace of a metrizable space is metrizable.*

**Corollary 3.8.25** *The unit interval* $\mathbf{I} = [0, 1]$ *is metrizable, since it is a subspace the real line space* $\mathbf{R}$.

### 3.8.4 Metrizability Is a Topological Property

A topological property is a property of a topological space which is shared by homeomorphic spaces. For example, Proposition 3.8.26 proves that metrizability is a topological property. This property plays an important role in classification of topological spaces.

**Proposition 3.8.26** *Metrizability is a topological property.*

**Proof** Let $(X, \tau)$ be a metrizable space. Then there exists a metric $d_X$ such that $\tau$ is induced by the metric $d_X$ on $X$. Let $(Y, \sigma)$ be a metrizable space homeomorphic to $(X, \tau)$. Then there exists a homeomorphism

$$f : (X, \tau) \to (Y, \sigma).$$

Consider the map

$$d_Y : Y \times Y, \ (y, y') \to d_X(f^{-1}(y), f^{-1}(y')).$$

Then $d_Y$ is a metric on $Y$. We claim that the metric $d_Y$ induces the same topology $\sigma$ on $Y$. Since every homeomorphism from $X$ onto $Y$ sends an open base for $\tau$ to an open base for $\sigma$, the homeomorphism $f$ sends the $\tau$-basis elements $B_{x_0}(\epsilon) = \{x \in X : d_X(x, x_0)) < \epsilon\}$ in $X$ to the set $B_{f(x_0)}(\epsilon) = \{y \in Y : d_Y(f(x), f(x_0)) < \epsilon\} \subset Y$. Denoting $f(x_0)$ by $y_0$ and $f(x)$ by $y$. It follows that all the subsets of $Y$ of the form

$$B_{f(x_0)}(\epsilon) = B_{y_0}(\epsilon) = \{y \in Y : d_Y(y, y_0) < \epsilon\}$$

forms an open base for the $\sigma$ on $Y$. It asserts that $\sigma$ is induced by the metric $d_Y$ on $Y$, and hence, it follows that the topological space space $(Y, \sigma)$ is also metrizable. This implies that metrizability is preserved under every homeomorphism, and hence, it is a topological property.

$\square$

## 3.8.5 Topologically Complete Metric Spaces

This subsection conveys the motivation of topologically complete metric spaces with illustrative examples. It is sometimes convenient to consider topologically complete metric spaces, specially, when a metric space is not complete. For example, there exist homeomorphic. spaces $(X, d)$ and $(X, \rho)$ such that $(X, \rho)$ is complete but $(X, d)$ is not so ( see Example 3.8.29).

.

**Definition 3.8.27** (*Topologically complete metric spaces*) Let $(X, d)$ be a metric space with its induced topology $\tau_d$, which is not complete. If there exists another metric $\rho$ on $X$ such that $(X, \rho)$ is complete and its induced topology $\tau_\rho$ coincides with the topology $\tau_d$, then the space $(X, \rho)$ is said to be topologically complete.

***Example 3.8.28*** The set **Q** of rational numbers under the metric inherited from the standard metric on **R** is not topologically complete.

Example 3.8.29 shows that at some situations, it is convenient to consider topologically complete metric spaces.

***Example 3.8.29*** Consider the metric space $X = (-1, 1) \subset \mathbf{R}$ equipped with standard metric $d$. Then the metric $d$ is given by

$$d:, X \times X \to \mathbf{R}, (x, y) \mapsto |x - y|.$$

The metric space $(X, d)$ is not complete. On the other hand, there exists a topologically equivalent metric

$$\rho: X \times X \to \mathbf{R}, (x, y) \mapsto (\frac{x}{(1 - x^2)^{1/2}}, \frac{y}{(1 - y^2)^{1/2}}).$$

The spaces $(X, d)$ and $(X, \rho)$ are homeomorphic. $(X, \rho)$ is complete but $(X, d)$ is not so.

***Example 3.8.30*** The set **Q** of rational numbers under the metric inherited from the standard metric on **R** is not topologically complete.

## 3.9  Topology Generated by a Family of Functions

There is a natural problem in topology: given a nonempty set $X$ and a family of topological spaces $\{(Y_a, \tau_a): a \in \mathbf{A}\}$ and a family of functions $\{f_a: X \to Y_a: a \in \mathbf{A}\}$, how to endow a topology $\sigma$ on $X$ such that the function $f_a: (X, \sigma) \to (Y_a, \tau_a)$ is continuous for each $a \in \mathbf{A}$? This section solves this problem by constructing a subbase for the topology $\sigma$.

**Definition 3.9.1** Let $X$ be a nonempty set and $\{(Y_a, \tau_a): a \in A\}$ be a family of topological spaces. If **S** is the family of subsets of $X$ defined by

$$\mathbf{S} = \bigcup \{f_a^{-1}(V): V \in \tau_a, \ a \in A\},$$

i.e., **S** is the set consisting of inverse image of every open subset of each $Y_a$ under $f_a$. Then **S** forms a subbase for a topology on $X$. The topology $\sigma$ generated by **S** is said to be the **topology generated by the family of functions** $\{f_a: X \to Y_a: a \in A\}$.

**Theorem 3.9.2** *Let $X$ be a nonempty set and $\{(Y_a, \tau_a): a \in A\}$ be a family of topological spaces. Let $\sigma$ be the topology on $X$ generated by $\mathbf{S} = \bigcup \{f_a^{-1}(V) : V \in \tau_a, a \in A\}$ given in Definition 3.9.1. Then*

*(i)  each function $f_a: (X, \sigma) \to (Y_a, \tau_a)$ is continuous;*

*(ii)  the topology $\sigma$ is the intersection of all the topologies on $X$ such that each $f_a: (X, \sigma) \to (Y_a, \tau_a)$ is continuous;*

*(iii)  the topology $\sigma$ on $X$ is the coarsest (weakest or smallest) topology on $X$ such that each $f_a$ is continuous.*

**Proof** It follows from the construction of **S**.                                □

**Definition 3.9.3** The set **S** given in Definition 3.9.1 is called the **defining subbase** for the topology $\sigma$ such that each function $f_a: X \to Y_a$ is continuous for each $a \in A$.

**Example 3.9.4** Let $\tau_\sigma$ be the topology on **R** generated by the family of all linear transformations

$$T : \mathbf{R} \to \mathbf{R}, x \mapsto cx + d, \ \forall c, d \in \mathbf{R}.$$

Then $\tau_\sigma$ is the weakest topology on **R** such that each linear transformation

$$T : (\mathbf{R}, \tau_\sigma) \to (\mathbf{R}, \sigma), x \mapsto cx + d, \ \forall c, d \in \mathbf{R}$$

is continuous and $\tau_\sigma$ is the usual topology $\sigma$ on **R**.

**Example 3.9.5** Consider the projection maps $p_1, p_2: \mathbf{R}^2 \to \mathbf{R}$ defined by $p_1(x, y) = x$, $p_2(x, y) = y$. Then the inverse images of an open interval $(a, b)$, i.e., $p_1^{-1}(a, b)$ and $p_2^{-1}(a, b)$, are infinite open strips in $\mathbf{R}^2$. Such strips constitute a subbase for the natural topology on $\mathbf{R}^2$, which is the smallest topology on $\mathbf{R}^2$ such that the projection maps $p_1$ and $p_2$ are continuous by Theorem 3.12.11.

**Example 3.9.6 (Projective topology)** Let $X$ be a linear space, $Y$ be a linear topological space and $F(X, Y)$ be a family of maps $f: X \to Y$. The weakest topolgy for $X$ is the weakest topology making each member $f \in F(X, Y)$ continuous is called the projective topolgy for $X$.

## 3.10   Kuratowski Closure Topology, Sierpinski Topology and Niemytzki's Disk Topology

This section discusses three special types of topologies such as Kuratowski closure topology, Sierpinski topology and Niemytzki's disk topology which are used in subsequent chapters. The unique topology determined in Theorem 3.10.2, called Kuratowski closure topology, satisfies the four axioms of closed sets **C(1)–C(4)** of Definition 3.1.28. So, a topology can be defined satisfying the axioms **OS(1)–OS(4)**, for open sets or the axioms for closed sets. It is found that there exist different approaches of defining a topological space. But the open set approach is the most natural. Other approaches are used according to the nature of the problem.

### 3.10.1   Kuratowski Closure Topology

This subsection addresses Kuratowski closure topology defined by K. Kuratowski. It is an interesting topology on the power set $\mathcal{P}(X)$ of a nonempty set $X$. In his honor, this topology is known as Kuratowski closure topology. More precisely, let $(X, \tau)$) be a topological space and $\mathcal{P}(X)$ be the power set $X$. If $\bar{A}$ (the intersection of all closed subsets containing $A$ in $(X, \tau)$) denotes the closure of $A$, then the closure operator operator

$$k: \mathcal{P}(X) \to P(X), \ A \mapsto \bar{A}$$

has the properties given in Theorem 3.10.1

**Theorem 3.10.1**  *Let $(X, \tau)$ be a topological space. For arbitrary subsets $A$, $B$ of $X$, the closure operator $k : \mathcal{P}(X) \to P(X)$, $A \mapsto \bar{A}$ has the properties:*

  *(i)*  **K(1)**  $A \subset \bar{A}$ ; *equivalently,* $A \subset k(A)$;
 *(ii)*  **K(2)**  $\overline{A \cup B} = \bar{A} \cup \bar{B}$; *equivalently,* $k(A \cup B) = k(A) \cup k(B)$;
*(iii)*  **K(3)**  $\bar{\bar{A}} = \bar{A}$; *equivalently,* $k(\bar{A}) = \bar{\bar{A}} = k(A)$;
 *(iv)*  **K(4)**  $\bar{\emptyset} = \emptyset$, *equivalently,* $k(\emptyset) = \emptyset$.

**Proof**  It follows from the definition of the closure of a set.    □

Theorem 3.10.2 asserts that Kuratowski's closure axioms on a nonempty set determine the unique topology, known as Kuratowski's closure topology on the set.

**Theorem 3.10.2**  *Let $X$ be a nonempty set and $\mathcal{P}(X)$ be its power set. If an operator $k: \mathcal{P}(X) \to \mathcal{P}(X)$, $A \mapsto \bar{A}$ is defined satisfying the properties*

  *(i)*  **K(1)**  $A \subset \bar{A}$;
 *(ii)*  **K(2)**  $\overline{A \cup B} = \bar{A} \cup \bar{B}$;
*(iii)*  **K(3)**  $\bar{\bar{A}} = \bar{A}$;
 *(iv)*  **K(4)**  $\bar{\emptyset} = \emptyset$,

*for arbitrary subsets A and B of X, then there exists a unique topology τ on X such that for every subset A of X, the set Ā (thus obtained) is the closure of A in this topological $(X, \tau)$.*

**Proof** Let $X$ be a nonempty set. Let a subset $A \subset X$ be called closed if $\bar{A} = A$ and **F** be the family of all closed sets defined in this way. Then **F** satisfies all the four axioms **C(1)–C(4)** of Definition 3.1.28 of closed sets by using the given properties **K(1)–K(4)**. Hence, a set $U$ is open in $X$ if $X - U$ is closed. Then the collection $\tau$ of all such open sets determines a unique topology on $X$. Hence, the closed sets of $(X, \tau)$ are the same as the members of **F**. Let $A^*$ denote the closure of $A$ in $(X, \tau)$, which is determined in terms of $\tau$. This implies that $A^*$ is the intersection of all closed sets in $(X, \tau)$ containing the set A, which is also the intersection of all those members of **F** containing the set A. Utilizing the properties **K(1)–K(4)**, it follows that $\bar{A} = A^*$. □

**Definition 3.10.3** The conditions **K(1)–K(4)** of Theorem 3.10.2 on the power set $\mathcal{P}(X)$ of $X$ are called the Kuratowski's closure axioms, and the resulting topology is known as **Kuratowski's closure topology** on $X$, which is unique by Theorem 3.10.2

**Example 3.10.4** In an arbitrary topological space $X$, the relation $\overline{A \cap B} = \bar{A} \cap \bar{B}$ is not necessarily true for its arbitrary subsets $A$ and $B$. For example, in the real line space $(\mathbf{R}, \sigma)$, for the subsets $A = \mathbf{Q}$ (set of rationals) and $B = \mathbf{T}$ (set of irrationals),

$$\overline{\mathbf{Q}} = \mathbf{R} = \overline{\mathbf{T}} \implies \overline{\mathbf{Q}} \cap \overline{\mathbf{T}} = \mathbf{R}.$$

On the other hand,

$$\mathbf{Q} \cap \mathbf{T} = \emptyset \implies \overline{\mathbf{Q} \cap \mathbf{T}} = \emptyset.$$

This shows that

$$\overline{\mathbf{Q} \cap \mathbf{T}} \neq \overline{\mathbf{Q}} \cap \overline{\mathbf{T}}.$$

## 3.10.2 Sierpinski Space

This subsection conveys the concept of Sierpinski space having only two distinct points. This space has certain properties, some of them are given in Proposition 3.10.6 and others are studied in subsequent chapters.

**Definition 3.10.5 Sierpinski space** is a finite topological space $(S, \tau_S)$ consisting of only two distinct points, abbreviated $S = \{0, 1\}$ such that only one of them is the closed set. If $\{0\}$ is the closed set, then the open sets of **Sierpinski topology** $\tau_S$ are precisely, $\{\emptyset, \{1\}, \{0, 1\}\}$ and its closed sets are precisely, $\{\emptyset, \{0\}, \{0, 1\}\}$. In the Sierpinski space $(S, \tau_S)$, if $\{0\}$ is closed, then $\{1\}$ is open. Again, if $\{1\}$ is closed, then $\{0\}$ is open by definition of Sierpinski topology.

Sierpinski space has some special properties proved in Proposition 3.10.6. For more properties of this space such as its separation, compactness, connectedness properties see Chaps. 4 and 5.

**Proposition 3.10.6** *Let* $(S, \tau_S)$ *be a Sierpinski space with* $\{0\}$ *as a closed set. Then* $\{1\}$ *is an open set.* $(S, \tau_S)$ *has the following properties:*

(i) $\{1\}$ *and* $\{0\}$ *are topologically distinct in* $(S, \tau_S)$.

(ii) *Let* $\mathbf{I} = [0, 1]$ *be the subspace of the real line space. Then the map*

$$f : I \to S, \ t \mapsto \begin{cases} 0, \ if \ t = 0 \\ 1, \ for \ all \ t > 0 \end{cases}$$

*is continuous in* $(S, \tau_S)$, *since* $f^{-1}(1) = (0, 1]$ *is open in* $\mathbf{I}$.

(iii) *The group of all homeomorphisms of the Sierpinski space* $(S, \tau_S)$ *is the trivial group. Because if* $f : (S, \tau_S) \to (S, \tau_S)$ *is any continuous map, then* $f$ *is the same as the identity map* $1_S : S \to S, x \mapsto x$, *or the constant map* $c_0 : S \to S, x \mapsto 0$, *or the constant map* $c_1 : S \to S, x \mapsto 1$. *This asserts that there exist only three distinct continuous maps from the space* $(S, \tau_S)$ *to itself. Hence, the group of all homeomorphisms of the space* $(S, \tau_S)$ *is the trivial group consisting of only the identity homeomorphism* $1_S$.

(iv) *The Sierpinski space* $(S, \tau_S)$ *is not metrizable (see Chap. 4).*

## 3.10.3 Niemytzki Topology

This subsection defines a special topology on the open upper half of the Euclidean plane $\mathbf{R}^2$, known as Niemytzki's disk space topology, For interesting properties of Niemytzki's tangent dick topology such as its Hausdorff and regularity properties, see Chap. 4 and for some other properties, see Chap. 6. This topology was first defined by Viktor Niemytzki (1900–1967).

**Definition 3.10.7** Let $\mathbf{R}_+^2 = \{(x, y) \in \mathbf{R}^2 : y \geq 0\}$ be the open upper half-plane endowed with the topology induced by the Euclidean metric. Let $\mathbf{R} = \mathbf{R}^1$ be the subset of $\mathbf{R}^2$ consisting of all points on the $x$-axis, which are precisely the points of $\mathbf{R}^2$ for which $y = 0$. Given $r > 0$, define

(i) for each point $x \in \mathbf{R}_+^2 - \mathbf{R}^1$, the set $U_x(r) = B_x(r) \cap \mathbf{R}^1$ where $B_x(r)$ denotes an open ball (disk) with center at $x$ and radius $r$, which lies entirely above the $x$-axis and

(ii) for each point $x \in \mathbf{R}^1$, the set $U_x(r) = \{x\} \cup B_x(r)$, where $x$ is a point in the Euclidean line $\mathbf{R}^1$ for which $y = 0$, i.e., $x$ is a point on the $x$-axis and $B_x(r)$ is an open disk in $\mathbf{R}_+^2$, which touches the $x$-axis at the point $x$.

Geometrically, $U_x(r)$ consists of all points inside the circumference of the arc of the disk $B_x(r)$ along with points on the $x$-axis for the case (i) and $U_x(r)$ does not contain

any point on the $x$-axis excepting the point $x$ of tangency of $B_x(r)$ with the $x$-axis for the case (ii).
For each point $x \in \mathbf{R}^2_+$, the family $\mathcal{F}_x = \{U_x(r)\}$ forms a nbd filter of the point $x$ (see Definition 3.2.24). If $\tau$ is the resulting nbd topology on $\mathbf{R}^2_+$, then the corresponding topological space $(\mathbf{R}^2_+, \tau)$ is called the **Niemytzki's space** and this topology $\tau$ is called **Niemytzki's tangent disk topology or Niemytzki's topology**

## 3.11 Two Countability and Separability Axioms

Topologies defined on an arbitrary set are weak in the sense that they fail to invite their deep study until certain additional condition or conditions are imposed on them. With this aim, this section initiates the concepts of first and second countable spaces and also separable spaces through an axiomatic approach. Their more study is available in Chap. 7. **Historically**, two axioms of countability were formulated by F. Hausdorff (1868–1942) in 1914 and the concept of separability was introduced by M. Fréchet in 1906. Such spaces are important for our subsequent study.

### 3.11.1 Countability Axioms: First and Second Countable Spaces

This subsection imposes the two axioms of countability to obtain two special classes of topological spaces such as first and second countable spaces.

**Definition 3.11.1** Let $(X, \tau)$ be a topological space. A collection of open sets $\mathbf{B}_x$ in $(X, \tau)$ such that each member of $\mathbf{B}_x$ contains a point $x \in X$ is said to an **open base (or local base)** for $\tau$ at $x$, if for every open set $U$ containing the point $x$, there exists a member $V_x \in \mathbf{B}_x$ such that $x \in V_x \subset U$.

**Definition 3.11.2** A topological space $(X, \tau)$ is said to satisfy the **first axiom of countability** if there is a countable open base for $\tau$ at every point of $X$. A topological space $(X, \tau)$ satisfying the first axiom of countability is said to be a **first countable (or locally separable) space**.

*Example 3.11.3* In the real line space $(\mathbf{R}, \sigma)$, an open interval $(a, b)$, where $a, b: a < b$, are both rational numbers is called an open rational interval and they (open rational intervals) form a countable open base for the natural topology $\sigma$ on $\mathbf{R}$. Hence, $(\mathbf{R}, \sigma)$ is a first countable space.

*Example 3.11.4* The Euclidean $n$-space $\mathbf{R}^n$ and the $n$-sphere $S^n$ have each a countable open base for their natural topology.

**Definition 3.11.5** A topological space $(X, \tau)$ is said to satisfy the **second axiom of countability**, if there exists a countable open base for the topology $\tau$. A topological space $(X, \tau)$ satisfying the second axiom of countability is said to be a **second countable (or strongly separable) space.**

***Example 3.11.6***  The real number space $(\mathbf{R}, \sigma)$, with the usual topology $\sigma$, is second countable, because, the empty set $\emptyset$ and the collection of open intervals $\{(a, b) : a < b, \text{ and } a, b \text{ are rational numbers }\}$ form a countable open base for the topology $\sigma$ on $\mathbf{R}$. This shows that $(\mathbf{R}, \sigma)$ is a second countable space.

***Example 3.11.7***  Every second countable space is first countable but its converse is not true. In support consider the examples:

(i)  Let $X$ be any metric space.Then it is first countable but there are some metric spaces which are not second countable. For example, if $(X, d)$ is an uncountable metric space with discrete metric

$$d: X \times X \to \mathbf{R}, \ (x, y) \mapsto \begin{cases} 0, \text{if} x = y \\ 1, \text{if } x \neq y \end{cases}$$

then it is not second countable, since the metric $d$ induces the discrete topology on $X$.

(ii)  Euclidean spaces $\mathbf{R}^n$ are second countable, because the family of open balls $\{B_x(\epsilon)\}$ with $x$ and $\epsilon$ both rational numbers, form a countable open base for the Euclidean topology on $\mathbf{R}^n$.

## *3.11.2  Separability Axioms: Separable Spaces*

This subsection imposes certain conditions to obtain separable spaces and compares these spaces with second countable spaces. The concept of separability of topological spaces introduced by M. Fréchet in 1906 is given below.

**Definition 3.11.8**  A topological space $(X, \tau)$ is said to be **separable,** if there is a countable subset $C$ in $X$ such that $\overline{C} = X$; i.e., if $X$ has a countable dense subset in $X$ **(separability axiom)**

***Example 3.11.9***  The real number space $(\mathbf{R}, \sigma)$, with the usual topology $\sigma$, is second countable, also first countable and separable,

***Example 3.11.10***  Let the set $C[0, 1]$ of real-valued continuous functions on $[0, 1]$ be endowed with sup-norm topology. Then this topological space is separable by using the result that a metruc space is separable if it has a countable dense set, because the set of all polynomials with rational coefficients forms a countable dense set in the space $C[0, 1]$, by **Weierstrass approximation theorem** saying that any continuous function on a closed and bounded interval can be approximated uniformly on the same interval by polynomials.

**Definition 3.11.11** Let $(X, d)$ be a metric space. It is said to be **separable,** if for every point $x \in X$, and any open ball $B_x(\epsilon)$, there is an element $y_n \in Y$ for some countable and everywhere dense set $Y$ in $X$, which can be represented as

$$Y = \{y_1, y_2, \dots\}.$$

**Proposition 3.11.12** *Every separable metric space satisfies the second axiom of countability.*

**Proof** $(X, d)$ be a separable metric space and $Y = \{y_1, y_2, \dots\}$ be a countable and everywhere dense set in $X$. Consider the family of open sets

$$\mathcal{F} = \{U_{n,\, t} = \{x \in X : d(x, y_n) < 1/t \text{ for } n,\ t = 1, 2, \dots\}.$$

Then $\mathcal{F}$ forms an open base for the topology $\tau_d$ induced by $d$ on $X$. $\qquad \square$

# 3.12 Sum and Product of Topological Spaces

This section constructs sum and product of a given family of topological spaces to obtain new topological spaces from the old ones. The concept of sums of topological spaces was first found in the book (Bourbaki 1940).

## 3.12.1 Sum of Topological Spaces

This subsection prescribes a method of construction of topological sum of a given family of topological spaces. Given a topological $X$, and its two disjoint subspaces $Y$ and $Z$ such that $X = Y \cup Z$, it not always possible to recover the topology of $X$ from the topologies of the subspaces $Y$ and $Z$. For example if $y \neq z$, then the set $X = \{y, z\}$ has 4 distinct topologies ( see Example 3.1.11); on the other hand, each of the sets $\{y\}$ and $\{z\}$ has exactly only one topology. To avoid such situation, the concept of topological sum of topological spaces is introduced.

**Definition 3.12.1** Let $X$ be a topological and $Y$ and $Z$ be its two disjoint subspaces such that $X = Y \cup Z$. Then $X$ can be topolozied by declaring a subset $U$ to be open in $X$ if $U \cap Y$ is open in $Y$ and if $U \cap Z$ is open in $Z$. With this topology, $X$ is called the **topological sum** of $Y$ and $Z$ and is written as $X = Y \sqcup Z$ or $X = Y + Z$.

Definition 3.12.2 gives a generalization of Definition 3.12.1 for an arbitrary family of topological spaces.

**Definition 3.12.2** Let $\{(X_i, \tau_i): i \in \mathbf{A}\}$ be a given family of topological spaces (finite or infinite) and $X = \bigcup_{i \in \mathbf{A}} X_i$ (set-theoretic union). Then the collection $\tau$ of subsets $U$ of $X$, defined by

$$\tau = \{U \subset X : U \cap X_i \in \tau_i, \, \forall i \in \mathbf{A}\}$$

forms a topology on $X$, called the **topological sum** or the **sum of the given family of topological spaces**, denoted by $(X, \tau) = \Sigma\{(X_i, \tau_i): i \in \mathbf{A}\}$.

Theorem 3.12.3 proves some properties of the topological sum of spaces.

**Theorem 3.12.3** *Let* $(X, \tau) = \Sigma\{(X_i, \tau_i): i \in \mathbf{A}\}$ *be the topological sum of a given family* $\{(X_i, \tau_i): i \in \mathbf{A}\}$ *of topological spaces and* $f_i: X_i \hookrightarrow X$ *be the inclusion map for each* $i \in \mathbf{A}$. *Then*

(i) *each* $f_i: (X_i, \tau_i) \hookrightarrow (X, \tau)$ *is continuous.*
(ii) *if the sets* $X_i$ *of the given family are pairwise disjoint, then each*

$$f_i: (X_i, \tau_i) \hookrightarrow (X, \tau)$$

*is both open and closed.*

**Proof** (i) Let $U$ be an arbitrary open set in $(X, \tau)$. Then $f_i^{-1}(U) = U \cap X_i \in \tau_i$ implies that $f_i$ is continuous.

(ii) Let the sets $X_i$ in the family $\{X_i : i \in \mathbf{A}\}$ be pairwise disjoint and $U$ be an arbitrary open set in $\tau_i$. Since $f_i$ is an inclusion map, $f_i(U) = U$. Moreover,

$$U \cap X_i = U \in \tau_i \text{ and } U \cap X_k = \emptyset \in \tau_k, \, \forall k \neq i \in \mathbf{A}\}$$

imply that $f_i(U) \in \tau, \, \forall \, U \in \tau$. This asserts that each $f_i$ is an open map. For the last part, let $B$ be an arbitrary closed set in $(X_i, \tau_i)$. Then $V = X - B$ is an open set in $(X_i, \tau_i)$. Hence

$$V \cap X_i = (X - B) \cap X_i = X_i - B \in \tau_i \text{ and } V \cap$$
$$X_k = (X - B) \cap X_k = X_k \in \tau_K$$

show that $B$ is a closed set in $(X, \tau)$, and hence, $f_i(B) = B$ is also closed in $(X, \tau)$. This implies that each $f_i$ is a closed map. $\square$

### 3.12.2  Product Space of a Finite Family of Topological Spaces

This subsection describes the construction process of product spaces, which can be viewed as a generalization of the usual geometric process of construction of the Euclidean spaces $\mathbf{R}^2$, $\mathbf{R}^3, \ldots$, $\mathbf{R}^n$ from the real line space $\mathbf{R}$. This subsection also describes a base for finite product topology to obtain all its open sets.

The Cartesian product $U \times V$ of two finite intervals U and V in $\mathbf{R}$ is an open rectangle in $\mathbf{R}^2$. The open rectangles form an open base for the natural topology on $\mathbf{R}^2$, which is called a product topology on $\mathbf{R}^2$. This technique is borrowed for construction of any finite product topology. M. Fréchet first studied a finite product of abstract topological spaces in 1910. Construction of the product space of an abitrary family of topological spaces and Tychonöff topology are available in Sect. 3.12.3.

**Definition 3.12.4**  Given topological spaces $X$ and $Y$, a topology $\tau$ is defined on their product set $X \times Y$ by declaring a subset $U \subset X \times Y$ to be open if $U$ is the union of the sets of the form $U_1 \times U_2$, where $U_1$ is open in $X$ and $U_2$ is open in $Y$. This topology $\tau$ is called the **product topology** on $X \times Y$.

**Example 3.12.5**  The product topology on $X \times Y$ is generated by the collection of sets $\{U \times V : U \subset X, \ V \subset Y \ are \ both \ open\}$ as a subbase. Because,

$$(U \times V) \cap (U' \times V') = (U \cap U') \times (V \times V').$$

**Remark 3.12.6**  The product topology given in Example 3.12.5 is now generalized in Proposition 3.12.7 for a finite product of topological spaces.

**Proposition 3.12.7**  *Let $(X_1, \tau_1), (X_2, \tau_2), \cdots, (X_n, \tau_n)$ be a finite number of topological spaces and $\mathbf{B}$ be a collection of subsets of the product set $X = X_1 \times X_2 \times \cdots \times X_n$ defined by*

$$\mathbf{B} = \{U_1 \times U_2 \times \cdots \times U_n : U_i \in \tau_i, \ \forall i = 1, 2, \ldots, n\}.$$

*Then $\mathbf{B}$ constitutes an open base for a topology $\tau$ on $X$.*

**Proof**  Clearly, $\emptyset \times \emptyset \times \cdots \times \emptyset = \emptyset \in \mathbf{B}$, since $\emptyset \in \tau_i$, $\forall i = 1, 2, \ldots, n$. Again, every point of $X$ is at least in one member of $\mathbf{B}$, since $X_1 \times X_2 \times \cdots \times X_n \in \mathbf{B}$. Finally, let $x \in U \cap V$ for some $U$, $V \in \mathbf{B}$. Hence, there exist $U_i$, $V_i \in \tau_i$ for $i = 1, 2, \ldots, n$ such that

$$U = U_1 \times U_2 \times \cdots \times U_n \text{ and } V = V_1 \times V_2 \times \cdots \times V_n.$$

Then

$$U \cap V = (U_1 \times U_2 \times \cdots \times U_n) \cap (V_1 \times V_2 \cdots \times V_n)$$
$$= (U_1 \cap V_1) \cap (U_2 \cap V_2) \cap \cdots \cap (U_n \cap V_n) \in \mathbf{B},$$

because, $U_i \cap V_i \in \tau_i \ \forall i = 1, 2, \ldots, n$. This asserts that if $x \in U$, $x \in V$ *and* $W = U \cap V$, then

$$x \in W \subset U \cap V.$$

Hence, it follows that given $\mathbf{B}$ forms an open base for a topology $\tau$ on $X$.

□

**Definition 3.12.8** The topology $\tau$ on $X$ generated by the open base $\mathbf{B}$ constructed in Proposition 3.12.7 is sometimes denoted by $\tau(\mathbf{B})$ and is called the **product topology** on $X$ and the resulting topological space $(X, \tau(\mathbf{B}))$ is called the topological **product space** of the finite family of topological spaces $(X_1, \tau_1), (X_2, \tau_2), \ldots, (X_n, \tau_n)$.

*Example 3.12.9* The Sorgenfrey plane $(\mathbf{R}^2, \sigma)$ is the product space defined by $(\mathbf{R}^2, \sigma) = (\mathbf{R}, \sigma_l) \times (\mathbf{R}, \sigma_l)$, where $(\mathbf{R}, \sigma_l)$ is the Sorgenfrey line. **Geometrically,** this topology $\sigma$ on $\mathbf{R}^2$ is generated by all rectangles of the form $[a, b) \times [c, d)$ as a basis, where $a, b, c, d \in \mathbf{R}$ with $a < b$ and $c < d$.

**Definition 3.12.10** Let $X$, $Y$ and $Z$ be topological spaces. Then a function $f : X \times Y \to Z$ from the product space $X \times Y$ to the space $Y$ considered as a function $f(x, y)$ of two variables, with values in $Z$ is said to be continuous if it is continuous jointly in both variables $x$ and $y$.

**Theorem 3.12.11** *Let $X$ and $Y$ be topological spaces. If $X \times Y$ is their product space and*

$$p_1 : X \times Y \to X, \ (x, y) \mapsto x \text{ and } p_2 : X \times Y \to Y, \ (x, y) \mapsto y$$

*are projection maps, then $p_1$ and $p_2$ are both continuous maps in the product topology, which is the weakest topology such that $p_1$ and $p_2$ are continuous.*

*Proof* Let $\mathbf{B}_X$ and $\mathbf{B}_Y$ be two bases for the topological spaces $X$ and $Y$, respectively. Given any open set $U_i \in \mathbf{B}_X$, the set $p_1^{-1}(U_i) = U_i \times Y$ is an open set in the product space $X \times Y$. However, the space $Y = \bigcup(V_j : V_j \in \mathbf{B}_Y)$ and

$$p^{-1}(U_i) = U_i \times Y = U_i \times \bigcup_j V_j = \bigcup_j (U_i \times V_j).$$

It asserts that $p^{-1}(U_i)$ is an open set in the product space $X \times Y$, and hence, $p_1$ is continuous. Similarly, $p_2$ is continuous. It proves the first part of theorem. To prove the second part, let $\sigma$ be any topology on $X \times Y$ such that $p_1$ and $p_2$ are continuous. Then the sets $p_1^{-1}(U_i) = U_i \times Y$ and $p_2^{-1}(V_i) = X \times V_j$ are both open in the product space $X \times Y$ with topology $\sigma$. Hence the sets

$$p_1^{-1}(U_i) \cap p_2^{-1}(V_j) = (U_i \times Y) \cap (X \times V_j) = U_i \times V_j$$

are open sets belonging to $\sigma$. This asserts that any topology on $X \times Y$ under which both the projection maps $p_1$ and $p_2$ are continuous contains all the basic open sets of the form $U_i \times U_j$ and also the topology generated by them. This shows that the topology $\sigma$ is stronger than the product topology on $X \times Y$. In other words, the product topology on $X \times Y$ is weaker than any other topology on $X \times Y$ such that the projection maps $p_1$ and $p_2$ are continuous.

$\square$

Theorem 3.12.11 is reformulated in Theorem 3.12.12.

**Theorem 3.12.12** *Let $(X, \tau)$ and $(Y, \sigma)$ be two topological space with $(X \times Y, \Sigma)$ be their topological product space. Then the projection maps*

*(i)* $p_1 : (X \times Y, \Sigma) \to (X, \tau)$, $(x, y) \mapsto x$ *and*
*(ii)* $p_2 : (X \times Y, \Sigma) \to (Y, \sigma)$, $(x, y) \mapsto y$
*are con both continuous.*
*(iii) The product topology $\Sigma$ is the smallest topology on $X \times Y$ such the projection maps $p_1$ and $p_2$ are continuous.*

### 3.12.3 Product Space of an Arbitrary Family of Topological Spaces and Tychonöff Topology

This subsection defines Tychonöff topology defined by Andrey Tychonöff (1906–1993) in 1930 with the help of product topology for any family (possibly, infinite) of topological spaces, which gives a generalization of the product topology for a finite family of topological spaces. An element of the product $X = \Pi_{a \in \mathbf{A}} X_a$ is a set $\{x_a: , x_a \in X_a, a \in \mathbf{A}\}$; i.e., the elements of $\Pi_{a \in \mathbf{A}} X_a$ are functions $x: \mathbf{A} \to \bigcup X_a$ such that $x(a) \in X_a$. In particular, if $\mathbf{A} = \{1, 2, \ldots, n\}$, then the product (finite) of $X_1, X_2, \ldots, X_n$ denoted by $X_1 \times X_2 \times \cdots \times X_n$ and its elements are denoted by the ordered $n$-tuples $(x_1, x_2, \ldots, x_n)$, where $x_i \in X_i$ for $i = 1, 2, \ldots, n$.

**Definition 3.12.13** (*Tychonöff topology*) Let $\{(X_a, \tau_a): a \in \mathbf{A}\}$, where $\mathbf{A}$ is an arbitrary (possibly, infinite) indexing set and let $X = \Pi_{a \in \mathbf{A}} X_a$ be the Cartesian product of the sets $X_a$, for all $a \in \mathbf{A}$. Consider the family $\mathbf{B}$ of subsets of $X$ defined by

$$\mathbf{B} = \{p_a^{-1}(U_a): U_a \in \tau_a, a \in \mathbf{A}\},$$

where $p_a : X = \Pi_{a \in \mathbf{A}} X_a \to X_a$ is the projection mapping onto $X_a$, for each $a \in \mathbf{A}$. Then $\mathbf{B}$ forms a subbase for a topology $\tau$ on $X = \Pi_{a \in \mathbf{A}} X_a$ called the product topology or the Tychonöff topology on $X$ and the resulting topological space $(X, \tau)$ is called the topological product space of the given family of topological spaces.

Let $\mathbf{G}$ be the base for the topology generated by the subbase $\mathbf{B}$ given in Definition 3.12.13. Then a subset $V$ of $X = \Pi_{a \in \mathbf{A}} X_a$ is in the base $\mathbf{G}$ iff $V$ is is the intersection of finitely many members of the subbase $\mathbf{B}$ prescribed in Definition 3.12.13.

**Definition 3.12.14** (*Base for Tychonöff topology*) A base for the Tychonöff topology on $X$ given in Definition 3.12.13 consists of the sets of the form

$$V = p_{a_1}^{-1}(U_{a_1}) \cap p_{a_2}^{-1}(U_{a_2}) \cap \cdots \cap p_{a_n}^{-1}(U_{a_n}),$$

where $a_1, a_2, \ldots, a_n$ are arbitrary finite number of elements in $\mathbf{A}$ and $U_{a_i}$ is an arbitrary member of the base $\mathbf{G}$. In particular, if the indexing set $\mathbf{A} = \{1, 2, \ldots, m\}$, then base $G$ consists of all products of the form $U_1 \times U_2 \times \cdots \times U_m$, where $U_i \in \tau_i$ for $i = 1, 2, \ldots, m$.

**Theorem 3.12.15** *Let $(X, \tau)$ and $(Y, \sigma)$ be two topological spaces with $(X \times Y, \Sigma)$ their topological product space. If $(W, \rho)$ is a given topological space and*

$$p_1 \colon (X \times Y, \Sigma) \to (X, \tau), \ (x, y) \mapsto x,$$

*and*

$$p_2 \colon (X \times Y, \Sigma) \to (Y, \sigma), \ (x, y) \mapsto y$$

*are projection maps, then a map*

$$f \colon (W, \rho) \to (X \times Y, \Sigma)$$

*is continuous iff the composite maps*

*(i) $p_1 \circ f \colon (W, \rho) \to (X, \tau)$ and*
*(ii) $p_2 \circ f \colon (W, \rho) \to (Y, \sigma)$*
*   are both continuous.*

***Proof*** First suppose that $f$ is continuous. Since $p_1$ and $p_2$ are both continuous by Theorem 3.12.12, it follows that the composites $p_1 \circ f$ and $p_2 \circ f$ are both continuous. Conversely, suppose that $p_1 \circ f$ and $p_2 \circ f$ are both continuous. Let $U \times V$ be an arbitrary basic open set in the product topology $\Sigma$ of $X \times Y$. To prove the continuity of $f$, it is sufficient to show that $f^{-1}(U \times V) \in \rho$. Now,

$$f^{-1}(U \times V) = (p_1 \circ f)^{-1}(U) \cap (p_2 \circ f)^{-1}(V) \in \rho,$$

since it is the intersection of two open sets of the topological space $W$ under continuity assumption of $p_1 \circ f$ and $p_2 \circ f$. This asserts that $f$ is continuous.   $\square$

## 3.13 Topological Product of Metrizable Spaces

This section proves the metrizability of product of metrizable spaces, which is used to prove the metrizabilty of the Euclidean space $\mathbf{R}^n$ and unit cube $\mathbf{I}^n \subset \mathbf{R}^n$. For more study of metrizability, see subsequent chapters. Theorem 3.13.1 proves that metrizability is preserved by product of countable family of metrizable spaces. An alternative proof of this Theorem is given in Theorem 3.13.6.

**Theorem 3.13.1** *The topological product of any countable family of metrizable spaces is metrizable.*

**Proof** Let $\mathcal{F} = \{(X_n, \tau_n)\}$ be a countable family of metrizable spaces and $(X, \tau)$ be their topological product space. We have to prove that $(X, \tau)$ is also metrizable.
**Case I**: Let $\mathcal{F}$ be finite and

$$(X, \tau) = (X_1, \tau_1) \times (X_2, \tau_1) \times \cdots \times (X_n, \tau_n).$$

Then there exists a metric $d_i$ on each $X_i$ such its induced topology $\tau_{d_i}$ coincides with $\tau_i$ for each $i = 1, 2, 3, \ldots, n$. Define a map

$$d : X \times X \to \mathbf{R}, \ (x, y) \mapsto [\Sigma_{i=1}^{i=n} (d_i(x_i, y_i))^2]^{\frac{1}{2}}$$

for every pair of points $x = (x_1, x_2, \ldots, x_n)$ and $y = (y_1, y_2, \ldots, y_n)$ in $X$. Then $d$ is a metric on $X$, and hence, it defines the topology $\tau_d = \tau$. This proves that the product space $(X, \tau)$ is also metrizable.

**Case II**: Let $\mathcal{F}$ be countable. Without loss of generality, assume by Exercise 61 of Sect. 3.20 (or see proof of Theorem 3.13.6) that for every positive integer $i$,

$$d_i(x_i, y_i) \leq 1, \ \forall (x_i, y_i) \in X_i \times X_i.$$

Then the map

$$d : X \times X \to \mathbf{R}, (x, y) \mapsto \Sigma_{i=1}^{\infty} \frac{1}{2^i} (d_i(x_i, y_i)) \ \forall x_i, y_i \in X_i, \ and \ \forall i > 0$$

for every pair of points $x = (x_1, x_2, \ldots)$ and $y = (y_1, y_2, \ldots)$ in $X$, is a metric. Hence, $d$ defines the topology $\tau_d = \tau$. This asserts that in this case, the product space $(X, \tau)$ is also metrizable.
Considering both the cases, the proof of the theorem is completed. $\square$

**Corollary 3.13.2** *Euclidean n-space $(\mathbf{R}^n, \sigma)$ with usual metric on $\mathbf{R}^n$ is metrizable*

**Proof** It follows from Theorem 3.13.1, since $\mathbf{R}$ is metrizable. $\square$

**Corollary 3.13.3** *The unit n-cube $\mathbf{I}^n$ with usual metric is metrizable.*

**Proof** The product space $\Pi_{n=1}^{n}(\mathbf{I}_n, \tau_n)$ is the unit $n$-cube. Hence, the Corollary follows from Theorem 3.13.1, since $\mathbf{I}$ is metrizable. □

## Metrizability of Hilbert Cube

This subsection defines Hilbert cube named after David Hilbert (1862–1943) and unit $n$-cube, which are important topological spaces and studies their metrizability property.

**Definition 3.13.4** Let the topological space $(\mathbf{I}_n, \tau_n)$ be homeomorphic to $[0, 1]$ for each positive integer $n$. Then the product space $\Pi_{n=1}^{\infty}(\mathbf{I}_n, \tau_n)$ is called the **Hilbert cube**, denoted by $\mathbf{I}^{\infty}$.

**Remark 3.13.5** An alternative statement of Theorem 3.13.1 with a proof in more precise form is given in Theorem 3.13.6.

**Theorem 3.13.6** *Let* $\{(X_n, \tau_n): n \in \mathbf{N}\}$ *be a countably infinite family of metrizable spaces. Then their product space* $\Pi_{n=1}^{\infty}(X_n, \tau_n)$ *is also metrizable.*

**Proof** Let $d_n$ be a metric on $X_n$, which induces the topology $\tau_n$ on $X_n$ for every $n \in \mathbf{N}$. Define a metric $m_n$ equivalent to the metric $d_n$ on $X_n$ for all $n \in \mathbf{N}$ by taking

$$m_n = \min\{1, d_n(x, y), \ \forall x, y \in X_n \text{ and } \forall n \in \mathbf{N}\}.$$

So, we may assume that

$$d_n(x, y) \leq 1, \ \forall x, y \in X_n \text{ and } \forall n \in \mathbf{N}.$$

Define the function

$$d: \Pi_{n=1}^{\infty} X_n \times \Pi_{n=1}^{\infty} X_n \to \mathbf{R}: (\Pi_{n=1}^{\infty} x_n, \Pi_{n=1}^{\infty} y_n) \mapsto \Sigma_{n=1}^{\infty} \frac{d_n(x_n, y_n)}{2^n}$$

Then the function $d$ is well-defined, because, each $d_n(x_n, y_n) \leq 1$ and so it is bounded above by $\Sigma_{n=1}^{\infty} \frac{1}{2^n} = 1$, and hence, the right-hand series $\Sigma_{n=1}^{\infty} \frac{d_n(x_n, y_n)}{2^n}$ is convergent. This function $d$ is a metric on $\Pi_{n=1}^{\infty} X_n$. Hence, it follows that the product space $\Pi_{n=1}^{\infty}(X_n, \sigma_n)$ is also metrizable. □

**Corollary 3.13.7** *The Hilbert cube* $\mathbf{I}^{\infty}$ *is metrizable.*

**Proof** It follows from Theorem 3.13.6, since $\mathbf{I}$ is metrizable. □

## 3.14 Continuous Maps into Product Spaces

This section studies the problems of continuity of maps from an arbitrary topological space to a product space. For this subsection, $\Pi_{a \in \mathbf{A}} X_a$ denotes the product space of the topological spaces $\{(X_a, \tau_a): a \in \mathbf{A}\}$, where $\mathbf{A}$ is an arbitrary (possibly, infinite) indexing set. Consider the map $f: X \to \Pi_{a \in \mathbf{A}} X_a$ from an arbitrary topological space $X$ into a product space $\Pi_{a \in \mathbf{A}} X_a$. If

$$p_a: \Pi_{a \in \mathbf{A}} X_a \to X_a$$

is the projection map, then the component mapping $f_a: X \to X_a$ of $f$ is such that $f_a = p_a \circ f$. Again given a set of mappings

$$\{f_a: X \to X_a: a \in \mathbf{A}\},$$

the mapping $f: X \to \Pi_{a \in \mathbf{A}} X_a$ is uniquely determined. This gives a bijective correspondence between the set of mappings

$$f: X \to \Pi_{a \in \mathbf{A}} X_a$$

and the set of the family of mappings $\{f_a: X \to X_a: a \in \mathbf{A}\}$.

**Theorem 3.14.1** *Let $\{(X_a, \tau_a): a \in \mathbf{A}\}$ be a family of topological spaces and $X$ be a given topological space. Then a mapping*

$$f: X \to \Pi_{a \in \mathbf{A}} X_a$$

*is continuous iff the component mapping $f_a: X \to X_a$ of $f$ is continuous for each $a \in \mathbf{A}$.*

***Proof*** Suppose $f_a: X \to X_a$ is continuous for each $a \in \mathbf{A}$. Let $U$ be any member of the open base for the product space $\Pi_{a \in \mathbf{A}} X_a$. Then $f^{-1}(U)$ is open in $X$. It asserts that $f$ is continuous. Its converse is left to an exercise. □

Theorem 3.14.1 is applied to prove Corollary 3.14.2, an important result in topology.

**Corollary 3.14.2** *Every diagonal map on a topological space is continuous.*

***Proof*** Let $X$ be a topological space and $X \times X$ be the product space. If

$$\Delta: X \to X \times X, x \mapsto (x, x)$$

is the diagonal map, then

$$\Delta(x) = (x, x) = (1_X(x), 1_X(x)), \ \forall x \in X$$

shows that the map $\Delta$ can be represented as the map $(1_X, 1_X)$. Since the identity map $1_X : X \to X$ is continuous, it follows by Theorem 3.14.1 that $\Delta$ is continuous.   $\square$

***Remark 3.14.3*** If $X = X_1 = X_2 = \cdots = X_n = \mathbf{R}$, then a mapping $f : \mathbf{R} \to \mathbf{R}^n$ is an $n$-tuple of numerical functions. A mapping $f : X \to \Pi_{a \in \mathbf{A}} X_a$ may be considered as its generalization.

## 3.15   Weak Topology and Construction of $S^\infty$, $\mathbf{RP}^\infty$ and $\mathbf{CP}^\infty$

This section presents the concept of weak topology which is used in construction of some important geometrical and topological spaces such as infinite dimensional sphere $S^\infty$, infinite dimensional real projective space $\mathbf{RP}^\infty$ and infinite dimensional complex projective space $\mathbf{CP}^\infty$.

### 3.15.1   Weak Topology (Union Topology)

This subsection introduces the concept of weak topology or union topology with an eye to construct some important spaces such as infinite dimensional sphere, infinite dimensional real and complex projective spaces.

**Definition 3.15.1** Let $\mathbf{A}$ be an indexing set (countable or noncountable) and $\{X_i : i \in \mathbf{A}\}$ be a family of subsets of a set X such that $X = \bigcup_{i \in \mathbf{A}} X_i$. Suppose

  (i)   each $X_i$ is a topological space;
  (ii)  for every pair of indexes $i, j \in \mathbf{A}$, the topologies on $X_i$ and $X_j$ agree on their intersection $X_i \cap X_j$;
  (iii) for every pair of indexes $i, j \in \mathbf{A}$, the intersection $X_i \cap X_j$ is closed in both $X_i$ and $X_j$.

Then the topology $\tau$ defined on $X$ by declaring a subset $A \subset \bigcup_{i \geq 1} X_i$ to be closed iff its intersection $A \cap X_i$ is closed in the topological space $X_i$ for each $i \in \mathbf{A}$. This topology $\tau$ is called the **union topology or weak topology** on $X$ determined by the family $\{X_i : i \in \mathbf{A}\}$.

***Example 3.15.2*** Important examples of topological spaces endowed with weak topology are available in Sect. 3.15.2.

### 3.15.2    Construction of $S^\infty$, $\mathbf{RP}^\infty$ and $\mathbf{CP}^\infty$ with Weak Topology

This subsection constructs the infinite dimensional spheres $S^\infty$, infinite dimensional real projective space $\mathbf{RP}^\infty$ and also infinite dimensional complex projective space $\mathbf{CP}^\infty$ in Proposition 3.15.3 by using the concept of weak topology. For construction of the $n$-sphere $S^n$ and $n$-dimensional real projective of $\mathbf{R}P^n$ see Sect. 3.16.3. More properties of weak topology are given in Exercise 30 of Sect. 3.20.

**Proposition 3.15.3** (Construction of $S^\infty$, $\mathbf{RP}^\infty$ and $\mathbf{CP}^\infty$)

(i) *(**Infinite dimensional sphere**): $S^\infty = \bigcup_{n=0}^{\infty} S^n$ with weak topology, where the $n$-sphere $S^n$ is defined by $S^n = \{x \in \mathbf{R}^{n+1}: ||x|| = 1\}$, which is the boundary of the $(n+1)$-disk $\mathbf{D}^{n+1}$ in $\mathbf{R}^{n+1}$. The space $S^\infty$ is called infinite dimensional sphere.*

(ii) *(**Infinite dimensional real projective space**): $\mathbf{RP}^\infty = \bigcup_{n=0}^{\infty} \mathbf{R}P^n$ with weak topology, where the n-dimensional real projective space $\mathbf{R}P^n$ is the space of all straight lines through the origin in Euclidean space $\mathbf{R}^{n+1}$ (see Example 3.16.20). The space $\mathbf{RP}^\infty$ is called infinite dimensional real projective space.*

(iii) *(**Infinite dimensional complex projective space**): $\mathbf{CP}^\infty = \bigcup_{n=0}^{\infty} \mathbf{C}P^n$ with weak topology, where the n- dimensional complex projective space $\mathbf{C}P^n$ is the space of all complex lines through the origin in complex space $\mathbf{C}^{n+1}$. The space $\mathbf{CP}^\infty$ is called infinite dimensional real projective space.*

*Example 3.15.4* If $X$ is endowed with weak topology determined by its subspaces $\{X_i: i \in \mathbf{A}\}$, then each $X_i$ is closed in $X$ and each subspace $X_i$ of $X$ has its original subspace topology. In particular, if the indexing set $\mathbf{A}$ is finite, there is only one topology on $X$, which is the weak topology.

## 3.16    Quotient Spaces: Construction of Geometrical Objects Through Topological Methods

The section is devoted to address the concept of quotient spaces or identification spaces and also presents the mathematical version of a geometric process to obtain new geometric objects by several methods to obtain quotient spaces described in this section. Many interesting topological spaces can be constructed from a simple topological space by identifying some subset (or points) of $X$. For example, a circle is obtained by gluing together the end points of a closed line segment. The concept of quotient spaces was introduced by Robert L. Moore in 1925 and was also independently, by Pavel Alexandroff (1896–1982).

## 3.16.1  Quotient Topology and Quotient Spaces

This subsection explains the formal definition of the quotient topology formalizing the geometric process of gluing and identification and also constructs quotient (factor) spaces by using the concept quotient topology (see Proposition 3.16.2).

**Definition 3.16.1  Identification topology or quotient topology** induced by a surjective map. Let $(X, \tau)$ be a topological space. Given a nonempty set $Y$ and a surjective map $f: X \to Y$, the family of subsets $\tau_f$, determined by

$$\tau_f = \{U \subset Y : f^{-1}(U) \in \tau\}$$

forms a topology, called the identification topology or quotient topology on $Y$, induced by the map $f$ and the new topological space $(Y, \tau_f)$ is called the **quotient space or identification space**. The process of identification is sometimes called **gluing or pasting**

**Proposition 3.16.2** *Let $(X, \tau)$ be a topological space. Given a nonempty set $Y$ and a surjective map $f: X \to Y$, the quotient topology $\tau_f$ on $Y$ is the strongest (largest) topology on $Y$ for which $f: X \to Y$ is continuous.*

**Proof** It follows from Proposition 3.6.5.                                                            □

**Example 3.16.3** (An immediate application of Proposition 3.16.2). Let $X$ be a given topological space, and $\mathcal{P}$ be a partition of $X$. A new topological space $Y$ is an identification space or a quotient space if the points of $Y$ are the members of $\mathcal{P}$ and if $p: X \to Y$ sends each point of $X$ to the subset of $\mathcal{P}$ containing it, the topology of $Y$ is the largest for which $p$ is continuous by Proposition 3.16.5. More precisely, let $(X, \tau)$ be a topological space and $\rho$ be an equivalence relation on $X$ generating the partition $\mathcal{P}$. Let $X/\rho$ be the quotient set and $[x]$ denote the class which contains the element $x \in X$. Then the map

$$p: X \to X/\rho, x \to [x]$$

called the **natural projection,** which is surjective. Let $\sigma$ be the family of subsets of $X/\rho$ given by

$$\sigma = \{U \subset X/\rho : p^{-1}(U) \in \tau\}.$$

Then $\sigma$ forms a topology by Proposition 3.16.2 on $X/\rho$, called the **quotient topology** or topology on $X/\rho$ determined by the equivalence relation $\rho$ and the topological space $(X/\rho, \sigma)$ thus obtained is called the **quotient space or the decomposition space or identification space, determined by** $\rho$.

**Example 3.16.4** Let $\mathbf{I}$ be the closed unit interval and $\rho$ be an equivalence relation on $\mathbf{I}$ such that

$$[0] = [1] = \{0, 1\} \text{ and } [x] = \{x\} \text{ for } 0 < x < 1.$$

Then $\mathbf{I}/\rho$ is the quotient space homeomorphic to the circle $S^1$. In other words, $S^1$ is obtained from $\mathbf{I}$ by identifying the end points 0 and 1 of $\mathbf{I}$.

**Proposition 3.16.5** *Given a topological space* $(X, \tau)$ *and an equivalence relation* $\rho$ *on* $X$, *the quotient topology* $\sigma$ *on* $X/\rho$ *is the strongest (largest) topology for which the projection*

$$p: X \to X/\rho, \ x \to [x], \ \text{the class which contains the element } x \in X,$$

*is continuous.*

***Proof*** **Proof I**: It follows from Proposition 3.16.2, since the projection map

$$p: X \to X/\rho, x \to [x]$$

is surjective.
**Proof II**: Let $\{V\}$ be a family of open sets forming a topology $\tau'$ on $X/\rho$ such that

$$p: X \to X/\rho, x \to [x]$$

is continuous. Then $p^{-1}(V)$ is open in $(X, \tau)$. Consequently, $V$ is open in the quotient space $X/\rho$, i.e., $V \in \tau_p \subset \sigma$. This implies that $V \in \sigma$. This means that the topology $\tau'$ is weaker than the topology $\sigma$. Consequently, the topology $\sigma$ on $X/\rho$ is the largest topology such that $p$ is continuous. □

***Remark 3.16.6*** The natural problem is: How to relate the topological spaces $(Y, \tau_f)$ and $(X/\rho, \sigma)$ constructed in Definition 3.16.1 and in Proposition 3.16.5? To solve this problem, let $(X, \tau)$ be a topological space and $Y$ be an arbitrary nonempty set. If $f: X \to Y$ be a surjective set function, then by Proposition 3.16.2, there exists a topology $\tau_f$, which is the largest topology on $Y$ such that the map $f: X \to Y$ is continuous. Again, the map $f: X \to Y$ induces an equivalence relation $\rho$ on $X$ defined by $(x, y) \in \rho$ iff $f(x) = f(y)$ such that the projection map $p: X \to X/\rho, x \to [x]$ is surjective. Then by Proposition 3.16.2, there exists a topology $\tau_p$, which is the largest topology on $X/\rho$ such that the projection map $p$ is continuous. Again, by Proposition 3.16.5 there exists a topology $\sigma$ on $X/\rho$, which is the largest topology such that the projection map

$$p: X \to X/\rho, x \to [x]$$

is continuous. It asserts that $\sigma = \tau_p$.

### 3.16.2  Quotient or Identification Maps

This subsection studies identification topology and identification map. For different applications of identification maps see the rest of this chapter as well as Chapters 5 and **Basic Topology, Volumes II and III** of the present book series..

**Definition 3.16.7** (*Quotient or identification map*) Let $(X, \tau)$ and $(Y, \sigma)$ be two topological spaces and $f: (X, \tau) \to (Y, \sigma)$ be a surjective map. If $\tau_f$ is the quotient topology on $Y$ induced by $f$, then the map $f$ is called a quotient or identification map, if $\sigma = \tau_f$.

**Proposition 3.16.8** *Let $(X, \tau)$ and $(Y, \sigma)$ be two topological spaces. Then a surjective map*

$$f: (X, \tau) \to (Y, \sigma)$$

*is a quotient map if either*

*(i) $f$ is both open and continuous or*
*(ii) $f$ is both closed and continuous.*

**Proof** To prove the proposition, it is sufficient to prove that $\sigma = \tau_f$.

(i) Let $f$ be both open and continuous. Since $f$ is continuous, then for every open set $U \in \sigma$, $f^{-1}(U) \in \tau$, and hence, $U \in \tau_f$. This shows that $\sigma \subset \tau_f$. Again, since $f$ is open, for every open set $V \in \tau_f$, the inverse image $f^{-1}(V) \in \tau$, and hence, $V = f(f^{-1}(V)) \in \sigma$ shows that $\tau_f \subset \sigma$. This asserts that $\sigma = \tau_f$.

(ii) Let $f$ be both closed and continuous. By continuity of $f$, it follows that $\sigma \subset \tau_f$. Conversely, for any open set

$$U \in \tau_f, \, f^{-1}(U) \in \tau \implies X - f^{-1}(U)$$

is a closed set in $(X, \tau)$. This asserts that $f(X - f^{-1}(U)) = Y - U$ is a closed set in $(Y, \sigma)$, and hence, $U \in \sigma$ shows that $\tau_f \subset \sigma$. This asserts that $\sigma = \tau_f$. $\square$

The conditions (i) and (ii) prescribed in Proposition 3.16.8 give a sufficient condition for a surjective map to be a quotient map but are not necessary (see Proposition 3.16.9).

**Proposition 3.16.9** *Let $(X, \tau)$. $(Y, \sigma)$ and $(Z, \mu)$ be three topological spaces. If $f: (X, \tau) \to (Y, \sigma)$ is a quotient map and $g: (Y, \sigma) \to (Z, \mu)$ is a given map, then $g$ is continuous iff*

$$g \circ f: (X, \tau) \to (Z, \mu)$$

*is continuous.*

**Proof** First let $g: (X, \tau) \to (Z, \mu)$ be continuous. Since any quotient map $f$ is continuous, then composite $g \circ f$ is also continuous. Conversely, let $g \circ f$ be continuous. Let $U$ be an open set in $Z$. Then $f^{-1}(g^{-1}(U)) = (g \circ f)^{-1}(U)$ is an open set in $(X, \tau)$, and hence, $g^{-1}(U) \in \sigma$, since $\sigma$ is the quotient topology $\tau_f$ induced by $f$. This implies that $g$ is continuous.

$\square$

**Remark 3.16.10** Theorem 3.16.11 solves the problem raised in the Remark 3.16.6 by proving that the topological spaces $(Y, \tau_f)$ and $(X/\rho, \tau_p)$ are homeomorphic. Notations used in Theorem 3.16.11 are explained in Remark 3.16.6.

**Theorem 3.16.11** *Let* $(X, \tau)$ *be a topological space, $Y$ be a nonempty set and* $f: X \to Y$ *be a given surjective set function. If $\tau_f$ is the identification topology on* $Y$, *induced by $f$, then topological space* $(Y, \tau_f)$ *is homeomorphic to the quotient space* $(X/\rho, \tau_\rho)$, *where $\rho$ is a equivalence relation on $X$ defined by $(x, y) \in \rho$ iff* $f(x) = f(y)$.

**Proof** Since $\rho$ is an equivalence on $X$, it determines a topological space $(X/\rho, \tau_\rho)$. By hypothesis, $f$ is surjective and hence the map $g: X/\rho \to Y$, $[x] \mapsto f(x)$ is a bijection. Let $p: (X, \tau) \to (X/\rho, \tau_\rho), x \mapsto [x]$ be the canonical projection mapping onto $X/\rho$. Since $g \circ p = f$ is continuous, it follows by Proposition 3.16.9 that $g$ is continuous. Again, since $g^{-1} \circ f = p$ is continuous it follows by Proposition 3.16.9 that $g^{-1}$ is also continuous. This asserts that $g: (X/\rho, \tau_p) \to (Y, \tau_f)$ is a homeomorphism.

□

**Theorem 3.16.12** *Given an identification map $p: X \to Y$ on $Y$ and an arbitrary topological space $Z$, a map*

$$f : Y \to Z$$

*is continuous iff the composite $f \circ p: X \to Z$ is continuous.*

**Proof** Given an open subset $V \subset Z$, $f^{-1}(V)$ is open in $Y$ iff $p^{-1}(f^{-1}(V))$ is open in $X$. This asserts that $p^{-1}(f^{-1}(V))$ is open in $X$ iff $\{p \circ p\}^{-1}(V)$ is open in $X$. This implies that $f$ is continuous iff $p^{-1}(f^{-1}(V))$ is open in $X$, i.e., iff $f \circ p: X \to Z$ is continuous.

□

**Proposition 3.16.13** *The quotient space of a quotient space is also a quotient space.*

**Proof** Let $X$, $Y$, $Z$ be three topological spaces such that $Y$ is endowed with quotient topology from $X$ and $Z$ is endowed with quotient topology from $Y$. Since for two continuous onto maps $f: X \to Y$ and $g: Y \to Z$, the quotient topology of $Z$ is induced by the composite map $g \circ f$, the proposition follows.

□

**Example 3.16.14** Define an equivalence $\sim$ on the real line space $\mathbf{R}$ by the rule $x \sim y$ iff $x - y$ is an integer. Then the quotient space $\mathbf{R}/\sim$ is homeomorphic to the circle $S^1$.

**Example 3.16.15** Define an equivalence $\sim$ on the real line space $\mathbf{R}$ by the rule $x \sim y$ iff $x - y$ is a rational number. Then the corresponding quotient space $\mathbf{R}/\sim$ is an indiscrete space.

Theorem 3.16.16 gives a criterion for an identification map.

**Theorem 3.16.16** *Let $X$ and $Y$ be topological spaces. If $f: X \to Y$ is a surjective map such that $f$ maps open sets of $X$ to open sets of $Y$ or $f$ maps closed sets of $X$ to closed sets of $Y$, then $f$ is an identification map.*

**Proof** Let $f$ be a map such that $f$ maps closed sets of $X$ to closed sets of $Y$. Suppose $A$ is a subset of $B$ such that $f^{-1}(A)$ is closed in $X$. As $f$ is onto by hypothesis, $f(f^{-1}(A)) = A$.. This asserts that $A$ must be closed in the given topology on $Y$. This topology is the largest topology for which $f$ is continuous and $f$ is an identification map. The proof is similar for open maps.                                          □

### 3.16.3   Construction of $S^n$ and $RP^n$ by Identification Method

This subsection constructs the $n$-sphere $S^n$ and real projective space $RP^n$ by **identification method**. In view of Theorem 3.16.11, the identification space and identification topology can be reformulated in the following form:

**Definition 3.16.17** Let $X$ be a given topological space, and $\mathcal{P}$ be a partition on $X$. A new topological space $Y$ is said to be an **identification space** if the points of $Y$ are the members of $\mathcal{P}$ and if $p: X \to Y$ sends each point of $X$ to the subset of $\mathcal{P}$ containing it, the topology of $Y$ is the largest for which $p$ is continuous. An identification space $Y$ is obtained from a topological space $(X, \tau)$ by identifying each member of the partition $\mathcal{P}$ on $X$ to a single point.

**Example 3.16.18** Let $\mathbf{I} = [0, 1]$ be the closed unit interval and $\rho$ be an equivalence relation $\sim$ on $\mathbf{I}$ such that $[0] = [1] = \{0, 1\}$ and $[x] = \{x\}$ for $0 < x < 1$. Then $\mathbf{I}/\rho$ is the quotient space homeomorphic to the circle $S^1$. In other words, $S^1$ is obtained from $I$ by identifying the end points 0 and 1 of $\mathbf{I}$.

**Example 3.16.19** (i) Let $\mathbf{D}^2 = \{(x, y) \in \mathbf{R}^2 : x^2 + y^2 = 1\}$ be the closed disk. Then its boundary $\partial \mathbf{D}^2$ is the circle $S^1 = \{(x, y) \in \mathbf{R}^2 : x^2 + y^2 = 1\}$. Define an equivalence relation $\rho$ on $S^1$: $(x_1, y_1)\rho(x_2, y_2)$ iff $x_1^2 + y_1^2 = 1 = x_2^2 + y_2^2$, i.e., iff $\rho$ identifies the points on $S^1$. The quotient space thus obtained is the 2-sphere $S^2$ written as $\mathbf{D}^2/S^1$. **Geometrically**, if we identify all the points of the circumference of a disk $\mathbf{D}^2$, then the resulting quotient space is homeomorphic to the sphere $S^2$. In general. the quotient space $\mathbf{D}^n/S^{n-1}$ is the $n$-sphere $S^n$.

**Theorem 3.16.20** (Real Projective Space $RP^n$): *Let* $X = \mathbf{R}^{n+1} - \{0\}$, *where* $0 = (0, 0, \ldots, 0) \in \mathbf{R}^{n+1}$ *and* $(X, \tau)$ *be the subspace of the Euclidean $n$-space* $\mathbf{R}^{n+1}$. *Let* "$\sim$" *be a binary relation on $X$ defined by*

$$(x_1, x_2, \ldots, x_{n+1}) \sim (y_1, y_2, \ldots, y_{n+1})$$

*holds iff $y_i = rx_i$ $(i = 1, 2, \ldots, n + 1)$ for some nonzero real number $r$. Then $\sim$ is an equivalence relation on $X$. The quotient space $X / \sim$ thus obtained is called the $n$-dimensional real projective space, denoted by* $RP^n$. *Hence, the points of* $RP^n$ *are the straight lines in* $\mathbf{R}^{n+1}$ *passing through its origin* $0$. *Let*

$$p: X \to RP^n$$

*be the natural projection. Since the n- sphere $S^n \subset X$, the restriction of $p$ to $S^n$,*

$$q = p|S^n: S^n \to \mathbf{RP}^n$$

*is surjective, because, any point $x = (x_1, x_2, \ldots, x_{n+1}) \in X$ can be normalized by multiplying with nonzero real number $r = \frac{1}{||x||}$, and hence, $q$ is an identification map. For any two points $x = (x_1, x_2, \ldots, x_{n+1})$ and $y = (y_1, y_2, \ldots, y_{n+1}) \in S^n$,*

$$q(x) = q(y)$$

*iff $x_i + y_i = 0$ $(i = 1, 2, \ldots, n + 1)$, i.e., iff $x$ and $y$ are antipodal points of $S^n$. This asserts that $\mathbf{RP}^n$ can be obtained from $S^n$ by identifying the antipodal points of $S^n$.*

**Remark 3.16.21** **Geometrical Interpretation of Theorem** 3.16.20: This theorem asserts that the real projective space $\mathbf{R}P^n$ obtained from $\mathbf{R}^{n+1} - \{0\}$ as a quotient space by an equivalence relation $\sim$ consists of all straight lines in $\mathbf{R}^{n+1}$ passing its origin. It is also obtained from the $n$-sphere $S^n$ by identifying its diametrically opposite points.

**Proposition 3.16.22** *Let $(Y, \sigma)$ be an identification space obtained from the topological space $(X, \tau)$ by a surjective map $f : X \to Y$. Given an arbitrary topological space $(Z, \mu)$, a map $g : Y \to Z$ is continuous iff the composite map*

$$g \circ f : (X, \tau) \to (Z, \mu)$$

*is continuous.*

**Proof** It follows from Proposition 3.16.9. □

### 3.16.4 Constructions of Spheres and Cones by Collapsing Method

This subsection constructs some interesting topological spaces such as sphere and cone by collapsing a subspace of a topological space to a point.

**Definition 3.16.23** Let $X$ be a topological space and $A$ be a given nonempty subspace of $X$. The set $A$ together and the individual points of $X - A$ constitute a decomposition of the space $X$. This defines a quotient space, called the topological space obtained from $X$ by **collapsing the subspace $A$ to a point**, and this process is known as **collapsing method**. In this method, all points of $A$ are identified to a single equivalence class and all points in $X - A$ are remained equivalent to themselves by this decomposition.

***Example 3.16.24*** **(Construction of the $n$-sphere $S^n$)** Let $X = \mathbf{D}^n$ be the closed unit disk in $\mathbf{R}^n$ and $A = S^{n-1} = \partial \mathbf{D}^n$ be the $(n-1)$-sphere in $\mathbf{R}^n$. The set $A$ being the boundary of $X$, it is a subspace of $X$ with relative topology. The unit $n$-sphere $S^n$ is homeomorphic to the quotient space obtained from $X$ by collapsing $A$ to a point. This implies that the $n$-sphere $S^n$ can be obtained as a quotient space from the closed $n$-disk $\mathbf{D}^n$ by collapsing its boundary $S^{n-1}$ to a point. This is sometimes written as $\mathbf{D}^n/S^{n-1} = S^n$ (up to homeomorphism). In particular, the 2-sphere $S^2$ is homeomorphic to the quotient space $\mathbf{D}^2/S^1$ obtained from the unit disk $\mathbf{D}^2$ by identifying its boundary $\partial \mathbf{D}^2 = S^1$ to the single equivalent class.

***Example 3.16.25*** **(Construction of Cone)** Let $A$ be an arbitrary space and $X = A \times \mathbf{I}$ be the product space of $A$ and the closed unit interval $\mathbf{I}$. If $T = A \times \{1\}$, then the quotient space obtained from $X$ by collapsing its top $T$ to a point, then the quotient space is called the cone over $A$, denoted by Con $(A)$. The point $v = A \times \{1\}$ is called the vertex of the cone Con $(A)$. The given space $A$ is considered as a subspace of Con $(A)$ by an embedding

$$f : A \to \mathrm{Con}(A), \ a \mapsto p(x, 0),$$

where $p : A \times \mathbf{I} \to \mathrm{Con}(A)$ is the natural projection.

### 3.16.5  Construction of New Spaces by Gluing Method

This subsection constructs some well-known surfaces in the Euclidean space obtained by gluing method. The concept of quotient spaces in topology and geometry formalizes mathematically the intuitive idea of the process of identification by gluing or pasting.

***Example 3.16.26*** The circle $S^1$ is obtained from the closed interval $I = [0, 1]$ by gluing (identifying) its end points 0 and 1, as shown in Example 3.16.18.

### 3.16.6  Attaching Map: Construction of Mapping Cylinder and Mapping Cone

This section concentrates on a special type of identification spaces, called **adjunct spaces**. The concept of attaching one space to another by a continuous map may be given by using identification space. The process of attaching a topological space to another space by a map plays a key role in topology which provides constructions of many important objects such as cone, cylinder and suspension space. Recall that given two disjoint spaces $X$ and $Y$, their disjoint union $X \sqcup Y$ (or $X + Y$) is the set $X \cup Y$ endowed with the topology defined: a subset $V \subset X \sqcup Y$ is open iff both

$V \cap X$ is open in $X$ and $V \cap Y$ is open in $Y$, is called the topological sum of $X$ and $Y$. Moreover, given a disjoint union $X \sqcup Y$, as $X \cap Y$ is the empty set, $X$ and $Y$ carry their own topologies and they are disjoint open sets in $X + Y$, Moreover, a subset $A \subset X + Y$ is closed if both $A \cap X$ is closed in $X$ and $A \cap Y$ is closed in $Y$. We now utilize these result in our next constructions. Definition 3.16.27 shows an important aspect of identification of topological spaces by formalizing the concept of attaching one topological space to another by a continuous map.

**Definition 3.16.27** (*Attaching map*) Given two disjoint topological spaces $X$ and $Y$, a closed subset $A$ of $X$, and a continuous map $f : A \to Y$, the topological sum $X + Y$ gives an equivalence relation $\rho$ obtained by identifying $a$ with $f(a)$ for each $a \in A$. More precisely, the subsets of the partition generated by $\rho$ are precisely,

  (i) the pairs of points $\{(a, f(a))\}$;
 (ii) the individual points in the set $X - A$:
(iii) the individual points in the set $Y - \mathrm{Imf}\,(f)$.

The quotient space $(X + Y)/\rho$ is denoted by $X \cup_f Y$. The map $f$ is called the **attaching map** $f$ and the space $X \cup_f Y$ is said that $X$ is attached to $Y$ by the attaching map $f$.

**Example 3.16.28** The 1-circle $S^1$ is obtained from the closed interval $\mathbf{I} = [0, 1]$ by attaching $\mathbf{I}$ to a point $x_0$ by the attaching map $f : \mathbf{I} \to \mathbf{R}^2$ such that $f(0) = f(1) = x_0$

**Remark 3.16.29** The space $X \cup_f Y$ obtained in Definition 3.16.27 is also called the **adjunction space** constructed by gluing $X$ to $Y$ defined by the equivalence relation $\rho : x \rho y$ iff $x \in A$ and $f(x) = y$. The equivalence classes of the disjoint union $X \sqcup Y$ under $\rho$ constitute $X \sqcup_f Y$. For $A = \emptyset$, the adjunction space $Y \sqcup_f X = Y \sqcup X$, and for $A = X$, the adjunction space $Y \sqcup_f X = Y$. In particular, for $X = Y$, the map $f = 1d$ and the adjunction space $Y \sqcup_{1d} X = X$.

**Example 3.16.30** The equivalence relation generated by all pairs

$$a \sim f(a) : a \in A$$

consists of the following pairs:

  (i) either for $x, y \in X$, $x \sim y$ if $x = y$
 (ii) or $x, y \in A$ and $f(x) = f(y)$
(iii) or $x \in A$ and $y = f(x) \in Y$ .

**Example 3.16.31** Let $f : S^1 \to S^1, z \mapsto z^2$. Then $f$ is an identification map identifying the points $z$, and $-z$ of $S^1$ to the point $z$. Hence, $S^1 \sqcup_f \mathbf{R}^2$ is homeomorphic to the quotient space obtained from $S^2$ with its diametrically opposite points identified. This produces the real projective space $\mathbf{R}P^2$.

Mapping cylinder and mapping cone are important objects in topology constructed is Example 3.16.32.

***Example 3.16.32*** **(Constructions of Mapping Cylinder and Mapping Cone)**

(i) Let $X$ and $Y$ be topological spaces and $f : X \to Y$ be continuous. Let $S = (X \times \mathbf{I}) \sqcup Y$ denote the disjoint union of topological spaces $X \times \mathbf{I}$ and $Y$. Then both $X \times \mathbf{I}$ and $Y$ are open sets of $(X \times \mathbf{I}) \sqcup Y$. If we define an equivalence relation $\rho$ on $(X \times \mathbf{I}) \sqcup Y$ by $(x, t) \rho y$ iff $y = f(x)$ and $t = 1$, then the quotient space $M_f = ((X \times \mathbf{I} \sqcup Y) / \rho$ is called the mapping cylinder of $f$. Thus, $M_f$ is the space obtained from $Y$ and $(X \times \mathbf{I})/(x_0 \times \mathbf{I})$ by identifying for each $x \in X$, the points $(x, 1)$ and $f(x)$ as shown in Fig. 3.2, in which the thick line is supposed to be identified to a point (the base point of $M_f$). Denote $(x, t)\rho$ in $M_f$ by $[x, t]$ and $y\rho$ in $M_f$ by $[y]$. Then $[x] = [x, 1] = [f(x)]$, $\forall x \in X$. **The space $Y$ is embedded in $M_f$** under the map $y \to [y]$. In particular, if $Y$ is a one-point space, then $f : X \to Y$ is a constant map and $M_f$ is $CX$, the cone over $X$.

(ii) **Alternative construction** of mapping cylinder and mapping cone: Let $X$ and $Y$ be topological spaces and $f : X \to Y$ be continuous. The individual points

$$\{(x, t) \in X \times \mathbf{I} : x \in X, \ 0 \leq t < 1\}$$

together with the subsets

$$\{(f^{-1}(y) \times 1) \ \cup \{y\} : y \in Y\}$$

of the topological sum $S = (X \times \mathbf{I}) + Y$ form a partition of $S$. The quotient space obtained in this way is the mapping cylinder of $f$, denoted by $M_f$ and the spaces $X$, $Y$ are considered as subspaces of $M_f$ by the embeddings

$$g : X \to M_f, \ x \mapsto p(x, 0) \text{ and } h : Y \to M_f, \ y \mapsto p(y),$$

where $p : S = (X \times \mathbf{I}) \sqcup Y \to M_f$ is the natural projection map. On the other hand, the quotient space obtained from the mapping cylinder $M_f$ by collapsing its subspace $X$ to a point $v_0$ is the mapping cone $C_f$ with vertex $v_0$. **The space $Y$ is embedded in $C_f$** by usual embedding.

***Example 3.16.33*** Let $X$ and $Y$ be topological spaces with base points $x_0$ and $y_0$, respectively. Given a continuous map $f : X \to Y$, the mapping cone $C_f$ as shown in Fig. 3.3.
is the quotient space obtained from $Y$ and the cone $CX$ over $X$ by identifying the point $[x, 1]$ of $CX$ with the point $f(x)$ of $Y$ for all $x \in X$. The base point of $C_f$ is the point to which $[x_0, t]$ and $y_0$ are identical for all $t \in I$.

***Example 3.16.34*** **(Wedge)** Let $(X, x_0)$ and $(Y, y_0)$ be two pointed topological spaces. Their wedge(or one-point union) $X \vee Y$ is the quotient space of their disjoint union $X \sqcup Y$ in which the base points are identified. In general, if $X_i$ is a collection of disjoint spaces, with base point $x_i \in X_i$, then their wedge (or one-point union) $\bigvee_{i \in I} X_i$ is the quotient space $X/X_0$, where $X = \bigsqcup_{i \in I} X_i$ and $X_0$ is the subspace of $X$

**Fig. 3.2** Mapping cylinder
$M_f$ of $f : X \to Y$

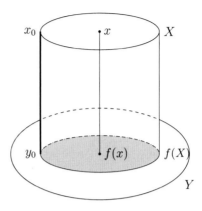

**Fig. 3.3** Mapping cone $C_f$
of $f : X \to Y$

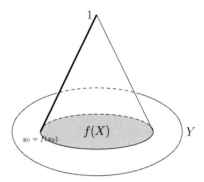

consisting of all base points $x_i$; the base point of $\bigvee_{i \in I} X_i$ is the point corresponding to $X_0$. In other words, $\bigvee_{i \in I} X_i$ is the space obtained from $X$ by identifying together the base point $x_i$.

**Example 3.16.35** (**Smash product**) Let $X$ and $Y$ be two pointed spaces with base points $x_0$ and $y_0$, respectively. Then their smash product (or reduced product) $X \wedge Y$ is defined to be the quotient space $X \times Y / (X \vee Y)$. We may think $X \wedge Y$ as a reduced version of $X \times Y$ obtained by collapsing $X \vee Y$ to a point. The smash product $X \vee Y$ is also written as $X \# Y$. The spaces $X$ and $Y$ are considered as subspaces of $X \vee Y$ by embedding

$$f : X \vee Y \to X \times Y, \quad x \mapsto \begin{cases} (x, y_0), \text{ if, } x \in X \\ (x_0, y), \text{ if } y \in Y, \end{cases} \ .$$

**Example 3.16.36** The smash product $X \vee Y$ depends on the base point. For example, if $0 \in [0, 1] = \mathbf{I}$ is taken as a base point of $\mathbf{I}$, then the smash product $\mathbf{I} \vee \mathbf{I}$ is homeomorphic to $\mathbf{I} \times \mathbf{I}$, where two adjacent sides of $\mathbf{I} \times \mathbf{I}$ are identified. On the other hand, if $\frac{1}{2} \in \mathbf{I}$ is taken as a base point of $\mathbf{I}$, then the smash product $\mathbf{I} \vee \mathbf{I}$ is homeomorphic to the wedge of four copies of $\mathbf{I} \times \mathbf{I}$.

***Example 3.16.37*** **(Reduced suspension space)** Let $X$ be a pointed topological space with base point $x_0$. Then the suspension of $X$, denoted by $\Sigma X$, is defined to be the quotient space of $X \times \mathbf{I}$ in which the subspace

$$(X \times 0) \cup (x_0 \times \mathbf{I}) \cup (X \times 1)$$

is identified to a single point. It is sometimes called the reduced suspension space. If $(x, t) \in X \times \mathbf{I}$, we use $[x, t]$ to denote the corresponding point of $\Sigma X$ under the identification map

$$f : X \times \mathbf{I} \to \Sigma X$$

such that $[x, 0] = [x_0, t] = [x', 1]$ for all $x, x' \in X$ and for all $t \in \mathbf{I}$. The point $[x_0, 0] \in \Sigma X$ is also denoted by $x_0$. Thus, $\Sigma X$ is a pointed space with base point $x_0$ and $\Sigma X = X \wedge S^1$. In particular, $\Sigma S^n$ is homeomorphic to $S^{n+1}$ for $n \geq 0$. Moreover, if $f : X \to Y$ is a base point preserving continuous map, then $\Sigma f : \Sigma X \to \Sigma Y$ is defined by $\Sigma f([x, t]) = [f(x), t]$.

**Proposition** 3.16.38 **expresses $\Sigma$ in the language of the category theory.**

**Proposition** 3.16.38 *Let* **Top**$_*$ *be the category of pointed topological spaces and base point preserving continuous maps. Then*

$$\Sigma : \mathbf{Top}_* \to \mathbf{Top}_*$$

*is a covariant functor.*

***Proof*** Define the object function

$$\Sigma : \mathbf{Top}_* \to \mathbf{Top}_*, \ X \mapsto \Sigma X$$

and define the morphism function for every morphism $f : X \to Y \in \mathbf{Top}_*$

$$\Sigma f : \Sigma X \to \Sigma Y, \ [x, t] \mapsto [f(x), t],$$

where $\Sigma X \in \mathbf{Top}_*$ for every object $X \in \mathbf{Top}_*$ by its construction and $\Sigma f \in \mathbf{Top}_*$ for every morphism $f \in \mathbf{Top}_*$ as described in Example 3.16.37.                            $\square$

***Example 3.16.39*** For the $n$-sphere $S^n$, its suspension space $\Sigma S^n = S^{n+1}$ and $\Sigma^{n-1}(S^1) = S^n$ (up to homeomorphism)

## 3.16.7 Construction of Cylinder, Möbius Band, Torus and Klein Bottle

This subsection describes construction of some geometrical objects such as the cylinder, Möbius band (Möbius strip) named after A. F. Möbius (1790–1868), torus and the Klein bottle named after F. Klein (1849–1925) from **the square by identification methods**. Historically, the Möbius band was constructed by Möbius in 1858 and the Klein bottle was constructed by Klein in 1882.

(i) **The cylinder** can be obtained as a quotient space obtained from the square $\mathbf{I} \times \mathbf{I}$ by identifying the point $(0, t)$ to the point $(1, t)$ for all $t \in \mathbf{I}$.

(ii) **The Möbius band (Möbius strip)** $M$ can be obtained as a quotient space obtained from the square $\mathbf{I} \times \mathbf{I}$ by identifying the point $(0, t)$ to the point $(1, 1 - t)$ for all $t \in \mathbf{I}$ as shown in Fig. 3.4. The identification topology on $M$ coincides with the subspace topology induced on $M$ by the usual topology on $\mathbf{R}^3$. Then Möbiu1s strip $M$ is embedded in $\mathbf{R}^3$. If

$$p: \mathbf{I} \times \mathbf{I} \to M$$

is natural projection, then the subspace

$$Y = p(\mathbf{I} \times \{0\} \cup (\mathbf{I} \times \{1\}) \subset M,$$

is called the edge of $M$, which is homeomorphic to the circle $S^1$. If $f: S^1 \to M$ is an embedding mapping $S^1$ homeomorphically onto $Y$, then the cone $C_f$ is homeomorphic to $\mathbf{R}P^2$. which can be embedded in $\mathbf{R}^4$.

(iii) The 2-**torus** $T^2$ **(or torus)** can be obtained as a quotient space obtained from the square $\mathbf{I} \times \mathbf{I}$ by identifying both pairs of opposite edges via identifying the point $(0, t)$ with the point $(1, t)$ and the point $(s, 0)$ with the point $(s, 1)$ for all $t, s \in \mathbf{I}$ as shown in Fig.3.5. The torus $T^2$ is homeomorphic to $S^1 \times S^1$ by an identification map

$$f: \mathbf{I} \times \mathbf{I} \to S^1 \times S^1, (t, s) \to (e^{2\pi i t}, e^{2\pi i s}).$$

(iv) **The Klein bottle** $K$ can be obtained as a quotient space obtained from the square $\mathbf{I} \times \mathbf{I}$ by identifying the point $(0, t)$ with the point $(1, 1 - t)$ and the point $(t, 0)$ with the point $(t, 1)$ for all $t \in \mathbf{I}$ as shown in Fig. 3.6. The Klein bottle $K$ is homeomorphic to the quotient space from the topological sum of two Möbius strips by identifying the corresponding points of the edges. It is not possible to visualize the Klein bottle $K$, since $K$ cannot be embedded in $\mathbf{R}^3$.

(v) **The real projective plane** $\mathbf{R}P^2$ can be obtained as the quotient space of the square $\mathbf{I} \times \mathbf{I}$ by identifying $(t, 0)$ with $(1 - t, 1)$ and $(0, t)$ with $(1, 1 - t)$ for all $t \in \mathbf{I}$.

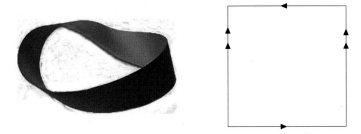

**Fig. 3.4** Möbius band obtained as the quotient space of unit square

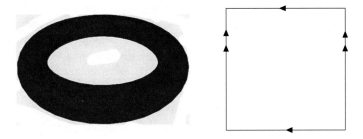

**Fig. 3.5** Torus obtained as the quotient space of unit square

**Fig. 3.6** Klein bottle
obtained as the quotient
space of the unit square

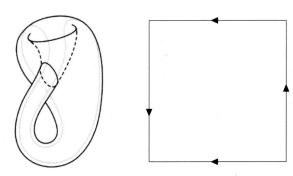

(vi) **The 2-sphere** $S^2$ can be obtained as the quotient space of the square $\mathbf{I} \times \mathbf{I}$ by identifying all its boundary points to a common point. In general, construction of $S^n$ by identification method is described in Example 3.16.24.

(vii) The general construction of the real projective space $\mathbf{R}P^n$ is done in two different ways:

**Method I**: $\mathbf{R}P^n$ is constructed from the $n$-sphere $S^n$ by identifying its antipodal points. For this identification, consider $S^n$ in $\mathbf{R}^{n+1}$ and a partition of $S^n$ into subsets which contain exactly two points lying at opposite ends of a diameter of $S^n$. Such points are called antipodal or diametrically opposite points of $S^n$.

**Method II**: It is described in Theorem 3.16.20.

## 3.17 Zariski Topology, Scheme and Zariski Space

This section describes the **Zariski topology** on an affine space and also Zariski topology on a commutative ring defined by Oscar Zariski (1899–1986) around 1950, which were born through algebraic geometry while studying algebraic varieties. This topology is an interplay among algebra, geometry and topology and is different from the metric topology because every metric topology is Hausdorff but every Zariski topology is not so. For example, the Zariski topology on the polynomial ring $\mathbf{R}[x]$ over $\mathbf{R}$ is not Hausdorff, and hence, this topology cannot be induced by a metric (see Chap. 4). Zariski topology on the $n$-dimensional complex space $\mathbf{C}^n$ has fewer open sets than its metric topology. This gives an advantage of Zariski topology over the metric topology. It is interesting that Zariski topology on $\mathbf{R}$ coincides with the cofinite topology on $\mathbf{R}$. (see Exercise 37 of Sect. 3.20). In this sense, Zariski topology on $\mathbf{R}^n$ is a generalization of cofinite topology on $\mathbf{R}$.

### 3.17.1 Zariski Topology on an Affine Space

This subsection studies Zariski topology on an affine space with the help of affine algebraic set. This specific topology provides an interplay between algebraic geometry and topology. It is interesting that the closed interval $\mathbf{I} = [0, 1]$ is closed under the Euclidean topology induced on $\mathbf{I}$ from $\mathbf{R}$ but it is not closed under Zariski topology. For more study on Zariski topology, see Chap. 4.

**Definition 3.17.1** Let $A$ be a field. Then an **affine $n$-space**

$$A^n = \{x = (x_1, x_2, \ldots, x_n) : x_i \in A, \ \forall\, i = 1, 2, \ldots, n\}$$

is a vector space of dimension $n$ over the field $A$.

**Example 3.17.2** The Euclidean $n$-space $\mathbf{R}^n$ is an affine $n$-space of dimension $n$.

**Definition 3.17.3** Given an affine space $A^n$, an $n$-tuple $x = (x_1, x_2, \ldots, x_n) \in A^n$ is said to be a **zero** of a polynomial $f(x) = f(x_1, x_2, \ldots, x_n) \in A[x_1, x_2, \ldots, x_n]$ (polynomial ring of $n$ determinates over $A$), if

$$f(x) = f(x_1, x_2, \ldots, x_n) = 0.$$

Given a subset $S \subset A[x_1, x_2, \ldots, x_n]$, the algebraic set $V(S)$ of zeros of $S$ is defined by

$$V(S) = \{x \in A^n : f(x) = 0, \ \forall f \in S\} \subset A^n.$$

A subset $X \subset A^n$ is said to be an affine algebraic set or simply, an **algebraic set** if there is a subset $S \subset A[x_1, x_2, \ldots, x_n]$ such that $X = V(S)$.

**Definition 3.17.4** Let $X$ be a subset of $A^n$ and the subset $I(X)$ of $A[x_1, x_2, \ldots, x_n]$ be defined by

$$I(X) = \{f \in A[x_1, x_2, \ldots, x_n]: f(x) = 0 \text{ for all } x \in A\},$$

is an ideal of $A[x_1, x_2, \ldots, x_n]$, called the ideal of $X$.

**Definition 3.17.5** An algebraic set $X$ in $A^n$ is said to be an **affine variety** if $I(X)$ is a prime ideal in the polynomial ring $A[x_1, x_2, \ldots, x_n]$.

Theorem 3.17.6 characterizes affine variety with the help of quotient ring.

**Theorem 3.17.6** *An algebraic set $X$ in $A^n$ is an affine variety iff the quotient (coordinate) ring $A[x_1, x_2, \ldots, x_n] / I(X)$ is an integral domain.*

**Proof** We use the result of algebra that a proper ideal of a commutative ring with identity is prime iff the quotient ring $R/I$ is an integral domain. Then the theorem follows from Definition 3.17.5 of affine variety.                                              □

We now use the concept of algebraic set given in Definition 3.17.3 to define Zariski topology.

**Definition 3.17.7** (*Zariski Topology on $A^n$*) The Zariski topology $\tau$ on an affine $n$-space $A^n$ for an arbitrary field $A$ is defined by declaring a subset $X \subset A^n$ to be closed if it is an algebraic set in $A^n$. A subset $U \subset A^n$ is open iff its complement $A^n - U$ is closed in the topology $\tau$.

*Example 3.17.8* Let $X$ be an algebraic set in $A^n$. If $X = V(S)$ for some subset $S \subset A[x_1, x_2, \ldots, x_n]$, then $X = V(I(X))$. If $K = I(X)$ is an ideal of the algebraic set $X$. then $K = I(V(K))$.

*Example 3.17.9* Let **C** be the field of complex numbers. Then it has only two ideals, viz. $\langle 0 \rangle$ and **C** itself. So the Zariski topology on **C** consists of only one open set $\langle 0 \rangle$.

*Example 3.17.10* For $n = 1$, in $A^n$, i.e., for $A = A^1$, the emptyset $\emptyset$, $A$ and the finite subset of $A$ are closed sets and hence $\emptyset$, $A$ and the complements of finite subsets of $A$ are open sets in the Zariski topology on $A$. If $A$ is an infinite field, then in the Zariski topology, any two nonempty open sets in $A$ have nonempty intersection.

*Example 3.17.11* The space **I** is not closed in Zariski topology on **R**. Because, if $A = \mathbf{R}$, then in Zariski topology on **R**, the closed sets are also closed sets under Euclidean topology on **R**, and Zariski open sets in topology on **R** are also open sets in Euclidean topology on **R**. But the closed interval $\mathbf{I} = [0, 1]$ is closed under the Euclidean topology induced on **I** from **R** but it is not closed under Zariski topology.

*Example 3.17.12* Let **C** be the field of complex numbers. Then it has only two ideals, viz. $< 0 >$ and **C** itself. So the Zariski topology on **C** consists of only one open set $< 0 >$.

## 3.17.2 Zariski Topology on the Spectrum of a Ring and Scheme

This subsection defines Zariski topology on the **spectrum spec R** of a commutative ring $R$. This specific topology provides an interplay between algebra and topology. This leads to the concept of the **"scheme"** defined by A. Grothendieck (1928–2014). "Fields Medal" was awarded to him in 1966 in recognition of his outstanding contribution in the theory of schemes.

**Definition 3.17.13** Given a commutative ring $R$ with identity, the prime spectrum of $R$ is defined to be the set of all prime ideals of $R$, i.e., prime spectrum or simply **spectrum of R**, abbreviated spec R is defined to be the set

$$spec\ R = \{P \colon P \text{ is a prime ideal in } R\}.$$

A topology is defined on the set *spec R* by declaring closed sets given in Definition 3.17.14.

**Definition 3.17.14** Let $P$ be a prime ideal of a commutative ring $R$ with identity. The closed sets in *spec R* are defined to be sets

$$V(P) = \{A \colon A \text{ is a prime ideal in } R \text{ containing } P\}.$$

The complements $c(P) = spec\ R - V(P)$ of $V(P)$ are called **open sets**. The topology defined in this way on *spec R* is called the **Zariski topology on *spec R***. The points in *spec R* corresponding to maximal ideals in *spec R* (see Chap. 5), are called **geometric points**.

**Definition 3.17.15** Given a commutative ring $R$ with identity, and a subset $X \subset R$, the set $V(X)$ is defined to be the set

$$V(X) = \{P \in spec\ R \colon X \subset P\}.$$

***Example 3.17.16*** Let I be the ideal generated by the subset $X \subset R$ of a commutative ring $R$, then $V(X) = V(I)$.

**Proposition 3.17.17** *For any subset $X \subset R$, the sets $V(X)$ in spec R have the following properties:*

*(i) $V(R) = \emptyset$;*
*(ii) $V(0) = spec\ R$;*
*(iii) For any two ideals $P$ and $Q$ in spec R, the union $V(P) \cup V(Q) = V(P \cap Q)$;*
*(iv) Given any family $\{V(X_a) \in spec\ R \colon a \in \mathbf{A}\}$ the intersection*

$$\bigcap_{a \in \mathbf{A}} V(X_a) = V\left(\bigcup_{a \in \mathbf{A}} V(X_a)\right).$$

**Proof** It follows from the definition of $V(X)$.                                                    □

**Definition 3.17.18** Given a commutative ring $R$, the **Zariski topology on** spec $R$ is defined by declaring open sets to be the unions and finite intersections of all sets of the form spec $R - V(P)$. The collection of all subsets of spec $R$ of the form $V(X)$, form a topology on spec $R$ in which the closed sets are of the form $V(X)$ for some $X \subset R$ by Proposition 3.17.17.

**Theorem 3.17.19** *Let $R$, $S$ be two rings with identity elements, spec$(R)$ and spec$(S)$ be topological spaces endowed with Zariski topology. Then every ring homomorphism $f : R \to S$ induces a continuous function*

$$spec\, f : spec\, S \to spec\, R, \ P \mapsto f^{-1}(P) for\ all\ P\ in\ spec\ S.$$

**Proof** Since $P$ is a prime ideal of $S$, the quotient ring $S/P$ is an integral domain. This asserts that $R/f^{-1}(P)$ is an integral domain. This implies that $f^{-1}(P)$ is a prime ideal of $R$. This shows that the function spec $f$ is well-defined. Moreover, for any $V(P)$, $(spec\, f)^{-1}(V(P)) = \{Q \in spec\ S : f(P) \subset Q\} = V(f(P))$                □

Grothendieck studied the scheme given in Definition 3.17.20.

**Definition 3.17.20** Given a commutative ring $R$, the *spec R* endowed with the Zariski topology is said to be a **scheme**.

The above discussion is now summarized in a basic and important Theorem 3.17.21 in the language of the category theory establishing a close relation between algebra and topology.

**Theorem 3.17.21** *Let **CRng** be the category of commutative rings with identity element and their homomorphisms and **Top** be the category of topological spaces and their continuous maps. Then*

$$spec : \mathbf{CRng} \to \mathbf{Top}$$

*is a contravariant functor.*

**Proof** The object function assigns to every object $R \in \mathbf{CRng}$, the object spec $R$ endowed with Zariski topology and hence spec $R \in \mathbf{Top}$. The morphism function assigns to every ring homomorphism $f : R \to S$, the continuous function

$$spec\, f : spec\, S \to spec\, R, \ P \mapsto f^{-1}(P) \ for\ all\ P\ in\ spec\ S$$

and hence spec $f$ is a morphism in the category **Top**. Hence, it follows that

$$spec : \mathbf{CRng} \to \mathbf{Top}$$

is a contravariant functor.

                                                                                                       □

### 3.17.3   Zariski Space Defined by Descending Chain of Closed Sets

This subsection discusses Zariski space having a particular topology defined by irreducible closed sets.

**Definition 3.17.22**  Let $X$ be a topological space. It is said to be **irreducible** if $X = A \cup B$ for closed sets $A$ and $B$ in $X$ implies either $X = A$ or $X = B$. A subspace $Y \subset X$ is irreducible if $Y$ is irreducible in the subspace topology. The topological space $X$ is said to be a **Zariski space** if given any descending chain

$$A_1 \supset A_2 \supset A_3 \supset \cdots$$

of closed sets in $X$, there exists an integer $n$ (depending on the sequence) such that

$$A_m = A_n, \ \forall \, m \geq n.$$

Every Zariski space $X$ can be expressed uniquely (up to order) as a finite union

$$X = X_1 \cup X_2 \cup \cdots \cup X_n,$$

where $X_i's$ are closed and irreducible subsets of $X$ such that $X_i$ is not a subset $X_j$ for $i \neq j$.

**Example 3.17.23**  Let **R** be the real line with the topology having the open sets $\emptyset$ together with the complements of finite subsets. Then **R** endowed with this topology is an irreducible Zariski space. Hence, **R** endowed with cofinite topology is an irreducible Zariski space.

**Proposition 3.17.24**  *Let $(X, \tau)$ be a topological space. Then the closure of every one-pointic set in $X$ is irreducible.*

***Proof***  Let $x \in X$ and $A = \overline{\{x\}} \subset X$. If $A = X_1 \cup X_2$, where both $X_1$ and $X_2$ are proper closed subsets of $A$. Hence, they are closed subsets of $X$ which are strictly contained in $X$. Then either $x \in X_1$ or $x \in X_2$. But it is not possible, because, $A$ is the smallest closed set containing the point x. It asserts that $A$ is closed in $(X, \tau)$, and hence, $A$ is irreducible by Definition 3.17.22.                                                                $\square$

## 3.18   Topological Applications

This section presents some interesting topological applications to communicate the importance and beauty of topology.

## 3.18.1 Topological Applications in Matrix Algebra

This subsection studies different classes of matrices from the viewpoint of topology. Let $M(n, \mathbf{R})$ be the set of all $n \times n$ matrices over $\mathbf{R}$ and $\mathbf{R}^{n^2}$ denote the Euclidean $n^2$-space. Then the map

$$f : M(n, \mathbf{R}) \to \mathbf{R}^{n^2},$$

$$(a_{ij}) \mapsto (a_{11}, a_{12}, \ldots, a_{1n}, a_{21}, a_{22}, \ldots, a_{2n}, \ldots, a_{n1}, a_{n2}, \ldots, a_{nn})$$

identifies $M(n, \mathbf{R})$ with the Euclidean $n^2$-space $\mathbf{R}^{n^2}$. As $\mathbf{R}$ has the usual topology, $\mathbf{R}^{n^2}$ has the product topology, called the usual topology on $\mathbf{R}^{n^2}$. Hence, $M(n, \mathbf{R})$ endowed with metric topology (usual topology) forms a topological space. This subsection begins a study of this topological space. A detailed study of this topological space of matrices is available in **Basic Topology, Volume 2** of the present series of books.

**Proposition 3.18.1** *The determinant function*

$$det : M(n, \mathbf{R}) \to \mathbf{R}, \ M \mapsto detM$$

*is continuous.*

**Proof** Since the determinant function

$$det : M(n, \mathbf{R}) \to \mathbf{R}, \ M \mapsto detM$$

is a polynomial function in $\mathbf{R}^{n^2}$, it follows that it is continuous.  □

**Proposition 3.18.2** *The general linear group* $GL(n, \mathbf{R}) = \{A \in M(n, \mathbf{R}) : detA \neq 0\}$ *is the set of all nonsingular matrices of order n over* $\mathbf{R}$. *The set* $GL(n, \mathbf{R})$ *is an open and a dense subset but it is not closed in* $M(n, \mathbf{R})$.

**Proof** Let $\mathbf{R}^* = \mathbf{R} - \{0\}$ be the set of nonzero real numbers. Then it is an open set in $\mathbf{R}$. Again, the determinant function

$$det : GL(n, \mathbf{R}) \to \mathbf{R}^*, \ A \mapsto det A$$

is continuous. It is surjective, since for any $r \in R^*$, the diagonal matrix $D$ with one diagonal entry r and the remaining $(n-1)$ diagonal entries 1 is such $D \in GL(n, \mathbf{R})$ with det $D = r$. Hence it follows that

$$GL(n, \mathbf{R}) = det^{-1}(\mathbf{R}^*)$$

is an open set in the given space, as it is the inverse image of an open set under a continuous map. Let $A \in M(n, \mathbf{R})$ be an arbitrary matrix. If det polynomial vanishes on some nbd of $A$ in $\mathbf{R}^{n^2}$, then there exists no nonsingular matrix in the nbd of $A$, since nonzero polynomial functions have countably many zeros. This asserts that det polynomial is identically the zero polynomial, which is not true. This shows that $GL(n,\ \mathbf{R})$ is dense in $M(n, \mathbf{R})$, since every nbd of $A$ contains an element of $GL(n,\ \mathbf{R})$. $\qquad\square$

**Proposition 3.18.3** *Let* $X = \{M \in M(n, \mathbf{R}): M$ *is singular*$\}$. *Then* $X$ *is nowhere dense in* $M(n, \mathbf{R})$.

*Proof* By hypothesis, $X = M(n, \mathbf{R}) - GL(n, \mathbf{R})$. Since $GL(n, \mathbf{R})$ is open in $M(n, \mathbf{R})$, the set $X$ is a closed set. To prove the proposition, it is sufficient to show that $Int(X) = \emptyset$. Let $M \in X$ be a matrix which is not identically zero. Then det $M$ is a nonzero polynomial function, called det function. Suppose det polynomial vanishes in some nbd of $M$. Then there exists no invertible matrix in this nbd, because nonzero polynomial functions have only finitely many zeros. This implies that det function vanishes identically by fundamental theorem of algebra (see Chap. 1). This is a contradiction of our supposition that $M$ is nonzero. This proves the proposition. $\qquad\square$

**Proposition 3.18.4** *Let* $GL(n, \mathbf{C})$ *be the set of all nonsingular complex matrices. The determinant function*

$$det: GL(n, \mathbf{C}) \to \mathbf{C}, M \mapsto detM$$

*is continuous.*

*Proof* Proceed as in Proposition 3.18.1.

$\qquad\qquad\qquad\qquad .$ $\qquad\square$

**Proposition 3.18.5** $X = \{M \in M(n, \mathbf{R}): M$ *is symmetric nonnegative definite matrices*$\}$. *Then the subspace* $X$ *is closed in* $M(n, \mathbf{R})$.

*Proof* Let $M \in X$ and $x \in \mathbf{R}^n$. Then the map

$$T_x: X \to \mathbf{R}, \ A \mapsto x^t A x$$

is linear, and hence, it is continuous. Consequently, $T_x^{-1}([0, \infty)$ is closed in $X$ and the set

$$X = \bigcap_{x \in \mathbf{R}^n} T_x^{-1}([0, \infty))$$

is closed, since it is the intersection of an arbitrary family closed sets in $M(n, \mathbf{R})$. $\qquad\square$

## 3.18.2    Uniform Convergence of Sequence of Functions to Metric Space

This subsection studies uniform convergent sequences from a topological space to a metric space with an eye to solve some continuity problems. The classical definition of uniform convergence in the language of $\epsilon - \delta$ method leads to its generalization in this subsection.

**Definition 3.18.6** Let $X$ be a topological space and $Y$ be a metric space with metric d. Then a sequence $\{f_n : X \to Y\}$ of functions is said to **converge uniformly** to a function $f : X \to Y$ if corresponding to a given $\epsilon > 0$, there exists a positive integer $n_0$ such that whenever $n \geq n_0$,

$$d(f_n(x) - f(x)) < \epsilon, \ \forall x \in X.$$

**Theorem 3.18.7** *Let $X$ be a topological space and $Y$ be a metric space with metric d. If a sequence of continuous functions $\{f_n : X \to Y\}$ converges uniformly to a function $f : X \to Y$, then $f$ is continuous.*

**Proof** Let the sequence $\{f_n : X \to Y\}$ of continuous functions converge uniformly to a function $f : X \to Y$. Hence corresponding to a given $\epsilon > 0$, there exists a positive integer $n_0$ such that whenever $n \geq n_0$,

$$d(f_n(x) - f(x)) < \epsilon/3, \ \forall x \in X.$$

Again, since each $f_n$ is continuous, given a point $a \in X$, there is a nbd $N_a$ of $a$ with the property that whenever $x \in N_a$,

$$d(f_{n_0}(x), f_{n_0}(a)) < \epsilon/3.$$

This shows that for every $x \in N_a$,

$$d(f(x), f(a)) \leq d(f(x), f_{n_0}(x)) + d(f_{n_0}(x), f_{n_0}(a))$$
$$+ d(f_{n_0}(a), f(a)) < \epsilon/3 + \epsilon/3 + \epsilon/3 = \epsilon.$$

This proves that $f$ is continuous, since the point $a \in X$ is arbitrary.     $\square$

### 3.18.3 Solution of Homeomorphism Problems in R by Cardinality

This subsection proves a necessary and sufficient condition for a certain class of subspaces of **R** to be homeomorphic with the help of cardinality of the corresponding underlying sets.

**Theorem 3.18.8** *Let X and Y be two finite subsets of the Euclidean line* **R**. *Then the subspace* **R** − X *and* **R** − Y *are homeomorphic iff card X = card Y.*

**Proof** Let $X = \{x_1, x_2, \ldots, x_n : x_1 < x_2 < \cdots < x_n\} \subset \mathbf{R}$ be a finite subset of **R**. Then its complement in **R**

$$\mathbf{R} - X = (-\infty, x_1) \cup (x_1, x_2) \cup (x_2, x_3) \cup \cdots \cup (x_n, \infty),$$

is homeomorphic to the disjoint union of $n + 1$ copies of the open interval $(0, 1)$. Hence, it follows that the subspaces **R** − X and **R** − Y are homeomorphic if card X = card Y, each being homeomorphic to the disjoint union of $n + 1$ copies of $(0, 1)$. Conversely, let card $X = n$ and card $Y = m$, where $m \neq n$. Without loss of generality, suppose $m < n$. If possible, the topological spaces **R** − X and **R** − Y are homeomorphic. Let $X = \{x_1, x_2, \ldots, x_n : x_1 < x_2 < \cdots < x_n\} \subset \mathbf{R}$ and $Y = \{y_1, y_2, \ldots, y_m : y_1 < y_2 < \cdots < y_m\} \subset \mathbf{R}$. Then

$$\mathbf{R} - X = (-\infty, x_1) \cup (x_1, x_2) \cup (x_2, x_3) \cup \cdots \cup (x_n, \infty),$$

and

$$\mathbf{R} - Y = (-\infty, y_1) \cup (y_1, y_2) \cup (y_2, y_3) \cup \cdots \cup (y_m, \infty),$$

This asserts **R** − X are homeomorphic to $(0, 1)^{n+1}$, which is the $(n + 1)$ copies of $(0, 1)$ and **R** − Y is homeomorphic to $(0, 1)^{m+1}$, which is the $(m + 1)$ copies of $(0, 1)$. But it is not possible, since $m \neq n$ by assumption. This forces to conclude that $m = n$.

$\square$

### 3.18.4 Cantor Space

This subsection studies **Cantor space**, defined by G. Cantor (1845–1918), which is the Cantor set endowed with the subspace topology inherited from the topological space $\mathbf{I} = [0, 1]$. It is interesting that the topological space $[0, 1]$ is a continuous image of the Cantor space and the Cantor space is also homeomorphic to a countably infinite product of two-point spaces. Recall that the Cantor set $C$ is obtained from

[0, 1] by successively deleting middle thirds and is defined by $C = \bigcap_{n=1}^{\infty} \mathbf{I}_n$ (see Chap. 1).

**Definition 3.18.9** **Cantor space** $(C, \tau_C)$ is the topological space with the subspace topology $\tau_C$ on the Cantor set $C$ induced on $\mathbf{I}$ relative to the usual topology $\tau$ on $\mathbf{R}$.

**Proposition 3.18.10** *Cantor space* $(C, \tau_C)$ *is a closed subset of* $\mathbf{I}$.

***Proof*** Since the Cantor set $C$ defined by

$$C = \bigcap_{n=1}^{\infty} \mathbf{I}_n$$

is the infinite intersection of closed sets $\mathbf{I}_n \colon \forall\, n = 1, 2, \ldots$, it follows that $(C, \tau_C)$ is a closed subset of $\mathbf{I} = [0, 1]$.

□

**Theorem 3.18.11** *Let the set* $\{0, 2\}$ *be endowed with the discrete topology. The Cantor space* $C$ *is homeomorphic to a countably infinite product space of* $\{0, 2\}$.

***Proof*** Let $X_n = \{0, 2\}$ and $\sigma_n$ be the discrete topology on $X_n$ for every $n \in \mathbf{N}$. Then $\{(X_n, \sigma_n)\}$ be a family of topological spaces and $\Pi_{n=1}^{\infty}(X_n, \sigma_n)$ be the product space of this family. Let $(C, \tau)$ be the Cantor space. Then the map

$$\psi \colon (C, \tau) \to \Pi_{n=1}^{\infty}(X_n, \sigma_n) \colon \Sigma_{n=1}^{\infty} \frac{x_n}{3^n} \mapsto (x_1, x_2, \ldots), \text{ where } x_n \text{ assumes the values } 0 \text{ or } 2$$

is a homeomorphism.

□

**Proposition 3.18.12** *The topological space* [0, 1] *is a continuous image of the Cantor space* $(C, \tau)$.

***Proof*** To prove the proposition, use the homeomorphism $\psi$ defined in Theorem 3.18.11 and consider the map

$$\phi \colon \Pi_{n=1}^{\infty}(X_n, \sigma_n) \to [0, 1] \colon (x_1, x_2, \ldots, x_n, \ldots) \mapsto \Sigma_{n=1}^{\infty} \frac{x_n}{2^{n+1}}$$

Then the map $\phi$ is continuous and onto. This asserts that the topological space [0, 1] is the continuous image of the Cantor space $(C, \tau)$ under the composite map $\phi \circ \psi \colon (C, \tau) \to [0, 1]$, which is continuous.

□

### 3.18.5  Application of Pasting Lemma for Functions from Product Spaces

This subsection gives an interesting application of Pasting Lemma 3.6.17 to examine continuity of functions from a product space in Proposition 3.18.13. For its use in homotopy theory (see **Basic Topology: Volume 3**) of the present series of books.

**Proposition 3.18.13** *Let $X$ and $Y$ be two topological spaces and $F, G: X \times I \to Y$ be two continuous maps from the product space $X \times I$ to the space $Y$ such that*

$$F(x, 1) = G(x, 0), \ \forall x \in X.$$

*Then the map $H: X \times I \to Y$, defined by*

$$H(x, t) = \begin{cases} F(x, 2t), & 0 \le t \le 1/2 \\ G(x, 2t - 1), & 1/2 \le t \le 1 \end{cases}$$

*is continuous.*

**Proof** Let $A = X \times [0, \frac{1}{2}]$ and $B = X \times [\frac{1}{2}, 1]$. Then $A$ and $B$ are closed sets in $X \times I$ such that
$$X \times I = A \cup B \text{ and } A \cap B = X \times \{\tfrac{1}{2}\}.$$

Since by hypothesis, $F(x, 1) = G(x, 0)$, $\forall x \in X$, it follows that

$$F(x.t) = G(x, t), \ \forall (x, t) \in A \cap B,$$

and hence by using pasting Lemma 3.6.17, it follows that $H$ is continuous.  □

## 3.19  Historical Note: Beginning of Topology Through the Work of Euler

This section gives a short historical note on **Euler seven bridge problem of Königsberg and Euler characteristic of a polyhedron**, named after L. Euler (1707–1783). These two papers of Euler involve no concept of distance, and they are considered as **beginning of the subject Topology** but as a well-defined mathematical discipline, it was systematically originated through the monumental work of Henri Poincare' (1854–1912) in Analysis Situs published during the period 1895–1904, but some isolated results can be traced back to the eighteenth century. Their study is given in **Basic Topology: Volume 3** of the present series.

### 3.19.1   Seven Bridge Problem of Königsberg

This subsection conveys **Seven Bridge Problem of Königsberg** posed by Euler which initiated the concept of a new geometry, now called topology, without the concept of distance. This problem is posed: Is it possible to cross each of the seven bridges of Konigsberg , once and only once on a walk through the town ? More precisely, Euler published a paper "Solutio problematic adgeometriam situs pertinentis" in 1736, where he studied the solution of a problem relating to geometry of position without the concept of distance. This problem is now called Seven Bridge Problem of Königsberg, displayed in Fig. 3.7. The diagram shows the original Konigsberg bridge problem, with two land areas on the opposite sides of Pregel River and two islands in the river, and also its graph-theoretic abstraction, in which the four land areas are represented by vertices and the seven bridges by edges. This problem arises the definition of an Eulerian graph. Its graphical representation is given in Fig. 3.8.

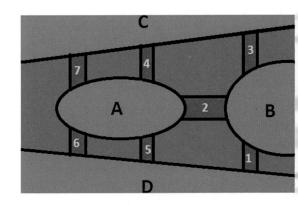

**Fig. 3.7**  City of Königsberg
with seven bridges

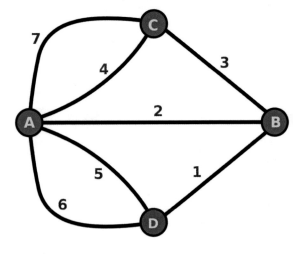

**Fig. 3.8**  Graph
corresponding to seven
bridge problem of
Königsberg

## 3.19.2   Euler Characteristic of a Polyhedron

This subsection conveys the concept of "Euler Characteristic of a Polyhedron," which is considered as the first topological (numerical) invariant. Euler characteristic establishes a relation between geometry and algebra. Euler sent a letter to C. Goldbach ( 1690–1764) in 1750 giving his formula for a connected graph $G$ on a two-dimensional sphere $S^2$, known as Euler formula: (number of vertices of $G$) − (number of edges of $G$) + (number of regions of the sphere divided by the graph $G$) = 2.

**Definition 3.19.1**  Let $G$ be finite graph with $\mathbf{V}$ vertices and $\mathbf{E}$ edges (number of 1-simplexes), then the **Euler characteristic** $\kappa(G)$ is defined to be the integer

$$\kappa(G) = \mathbf{V} - \mathbf{E}.$$

On the other hand, Euler theorem on a polyhedron is stated in Theorem 3.19.2.

**Theorem 3.19.2**  (Euler's theorem) *If $P$ is any polyhedron homeomorphic to the 2-sphere $S^2$, then*

$$\kappa(P) = \mathbf{V} - \mathbf{E} + \mathbf{F} = 2,$$

*where $\mathbf{V}$ is the number of vertices, $\mathbf{E}$ is the number of edges and $\mathbf{F}$ is the number of faces of the polyhedron $P$ and $\kappa(P)$ is independent of the choice of the polyhedron provided $P$ is homeomorphic to $S^2$.*

**Remark 3.19.3**  Euler characteristic is an integral invariant. As $\mathbf{Z}$ is an algebraic object, the concept of Euler characteristic establishes a relation between geometry and algebra. Euler's Theorem 3.19.2 is considered the first basic result conveying the geometric properties of a polyhedron without using the concept of distance, though both Archimedes (287 BC–212 BC) and R. Descartes (1556–1650) did extensive work on polyhedron. A detailed study of the Euler characteristic of a polyhedron is available in **Basic Topology, Volume 3** of the present series of books.

## 3.20   Exercises

1. Show that the open $n$-disk $\mathbf{D}^n = \{x \in \mathbf{R}^n : ||x|| < 1\}$ is homeomorphic to the Euclidean $n$-space $\mathbf{R}^n$ for all $n \geq 1$.
2. Show that the circle $S^1$ is homeomorphic to the square $\mathbf{I}^2$ in $\mathbf{R}^2$ with usual topology.
   [Hint: The map $f : \mathbf{I}^2 \to S^1$, $(x, y) \mapsto (\frac{x}{r}, \frac{y}{r}) : r = (x^2 + y^2)^{\frac{1}{2}}$ is a homeomorphism.]
3. Let $(\mathbf{R}, \sigma)$ be the real line space with natural topology $\sigma$. Show that

   (i)  every open subset of $\mathbf{R}$ is a union of disjoint open intervals;
   (ii) the topology $\sigma$ is generated by the metric $d : \mathbf{R} \times \mathbf{R} \to \mathbf{R}$, $(x, y) \mapsto |x - y|$.

4. In the real line space $\mathbf{R}$, show that

   (i) a subset $U \subset \mathbf{R}$ is open iff $U$ is a countable union of disjoint open intervals;
   (ii) the only subsets of $\mathbf{R}$ which are both open and closed are the emptyset $\emptyset$ and the whole set $\mathbf{R}$.

5. Show that every topological space satisfying the second axiom of countability is separable.

6. Show that the standard unit $n$-simplex $s_n = \{(x_1, x_2, \ldots, x_{n+1}) \in \mathbf{R}^{n+1} : x_i \geq 0,$ $\forall\, i = 1, 2, \ldots, n + 1$ and $x_1 + x_2 + \cdots + x_{n+1} = 1\}$ in $\mathbf{R}^{n+1}$ is homeomorphic to the both

   (i) unit $n$-cube $\mathbf{I}^n$ in $\mathbf{R}^n$ and
   (ii) closed unit $n$-disk $\overline{\mathbf{D}^n} = \{x \in \mathbf{R}^n : ||x|| \leq 1\}$ in $\mathbf{R}^n$.

7. Let $(X, d)$ be a metric space and $A$ be a subset of $X$. Show that its closure $\overline{A}$ coincides with the set limits in $X$ of the sequences of points lying in $A$.
   [Hint: If $x$ is the limit point of a sequence of points in $A$, then $x \in \overline{A}$ and conversely if $x \in \overline{A}$ and $n \in \mathbf{N}$, then $B_x(1/n)$ contains a point $x_n \in A$ and $lim_{n \to \infty} x_n = x$.]

8. In the real line space $\mathbf{R}$, prove that

   (i) $\overline{\mathbf{Q}} = \mathbf{R}$;
   (ii) $\overline{[0, 1)} = [0, 1]$;
   (iii) $\overline{(\mathbf{R} - \mathbf{Q})} = \mathbf{R}$.

9. Let $f : \mathbf{R}^n \to \mathbf{R}$ be a continuous function and $r$ be any real number. Then the sets

$$X_r = \{x \in \mathbf{R}^n : f(x) < r\} \subset \mathbf{R}^n;$$

$$Y_r = \{x \in \mathbf{R}^n : f(x) \leq r\} \subset \mathbf{R}^n;$$

and

$$Z_r = \{x \in \mathbf{R}^n : f(x) = r\} \subset \mathbf{R}^n$$

are Lebesgue sets of the function $f$ corresponding to $r$. Give

   (i) an example showing that $\overline{X_r} = Y_r$;
   (ii) an example showing that $\partial X_r = Z_r$;
   (iii) an example showing that $\overline{X_r} \neq Y_r$ and $\partial X_r \neq Z_r$.

10. Show that a finite topological space is metrizable iff it is discrete.

11. Let $(X, \tau)$ be a topological space and $X = A \cup B$ be the union of its two closed sets $A$ and $B$. If $f : X \to Y$ is a function such that its restrictions $f|_A$ and $f_B$ are both continuous, show that $f$ is continuous.
   [Hint: Use Pasting Lemma 3.6.17.]

12. Let $(X, \tau)$ and $(Y, \sigma)$ be topological spaces and $f : (X, \tau) \to (Y, \sigma)$ be a map. Show that the following statements are equivalent:

    (i) $f$ is continuous;
    (ii) for every subset $A$ of $X$, the set $f(\overline{A})$ is a subset of $\overline{f(A)}$;
    (iii) for every closed set $B$ of $(Y, \sigma)$, the set $f^{-1}(B)$ is closed in $(X, \tau)$.

13. Let $(X, \tau)$ and $(Y, \sigma)$ be topological spaces and $f : (X, \tau) \to (Y, \sigma)$ be a bijective map. Show that $f$ is a homeomorphism iff $f$ satisfies any one of the following equivalent conditions:

    (i) the map $f$ and its inverse $f^{-1}$ are both continuous;
    (ii) the map $f$ and its inverse $f^{-1}$ are both open or both closed;
    (iii) the map $f$ is both continuous and closed;
    (iv) $f(\overline{A}) = \overline{f(A)}$ for every subset $A \subset X$;
    (v) $f(\mathrm{Int}(A)) = \mathrm{Int} f(A)$ for every subset $A \subset X$.

14. Let $(X, \sigma)$ be a nonempty closed subspace of the Cantor space $(C, \tau)$. Show that there exists a continuous onto map

$$f : (C, \tau) \to (X, \sigma).$$

15. Show that the topology $\tau_1$ on a nonempty set $X$, generated by a base $\mathcal{B}_1$ is finer than the topology $\tau_2$ on $X$, generated by a base $\mathcal{B}_2$ iff given any point $x \in X$ and any $U_2$ in $\mathcal{B}_2$ with $x \in U_2$, there is some $U_1$ in $\mathcal{B}_1$ such that $x \in U_1 \subset U_2$.

16. If $A$ is a closed subset of a topological space $X$, show that $A$ is nowhere dense iff its complement is everywhere dense.

17. Show that the Cantor space is nowhere dense.

18. Show that any infinite subset of $\mathbf{R}$ with cofinite topology is dense in $\mathbf{R}$. Use this result to show that $\mathbf{Z}$ is dense in this topological space.

19. Let $\mathcal{T}$ be the set of all triangular regions in the Euclidean plane $\mathbf{R}^2$. Show that $\mathcal{T}$ forms a base of the standard topology on $\mathbf{R}^2$.

20. Let $X$ be an arbitrary topological space and $f : X \to \mathbf{R}$ be a continuous function. Show that for every real number $r \in \mathbf{R}$, $f^{-1}(r)$ is a closed subset in $X$. Hence show that the sphere $S^n$ is a closed subset in $\mathbf{R}^{n+1}$.
    [Hint: Use the facts that the map $f : \mathbf{R}^{n+1} \to \mathbf{R}, x \mapsto ||x||$ is continuous, $1 \in \mathbf{R}$ and $S^n = f^{-1}(1)$].

21. Examine whether the two bases $\mathcal{B}_n$ and $\mathcal{B}$ in $\mathbf{R}^n$ defined by

    (i) $\mathcal{B}_n = \{(x_1, x_2, \ldots, x_n) \in \mathbf{R}^n : a_i < x_i < b_i, c < y < d\} \subset \mathbf{R}^n$ with sides parallel to axes, and
    (ii) $\mathcal{B}$, the collection of open balls in $\mathbf{R}^n$

    induce the same topology on $\mathbf{R}^n$.

22. **(Embedding Theorem)** Let $(X, \tau)$ be a topological space and $\{(Y_n, \tau_n) : \forall n \in \mathbf{N}\}$ be a countable family of topological spaces and

$$f_n: (X, \tau) \to (Y_n, \tau_n): \forall n \in \mathbf{N}$$

be a countable family of maps. If

$$e: (X, \tau) \to \Pi_{n=1}^{\infty}(Y_n, \tau_n), \quad x \mapsto \Pi_{n=1}^{\infty} f_n(x)$$

is the evaluation map, show that

(i) $e$ is a homeomorphism of $(X, \tau)$ onto the subspace $(e(X), \sigma)$; i.e., $e$ is an embedding of $(X, \tau)$ in the product space $\Pi_{n=1}^{\infty}(Y_n, \tau_n)$, where $\sigma$ is the subspace topology on $e(X)$;

(ii) each $f_n$ is continuous;

(iii) the countable family $\{f_n\}$ of continuous maps separates points of $X$ in the sense that for every pair of distinct points $x, y \in X$, their images $f_n(x)$ and $f_n(y)$ are distinct for the same value of $n$;

(iv) the countable family $\{f_n\}$ of continuous maps separates points of $X$ and closed sets in the sense that for any point $x \in X$ and any closed set $A$ of $(X, \tau)$, not containing the point $x$, $f_n(x) \notin \overline{f_n(A)}$ for some $n$.

23. Let $(X, d)$ be a metric space. If $A \subset X, x \in X$ and

$$d(x, A) = \inf\{d(x, a): a \in A\},$$

show that $x \in \bar{A}$ iff $d(x, A) = 0$.

24. Let $(X, d)$ be a metric space and $A \subset X$. Show that

(i) the collection of all open balls $\{B_x(r): r > 0\}$ forms a basis for a topology $\tau_d$ on $X$:

(ii) the metric topology on $A$ is the same as the subspace topology induced from the topology $\tau_d$ on $X$.

[Hint: The metric $d: X \times X \to \mathbf{R}$ induces a metric $d|_A: A \times A \to \mathbf{R}$.]

25. Let $X$ and $Y$ be two topological spaces and $p: X \times Y \to X$ and $q: X \times Y \to Y$ be canonical projections. Show that

(i) corresponding to any pair of continuous maps $f: \mathbf{I} \to X$ and $g: \mathbf{I} \to Y$, there is a continuous map

$$(f, g): \mathbf{I} \to X \times Y, \quad t \mapsto (f(t), g(t));$$

(ii) conversely, corresponding to any continuous map $\alpha: \mathbf{I} \to X \times Y$, there is a pair of continuous maps $(f, g)$ such that

$$f = p \circ \alpha: \mathbf{I} \to X, \quad g = q \circ \alpha: \mathbf{I} \to X.$$

26. Let $a, b \in \mathbf{Q}$ and $a < b$. If $S$ is the family of such closed intervals $[a, b]$, Show that the set $B = S \cup \{x : x \in \mathbf{Q}\}$ forms a base for the topology $\sigma$ generated by $S$ on the real line $\mathbf{R}$.

27. Let $Y$ and $Z$ be disjoint subspaces of a topological space $X$ such that $X = Y \cup Z$. Show that the following statements are equivalent:

   (i)   $X = Y \cup Z$ (topological sum);
   (ii)  $Y$ and $Z$ are both open sets in $X$;
   (iii) $Y$ is open and closed in $X$;
   (iv)  $\bar{Y} \cap Z = \emptyset$ and $Y \cap \bar{Z} = \emptyset$.

28. Let $X = Y + Z$ be the topological sum of two disjoint subspaces $Y$ and $Z$ of a topological space $X$ and $i : Y \to X$, $j : Z \to X$ be inclusion maps. If $f : Y \to W$ and $g : Z \to W$ are continuous maps, show that there exists a unique continuous map $h : X \to W$ such that $h \circ i = f$, $h \circ j = g$.

29. Let $X$ and $Y$ be topological spaces and given a closed subspace $A$ of $X$, let $f : X \to Y$ be a continuous map. If $Y \cup_f X$ is the quotient space obtained from the disjoint union $X + Y$ by the equivalence relation generated by identifying $a \sim f(a)$, $\forall a \in A$, show that

   (i)   for $Y = \{y\}$, $Y \cup_f X = X/A$;
   (ii)  the canonical map $p : Y \to Y \cup_f X$ is an embedding onto a closed subspace of $Y \cup_f X$;
   (iii) the canonical map $p : X - A \to Y \cup_f X$ is an embedding onto an open subspace of $Y \cup_f X$.

30. Let $X$ be the topological space endowed with the weak topology determined by subspaces $\{X_i : i \in J\}$. Show that

   (i)   a subset $U \subset X$ is open iff $U \cap X_i$ is open in $X_i$ for every $i \in J$;
   (ii)  if $B$ is a closed subspace of $X$, then $B$ has also the weak topology determined by the subspaces $\{X_i \cap B : i \in J\}$.

31. Let $(X, \tau)$ and $(Y, \sigma)$ be two topological spaces and $S$ be a subbase for the topology $\sigma$. Show that a function

$$f : (X, \tau) \to (Y, \sigma)$$

is continuous iff the inverse image of every member of $S$ under $f$ is an open set in $(X, \tau)$.

32. Let $X_0 = \{0, 1\}$ and $X_n = \{m/2^n : m \in \mathbf{N}, \ m \text{ is odd, and } 0 < m < 2^n\}$ for all $n \geq 1$. The set $X = \bigcup_{n=0}^{\infty} X_n$ is called **the set of dyadic rational numbers in the closed interval** $[0, 1]$. Show that the set $X$ of dyadic rational numbers in the closed interval $[0,1]$ is dense in $[0,1]$.

33. Let $f : X \to Y$ be a map between two topological spaces $X$ and $Y$. If $G = \{(x, f(x)\}$ is the graph of $f$, endowed with product topology (inherited fron $X \times Y$). Show that the map

$$\psi : X \to G, x \mapsto (x, f(x))$$

is a homeomorphism iff $f$ is continuous.

34. Let $(\mathbf{R}, d)$ be the Euclidean line and $\mathbf{Z} \subset \mathbf{R}$ be set of integers endowed with metric induced from the Euclidean metric $d$ from $\mathbf{R}$. Show that the set of open balls in $\mathbf{Z}$ is the set of all subsets of $\mathbf{Z}$ that consists of an odd number of consecutive integers with the center of the ball at its middle position. y consists of precisely the set of all arithmetic progressions of common difference one and of the form

$$A(n, k) = \{n - k, \ldots, n, \ldots, n + k\}$$

for some nonnegative integer $k$.

[Hint:The open ball $B_n(r)$ in $\mathbf{Z}$ at center $n$ with radius $r > 0 \in \mathbf{R}$ is given by

$$B_n(r) = \{a \in \mathbf{Z} : d(n, a) = |n - a| < r\}$$

It is of the form

$$A(n, k) = \{n - k, \ldots, n, \ldots, n + k\}$$

for some nonnegative integer $k$ depending on $r$.

Let p = [r] denote the greatest nonnegative integer such that $p \leq r$. Then $k = p$, if $r > p$ and $k = p - 1$, if $r = p$. For example,

$$d_1(\pi) = \{-2, -1, 0, 1, 2, 3, 4\}.$$

35. Let $\Omega$ be the family of subsets of $\mathbf{N}$ consisting of the empty set $\emptyset$ and all those subsets $X_n$ of $\mathbf{N}$ which are expressible in the form

$$X_n = \{n, n + 1, n + 2, n + 3, \ldots : n \in \mathbf{N}\}.$$

Show that $\Omega$ forms a topology on $\mathbf{N}$.

[Hint: Use that $X_1 = \mathbf{N}$ to show that $\mathbf{N} \in \Omega$. Again, since $\Omega$ is totally ordered by set inclusion, the intersection of any two sets in $\Omega$ is also in $\Omega$. Let $\mathbf{S}$ be a subfamily of $\Omega - \{\emptyset, \mathbf{N}\}$ in the sense that $\mathbf{S} = \{X_n : n \in J \subset \mathbf{N}\}$. As $J$ is a subset of positive integers, it has a smallest positive integer $p$. Hence

$$\bigcup \{X_n : n \in J\} = \{p, p + 1, p + 2, p + 3, \ldots\} = X_p \in \Omega.]$$

36.    (i) Given a nonempty subset $X \subset \mathbf{N}$, let there exist a positive integer $n_X$ such that $X$ contains no arithmetic progression (AP) of length greater than $n_X$. Show that the family of subsets of $\mathbf{N}$ having this property together with $\emptyset$ and the set $\mathbf{N}$ form a collection of closed sets for some topology on $\mathbf{N}$.

[Hint: Use Van der Waerden's theorem which asserts that given an integer $n \in \mathbf{N}$, there is an integer $n_0$ such that for any subset $X \subset \{1, 2, \ldots, n_0\} = Y$, either $X$ or $Y - X$ contains an AP of length $n$.]

(ii) Show that the collection of all infinite AP's in $\mathbf{N}$ forms a base for some topology on $\mathbf{N}$.
    [Hint: Use the result that the finte intersection of AP's in $\mathbf{N}$ is also an AP.]

(iii) Using this topology on $\mathbf{N}$ show that the set of prime integers is infinite.
    [Hint: The sets $A(k, d) = \{k, k + d, k + 2d, \ldots : k = 1, 2, \ldots, d\}$ are open, pairwise disjoint, and form a covering of $\mathbf{N}$. Hence, it follows that each of them is closed. As a particular situation, for each prime integer p, the sets of the form $\{p, 2p, 3p, \ldots\}$ forms a covering of $\mathbf{N} - \{1\}$. This shows that the set of prime integers cannot be finite, otherwise, if the set were finite, then the set $\{1\}$ would be open. This shows that it is not a union of arithmetic progressions. This concludes that the set of prime integers cannot be finite, and hence, it is infinite.

(iv) Show that there exists infinite primes in $\mathbf{Z}$.
    [Hint: An independent proof: For any pair of integers $d > 0, 0 \leq k < b$ define the sets

$$A(k, d) = \{x \in \mathbf{Z} : x \equiv k \quad \mathrm{mod} \ (d)\} = \{k, \ k \pm d, \ k \pm 2d, \ k \pm 3d, \ldots\} \subset \mathbf{Z}.$$

Hence, $\{A(k, d) : d > 0, 0 \leq k < b\}$ forms a family of infinite arithmetic progressions in $\mathbf{Z}$. If $x \in A(k, d) \cap A(a, b)$, then $x \in A(x, db) \subset A(k, d) \cap A(a, b)$. Hence, the family of subsets $\{A(k, d)\}$ forms a subbase generating a base $\mathcal{B}$ for a topology $\tau$ on $\mathbf{Z}$. An element $B$ of $\mathcal{B}$ is $B = A(k_1, d_1) \cap A(k_2, d_2) \cap \cdots \cap A(k_n, d_n)$ Then $B$ is either $\emptyset$ or infinite by Chinese remainder theorem.]

37. Let $\mathbf{R}$ be the field of real of real numbers and $x = (x_1, x_2, x_3, \ldots, x_n) \in \mathbf{R}^n$ be an arbitrary point. Given any polynomial $f \in R$, in the polynomial ring $R = \mathbf{R}[x_1, x_2, x_3, \ldots, x_n]$, define

$$V(f) = \{x \in \mathbf{R}^n : f(x) = 0\} \subset \mathbf{R}^n.$$

Show that

(i) the collection

$$\mathcal{B} = \{\mathbf{R}^n - V(f) : f \in \mathbf{R}[x_1, x_2, x_3, \ldots, x_n]\}$$

forms a base for a topology $\tau$ on $\mathbf{R}^n$, called **Zariski topology** on $\mathbf{R}^n$;

(ii) the Zariski topology $\tau$ on $\mathbf{R}$ is the cofinite topology on $\mathbf{R}$

38. Let $X$ be a subset of the affine space $A^n$, then the subset $I(X)$ of $A[x_1, x_2, \ldots, x_n]$ defined by

$$I(X) = \{f \in A[x_1, x_2, \ldots, x_n] : f(x) = 0 \text{ for all } x = (x_1, x_2, \ldots, x_n) \in X\}$$

is an ideal of $A[x_1, x_2, \ldots, x_n]$, **called the ideal of** $X$ and $V(I(X))$ is called **the algebraic set** in $A^n$ corresponding to the ideal $I(X)$. Let $\mathcal{I}$ be the set if all ideals of $A[x_1, x_2, \ldots, x_n]$, defined above and $\mathcal{V}$ be the set of all algebraic sets in $A^n$. Show the mapping

$$V : \mathcal{I} \to \mathcal{V}, \ I \mapsto V(I)$$

has the following properties

(i) $V(0) = A^n$;

(ii) $V(A[x_1, x_2, \ldots, x_n]) = \emptyset$;

(iii) $V(I \cap J) = V(I) \cup V(J)$;

(iv) $V(\Sigma_k I_k) = \bigcap_k V(I_k)$

Hence show that the above properties define the Zariski topology on $A^n$.

39. Let $X$ be a Zariski space. Show that $X$ can be uniquely represented (up to order) as a finite union of closed and irreducible sets $X_i$ such that

$$X = X_1 \cup X_2 \cup X_3 \cup \cdots \cup X_n,$$

where $X_i$ is not contained in $X_j$ for any $i, j$ with $i \neq j$.

40. Find the Zariski topology on the ring $\mathbf{Z}$ of integers.

41. Prove that the collection of open disks forms an open base for the usual topology on the Euclidean space $\mathbf{R}^2$.

42. Show that the Sorgenfrey line is not metrizable.

43. Let $X \subset [0, 1]$ be the set of dyadic fractions

$$X = \left\{ \frac{1}{2}, \frac{1}{4}, \frac{3}{4}, \frac{1}{8}, \frac{3}{8}, \frac{5}{8}, \frac{7}{8}, \frac{1}{16}, \ldots, \frac{15}{16}, \ldots \right\},$$

(i.e., the set of proper fractions having denominators powers of 2). Show that $X$ is dense in $[0, 1]$.

[Hint: Any open interval $(a - \epsilon, a + \epsilon)$ for every $a \in [0, 1]$ contains a point of $X$. Moreover, $\lim_{n \to \infty} \frac{1}{2^n} = 0$. Hence, it follows that $\bar{X} = [0, 1]$.]

44. Let $\mathcal{H} = \{(x, y) \in \mathbf{R}^2 : x > a, \text{ or } x < a \text{ or } y > a \text{ or } y < a\}$ be the family of all open half planes. Show that the topology on $\mathbf{R}^2$ generated by $\mathcal{H}$ is the usual topology on $\mathbf{R}^2$.

45. Show that the family $\mathcal{S} = \{(x, 1] \cup [0, y) : 0 < x, y < 1\}$ of all half-open intervals form a subbase for the topology induced on $[0, 1]$ by the usual topology on $\mathbf{R}$.

46. Let $X$ be a complete metric space with metric $d$ and $Y$ be an arbitrary topological space. Let $T : X \times Y \to X$ be continuous map such that

$$d(T(x_1), y), T(x_2), y)) \leq r \, d(x_1, x_2) \ (r < 1), \ \forall x_1, x_2 \in X, \ \forall y \in Y.$$

Show that for a given $y \in Y$, the map

$$T : X \to X, \ x \mapsto T(x, y)$$

has a unique fixed point $f(y)$, say and the map $\psi: Y \to X$, $y \mapsto f(y)$ is continuous.

47. Show that every open (closed) subspace of a topologically complete space is also topologically complete.
    [Hint Let $(X, d)$ be a complete metric space and $U \subset X$ be open. Define a map

$$\psi: U \to \mathbf{R}, \; x \mapsto \frac{1}{d(x, X - U)}.$$

Then there exists an embedding

$$f: U \to X \times \mathbf{R}, x \mapsto x \times \psi(x)].$$

48. Show that the area of a region in Euclidean plane $\mathbf{R}^2$ is not a topological property.
    [Hint: Let $(r, \theta)$ be the polar coordinates of a point in $\mathbf{R}^2$, and $\mathbf{D}_1, \mathbf{D}_2$ be two open disks in $\mathbf{R}^2$ defined by

$$\mathbf{D}_1 = \{(r, \theta): r < 1\} \text{ and } \mathbf{D}_2 = \{(r, \theta): r < 2\}.$$

Then the map $h: \mathbf{D}_1 \to \mathbf{D}_2$, $(r, \theta) \mapsto (2r, \theta)$ is a homeomorphism.]

49. Let $\sigma$ be the topology on the real line $\mathbf{R}$ generated by the closed-open intervals $[c, d)$ and $\rho$ be the topology on $\mathbf{R}$ induced by collection of all the linear maps $T: \mathbf{R} \to \mathbf{R}, x \mapsto cx + d$, $\forall c, d \in \mathbf{R}$. Show thar $\rho$ is the discrete topology on $\mathbf{R}$.

50. Let $(0. \infty)$ and $[-1, 1]$ have the relative topology induced from the usual topology on $\mathbf{R}$. Show that the map

$$f: (0.\infty) \to [-1, 1], \; x \mapsto \sin(1/x)$$

is neither open nor closed but it is continuous.

51. Let $\kappa_A: X \to \mathbf{R}$ be the characteristic function for a subset $A$ of a topological space X. Show that it is continuous at a point $a \in X$ iff the point $a$ is not a point of the boundary of $A$.
    [Hint: Use the fact that $\kappa_A(x) = 1$ if $x \in A$ and is 0, otherwise.]

52. Let $\mathbf{B} = \{(\frac{a}{b}, \frac{c}{d}): a, b, c, d \in \mathbf{Z} \text{ and } b, d \neq 0\}$. Show that $\mathbf{B}$ constitutes an open base for the Euclidean topology on $\mathbf{R}$.

53. Let $\rho$ be the equivalence relation defined on $\mathbf{R}$ by the rule $x \rho y$ iff $x - y \in \mathbf{Q}$ i.e., $(x, y) \in \rho \iff x - y \in \mathbf{Q}$. Show that $\mathbf{R}/\rho$ has uncountably many points, but it has the trivial topology.

54. Let X and Y are two topological spaces. Show that the twist map $T: X \times Y \to Y \times X$, $(x, y) \mapsto (y, x)$ is continuous.
    [Hint: Let $p_1: X \times Y \to X$, $(x, y) \mapsto x$ and $p_2: X \times Y \to Y$, $(x, y) \mapsto y$ be the projection maps. Then they are continuous and $T$ is $(p_2, p_1)$. Hence, $T$ is continuous.]

55. Show that the bijective map $p: [0, 1] \to S^1$, $t \mapsto (cos 2\pi t, \sin 2\pi t)$ is continuous but not a homeomorphism.

[Hint: The map $p$ is a bijection. Since the components of p are continuous, p is continuous but $p^{-1}$ is not continuous because $N_0 = [0, \frac{1}{2}]$ is a nbd in $(0,1)$ of 0, but $p(N_0)$ is not a nbd in $S^1$ of p(0)=(1,0)]

56. **(Structure of open sets in R)** Let **R** be the real line space. Show that a nonempty subset of **R** is open iff it is a countable union of pairwise disjoint open intervals.

57. Show that the boundary of a closed set in a topological space is nowhere dense.

58. **(Characterization of nowhere dense sets)** Let $(X, \tau)$ be a topological space. Show that

   (i) a nonempty subset of $A$ of $X$ is nowhere dense iff each nonempty open set has a nonempty open subset $U$ such that $U \cap A = \emptyset$;
   (ii) a closed set $A$ of $X$ is nowhere dense iff its complement is everywhere dense.

59. Let $(X, \tau)$ be a topological space and $A$ be a nonempty subset of $X$. Show that $X$ is the disjoint union of Int $(A)$, boundary $\partial A$ and $X - \overline{A}$.

60. **(Invariance of domain)** (Brouwer) Let $A$ and $B$ be subsets of $\mathbf{R}^n$ such that there exists a homeomorphism $f : A \to B$. Show that

$$f(\text{Int}(A)) \subset \text{Int}(B).$$

Use this result to prove that if $f : \mathbf{R}^m \to \mathbf{R}^n$ is a homeomorphism, then $m = n$.

61. A metric $d : X \times X \to \mathbf{R}$ is said to be bounded if there exists a real number $M$ such that $d(x, y) < M$, $\forall (x, y) \in X \times X$. Show that every metric on $X$ is equivalent to a bounded metric on $X$.
   [Hint: Given a metric $d'$ on $X$, define a map

$$d : X \times X \to \mathbf{R}, \ (x, y) \mapsto \min \{1, d'(x, y)\}.$$

Then $d$ is a bounded metric on $X$, which is equivalent to the given metric $d'$.]

62. Let $X = \Pi_n^\infty M_n$ be the Tychonoff product of a countable number of metric spaces $(M_n, d_n)$ endowed with the metric

$$d : X \times X \to \mathbf{R}, \ ((x_1, x_2, \ldots, x_n, \ldots), \ (y_1, y_2, \ldots, y_n, \ldots))$$
$$\mapsto \Sigma_{n=1}^\infty \frac{1}{2^n} \cdot \frac{d_n(x_n, y_n)}{1 + d_n(x_n, y_n)}.$$

   (i) Show that the topology induced by $d$ on $X$ is the Tychonoff topology and hence prove that
   (ii) the countable product of the line segment **I** (called **Hilbert cube**) is a metrizable topology space.

## Multiple Choice Exercises

Identify the correct alternative (s) (there may be none or more than one) from the following list of exercises:

1. Let $M(n, \mathbf{R})$ be the set of all $n \times n$ matrices over $\mathbf{R}$ identified with the Euclidean $n^2$-space $\mathbf{R}^{n^2}$ (endowed with its usual topology). Then the general liner group $GL(n, \mathbf{R}) = \{M \in M(n, \mathbf{R}): \det M \neq 0\}$ is

   (i) open but not closed in $M(n, \mathbf{R})$;
   (ii) closed but not open in $M(n, \mathbf{R})$;
   (iii) dense in $M(n, \mathbf{R})$.

2. The subspace $Z = \{(x, y) \in \mathbf{R}^2 : y = mx\} - \{(0, 0)\}$ of the Euclidean plane $\mathbf{R}^2$ is

   (i) open in $\mathbf{R}^2$;
   (ii) neither open nor closed in $\mathbf{R}^2$;
   (iii) dense in $\mathbf{R}^2$.

3. Let $(X, \tau)$ be a topological space.

   (i) Int $(A \cup B) = $ Int $A \cup$ Int $B$ for any two subsets $A, B$ of $X$;
   (ii) Int $(A \cap B) = $ Int $A \cap$ Int $B$ for any two subsets $A, B$ of $X$;
   (iii) Int $A = \overline{A^c}^c$, for any subset $A$ of $X$, where $A^c = X - A$ denotes the complement of $A$ in $X$.

4. Let $(X, \tau)$ be a topological space.

   (i) $\overline{A \cup B} = \overline{A} \cup \overline{B}$, $\forall A, B \subset X$.
   (ii) $\overline{A \cap B} = \overline{A} \cap \overline{B}$, $\forall A, B \subset X$.
   (iii) If $A = \{(x, y): y = 0\}$, then Int $(A) = \emptyset$.

5. let $(X, \tau)$ be a topological space.

   (i) If $A$ is an arbitrary dense subset in $(X, \tau)$, then its complement $X - A$ is nowhere dense in $(X, \tau)$.
   (ii) If $A$ is an arbitrary nowhere dense subset in $(X, \tau)$, then its complement $X - A$ is dense in $(X, \tau)$.
   (iii) The set $\mathbf{R}$ identified with the $x$-axis of the Euclidean plane is nowhere dense.

6. Let $(X, \tau)$ be a topological space and $p \in X$. If $A = X - \{p\}$ is endowed with relative topology induced from $X$ on $A$, then the subspace $A$ is

   (i) open in $(X, \tau)$;
   (ii) closed in $(X, \tau)$;
   (iii) dense in $(X, \tau)$.

7. (i) Let $(X, \tau)$ and $Y, \sigma)$ be two topological space. If $f: X \to Y$ is a continuous map and $G_f = \{(x, f(x))\}$ endowed with product topology (inherited fron $X \times Y$) is its graph. then the map

$$\psi: X \rightarrow G_f, \ x \mapsto (x, f(x))$$

is continuous but not necessarily a homeomorphism.

(ii) The open ball $B = \{x = (x_1, x_2) \in \mathbf{R}^2 : \|x\| < 1\}$ is homeomorphic to the whole plane $\mathbf{R}^2$.

(iii) The cone $A = \{(x, y, z) \in \mathbf{R}^3 : x^2 + y^2 = z^2, z > 0\}$ is homeomorphic to the plane $\mathbf{R}^2$.

8. Let $M(n, \mathbf{R})$ be the set of all $n \times n$ matrices over $\mathbf{R}$ and be identified with $\mathbf{R}^{n^2}$ and endowed with its usual topology. If $\mathbf{S} = \{M \in M(n, \mathbf{R}) : \text{trace } M = 0\}$, then $\mathbf{S}$

(i)   is nowhere dense in $M(n, \mathbf{R})$;
(ii)  is dense in $M(n, \mathbf{R})$;
(iii) is a closed set in $M(n, \mathbf{R})$.

# References

Adhikari, A., Adhikari, M.R.: Basic Topology, Volume 2: Topological Groups, Topology of Manifolds and Lie Groups. Springer, India (2022)

Adhikari, M.R.: Basic Topology, Volume 3: Algebraic Topology and Topology of Fiber Bundles. Springer, India (2022)

Adhikari, M.R.: Basic Algebraic Topology and its Applications. Springer, India (2016)

Adhikari, M.R., Adhikari, A.: Basic Modern Algebra with Applications. Springer, New Delhi, New York, Heidelberg (2014)

Alexandrov, P.S.: Introduction to Set Theory and General Topology. Moscow (1979)

Armstrong, M.A.: Basic Topology. Springer-Verlag, New York (1983)

Bredon, G.E.: Topology and Geometry. Springer-Verlag, New York (1983)

Borisovich, Y.C.U., Blznyakov, N., Formenko, T.: Introduction to Topology. Mir Publishers, Moscow, Translated from the Russia by Oleg Efimov (1985)

Bourbaki, N.: General Topology, 1st edn. (1940)

Chatterjee, B.C., Ganguly, S., Adhikari, M.R.: Introduction to Topology. Asian Books, New Delhi (2002)

Dugundji, J.: Topology. Allyn & Bacon, Newton, MA (1966)

Hilbert, D., Cohn-Vossen, S.: Geometry and Imagination. Chelsea, New York (1952)

Hu, S.T.: Introduction to General Topology. Holden-Day, San Francisco (1966)

Kelly, J.L.: General Topology, Van Nostrand, New York, 1955. Springer- Verlag, New York (1975)

Lipschutz, S.: Schaum's Outline of General Topology. McGraw-Hill (1988)

Mendelson, B.: Introduction to Topology College Mathematical Series. Allyn and Bacon, Boston (1962)

Munkres, J.R.: Topology. Prentice-Hall, New Jersey (2000)

Patterson, E.M.: Topology. Oliver and Boyd (1959)

Prasolov, V.V.: Intuitive Topology. Universities Press (India) (1995)

Singer, I.M., Thorpe, J.A.: Lecture Notes on Elementary Topology and Geometry. Springer-Verlag, New York (1967)

Stephen, W.: General Topology. Addison-Wesley (1970)

# Chapter 4
# Separation Axioms

This chapter studies topological spaces by imposing certain conditions, called **separation axioms** on these spaces in terms of their points and open sets, specially, where there is possibly no concept of distance. The additional conditions are needed, because the defining axioms for a topological space are extremely general and they are too weak to study them in depth. These axioms are natural restrictions on topological structure to make the structure nearer to metrizable structure, and they are called $T_i$-axioms for $i = 0, 1, 2, 3, 4, 5$. The corresponding topological spaces are known as $T_i$-**spaces**. These spaces are interrelated to some extent (see Exercise 17 of Sect. 4.7).

**Historically**, the symbol "$T$" in their designations comes from the German word **Trennungs axiom**, meaning separation axiom. These axioms facilitate to classify topological spaces. Most of the topological spaces of our interest carry more structure (not necessarily a metric), but they have separation properties which provide a rich supply of continuous functions. There are topological spaces having only two open sets such as the empty set and the whole set, on which the constant functions are the only continuous functions. On the other hand, every subset is open in a discrete topological space on which every function is continuous. Most of the important spaces of analysis and geometry lie between these two extremes which are studied in this chapter by separation properties to obtain interesting topological spaces providing enough supply of continuous functions.

Various interesting applications of separation properties of topological spaces are also available in Sect. 4.6. Several separation axioms are known, but this chapter addresses only $T_i$-**axioms** for $i = 0, 1, 2, 3, 4, 5$ and the corresponding $T_i$-**spaces**. The interest of this address is caused by the fact that continuous functions play a key role in topology, and the supply of open sets of such a space is closely linked to its supply of continuous functions. For example, normal spaces and completely regular spaces include metric spaces, and they are closely linked with the real-valued continuous functions. More studies on these topics are made in Chap. 6.

© The Author(s), under exclusive license to Springer Nature Singapore Pte Ltd. 2022    233
A. Adhikari and M. R. Adhikari, *Basic Topology 1*,
https://doi.org/10.1007/978-981-16-6509-7_4

Motivation of separation axioms was born through the observation that any two points in a metric space are separated if they have a strictly positive distance. But there exist many topological spaces satisfying a set of certain conditions in addition to the axioms defining topological spaces which can recover many significant properties of metric spaces lost to arbitrary topological spaces. Such conditions defined in terms of points and open sets lead to the so-called separation axioms on the topology, initially used by Alexandroff (1896–1982) and Hopf (1894–1971). Such spaces $X$ are important objects in topology as many important topological properties can be characterized with the help of separation axioms by distributing the open sets in the space $X$ and imposing natural conditions on $X$ such that $X$ behaves like a metric space. Any structure like a metric is not studied in this chapter, and instead, certain conditions are completely described in terms of the points and open sets of the topological spaces. For example, every metrizable space is a $T_4$ space, called a normal space, which is characterized by Minor Urysohn Lemma 4.6.1. The importance of separation axioms is reflected throughout the three volumes of the present book series.

For this chapter, the books Chatterjee et al. (2002), Conway (2014), Dugundji (1966), Kelly (1975), Munkres (2000), Adhikari (2016, 2022), Adhikari and Adhikari (2014, 2022), Bredon (1983), Borisovich et al. (1985), Brown (1988), Fuks and Rokhlin (1984), Hu (1966), Patterson (1959), Singer and Thorpe (1967), Stephen (1970) and some other books are referred in Bibliography.

## 4.1   Separation by Open Sets and $T_i$-Spaces

There is a natural question: can any two distinct points or distinct subsets in a topological space be separated by open sets? The answer is positive for the real number space $\mathbf{R}$ under usual topology. But the answer is not positive for an arbitrary topological space. So a search is made for finding suitable conditions to have a positive answer. The two concepts such as weakly and strongly separated subsets of a topological space given in Definition 4.4.1 play a key role in the study of separation properties of topological spaces formulating $T_i$-axioms followed by $T_i$-spaces, for $i = 0, 1, 2, 3, 4, 5$.

### 4.1.1   Separation by Open Sets

This section introduces the concept of separation by open sets to formalize the separation axioms.

**Definition 4.1.1** Let $(X, \tau)$ be a topological space. Two nonempty subsets $A$ and $B$ in $X$ are said to be

(i) **weakly separated or (simply) separated** in $(X, \tau)$, if there exist two open sets $U$ and $V$ in $(X, \tau)$ such that

$$A \subset U, B \subset V, A \cap V = \emptyset \text{ and } B \cap U = \emptyset;$$

(ii) **strongly separated** in $(X, \tau)$, if there exist two open sets $U$ and $V$ in $(X, \tau)$ such that
$$A \subset U, \ B \subset V \text{ and } U \cap V = \emptyset.$$

***Example 4.1.2*** Let $(X, \tau)$ be a topological space. If two nonempty subsets $A$ and $B$ in $X$ are strongly separated in $(X, \tau)$, then they are also weakly separated in $(X, \tau)$, but its converse is not necessarily true. For example, consider the space $(\mathbf{R}, \tau)$ with cofinite topology $\tau$ on $\mathbf{R}$.

## 4.1.2 Separation Axioms and $T_i$-Spaces

This section imposes certain conditions prescribed in Definition 4.1.3, known as **separation axioms** on a topological space. They are called separation axioms, because they generate a separation of certain types of sets from each other by disjoint open sets. Some separation axioms are introduced, and the corresponding topological spaces are studied.

**Definition 4.1.3** A topological space $(X, \tau)$ is said to be a

(i) $T_0$-**space (due to Kolmogoroff)** if it satisfies $T_0$-axiom: for every pair of distinct points $x, y \in X$, there exists an open set which contains only one of the points $x$ and $y$; equivalently, $\overline{\{x\}} \neq \overline{\{y\}}$. Andrey Kolmogorov (1903–1987) introduced $T_0$-spaces around 1930;

(ii) $T_1$-**space (due to Fréchet)** if it satisfies $T_1$-axiom: for every pair of distinct points $x, y \in X$, there exist two open sets $U$ and $V$ in $X$ such that

$$x \in U, y \in V, x \notin V, \text{ and } y \notin U,$$

i.e., every pair of distinct points is weakly separated in $X$; equivalently, for every pair of distinct points $x, y \in X$, there exists a neighborhood of $x$ which does not contain $y$, and a neighborhood of $y$ which does not contain $x$. Some authors say that the concept of $T_1$-spaces was given by Frigyes Riesz (1880–1956) in 1907;

(iii) **Hausdorff space (due to Hausdorff)** if it satisfies $T_2$-axiom: any two distinct points are strongly separated in $X$; equivalently, distinct points have disjoint neighborhoods, i.e., if $x, y$ are any two distinct points of $X$, then there exist two disjoint open sets $U$ and $V$ in $X$ such that $x \in U, y \in V$. This class of topological spaces introduced by Felix Hausdorff (1868–1942) in 1914 is commonly known as Hausdorff spaces. On the other hand, other classes of topological spaces are rarely named after their inventors;

(iv) **regular space (due to Vietoris)** if it satisfies $T_3$-axiom: any closed set $F$ and any point $p \notin F$ are always strongly separated in $X$. This class of topological spaces was introduced by Leopold Vietoris (1891–2002) in 1921;

(v) **normal space (due to Tietze)** if it satisfies $T_4$-axiom: any two disjoint closed sets are strongly separated in $X$; equivalently, each pair of disjoint closed sets in $X$ have disjoint neighborhoods. This class of topological spaces was introduced by Tietze in 1923 and also independently by Pavel Alexandroff (1896–1982) and Pavel Urysohn (1898–1924) in 1929;

(vi) $T_5$-**space** if it satisfies $T_5$-axiom: if any two sets are weakly separated, then they are also strongly separated in $X$;

(vii) **completely normal space** if every subspace of $X$ is normal.

**Remark 4.1.4** (i) The $T_0$-axiom on a topological space $(X, \tau)$ asserts that every pair of distinct points in $X$ can be separated by the open sets in $(X, \tau)$, which contain only one of them.

(ii) Every topological space can be made into a $T_0$-space by identifying points having identical closures.

(iii) The $T_1$-axiom on a topological space $(X, \tau)$ asserts that for every $x \in X$, the one-pointic set $\{x\}$ is closed in $(X, \tau)$. The reason is that for every $y \in X$ different from the point $x$, there exists by $T_1$-axiom, an open set $U_y$ containing $y$ but not containing $x$. Then $X - \{x\} = \bigcup_{x \neq y \in X} U_y$ being the union of open sets, $X - \{x\}$ is an open set. On the other hand, if $\{x\}$ is closed in $X$, the open set $X - \{x\}$ can be chosen by the $T_1$-axiom as an open set containing any point other than $x$.

(iv) Every discrete space and every indiscrete space are both normal, and hence, a normal space may not satisfy $T_1$-axiom, the first or the second countability axioms for topological spaces. Since singleton sets in a $T_1$-space are closed, it follows that every normal $T_1$-space is a regular space. It is also Hausdorff by Exercise 23 of Sect. 4.7.

**Remark 4.1.5** Topologies defined in Examples 4.1.6, 4.1.7, and 4.1.8 are interesting, and they are used throughout the book.

**Example 4.1.6** Let $\tau$ be the natural topology, $\tau_1$ be the left-hand (or right-hand) topology, and $\tau_2$ be the lower-limit (or upper-limit) topology on **R**. Then,

(i) $(\mathbf{R}, \tau)$ is a Hausdorff space by Proposition 4.2.4. It is also a $T_5$-space by Exercise 30 of Sect. 4.7.

(ii) $(\mathbf{R}, \tau_1)$ is not Hausdorff by Proposition 4.2.13.

(iii) $(\mathbf{R}, \tau_1)$ is a normal and is also a completely normal space by Proposition 4.2.15.

(iv) $(\mathbf{R}, \tau_2)$ is a $T_5$-space by Proposition 4.2.18.

**Example 4.1.7** The **Sierpinski space** $(S, \tau_S)$, say, $S = \{0, 1\}$ and $\tau_S = \{S, \emptyset, \{1\}\}$ (see Chap. 3). Its concrete example is given in Example 4.2.10. The space $(S, \tau_S)$,

(i) is $T_0$, since $\{1\}$ is an open set containing only one of the points 0 or 1.

(ii) is not $T_1$, since $\{1\}$ is not closed in $(S, \tau_S)$.

(iii) is not Hausdorff, since the points 0 and 1 are not strongly separated by open sets in $(S, \tau_S)$.

(iv) is neither regular nor completely regular, because the point 1 and the closed set $\{0\}$ are not strongly separated by open sets in $(S, \tau_S)$.

**Example 4.1.8** Consider all the four the topologies $\tau_1$, $\tau_2$, $\tau_3$, $\tau_4$ on $X = \{x, y\}$ consisting of two distinct elements $x$ and $y$ only.

(i) $\tau_1 = \{\emptyset, \{x\}, \{y\}, \{x, y\}\}$ (discrete topology);
(ii) $\tau_2 = \{\emptyset, \{x\}, \{x, y\}\}$;
(iii) $\tau_3 = \{\emptyset, \{y\}, \{x, y\}\}$; and
(iv) $\tau_4 = \{\emptyset, \{x, y\}\}$ (trivial topology).

**Remark 4.1.9** The topology $\tau_1$ is Hausdorff, but the other three topologies are not so. On the other hand, all the four topologies are normal or (vacuously) normal in the sense that there exists no pair of nontrivial disjoint closed sets in $\tau_2$, $\tau_2$ and $\tau_3$.

### 4.1.3 Characterization of $T_1$-Spaces

This section gives a characterization of $T_1$-spaces in Theorem 4.1.10 and studies its immediate consequences. This characterization and its consequences assert that $T_1$ spaces are precisely the topological spaces in which points are closed sets. This characterization fulfills a natural requirement that each point of a topological space is a closed set. Characterizations of $T_1$-spaces in some other forms follow as a direct consequence of Theorem 4.1.10.

**Theorem 4.1.10** *A topological space $(X, \tau)$ is a $T_1$-space iff every one-pointic subset of $X$ is a closed set.*

**Proof** Let $(X, \tau)$ be a $T_1$-space and $\{x\}$ be a one-pointic subset of $X$. If $y \in X$ is any point different from $x$, then there exists an open set $V$ in $X$ such that $y \in V$ but $x \notin V$ by $T_1$ axiom. This shows that $y$ is not an accumulation point of the set $\{x\}$, and hence, $\{x\}$ is a closed set. Conversely, suppose $X$ is a topological space such that every one-pointic subset of $X$ is closed. If $x, y \in X$ and $x \neq y$, then the points $x$ and $y$ are weakly separated by the open sets $X - \{y\}$ and $X - \{x\}$ in $X$. This asserts that $X$ is a $T_1$-space. $\quad\square$

**Remark 4.1.11** An **alternative proof** of Theorem 4.1.10 is already given in Remark 4.1.4 (iii).

**Some immediate consequences of Theorem** 4.1.10.

**Corollary 4.1.12** *A topological space $(X, \tau)$ is a $T_1$-space iff every point $x \in X$, considered as a one-pointic set is a closed set.*

**Proof** It follows from Theorem 4.1.10. $\quad\square$

**Corollary 4.1.13** *A topological space* $(X, \tau)$ *is a* $T_1$-*space iff every finite subset set of* $X$ *is a closed set.*

*Proof* Let $(X, \tau)$ be a $T_1$-space. Then every finite subset set of $X$ is a closed set, because every one-pointic set in a $T_1$-space is closed and finite unions of closed sets are closed sets. Again if every finite subset of $X$ is a closed set, then every one-pointic subset of $X$ is closed in $X$. This asserts that $X$ be a $T_1$-space.                $\Box$

**Corollary 4.1.14** *Let* $(X, \tau)$ *be a topological space. Then it is a* $T_1$-*space iff* $\tau$ *contains the cofinite topolgy on* $X$.

*Proof* It follows from Theorem 4.1.10, because every finite union of closed sets in $(X, \tau)$ is a closed set.                $\Box$

The above discussion is summarized in a basic result embodied in Theorem 4.1.15.

**Theorem 4.1.15** $T_1$-*spaces are precisely the topological spaces* $(X, \tau)$ *in which singleton subsets* $\{x\}$ *of* $X$ *are closed sets.*

## 4.2   Hausdorff Spaces

This section addresses Hausdorff spaces which form an important family of topological spaces, because

(i) Every metric space is a Hausdorff spaces by Proposition 4.2.4.
(ii) Every metrizable space is also Hausdorff by Corollary 4.2.5.
(iii) Compact Hausdorff spaces studied in Chap. 5 form a very important class of topological spaces.
(iv) Locally Euclidean spaces discussed in Sect. 4.6.5 form an important class of Hausdorff spaces, specially, in the study of manifold theory (see Adhikari and Adhikari (2022) of the present series of book series), and
(v) Many other important results involving Hausdorff spaces.

*Remark 4.2.1* The Hausdorff property is a topological property in the sense that any topological space homeomorphic to a Hausdorff space is also Hausdorff. This property is stronger than that of a $T_1$-space and every pair of distinct points in a Hausdorff space have disjoint nbds. This property is utilized to prove the uniqueness of limit of a convergent sequence in a Hausdorff space in Theorem 4.2.2.

## 4.2.1   Basic Properties of Hausdorff Spaces

This section proves some basic properties enjoyed by Hausdorff spaces. Every Hausdorff space is a $T_1$-space. We now consider some other properties of Hausdorff spaces. Moreover, Proposition 4.2.4 and Corollary 4.2.5 provide a vast supply of Hausdorff spaces.

**Theorem 4.2.2** *Every convergent sequence in a Hausdorff space converges to a unique limit.*

**Proof** Let $(X, \tau)$ be a Hausdorff space and $\{x_n\} = \{x_1, x_2, x_3, \ldots\}$ be a convergent sequence in $X$ and converge to a point $a \in X$. If $b \neq a$ is a point in $X$, then by $T_2$-axiom, there exist two open sets $U, V$ in $X$ such that $a \in U, b \in V$ and $U \cap V = \emptyset$. Since $a$ is a limit point of $\{x_n\}$ and $U$ is an open set containing the point $a$, there exists a positive integer $n_0$ such that $x_n \in U$, $\forall n \geq n_0$. Since $U \cap V = \emptyset$, it follows that $V$ can contain at most $n_0$ elements of the sequence $\{x_n\}$. This asserts that $b$ cannot be a limit point of $\{x_n\}$. □

**Example 4.2.3** The converse of Theorem 4.2.2 is not true in general, because there is a topological space $(X, \tau)$ in which every convergent sequence has a unique limit in it, but the topological space is not Hausdorff. For example, if $X$ is an uncountable set endowed with the topology $\tau$ of countable complements, then every convergent sequence in this topological space $(X, \tau)$ has a unique limit, but the space $(X, \tau)$ is not Hausdorff (see Exercise 9 of Sect. 4.7). On the other hand, the converse of Theorem 4.2.2 is true with an additional condition: for a first countable topological space $(X, \tau)$, if every convergent sequence in $X$ has a unique limit, then topological space is Hausdorff (see Chap. 7).

**Proposition 4.2.4** *Every metric space is Hausdorff.*

**Proof** Let $(X, d)$ be a metric space and $x, y \in X$ be any two distinct points. Supposes $\epsilon = d(x, y)/2$. Then the $\epsilon$-neighborhoods $B_x(\epsilon)$ of $x$ and $B_y(\epsilon)$ of $y$ are disjoint, otherwise, if $z \in B_x(\epsilon) \cap B_y(\epsilon)$, then $d(x, z) < d(x, y)/2$ and $d(y, z) < d(x, y)/2$ would jointly imply that $d(x, z) + d(z, y) < d(x, y)$, which contradicts the triangle inequality condition of a metric space. Since $B_x(\epsilon)$ and $B_y(\epsilon)$ are open sets for the metric topology in $X$, it follows that the metric $(X, d)$ is Hausdorff. □

**Corollary 4.2.5** *Every metrizable space is a Hausdorff space.*

**Proof** Let $(X, \tau)$ be a metrizable space. Then there exists a metric $d$ on $X$ such that its induced topology $\tau_d$ on $X$ coincides with $\tau$. Hence the corollary follows from Proposition 4.2.4. □

**Proposition 4.2.6** *Let $(X, d)$ and $(Y, \rho)$ be two metric spaces and $D \subset X$ be dense in $X$. If $f, g : X \to Y$ are two continuous maps such that $f(x) = g(x)$, $\forall x \in D$. Then $f(x) = g(x)$, $\forall x \in X$.*

**Proof** Suppose $f(x) \neq g(x)$, for some $x \in X$. Then there exists at least one point $x_0 \in X$, such that $f(x_0) \neq g(x_0)$. Since $Y$ is Hausdorff by Proposition 4.2.4, then for this distinct pair of elements $f(x_0)$ and $g(x_0)$ in $Y$, there exist disjoint open sets $U$ and $V$ in $Y$, such that $f(x_0) \in U$ and $g(x_0) \in V$. Then $U_1 = f^{-1}(U)$ is an open set in $X$ containing $x_0$, and $V_1 = f^{-1}(V)$ is an open set in $X$ containing $x_0$, and hence $x_0 \in U_1 \cap V_1$. Since by hypothesis $D$ is dense in $X$ and $f(x) = g(x)$, $\forall x \in D$, there exists a point $x_1 \in D \cap U_1 \cap V_1$ such that $f(x_1) = g(x_1)$ and hence $f(x_1) = g(x_1) \in U \cap V$, which contradicts that $U \cap V = \emptyset$. This contradiction asserts that $f(x) = g(x)$, $\forall x \in X$. □

Corollary 4.2.7 gives an extension of Proposition 4.2.6 for topological spaces with an alternative condition.

**Corollary 4.2.7**  *Let* $(X, \tau)$ *be a topological space and* $(Y, \sigma)$ *be a Hausdorff space. If* $D \subset X$ *is dense in* $X$ *and* $f, g : X \to Y$ *are two continuous maps such that* $f(x) = g(x), \forall x \in D$. *Then*

$$f(x) = g(x), \forall x \in X.$$

*Proof*  Proceed as in Proposition 4.2.6.                                             $\square$

**Remark 4.2.8**  For alternative proof of Proposition 4.2.4 and its Corollary 4.2.5 see Chap. 6.

**Example 4.2.9**  Quotient space of a Hausdorff space may not be Hausdorff. For example, **R** endowed with the natural topology is Hausdorff by Proposition 4.2.4 but the quotient space $\mathbf{R}/\sim$, where $x \sim y$ iff $x - y$ is a rational number, is an uncountable set having trivial topology. Hence $\mathbf{R}/\sim$ is not a Hausdorff.

**Example 4.2.10**  Quotient space of even a compact Hausdorff space may not be Hausdorff. For example, consider the quotient space $X/A$, where $X = \mathbf{I} = [0, 1]$ and $A = [0, 1)$ are subspaces of **R** with subspace topology of **R** and the canonical map $p : X \to X/A, x \mapsto [x]$, where $X/A$ is the quotient space corresponding to the equivalence relation $\sim$, which identifies every pair of elements in $A$ and no other pair of points is continuous and surjective. Since $p^{-1}([0]) = [0, 1)$ is open, the point $[0] \in X/A$ is open. On the other hand, since $p^{-1}([1]) = \{1\}$ is not open, the point $[1] \in X/A$ is not open. This implies that the quotient space $X/A$ is Sierspenski (see Chap. 3), and hence, the quotient space $X/A$ is not Hausdorff. This gives an application of Sierspenski space.

### 4.2.2  Separation Property of Left-Hand (Right-Hand) Topology on R

This section proves some special properties (structures) of the left-hand and right-hand topologies on **R** and their deviation from natural topology on **R**, defined in Chap. 3. For example, let $\tau$ be the natural topology, $\tau_1$ be the left-hand (or right-hand) topology. Then

(i)  $(\mathbf{R}, \tau)$ is a Hausdorff space by Proposition 4.2.4. It is also a $T_5$-space by Exercise 30 of Sect. 4.7.

(ii)  $(\mathbf{R}, \tau_1)$ is not Hausdorff by Proposition 4.2.13.

(iii)  $(\mathbf{R}, \tau_1)$ is a normal and completely normal space by Proposition 4.2.15.

**Remark 4.2.11**  All the results evolving left-hand topology are also valid for right-hand topology and vice-versa. So, it is sufficient to study only one of them.

***Example 4.2.12*** **Hausdorff property** of a topological space $(X, \tau)$ depends on the topology of the set $X$. For example, $\mathbf{R}$ endowed with the natural topology is Hausdorff by Proposition 4.2.4, but $\mathbf{R}$ endowed with the left-hand topology is not so by Proposition 4.2.13.

Proposition 4.2.13 proves nonHausdorff property of $\tau_l$ on $\mathbf{R}$, and hence, the space $(\mathbf{R}, \tau_l)$ is not metrizable.

**Proposition 4.2.13** *Let $\tau_l$ be the left-hand topology on $\mathbf{R}$. Then the space $(\mathbf{R}, \tau_l)$ is not Hausdorff.*

***Proof*** The open sets in $\tau_l$ consist of all the subsets $U = \{x \in \mathbf{R} : x < a, \, \forall a \in \mathbf{R}\}$, the empty set $\emptyset$ and the whole set $\mathbf{R}$, and hence the closed sets are the subsets $\{x \in \mathbf{R} : a \leq x, \, \forall a \in \mathbf{R}\}$, the whole set $\mathbf{R}$ and the empty set $\emptyset$. For any two distinct points $a, b \in \mathbf{R}$, with $a < b$, the open set $V = \{x :\in \mathbf{R} : x < b\}$, contains $a$, but does not contain $b$. This asserts that the space $(\mathbf{R}, \tau_l)$ can not be Hausdorff and hence it not metrizable by Corollary 4.2.5. $\qquad\square$

**Corollary 4.2.14** *Let $\tau_l$ be the left-hand topology on $\mathbf{R}$. Then the space $(\mathbf{R}, \tau_l)$ is not metrizable.*

***Proof*** The space $(\mathbf{R}, \tau_l)$ is not Hausdorff by Proposition 4.2.13, and hence it is not metrizable by Corollary 4.2.5. $\qquad\square$

**Proposition 4.2.15** *Let $\tau_l$ be the left-hand topology on $\mathbf{R}$. Then the space $(\mathbf{R}, \tau_l)$ is normal and also completely normal.*

***Proof*** The topology $\tau_l$ consists of $\mathbf{R}, \emptyset$ and the family of subsets $U_a = \{x \in \mathbf{R} : x \in \mathbf{R}, \, x < a\} \subset \mathbf{R}$ for all $a \in X$. Hence the closed sets in $(X, \tau_l)$ are $\emptyset, \mathbf{R}$ and the family of subsets $\{x \in \mathbf{R} : x \in \mathbf{R}, \, a \leq x\} \subset \mathbf{R}$ for all $a \in X$. Consequently, for any two distinct nonempty open sets one open set will lie in the other. This shows that no two nonempty subsets of $\mathbf{R}$ are weakly separated in $(X, \tau_l)$. Again, no two nonempty closed sets are disjoint in $(X, \tau_l)$. This asserts that the topology $\tau_l$ satisfies $T_4$ and $T_5$ axioms on $\mathbf{R}$ vacuously. This proves that the space $(\mathbf{R}, \tau_l)$ is normal, and it is also completely normal. $\qquad\square$

***Example 4.2.16*** There is a natural problem: does every topological space admit a metric structure? The answer is negative. Because, if a topological space $(X, \tau)$ is such that it not Hausdorff, then $(X, \tau)$ can not be a metrizable by Corollary 4.2.5, and hence there exists no metric $d$ on $X$ such that its induced topology $\tau_d$ coincides with $\tau$. For example, as the topological space $(\mathbf{R}, \tau_l)$ with the left-hand topology $\tau_l$ on $\mathbf{R}$, is not metrizable by Corollary 4.2.14, there exists no metric $d$ on $X$ such that its induced topology $\tau_d$ coincides with $\tau_l$.

### 4.2.3   Separation Property of Lower-Limit (Upper-Limit) Topology on **R**

This section proves some special properties (structures) of the lower-limit and upper-limit topologies on **R** and their similarity with natural topology on **R**, defined in Chap. 3. For example, let $\tau$ be the natural topology and $\tau_2$ be the lower-limit (or upper-limit) topology on **R**. Then

(i) $(\mathbf{R}, \tau)$ is a Hausdorff space by Proposition 4.2.4. It is also a $T_5$-space by Exercise 30 of Sect. 4.7;

(ii) $(\mathbf{R}, \tau_2)$ is also a Hausdorff space by Proposition 4.2.19.

(iii) $(\mathbf{R}, \tau_2)$ is also a $T_5$-space by Proposition 4.2.18.

***Remark 4.2.17***   All the results evolving lower-limit topology are also valid for upper-limit topology and vice-versa. So, it is sufficient to study only one of them.

**Proposition 4.2.18**   *Let $\tau_2$ be the lower-limit (upper-limit) topology on **R**. Then $(\mathbf{R}, \tau_2)$ is a $T_5$ space.*

***Proof***   Let $\tau_2$ be the the lower-limit topology on **R** generated by the closed–open intervals $[z, w)$. Let $x, y$ $(x < y) \in \mathbf{R}$ be any pair of distinct points. Then they are strongly separated by two disjoint open sets $[x, x + t)$ and $[y, y + t)$, where $t = y - x$. This implies that $(\mathbf{R}, \tau_2)$ is a Hausdorff space. Let $A$ and $B$ be an arbitrary pair of weakly separated subsets in $(\mathbf{R}, \tau_2)$. Then for any point $a \in A$, there exists a real number $r_a > a$ such that $[a, r_a) \cap B = \emptyset$, because $a$ is not an accumulation point of $B$. Similarly, for any point $b \in B$, there exists a real number $r_b > b$ such that $[b, r_b) \cap A = \emptyset$. This implies that $[a, r_a) \cap [b, r_b) = \emptyset$, otherwise, either $a \in [b, r_b)$ or $b \in [a, r_a)$ but it is not possible. Consider two open sets

$$U = \bigcup_{a \in A} [a, r_a) \text{ and } V = \bigcup_{b \in b} [b, r_b).$$

Then $U$ and $V$ are two disjoint open sets in $(\mathbf{R}, \tau_2)$ such that the pair of weakly separated sets $A$ and $B$ are also strongly separated by $U$ and $V$ in $(\mathbf{R}, \tau_2)$. This proves that $(\mathbf{R}, \tau_2)$ is a $T_5$ space.

Similar result holds for lower-limit topology.                                            □

**Proposition 4.2.19**   *Let $\sigma$ be the the upper-limit (lower-limit) topology on **R**. Then $(\mathbf{R}, \sigma)$ is a Hausdorff space.*

***Proof***   Let $\sigma$ be the the upper-limit topology on **R** generated by the open–closed intervals $(z, w]$ and $x, y \in \mathbf{R}$ be two elements such that $x \neq y$. Without loss of generality, assume that $x < y$. Take $U = (x - 1, x]$ and $V = (x, y]$. Then $U, V \in \sigma$ are such that $x \in U$, $y \in V$ and $U \cap V = \emptyset$. This shows that the space $(\mathbf{R}, \sigma)$ is Hausdorff.                                            □

### 4.2.4 First Countable Hausdorff Spaces

This section studies first countable Hausdorff spaces and characterizes such spaces with the help of convergent sequences on them. Theorem 4.2.2 asserts that every convergent sequence in a Hausdorff space converges to a unique limit. On the other hand, Exercise 9 of Sect. 4.7 shows that its converse is not true in general but is true for a certain class of topological spaces (see Theorem 4.2.21).

**Definition 4.2.20** Let $(X, \tau)$ be a first countable topological space and $\mathcal{B}_x = \{B_1, B_2, \ldots\}$ be a countable local base at the point $x \in X$. Then $\mathcal{B}_x$ is said to be a **nested local base** at the point $x \in X$ if

$$B_1 \supset B_2 \supset B_3 \supset \cdots .$$

Theorem 4.2.21 characterizes Hausdorff property of first countable topological spaces by convergent sequences using the concept of nested local base.

**Theorem 4.2.21** *A first countable topological space $(X, \tau)$ is Hausdorff iff every convergent sequence on $X$ has a unique limit.*

**Proof** Let $(X, \tau)$ be a Hausdorff space and $\{x_n\} = \{x_1, x_2, x_3, \ldots\}$ be a convergent sequence in $X$. Then by Theorem 4.2.2, the sequence $\{x_n\}$ has a unique limit point in $X$. Conversely, let $(X, \tau)$ be a first countable space such that every convergent sequence in $X$ has a unique limit point in $X$. Then $(X, \tau)$ must be a Hausdorff space. Otherwise, there exist distinct points $x, y \in X$ such that every open set containing the point $x$ has a nonempty intersection with each open set containing the point $y$ in $X$. Let $\{U_n\}$ be a nested local base at the point $x$ and $\{V_n\}$ be a nested local base at the point $y$. Then $U_n \cap V_n \neq \emptyset$, otherwise, for every positive integer $n$, there exists a sequence $\{x_n\}$ in $X$ such that

$$x_1 \in U_1 \cap V_1, \ x_2 \in U_2 \cap V_2, \ \ldots .$$

This implies that the sequence $\{x_n\}$ converges to both the points $x$ and $y$, which is not possible, since $x \neq y$. This contradiction proves that the topological space $(X, \tau)$ is Hausdorff.

$\square$

## 4.3 Structures of Normal and Completely Normal Spaces

This section studies normal and completely normal spaces to determine their additional topological structures. Metric and metrizable spaces form important classes of normal spaces. On normal spaces, in particular, on metric spaces, there always exist nonconstant real-valued continuous functions by Urysohn lemma, named after Urysohn (1998–1924) (see Chap. 6).

Since an arbitrary subspace of a normal space is not normal, the concept of a completely normal space is introduced in Definition 4.3.12 to avoid the situation for failure of normality criterion with regard to its subspaces in the sense that a completely normal space is a topological space such that its every subspace is normal. Every metric space is also completely normal. Every completely normal space is normal by its defining property. But Example 4.3.15 asserts that its converse is not true. Theorem 4.3.1 gives a characterization of normal spaces, known as normality criterion of Urysohn. On the other hand, Exercise 32 of Sect. 4.7 gives a characterization of completely normal spaces. Theorem 4.3.3 and Corollary 4.3.5 provide a rich supply of normal spaces.

### 4.3.1  Normal Spaces and Normality Criterion of Urysohn

This section studies normal spaces and **proves normality criterion of Urysohn in Theorem** 4.3.1, which is one of the outstanding results for characterization of normal spaces. There exist several characterizations of normal spaces. For example, Theorem 4.3.1 and Exercise 28 of Sect. 4.7 give different characterizations of normal spaces. Two other characterizations of normal spaces formulated by real-valued functions and proved in Urysohn's lemma and Tietze's theorem (see Chap. 6) are basic results in topology. It has been proved in this section that metric spaces and metrizable spaces are normal spaces. Moreover, compact Hausdorff spaces form also an important class normal of spaces (see Chap. 5)

**Theorem 4.3.1** *(Normality Criterion of Urysohn) Let $(X, \tau)$ be a topological space. Then it is normal iff for every closed set $A \subset X$ and any open set $U$ containing $A$, there exists an open set $V$ such that $A \subset V$, $\bar{V} \subset U$.*

**Proof** Let the given condition hold for the topological space $(X, \tau)$. We claim that it is normal. Let $A$ and $B$ be a pair of disjoint closed sets in $(X, \tau)$. Then $A$ is contained in the open set $X - B$. This asserts by hypothesis that there exists an open set $V$ such that

$$A \subset V, \bar{V} \subset X - B.$$

It implies that the pair of closed sets $A$ and $B$ are strongly separated by disjoint open sets $V$ and $X - \bar{V}$ in $(X, \tau)$. This implies that the space $(X, \tau)$ is normal.

Conversely, let the space $(X, \tau)$ be normal and $U$ be an open set containing a closed set $A$. Then the pair of disjoint closed sets $A$ and $X - U$ are strongly separated in $(X, \tau)$, and hence for this pair of closed sets, there exist a pair of open sets $V$ and $W$ such that

$$A \subset V, \ X - U \subset W, \ V \cap W = \emptyset.$$

Since $V \cap W = \emptyset$, it follows that $V \subset X - W$ and hence

$$\bar{V} \subset \overline{X - W} = X - W \subset U.$$

This asserts that

$$A \subset V \text{ and } \bar{V} \subset U,$$

which proves the necessity of the condition.

□

**Corollary 4.3.2** *A topological space* $(X, \tau)$ *is normal iff either of the following conditions is satisfied:*

(i) *For every pair of disjoint closed sets $A$ and $B$ in $(X, \tau)$, there exists an open set $U$ such that $A \subset U$ and $B \cap \bar{U} = \emptyset$, or*

(ii) *Every pair of disjoint closed sets have nbds with disjoint closures.*

**Proof** Left as an exercise.

□

**Theorem 4.3.3** *Every metric space is normal.*

**Proof** Let $X$ be a metric space with a metric $d$ and $A$ be a nonempty subset in $X$. Given $x \in X$, define $d(x, A)$ to be the glb of the distances $d(x, a)$ for $a \in A$, i.e., $d(x, A) = \text{glb}\,\{d(x, a) : a \in A\}$. If $A$ is closed, then $d(x, A) = 0$ iff $x \in A$. Because, if $x \in A$, then $d(x, A) = 0$. Moreover, if $d(x, A) = 0$, then corresponding to a given $\epsilon > 0$, there exists an $a \in A$ such that $d(x, a) < \epsilon$, which shows that $x \in \bar{A} = A$, as $A$ is a closed set in $X$. Let $A, B$ be two disjoint closed sets in a metric space $X$ and $Y$ be the subset of $X$ defined by $Y = \{x \in X : d(x, A) < d(x, B)\} \subset X$. Given $y \in Y$, there exists a $\delta > 0$ (depending on $y$) such that $d(y, A) = d(y, B) - \delta$. If $z \in N_{\frac{\delta}{2}}$, an $\frac{\delta}{2} - nbd$ of $y$, then $d(y, z) < \frac{\delta}{2}$. Again, by triangle inequality,

$$d(z, a) \leq d(z, y) + d(y, a).$$

It implies that $d(z, A) \leq d(z, y) + d(y, A)$ and hence

$$d(z, A) \leq d(z, y) + d(y, B) - \delta.$$

Again, $d(y, B) \leq d(y, z) + d(z, B)$ and hence

$$d(z, A) \leq 2d(y, z) - \delta + d(z, B).$$

Now, $d(z, A) < d(z, B)$, since $d(y, z) < \frac{\delta}{2}$. This shows that the set of points $y \in Y$ such that $d(y, A) < d(y, B)$ is an open set containing $A$. In a similar way, the set of points $y \in Y$ satisfying $d(y, B) < d(y, A)$ is an open set containing $B$. As these two sets are disjoint, the metric space $X$ is normal.

□

**Remark 4.3.4** An alternative proof of Theorem 4.3.3 by using Urysohn lemma is available in Chap. 6.

**Corollary 4.3.5**  *Every metrizable space is normal.*

**Proof** Let $(X, \tau)$ be a metrizable space. Then there exists a metric $d$ on $X$ such the $\tau = \tau_d$ (topology induced on $X$ by $d$). Hence the corollary follows from Theorem 4.3.3.                                                                                          □

**Corollary 4.3.6**  *The real line-space* **R** *is normal.*

**Proof** It follows from Theorem 4.3.3.                                                      □

**Example 4.3.7**  Every metrizable space is normal by Corollary 4.3.5, but its converse is not true. For example, if $\tau_l$ is the left-hand topology on **R**, then the space $(\mathbf{R}, \tau_l)$ is normal by Proposition 4.2.15, but it is not metrizable by Corollary 4.2.14.

**Example 4.3.8**  An arbitrary subspace of a normal space may not be normal. In support, consider Example 4.3.15. On the other hand, Proposition 4.3.9 asserts that every closed subspace of a normal space is normal.

**Proposition 4.3.9**  *Every closed subspace of a normal space is normal.*

**Proof** Let $(X, \tau)$ be a normal space and $Y$ be a closed subspace of $X$. Consider the the subspace $(Y, \tau_Y)$ of $(X, \tau)$. Let $A$ and $B$ be any two disjoint closed sets in $(Y, \tau_Y)$. Then they are also disjoint closed sets in $(X, \tau)$. By hypothesis, $(X, \tau)$ is normal. Hence $A$ and $B$ are strongly separated by two open sets $U$ and $V$ in $(X, \tau)$. This implies that $A$ and $B$ are strongly separated by two open sets $Y \cap U$ and $Y \cap V$ in $(X, \tau_Y)$. This asserts that the subspace $(X, \tau_Y)$ is normal.
                                                                                              □

**Corollary 4.3.10**  *Let $(X, \tau)$ and $(Y, \sigma)$ be two topological spaces such that their product space $(Z, \alpha)$ is normal. Then each of the topological spaces $(X, \tau)$ and $(Y, \sigma)$ is normal.*

**Proof** By hypothesis, product space $(Z, \alpha)$ of the spaces $(X, \tau)$ and $(Y, \sigma)$ is normal. Then each of the factor spaces $(X, \tau)$ and $(Y, \sigma)$ is closed in $(Z, \alpha)$, and hence the Corollary follows from Proposition 4.3.9.                                                     □

**Example 4.3.11**  The topological product space of normal spaces may not be normal. For example, consider the topological product space $(X, \sigma) = (\mathbf{R}, \tau_l) \times (\mathbf{R}, \tau_l)$ of Sorgenfrey line $(\mathbf{R}, \tau_l)$. The space $(\mathbf{R}, \tau_l)$ is normal, but $(X, \sigma)$ is not normal.

### 4.3.2   Completely Normal Spaces

This section continues the study of a completely normal spaces initiated in Sect. 4.1.2. This space is different from a normal space. Example 4.3.15 shows that every subspace of a normal space may not be normal. This motivates to define a completely normal space, which asserts that its every subspace is normal. A characterization of a completely normal space in terms of disjoint closed sets containing separated sets is given in Exercise 32 of Sect. 4.7.

**Definition 4.3.12** A topological space $(X, \tau)$ is said to be **completely normal** if every subspace of $(X, \tau)$ is normal.

**Example 4.3.13** The set **R** equipped with left-hand topology is completely normal. by Proposition 4.2.15.

**Example 4.3.14** Every metric space is completely normal.

**Example 4.3.15** The concepts of normal and completely normal spaces are different, because every subspace of a completely normal space is normal by its defining property. On the other hand, every subspace of a normal space may not be normal. For example, consider the set $X = \{x, y, z, t\}$ and its family of subsets $\tau = \{\emptyset, \{t\}, \{y, t\}, \{z, t\}, \{y, z, t\}, X\}$. Then $\tau$ forms a topology on $X$. The closed sets of $(X, \tau)$ are

$$X, \ \{x, y, z\}. \ \{x, z\}, \ \{x, y\}, \ \{x\} \text{ and } \emptyset.$$

Then $(X, \tau)$ is (vacuously) a normal space, since there exists no pair of nontrivial disjoint closed sets in $(X, \tau)$. On the other hand, if $Y = \{y, z, t\}$, then the subspace $(Y, \tau_Y)$ of the normal space $(X, \tau)$ with relative topology $\tau_Y$ on $Y$ is not normal, because the pair of disjoint closed sets $\{y\}$ and $\{z\}$ can not be strongly separated in $(Y, \tau_Y,)$ since $\tau_Y = \{\emptyset, \{t\}, \{y, t\}, \{z, t\}, Y\}$ and the closed sets in $(Y, \tau_Y)$ are

$$Y, \ \{y, z\}, \ \{z\}, \ \{y\} \text{ and } \emptyset.$$

This shows that $(X, \tau)$ is normal, but its subspace $(X, \tau_Y)$ is not normal, and hence, $(X, \tau)$ is not completely normal.

## 4.4 Structures of Regular and Completely Regular Spaces

This section studies the structures of regular and completely regular spaces and proves regularity criterion of Tychonoff in Theorem 4.4.5. Every completely regular space is closely linked with real-valued continuous functions by its Definition 4.4.13. Metric spaces provide a rich supply of completely regular spaces by Corollary 4.4.19.

### 4.4.1 Regular Spaces

This section proves "Regularity Criterion of Tychonoff" in Theorem 4.4.5 providing a characterization of regular spaces and also establishes some other properties of regular spaces.

**Definition 4.4.1** A topological space $(X, \tau)$ is said to be **regular at a point** $a \in X$ if the one-pointic set $\{a\}$ and any closed $A$ in $X$ not containing the point $a$ are strongly separated. It is said to be a **regular space** if it is regular at every point of $X$.

**Remark 4.4.2** It follows from Definition 4.4.1 that a topological space $(X, \tau)$ is said to be regular if given any closed set $A$ in $(X, \tau)$ and a point $y \in X$, *but* $y \notin A$, there exist disjoint open sets $U$ and $V$ such that

$$A \subset U \text{ and } y \in V.$$

*Example 4.4.3* A regular space may not be $T_1$. In support, let $X = \{x, y, z\}$ and the topology on $X$ be $\tau = \{\emptyset, X, \{x\}, \{y, z\}\}$. Then the closed sets of $(X, \tau)$ are precisely

$$\{X, \emptyset, \{y, z\}, \{x\}\}.$$

This shows that $(X, \tau)$ is a regular space, but it is not a $T_1$ space, because there exists a finite set $\{y\}$, which is not closed.

**Remark 4.4.4** Theorem 4.4.5 gives a characterization known as regularity criterion of Tychonoff. Again, Corollary 4.4.6 gives another characterization of regular spaces formulated in an equivalent form of regularity criterion of Tychonoff.

**Theorem 4.4.5** *(Regularity Criterion of Tychonoff) A topological space $(X, \tau)$ is regular iff for every point $a \in X$, and every nbd $U$ of $a$, there exists a nbd $K$ of $a$ such that $\bar{K} \subset U$.*

*Proof* Let $(X, \tau)$ be a regular space and $U$ be a nbd of the point $a \in X$. Then there exists an open set $V$ in $X$ such that $a \in V \subset U$. Hence the closed set $X - V$ can not contain the point $a$. This shows that $X - V$ and $\{a\}$ are strongly separated in $(X, \tau)$, since $(X, \tau)$ be a regular space. This asserts that there exist two disjoint open sets $S, T$ in $X$ such that

$$X - V \subset S, \ a \in T \text{ and } S \cap T = \emptyset.$$

But $S \cap T = \emptyset \implies T \subset X - S$. Since $X - S$ is closed, it follows that

$$\bar{T} \subset \overline{X - S} \subset X - S \subset V \subset U.$$

Taking $T = K$, it shows that $K$ is an open nbd of the point $a$ such that $\bar{K} = \bar{T} \subset U$. Conversely, let a topological space $(X, \tau)$ satisfy the given conditions and $A$ be a closed set such that the given point $a$ lies outside $A$. This implies that $U = X - A$ is an open nbd of the point $a$ and hence by hypothesis, there exists a nbd $K$ of $a$ such that $\bar{K} \subset U = X - A$. Let $W$ be an open set such that $a \in W \subset K$. This shows that $\{a\}$ and $A$ are strongly separated by two disjoint open sets $W$ and $X - \bar{K}$. This implies that the topological space $(X, \tau)$ is regular.

$\square$

Corollaries 4.4.6 and 4.4.7 give regularity criterion of Tychonoff formulated in two equivalent forms.

**Corollary 4.4.6** *A topological space* $(X, \tau)$ *is regular iff for every point* $a \in X$ *and for any open set* $U$ *containing the point* $a$, *there exists an open set* $V$ *such that*

$$a \in V \subset \overline{V} \subset U.$$

*Proof* It follows from Theorem 4.4.5. $\quad\square$

**Corollary 4.4.7** *A topological space* $(X, \tau)$ *is regular iff for any nbd* $U_a$ *containing the point* $a \in X$, *there exists an open nbd* $N$ *such that*

$$a \in N \subset \overline{N} \subset U.$$

*Proof* It follows from Theorem 4.4.5. $\quad\square$

**Proposition 4.4.8** *Let* $(X, \tau)$ *be a regular space and* $x, y$ *be two distinct points in* $X$. *Then the subsets* $\overline{\{x\}}$ *and* $\overline{\{y\}}$ *are either identical or they are disjoint.*

*Proof* If both $x \in \overline{\{y\}}$ and $y \in \overline{\{x\}}$ hold, then

$$\overline{\{x\}} \subset \overline{\overline{\{y\}}} = \overline{\{y\}} \subset \overline{\overline{\{x\}}} = \overline{\{x\}}$$

asserts that $\overline{\{x\}} = \overline{\{y\}}$. Again, if at least one relation is not true, say $y \notin \overline{\{x\}}$, then the point $y$ lies outside the closed set $\overline{\{x\}}$, and hence, the closed set $\overline{\{x\}}$ and the point $y$ lying outside the closed set $\overline{\{x\}}$ are strongly separated in $(X, \tau)$. This implies that there exists an open set $V$ in $(X, \tau)$ such that $\overline{\{x\}} \subset V$ and $y \in X - V$. This shows that $\overline{\{y\}} \subset \overline{X - V} = X - V$. This implies that $\overline{\{y\}} \cap \overline{\{x\}} = \emptyset$, since $\overline{\{x\}} \subset V$ and $\overline{\{y\}} \subset X - V$. This proves that either $\overline{\{x\}} = \overline{\{y\}}$ or they are disjoint. $\quad\square$

## 4.4.2 Independence of Regularity and Hausdorf Properties

This section shows that regularity and Hausdorff properties of topological spaces are independent in the sense one property does not guarantee the holding of other property. In support consider the Examples 4.4.3 and 4.4.10.

*Example 4.4.9* A regular space may not be Hausdorff. For example, the topological space $(X, \tau)$ given in Example 4.4.3 is regular but not Hausdorff.

*Example 4.4.10* A Hausdorff space may not be regular.

(i) Let **R** be the sets reals and **Q** be the set of rationals. If **B** is the family of all open intervals $(a, b) \subset \mathbf{R}$ together with the subset **Q**, then **B** forms an open base for topology $\tau$ on $\mathbf{R}$. If $\mathbf{R}$ is the usual topology $\sigma$ on $\mathbf{R}$, then the topology $\tau$ is stronger than the usual topology on $\sigma$. Since $(\mathbf{R}, \sigma)$ is Hausdorff, it follows that $(\mathbf{R}, \tau)$ is also Hausdorff. But the space $(\mathbf{R}, \tau)$ is not regular, because the closed set of irrationals which is $\mathbf{R} - \mathbf{Q}$ and the point 1 staying outside $\mathbf{R} - \mathbf{Q}$ cannot be strongly separated by the disjoint open sets in the space $(\mathbf{R}, \tau)$.

(ii)  Consider **Niemytzki's space** defined in Chap. 3. This space may be reformulated: Let $X^{+} = \{p = (x, y) \in \mathbf{R}^2 : y \geq 0\}$, which is the closed upper-half of the Euclidean plane $\mathbf{R}^2$. Let the $x$-axis for which $y = 0$ be denoted by $\mathbf{R}^1$ and open upper-half plane $X^{+} - \mathbf{R}^1$ by $X^*$. Define $\mathbf{B}_1$ and $\mathbf{B}_2$ as

$$\mathbf{B}_1 = \{B_p(\epsilon) : p = (x, y) \in X^* \text{ and } \epsilon < y\},$$

$$\mathbf{B}_2 = \{(B_p(\epsilon) \cap X^*) \cup \{x\} : p = (x, 0) \text{ and } x \in \mathbf{R}^1\},$$

where $B_p(\epsilon)$ denotes the open ball having center $p$ and radius $\epsilon$ with respect to the Euclidean metric on $\mathbf{R}^2$. Then $\mathbf{B} = \mathbf{B}_1 \cup \mathbf{B}_2$ forms an open base for a topology $\tau$ on $X^{+}$, which is different from the usual topology $\sigma$ induced on $X^{+}$ by the Euclidean topology on $\mathbf{R}^2$. The topological space $(X^{+}, \tau)$ is Hausdorff. It is not regular, because one-pointic sets in this space $(X^{+}, \tau)$ are closed by its Hausdorff property and the set $A = \mathbf{R}^1 - \{0\}$ is also closed in $(X^{+}, \tau)$, since its complement $X^* \cup \{0\}$ is open in $(X^{+}, \tau)$. The closed set $A$ and the point 0 which lies outside $A$ can not be separated by open sets, because if $U$ is an open set containing the point 0, then there exists some $\epsilon > 0$ such that the basis element $(B_0(\epsilon) \cap X^*) \cup \{0\} \subset U$. This implies that any open set containing the point $(\epsilon/2, 0)$ intersects $U$. This shows that space $(X^{+}, \tau)$ is not regular.

### 4.4.3    Completely Regular Spaces

This section conveys the concept of completely regular spaces introduced by Andrey L. Tychonoff (1906–1993) (also spelled Tikhonov) in 1930. This class includes metric spaces. A completely regular spaces is closely linked with real-valued continuous functions in the sense that one can separate a point from a closed set by a real-valued continuous function in this space. On the other hand, Urysohn lemma asserts the existence of a real-valued continuous function for every pair of disjoint closed sets in a normal space. Since every completely regular space is regular by Proposition 4.4.17 and every normal space is completely regular by Uryshon lemma, it has become necessary a new separation axiom to define a completely regular space. A detailed study on Urysohn lemma and completely regular spaces is available in Chap. 6.

**Definition 4.4.11**  Let $(X, \tau)$ be a topological space. A pair of subsets $A$ and $B$ in $X$ are said to be separated by a real-valued continuous function if there exists a continuous function $f : X \to \mathbf{R}$ such that

$$f(x) = \begin{cases} 0, \text{ for all } x \in A \\ 1, \text{ for all } x \in X - U, \end{cases}$$

and

$$0 \le f(x) \le 1.$$

**Remark 4.4.12**  **Urysohn Lemma** separates every pair of disjoint closed sets of a normal space $X$ by a real-valued continuous function on $X$. This lemma characterizes a normal space by a real-valued continuous function asserting that a topological space $(X, \tau)$ is normal iff for every pair of disjoint closed sets $A$ and $B$ in $X$, then there exists a continuous real-valued function $f : X \to \mathbf{R}$ such that

$$f(x) = \begin{cases} 0, \text{ for all } x \in A \\ 1, \text{ for all } x \in B, \end{cases}$$

and

$$0 \le f(x) \le 1 \text{ for all } x \in X$$

The proof of Urysohn Lemma is deferred to Chap. 6.

**Definition 4.4.13**  A Hausdorff space $(X, \tau)$ is said to be **completely regular** if given any closed set $A$ in $(X, \tau)$ and any point $a \in X$ but not lying in $A$, (i.e., $a \notin A$) there is a real-valued continuous function $f : X \to \mathbf{R}$ such that

$$f(x) = \begin{cases} 0, \text{ for } x = a \\ 1, \text{ for all } x \in A \end{cases}$$

and

$$0 \le f(x) \le 1 \text{for all } x \in X,$$

i.e., if $A$ and $\{a\}$ are separated by a continuous real-valued function on $X$.

By replacing $f(x)$ by $1 - g(x)$ in Definition 4.4.13, its equivalent definition is now given:

**Definition 4.4.14**  A topological space $(X, \tau)$ is is said to be **completely regular** if every one-pointic set is closed and given a closed set $A$ and any point $x_0 \in X - A$, there exists a continuous function $g : X \to [0, 1]$ such that

$$g(x_0) = 1, \text{ and } g(x) = 0, \forall x \in A.$$

**Example 4.4.15**    (i) Every normal space is completely normal by Urysohn lemma (see Chap. 6).

(ii) Every regular normal space is completely regular by Proposition 4.4.16.

(iii) Every metric space is completely regular by Corollary 4.4.19.

**Proposition 4.4.16** *Every regular normal space is completely regular.*

**Proof** Let $(X, \tau)$ be a regular and normal space and $A$ be closed in $(X, \tau)$. Take a point $y \in X$ such that $y \notin A$. Then $X - A$ is an open set in $(X, \tau)$, which contains the point $y$. Since by hypothesis, $(X, \tau)$ is regular, there exists an open set $U \in \tau$ such that

$$y \in U, \ \overline{U} \subset X - A.$$

Again, since $(X, \tau)$ is normal, there exists a continuous function $f : X \to I$ such that

(i) $f(x) = 0, \ \forall x \in \overline{U}$;

(ii) $f(x) = 1, \ \forall x \in A$ and

(iii) $0 \le f(x) \le 1, \ \forall x \in X$.

This shows that $f(y) = 0$, since $y \in \overline{U}$ and hence $(X, \tau)$ is completely regular.

$\square$

**Proposition 4.4.17** *Every completely regular space $(X, \tau)$ is regular.*

**Proof** Let $(X, \tau)$ be completely regular. Then it is also Hausdorff by definition. We claim that it is regular. To show it, let $A$ be a given closed subset of $(X, \tau)$ and $a \in X$ be a point such that $a \notin A$. Then the closed sets $A$ and $\{a\}$ are separated by a real function $f$ on $X$, i.e., there exists a real function $f : X \to \mathbf{R}$ such that

$$f(x) = \begin{cases} 0, \text{ for } x = a \\ 1, \text{ for all } x \in A \end{cases}$$

and

$$0 \le f(x) \le 1 \text{ for all } x \in X.$$

Consider two open sets $U = f^{-1}([0, \frac{1}{2}))$ and $V = f^{-1}((\frac{1}{2}, 1])$ in $(X, \tau)$. The lower-limit and upper-limit topologies on $\mathbf{R}$ are both strictly stronger than the usual topology on $\mathbf{R}$. This asserts that the closed sets $A$ and $\{a\}$ are strongly separated by disjoint open sets $V$ and $U$ in $X$. This shows that the space $(X, \tau)$ is regular.

$\square$

**Proposition 4.4.18** *Let $(X, \tau)$ be a topological space and $A$ be any closed set in $X$. If for every point $x \in X - A$, there is a continuous function $f : X \to \mathbf{R}$ such that $f(y) = 0, \ \forall y \in A$ but for every $x \in X - A, \ f(x) \ne 0$. Then the space $X$ is completely regular.*

**Proof** Let $(X, \tau)$ be a topological space and $A$ be any closed set in $X$. Suppose that for every point $x \in X - A$, there is a function $g : X \to \mathbf{R}$ such that $g(y) = 0, \ \forall y \in A$ but $g(x) \ne 0 \in \mathbf{R}$. Define a continuous function

$$f : X \to \mathbf{R}, y \mapsto g(x)^{-1}g(y).$$

This function shows that there is a continuous function $f$ such that

$$f : X \to \mathbf{R}, \ y \mapsto \begin{cases} g(y), & \text{for all } y \in A \\ g(x)^{-1}g(y), & \text{for all } y \in X - A, \end{cases}$$

Then

$$0 \le f(y) \le 1 \text{ for all } y \in X.$$

This asserts that $f$ is a real-valued continuous function such that

$$f : X \to \mathbf{R}, y \mapsto \begin{cases} 0, \text{ for all } y \in A \\ 1, \text{ for all } x \in X - A, \end{cases}$$

and

$$0 \le f(x) \le 1 \text{ for all } x \in X.$$

This implies that $(X, \tau)$ is completely regular.    □

**Corollary 4.4.19** *Every metric space is completely regular.*

**Proof** Let $(X, d)$ be a metric space, $A$ be any closed set in $X$ and $x \in X - A$ be an arbitrary point. Then

$$f : X \to \mathbf{R}, \ y \mapsto d(y, A)$$

is a continuous function such that $f(y) = 0$, $\forall y \in A$ and $f(x) \ne 0$, $\forall x \in X - A$. This asserts by Proposition 4.4.18 that $(X, d)$ is a completely regular space.    □

**Remark 4.4.20** Corollary 4.4.19 raises the problem: is every normal space is completely regular? Its answer by using Urysohn function is available in Chap. 6.

## 4.5   Homeomorphisms of $T_i$-Spaces and Topological Property

This section proves that homeomorphic images of $T_i$-spaces are also $T_i$-spaces for $i = 0, 1, 2, 3, 4$, and 5 in Theorem 4.5.4 asserting that the property of a topological space of being a $T_i$-space is a **topological property** in the sense that homeomorphic images of $T_i$-spaces are also $T_i$-spaces for each $i = 0, 1, 2, 3, 4$ and 5, which is the main result of this section.

**Theorem 4.5.1** *Let* $f : (X, \tau) \to (Y, \sigma)$ *be a map.*

(i) *If f is closed and onto, and the space $(X, \tau)$ is $T_1$, then the space $(Y, \sigma)$ is also $T_1$.*

(ii) *If f is closed and bijective, and the space $(X, \tau)$ is $T_2$, then the space $(Y, \sigma)$ is also $T_2$.*

(iii) *If f is closed, onto and continuous, and the space $(X, \tau)$ is $T_i$, then the space $(Y, \sigma)$ is also $T_i$ for $i = 4$ and 5.*

**Proof**    (i) Let $(X, \tau)$ be a $T_1$-space and $f : (X, \tau) \to (Y, \sigma)$ be a closed and onto map. Then for a point $y \in Y$, there exists a point $x \in X$ such that $f(x) = y$. Since one-pointic set $\{x\}$ is closed in $X$, and $f$ is closed by hypothesis, it follows that the one-pointic set $\{f(x)\} = \{y\}$ is also closed in $Y$. Since the point $x \in X$ is arbitrary, hence (i) follows.

(ii) Let $(X, \tau)$ be a $T_2$-space and $f : (X, \tau) \to (Y, \sigma)$ be a closed and bijective map. Then $f$ is also an open map, by using the result that a bijective map $f$ is open iff it is also closed. Since $f$ is bijective, for any two distinct point $y, z \in Y$, points $f^{-1}(y)$ and $f^{-1}(z)$ are also two distinct points in $X$. Since by hypothesis $(X, \tau)$ be a $T_2$-space, then there exist two disjoint open sets $U$ and $V$ in $(X, \tau)$ such that $f^{-1}(y) \in U$ and $f^{-1}(z) \in V$. Since $f$ is a bijective open map, it follows that $f(U)$ and $f(V)$ are two disjoint open sets in $(Y, \sigma)$, which strongly separate the points $y$ and $z$. This proves (ii).

(iii) Let $(X, \tau)$ be a $T_4$-space and $f : (X, \tau) \to (Y, \sigma)$ be a closed, onto and continuous map. Since $(X, \tau)$ is also a $T_1$-space, it follows by (i) that $(Y, \sigma)$ is also a $T_1$-space. Let $A$ and $B$ be two disjoint subsets in $(Y, \sigma)$. Since by hypothesis, $f$ is continuous, and onto, it follows that $f^{-1}(A)$ and $f^{-1}(B)$ are two disjoint closed sets in $(X, \tau)$. By normality property of $(X, \tau)$, it follows that there exist two disjoint open sets $U$ and $V$ in $(X, \tau)$ such that $f^{-1}(A) \subset U$ and $f^{-1}(B) \subset V$. Take $G = Y - f(X - U)$ and $H = Y - f(X - V)$. Then, $G$ and $H$ are two disjoint open sets in $(Y, \sigma)$ such that $A \subset G$ and $B \subset H$. This implies that $A$ and $B$ are strongly separated by $G$ and $H$ in $(Y, \sigma)$. Hence $(Y, \sigma)$ is a normal, i.e., a $T_4$-space. The proof for the case of $T_5$ is similar.

□

**Corollary 4.5.2**  *If $f : (X, \tau) \to (Y, \sigma)$ is a homeomorphism, and $(X, \tau)$ is a $T_i$-space, for $i = 0, 1, 2, 3, 4,$ and 5. then $(Y, \sigma)$ is also a $T_i$-space, for $i = 0, 1, 2, 3, 4,$ and 5.*

**Proof** Let $f : (X, \tau) \to (Y, \sigma)$ be a homeomorphism. Then $f$ is a bijective, closed and continuous map. Hence the corollary follows from Theorem 4.5.1, for $i = 1.2, 4, 5$. It is left as an exercise for $i = 0, 3$.    □

**Remark 4.5.3** Theorem 4.5.4, proves the **topological property of $T_i$-spaces**, for $i = 0, 1, 2, 3, 4$ and 5, which is a basic result of topology. The same result also follows from Corollary 4.5.2.

**Theorem 4.5.4** *Let $f : (X, \tau) \to (Y, \sigma)$ be a homeomorphism. If $(X, \tau)$ is a $T_i$-space, then its homeomorphic image $(Y, \sigma)$ under $f$ is also a $T_i$-space , $\forall i = 0, 1, 2, 3, 4, 5$.*

**Proof** We consider the following cases:

(i) $a$ and $b$ are distinct points in $Y$.
(ii) $a \in Y$ is a point and $B$ is a closed set in $(Y, \sigma)$, not containing the point $a$.
(iii) $A$ and $B$ are disjoint closed sets in $(Y, \sigma)$.
(iv) $A$ and $B$ are weakly separated in $(Y, \sigma)$.

By hypothesis $f$ is a homeomorphism. Hence its inverse $f^{-1}$ is also a homeomorphism, and it is a bijective and closed map. Consider the above four cases:

(i) In this case, $f^{-1}(a)$ and $f^{-1}(b)$ are two distinct points in $X$. If $U$ is an open set containing only one of the points $f^{-1}(a)$ and $f^{-1}(b)$ in $X$, then $f(U)$ is an open set, which contains one only of the points $a$ and $b$.
(ii) In this case, $f^{-1}(a) \in X$ is a point and $f^{-1}(B)$ is a closed in $(X, \tau)$, not containing the point $f^{-1}(a)$. If $f^{-1}(a)$ and $f^{-1}(B)$ are strongly separated by open sets $U$ and $V$ in $(X, \tau)$, then $f(U)$ and $f(V)$ are open sets and strongly separate $\{a\}$ and $B$ in $(Y, \sigma)$, since $f$ is an open mapping.
(iii) In this case, $f^{-1}(A)$ and $f^{-1}(B)$ are disjoint closed in $(X, \tau)$. If $f^{-1}(A)$ and $f^{-1}(B)$ are strongly separated by open sets $U$ and $V$ in $(X, \tau)$, then $f(U)$ and $f(V)$ are open sets and strongly separate $A$ and $B$ in $(Y, \sigma)$, since $f$ is an open mapping.
(iv) In this case, $f^{-1}(A)$ and $f^{-1}(B)$ form a pair of weakly separated sets in $(X, \tau)$. If $f^{-1}(A)$ and $f^{-1}(B)$ are weakly separated by open sets $U$ and $V$ in $(X, \tau)$, then $f(U)$ and $f(V)$ are open sets and weakly separate $A$ and $B$ in $(Y, \sigma)$, since $f$ is an open mapping.
Considering all of the above cases, it follows that for any homeomorphism $f : (X, \tau) \to (Y, \sigma)$, if $(X, \tau)$ is a $T_i$ space, then its homeomorphic image $(Y, \sigma)$ is also a $T_i$-space, $\forall i = 0, 1, 2, 3, 4, 5$.

$\square$

**Corollary 4.5.5** *The property of a topological space of being a $T_i$-space is a **topological property** for $i = 0, 1, 2, 3, 4$, and 5.*

**Proof** It follows from Theorem 4.5.4. $\square$

## Continuity of Functions on Normal Spaces

This section studies continuity of functions on normal spaces. Continuous image of an arbitrary normal space may not be normal (see Example 4.5.6). But Proposition 4.5.7 gives a sufficient condition asserting that the image of every closed continuous map of a normal space is normal. For more study of continuity of functions on normal spaces, see Exercise 27 of Sect. 4.7.

*Example 4.5.6* Continuous image of an arbitrary normal space may not be normal. The identity map on the same set endowed with different topologies may not be continuous. For example, let $(\mathbf{R}, \sigma)$ be the Euclidean line space and $(\mathbf{R}, \tau)$ be the topological space with trivial topology $\tau$ on $\mathbf{R}$. Then the identity map

$$1_d : (\mathbf{R}, \tau) \to (\mathbf{R}, \sigma), \ x \mapsto x$$

is not continuous, because its inverse image $1_d^{-1}(a, b)$ of the open interval $(a, b) \in \sigma$, which is also $(a, b)$ but $(a, b) \notin \tau$. On the other hand, the identity map

$$1_d : (\mathbf{R}, \sigma) \to (\mathbf{R}, \tau), \ x \mapsto x$$

is continuous but its image is not Hausdorff, because the topological space $(\mathbf{R}, \tau)$ is not Hausdorff.

**Proposition 4.5.7** *Let $(X, \tau)$ be a normal space and $f : (X, \tau) \to (Y, \sigma)$ be a continuous closed surjective map. The the space $(Y, \sigma)$ is also normal.*

**Proof** It follows from Theorem 4.5.1.                                        □

## 4.6 Applications

This section studies different applications of separation axioms such as Minor Urysohn Lemma 4.6.1 characterizing normal spaces by open sets, Hausdorff Property of $\mathbf{R}^n$ and $S^n$ in Sect. 4.6.3, retraction property of Hausdorff spaces in Sect. 4.6.4 and locally Euclidean property of the real projective space $\mathbf{R}P^n$, that of complex projective space $\mathbf{C}P^n$ and that of general linear group $GL(n, \mathbf{R})$ in Sect. 4.6.5, which are used in subsequent development of topology.

### 4.6.1 Minor Urysohn Lemma for Normal Spaces

This section proves Minor Urysohn Lemma 4.6.1 characterizing normal spaces, which can be used as an equivalent definition of normal spaces. On the other hand, Urysohn Lemma 6.3.9 proved in Chap. 6 characterizes normal spaces by real-valued continuous functions. This characterization provides an equivalent definition of normal spaces.

**Lemma 4.6.1 (*Minor Urysohn lemma*)** *A topological space $(X, \tau)$ is normal iff for any open set $U$ containing a given closed set $A$, there exists an open set $V$ such that*

$$A \subset V \subset \bar{V} \subset U.$$

***Proof*** Let $(X, \tau)$ be a normal space and $U$ be an open set containing a given closed set $A$ in $(X, \tau)$. Then $B = X - U$ is a closed set such that $A \cap B = \emptyset$. Since $X$ is normal, it follows that there exist two open sets $V$ and $V'$ such that

$$A \subset V, B \subset V' \text{ and } V \cap V' = \emptyset.$$

But $V \cap V' = \emptyset \implies V \subset X - V'$ and $X - U = B \subset V' \implies X - V' \subset U$. Since $X - V'$ is closed, it follows that

$$A \subset V \subset \bar{V} \subset X - V' \subset U.$$

Consequently, $A \subset V \subset \bar{V} \subset U$.

Conversely, let $(X, \tau)$ be a topological space and $A$, $B$ be two disjoint closed sets in $(X, \tau)$. Then $A \subset X - B$ and $X - B$ is an open set. Hence by hypothesis, there exists an open set $V$ such that $A \subset V \subset \bar{V} \subset X - B$. Consequently, it follows that $B \subset X - \bar{V}$, $V \cap (X - \bar{V}) = \emptyset$. Hence $A \subset V$ and $B \subset X - \bar{V}$, which shows that the disjoint closed sets $A$ and $B$ are strongly separated by two disjoint open sets by $V$ and $X - \bar{V}$. Since $A$ and $B$ is an arbitrary pair of closed sets in $(X, \tau)$, it proves that $(X, \tau)$ is normal. □

**Definition 4.6.2** Let $(X, \tau)$ be a topological and A be a subset of $X$. An open set $U_A$ in $(X, \tau)$ containing $A$ is said to be a nbd of $A$.

**Corollary 4.6.3** *Any two disjoint closed sets $A$ and $B$ in a normal space $(X, \tau)$ have open nbds $U_A$ and $U_B$ containing $A$ and $B$, respectively, such that $\bar{U}_A \cap \bar{U}_B = \emptyset$.*

***Proof*** It follows from Urysohn Lemma 4.6.1. □

***Example 4.6.4*** Normality property of a topological space is not hereditary in the sense that a subspace of a normal space is not in general a normal space. But under certain conditions, it is hereditarily normal (see Exercise 31 of Sect. 4.7).

## 4.6.2 Link Between Hausdorff Property and Continuity of Real Functions

This section studies Hausdorff spaces with the help of real-valued continuous functions by providing a sufficient condition for a space to be Hausdorff in Proposition 4.6.5.

**Proposition 4.6.5** *Let $(X, \tau)$ be a topological space such that for every pair of distinct points $x, y \in X$, there exists a real-valued continuous function*

$$f : X \to \mathbf{R}, \text{ such that } f(x) \neq f(y).$$

*Then $(X, \tau)$ is Hausdorff.*

***Proof*** Let $(X, \tau)$ be a topological space and $x, y \in X$ be a pair of distinct points satisfying the given condition. To show that $(X, \tau)$ is Hausdorff, let $U$ and $V$ be two disjoint open sets in the real line space $\mathbf{R}$ such that $f(x) \in U$ and $f(y) \in V$. Then

$$x \in f^{-1}(U), \ y \in f^{-1}(V) \text{ and } f^{-1}(U) \cap f^{-1}(V) = \emptyset,$$

otherwise, if $z \in f^{-1}(U) \cap f^{-1}(V)$, then $f(z) \in U$ and $f(z) \in V$ would imply that $U \cap V \neq \emptyset$. Hence it follows that $(X, \tau)$ is Hausdorff. $\qquad\square$

### 4.6.3 Hausdorff Property of $\mathbf{R}^n$, $S^n$, and Hilbert Cube

The section proves the **Hausdorff property of $\mathbf{R}^n$, $S^n$, and Hilbert cube** which are important objects in the study of analysis, geometry, algebra and topology.

**Proposition 4.6.6** *The n-dimensional Euclidean space $\mathbf{R}^n$ is Hausdorff.*

***Proof*** Since $\mathbf{R}$ is a Hausdorff space, it follows by Exercise 22 of Sect. 4.7 that its topological product space $\mathbf{R}^n$ is also Hausdorff. $\qquad\square$

**Corollary 4.6.7** *The n-sphere is Hausdorff.*

***Proof*** The $n$-sphere $S^n = \{x \in \mathbf{R}^{n+1} : ||x|| = 1\}$ is a subspace of the $(n+1)$-dimensional Euclidean space $\mathbf{R}^{n+1}$. Since any subspace of a Hausdorff space is Hausdorff, it follows from Proposition 4.6.6 that the $n$-sphere $S^n$ is Hausdorff. $\qquad\square$

***Remark 4.6.8*** Proposition 4.6.9 proves that the real Hilbert space $H$ is Hausdorff. The Hausdorff property of the Hilbert space also follows from Exercise 22 of Sect. 4.7.

**Proposition 4.6.9** *The real Hilbert space $H$ is Hausdorff.*

***Proof*** Since the real Hilbert space $H$ is the set of all sequences $\{x = (x_1, x_2, \ldots, x_n, \ldots) \in \mathbf{R}^\infty : \Sigma x_n^2$ is a convergent series$\}$, $H$ is a metric space with a metric

$$d : H \times H, \ (x, y) \mapsto [\Sigma_{n+1}^\infty (x_n - y_n)^2]^{\frac{1}{2}}.$$

Hence $H$ is Hausdorff. $\qquad\square$

***Remark 4.6.10*** The space $\mathbf{R}^\infty$ is metrizable and can be embedded in metric space. Because, as the space $\mathbf{R}$ is homeomorphic to the open interval $(0, 1)$, it follows that the space $\mathbf{R}^\infty$ can be embedded into the product space $\mathbf{I}^\infty = \Pi_{n=1}^\infty \mathbf{I}_n = [0, 1]$.

**Definition 4.6.11** The product space $\mathbf{I}^\infty = \Pi_{n=1}^\infty \mathbf{I}_n = [0, 1]$, which is the countable product of the line segment $\mathbf{I}$, is called the Hilbert cube. In particular, $\mathbf{I}^m = \Pi_{n=1}^m \mathbf{I}_n = [0, 1]$ is called the unit $m$-cube.

**Proposition 4.6.12** *The Hilbert cube* $\mathbf{I}^\infty$ *is Hausdorff.*

**Proof** Let $H$ be the Hilbert space. Consider the map

$$f : \mathbf{I}^\infty \to H, \ (x_1, x_2, \ldots, x_n, \ldots) \mapsto \left( x_1, \frac{x_2}{2}, \ldots, \frac{x_n}{n}, \ldots \right)$$

$f$ is well-defined, because

$$\{ [x_1^2 + x_2^2 + \ldots + x_n^2]^{\frac{1}{2}} \}_{n=1}^\infty$$

converges for all $x_n \in \mathbf{I}$. Clearly, $f$ is continuous and injection such that $\mathrm{Imf}(f) = \Pi_{m=1}^\infty [0, 1/m]$. Since $\mathbf{I}$ is compact, $\mathbf{I}^\infty$ is also compact by Tychonoff product theorem asserting that the topological product of any family of compact spaces is also compact (see Chap. 5). Moreover, $H$ is Hausdorff by Proposition 4.6.9. This shows that $f$ is a homeomorphism onto a subspace of $H$ by the result saying that every bijective continuous map from a compact space to a Hausdorff space is a homeomorphism (see Chap. 5).

$\square$

### 4.6.4 Retraction of a Hausdorff Space

This section studies the concept of retraction of a topological space from the viewpoint of Hausdorff property, which is an important concept in topology. It has wide applications in homotopy and homology theories to study topological problems by algebraic methods, discussed in **Basic Topology, Volume 3** of the present series of books.

**Definition 4.6.13** Let $(X, \tau)$ be a topological space and $A \subset X$ be a subspace of $X$. Then $A$ is said to be a **retract** of $X$, if there exists a continuous map

$$r : X \to A \text{ such that } r|_A = 1_A,$$

i.e., for the inclusion map $i : A \hookrightarrow X$,

$$r \circ i = 1_A.$$

This map $r$ is called a **retraction**.

**Example 4.6.14** Let $(X, \tau)$ be a topological space. Then

(i) $(X, \tau)$ is a retract of itself.
(ii) Every singleton subspace $\{x\}$ of $(X, \tau)$ is a retract of $X$.

**Proposition 4.6.15** *Let* $\mathbf{R}$ *be the Euclidean line and* $\mathbf{I} = [0, 1]$ *be the unit interval of* $\mathbf{R}$ *with subspace topology of* $\mathbf{R}$. *Then* $\mathbf{I}$ *is a retract of* $\mathbf{R}$.

**Proof** Define the map

$$r : \mathbf{R} \to \mathbf{I}, \ t \mapsto \begin{cases} 0, & \text{if } t \leq 0, \\ t, & \text{if } t \in \mathbf{I}, \\ 1, & \text{if } t \geq 1. \end{cases}$$

Then $r$ is a continuous map such that its restriction to $\mathbf{I}$ is the identity map on $\mathbf{I}$. This proves the proposition.

$\square$

Proposition 4.6.16 relates retraction of a subspace $A$ of a Hausdorff space $X$ by proving that $A$ is closed set of $X$.

**Proposition 4.6.16** *Let* $(X, \tau)$ *be a Hausdorff space and* $A$ *be a retract of* $X$. *Then* $A$ *is closed in* $(X, \tau)$.

**Proof** Let $(X, \tau)$ be a Hausdorff space and $A$ be a retract of $X$. Then there exists a retraction $r : X \to A$ such that $r(x) = x$, $\forall\, x \in A$. To prove the proposition, it is sufficient to show that $X - A$ is an open set in $(X, \tau)$. If $X - A = \emptyset$, then there is nothing to prove. So, assume that $X - A \neq \emptyset$. Then there exists a point $x \in X - A$. This point $x \notin A$ but its image point $r(x) = y \in A$ shows that $x \neq y$. Since $(X, \tau)$ is Hausdorff, it follows that there exist open sets $U$ and $V$ in $(X, \tau)$ such that

$$x \in U, \ y \in V \text{ and } U \cap V = \emptyset.$$

Since $r$ is continuous and $A \cap V$ is open in $A$, it follows that $r^{-1}(A \cap V)$ is an open set in $(X, \tau)$ containing the point $x$. Then $W = U \cap r^{-1}(A \cap V)$ is an open set in $(X, \tau)$ such that $W \subset X - A$. This implies that $X - A$ is an open set in $(X, \tau)$. This proves that $A$ is closed in $(X, \tau)$.

$\square$

Corollary 4.6.17 gives an alternative proof of the well-known result asserting that every one-pointic set in a Hausdorff space is closed.

**Corollary 4.6.17** *Every one-pointic set in a Hausdorff space is closed.*

**Proof** Let $(X, \tau)$ be a Hausdorff and $x \in X$ be an arbitrary point. Consider the one-pointic subspace $\{x\}$ of $X$. Then the corollary follows from Proposition 4.6.16, by taking $A = \{x\}$.

$\square$

### 4.6.5  Locally Euclidean Spaces

This section studies a special class of Hausdorff spaces, called **locally Euclidean spaces,** which are very important spaces in geometry, specially, in the study of manifolds in **Basic Topology, Volume 2** of the present series of books.

**Definition 4.6.18** A Hausdorff space $(X, \tau)$ is said to be **locally Euclidean** of dimension $n$ if for every point $x \in X$, there exists a homeomorphism $\psi_x$ such that it maps some open set in $(X, \tau)$ containing $x$ onto an open subset in the Euclidean space $\mathbf{R}^n$.

**Example 4.6.19** Consider the following Hausdorff spaces.

(i) The Euclidean space $n$-space $\mathbf{R}^n$ is a locally Euclidean space of dimension $n$. Here for each point $x \in \mathbf{R}^n$, the homeomorphism $\psi_x$ is taken to be the identity map on $\mathbf{R}^n$.

(ii) The $n$-sphere $S^n$ is a locally Euclidean space of dimension $n$. Because, for every pair of distinct points $x, y \in S^n$, the stereographic projection

$$f : S^n - N \to \mathbf{R}^n, x \mapsto \frac{1}{1 - x_{n+1}}(x_1, x_2, \ldots, x_n),$$
$$\forall x = (x_1, x_2, \ldots, x_{n+1}) \in S^n - N.$$

$$\psi_y : S^n - \{y\} \to \mathbf{R}^n, , x \mapsto \frac{1}{1 - x_{n+1}}(x_1, x_2, \ldots, x_n),$$
$$\forall x = (x_1, x_2, \ldots, x_{n+1}) \in S^n - \{y\}$$

is a homeomorphism.

(iii) Locally Euclidean property of the real projective space. The real projective space $\mathbf{R}P^n$ consisting of all straight lines passing through the origin of $\mathbf{R}^n$ is a locally Euclidean space of dimension $n$, because, $\mathbf{R}P^n$ is covered by $S^n$ and each point $x \in \mathbf{R}P^n$ is contained in an open set homeomorphic to an open set in $S^n$, which is homeomorphic to an open set in $\mathbf{R}^n$.

(iv) Locally Euclidean property of the complex projective space $\mathbf{C}P^n$. It consists of all complex lines passing through the origin of $\mathbf{C}^n$ is a locally Euclidean space of (complex) dimension $n$.

(v) Locally Euclidean property of $GL(n, \mathbf{R})$. The set $M(n, \mathbf{R})$ of all $n \times n$ matrices over $\mathbf{R}$ can be identified with $\mathbf{R}^{n^2}$ by the map

$$f : M(n, \mathbf{R}) \to \mathbf{R}^{n^2}, a_{i,j}$$
$$\mapsto (a_{11}, a_{12j}, \ldots, a_{1n}, a_{21}, a_{22}, \ldots, a_{2n}, \ldots, a_{n1}, a_{n2}, \ldots, a_{nn}).$$

The determinant function

$$\det : M(n, \mathbf{R}) \to \mathbf{R}, M \to \det M$$

is just a polynomial in the matrix coefficients and hence it is continuous. Then it follows that $GL(n, \mathbf{R}) = M(n, \mathbf{R}) - \det^{-1}\{0\}$ is open, because the set $\{0\})$ is closed in $\mathbf{R}$ and hence $\det^{-1}\{0\}$ is closed in $M(n, \mathbf{R})$. This asserts that for each point $x \in GL(n, \mathbf{R})$, there exists a nbd of $x$, homeomorphic to an open set in $\mathbf{R}^{n^2}$. It asserts that $GL(n, \mathbf{R}) = M(n, \mathbf{R}) - \det^{-1}\{0\}$ is locally Euclidean. More study on $GL(n, \mathbf{R})$ is found in **Basic Topology, Volume 2** of the present series of books.

**Definition 4.6.20** An **n-dimensional (topological) manifold** or an **n-manifold** $M$ is a Hausdorff space with a countable base such that for each point $x \in M$, there exists a homeomorphism $\psi_x$ mapping some neighborhood of the point $x$ onto an open subset of $\mathbf{R}^n$. **A one-dimensional manifold is called a curve and a two-dimensional manifold is called a surface.**

*Example 4.6.21* Every $n$-dimensional manifold is a locally Euclidean space of dimension $n$ satisfying the second axiom of countability (see **Basic Topology, Volume II** of the present series of books for its study).

## 4.7  Exercises

1. Let $(X, \tau)$ be a topological space. Show that the following statements are equivalent

    (i) $X$ is a $T_1$-space;
    (ii) Every one-pointic subset in $X$ is closed in $X$;
    (iii) Every finite subset of $X$ is closed in X;
    (iv) The derived set $\{x\}'$ of every point $x \in X$ is $\emptyset$;
    (v) Given a subset $S \subset X$, the intersection of all open sets in $X$ containing $S$ is the set $S$;
    (vi) Every open set containing an accumulation point $a$ of a subset $S$ of $X$ intersects $S$ in a countably infinite collection of points (called an $\omega$-accumulation points of $S$).

2. Let $(X, \tau)$ be a topological space. Show that two nonempty subsets $A$ and $B$ of $X$ are weakly separated in $X$ iff either of the following conditions is satisfied

    (i) $(A \cap \bar{B}) \cup (\bar{A} \cap B) = \emptyset$ **(Hausdorff-Lennes condition)**;
    (ii) $A \cap B = \emptyset$, $A' \cap B = \emptyset$, and $A \cap B' = \emptyset$.

3. The set $\mathbf{R}$ with finite-complement topology is not Hausdorff.
   [Hint: Use the fact that any two nonempty open sets in this space are overlapping.]

4. Find the separation axioms under which there is a smallest topology on the set $\mathbf{R}$ such that every singleton is closed.

5. Let $X$ be a nonempty set and $\mathbf{B}_x$ be a family of subsets of $X$ forming a nbd basis of a point $x \in X$ in a topological space $(X, \tau)$. Then the family $\mathbf{B}_x$ is called a **nbd filter** of the point $x$.
   Let $(X, \tau)$ be a topological space. Show that the space $(X, \tau)$ is Hausdorff iff

    (i) either given any point $x \in X$, $\{x\} = \cap\{\bar{B}_x : B_x \in \mathbf{B}_x\}$, where $\mathbf{B}_x$ denotes the nbd filter of the point $x \in X$;

(ii) or the diagonal set $\Delta = \{(x, x) : x \in X\} \subset X \times X$ is a closed set in the product space $(X \times X, \tau \times \tau)$.

6. Let $(X, \tau)$ be a a Hausdorff space. Show that $(X, \tau)$ is regular iff the closed nbds of any point $x \in X$, constitute a nbd basis of the point $x$.

7. Show that every subspace of a Hausdorff space is also Hausdorff.

8. Show that every subspace of a regular space is regular.
   [Hint: Use Exercise 6.]

9. (A nonHausdorff space in which every convergent sequence has a unique limit) Let $X$ be an uncountable set endowed with the topology $\tau$ of countable complements. Show that

   (i) The topology $\tau$ is $T_1$;
   (ii) The topology $\tau$ is not Hausdorff;
   (iii) Every convergent sequence in this topological space $(X, \tau)$ has a unique limit.

0. Let $(X, \tau)$ be a Hausdorff space and $Y, Z$ be two subspaces of $X$ consisting of points of two convergent (infinite) sequences in $X$. Show that the subspaces $Y$ and $Z$ are homeomorphic.

1. Let $(X, \tau)$ be a **Zariski space** defined in Chap. 3. If $X$ is Hausdorff, show that it is finite.

2. Let $A$ be an infinite field. Show that the **Zariski topology** defined in Chap. 3 on $A = A^1$ is not Hausdorff

3. Let $X$ be the set whose point set is the Euclidean plane $\mathbf{R}^2$ and whose open sets described by the basis consisting of the usual open sets in $\mathbf{R}^2$ together with the sets defined by $\{(x, y) \in \mathbf{R}^2 : x^2 + y^2 < r, \ \forall r > 0, \ y \neq 0\} \cup \{(0, 0)\}$. Show that $X$ is not regular but it is Hausdorff.

4. Let $X = [0, 1] \subset \mathbf{R}$ be given relative topology induced from the usual topology on $\mathbf{R}$. If $\rho$ is an equivalence relation on $X$ defined by the rule $(x, y) \in \rho$ iff $x - y$ is a rational number, show that the quotient space $X/\rho$ is

   (i) not Hausdorff;
   (ii) is not metrizable.

5. Let $(X, \tau)$ be a topological space. Show that it is Hausdorff iff its diagonal $\Delta = \{(x, x) : x \in X\} \subset X \times X$ is closed in the product topology on $X \times X$.

6. Let $(X, \tau)$ be a regular space, $A$ be a closed set in $(X, \tau)$ and $p$ be a point in $X$ but lying outside $A$. Show that there exist open sets $U$ and $V$ in $X$ such that

$$p \in U, \ A \subset V, \text{ and } \bar{U} \cap \bar{V} = \emptyset.$$

7. Show that

   (i) Every $T_1$-space is a $T_0$-space.
   (ii) Every Hausdorff space is both a $T_1$ and a $T_2$-space;
   (iii) Every completely normal space is a normal space;

(iv) Every $T_5$-space is also a $T_4$-space;

(v) A normal space may not be a regular space;

(vi) A topological space $(X.\tau)$ is $T_1$ iff every one-pointic set is closed in $(X, \tau)$;

(vii) Every $T_i$ space is a $T_{i-1}$-space for every $i = 1, 2, 3, 4$, and5 but its converse is not necessarily true.

18. Let **R** be the real line endowed with the topology generated by the base consisting of all rays of the form $a < x < +\infty$. Show that this topological space is $T_0$ but it is not $T_1$.

19. (Example of a $T_1$-space which is not Hausdorff) The closed interval $\mathbf{I} = [0, 1]$ endowed with the topology $\tau$ whose open sets are $\emptyset$ and all the subsets of $\mathbf{I}$ obtained by deleting either a finite or a countable number of points from $\mathbf{I}$. Show the resulting topological space $(\mathbf{I}, \tau)$ is $T_1$ but it is not Hausdorff.

20. Show that the set **R** endowed with a right-hand or left-hand topology is a $T_0$-space, a normal and completely normal space but is not a regular space.

21. Show that the topological sum of any nonempty family of Hausdorff spaces is also Hausdorff.

22. Show that the topological product of any nonempty family of Hausdorff spaces is also Hausdorff.
    [Hint: Let $\{X_i, \tau_i\}$ be a nonempty family Hausdorff spaces and $(X, \tau) = \Pi_i(X_i, \tau_i)$ be their product space. Suppose $x = \{x_i\}$ and $y = \{y_i\}$ are two distinct points of $X$. Then there exists at least one index $i_k$ such that $x_{i_k} \neq y_{i_k}$. Use the Hausdorff property of the space $X_{i_k}$.]

23. Show that every regular $T_1$-space is Hausdorff.

24. Let $(X, \tau)$ be a completely regular space $T_1$ space and $C(X, \mathbf{R})$ be the set of all real-valued continuous functions on $X$. Show that the space $C(X, \mathbf{R})$ separates points of $X$ in the sense that if $a, b$ are two distinct points in $X$, then there exists a function $f \in C(X, \mathbf{R})$ such that $f(a) \neq f(b)$.

25. Consider the nbds of an arbitrary point in $\mathbf{I} = [0.1]$, except 0, consisting of ordinary nbds and the nbd of 0, defined by all the possible half-intervals $[0, x)$ with the deleted points $\{1/n, n = 1, 2, 3, \ldots\}$ Show that the resulting space $(X, \tau)$ is Hausdorff but not regular.
    [Hint: The point $0 \in X$ and the closed set $\{0\}$ and $A = \{1/n, n = 1, 2, \ldots\}$ are not strongly separated by open nbds.]

26. Let $(X, \tau)$ be the Niemytzki's space endowed with Niemytzki's tangent dick topology (see Chap. 3). Show that this space $(X, \tau)$

    (i) is Hausdorff;

    (ii) is not normal;

    (iii) is not completely normal.

27. Let $(X, \tau)$ be a normal space and $(Y, \sigma)$ be an arbitrary topological space. Show that for every continuous closed onto mapping

$$f : (X, \tau) \to (Y, \sigma).$$

its image $(Y, \sigma)$, is also normal.

8. Let $(X, \tau)$ be a topological space, $\mathcal{U} = \{U_\alpha : \alpha \in \mathbf{A}\}$ and $\mathcal{V} = \{V_\alpha : \alpha \in \mathbf{A}\}$ be two open coverings of $X$. If $\overline{V}_\alpha \subset U_\alpha$, $\forall \alpha \in \mathbf{A}$, then $\mathcal{U}$ is said to be **shrinkable** and $\mathcal{V}$ is said to be **shrinking** of $\mathcal{U}$. The covering $\mathcal{U}$ is said to be point-finite if every point $x \in X$ is in only finitely many $U_\alpha$.
   Show that a topological space $(X, \tau)$ is normal iff

   (i) either every finite open covering of $X$ is shrinkable;
   (ii) or every point-finite open covering of $X$ is shrinkable.

9. (**Jones lemma**) Let $(X, \tau)$ be a topological space such that it contains a dense subset $D$ and a closed discrete subspace $C$. If card$(D) = \beta$ and card$(C) \geq 2^\beta$, show that the space $(X, \tau)$ is not normal.
   [Hint: If $X$ is normal, show that card$(\mathcal{P}(C)) \leq$ card$(\mathcal{P}(D))$, which is not possible, since by hypothesis card$(C) \geq 2^\beta$.]

10. Show that

    (i) Every metric space is completely normal.
    (ii) Every metrizable space is completely normal.

11. (**Urysohn**) Two sets $A$ and $B$ in a topological space are said to be **separated** 1 if $\overline{A} \cap B = A \cap \overline{B} = \emptyset$. Prove that a topological space $(X, \tau)$ is hereditarily normal iff any two separated sets of $(X, \tau)$ have disjoint nbds.

12. Let $(X, \tau)$ be a topological space. Show that $X$ is completely normal iff for any two subsets $A, B \subset X$ satisfying the condition

$$A \cap \overline{B} = \emptyset, \text{ and } \overline{A} \cap B = \emptyset,$$

there exist two disjoint open sets $U$ and $V$ in $(X, \tau)$ such that

$$A \subset U \text{ and } B \subset V.$$

## Multiple Choice Exercises

Identify the correct alternative (s) (there may be none or more than one) from the following list of exercises:

1. (i) Let $(X, \tau)$ be a Hausdorff space. Then every retract of $X$ is open in $(X, \tau)$.
   (ii) The Euclidean plane $\mathbf{R}^2$ is Hausdorff.
   (iii) The standard sphere $S^2$ is Hausdorff.

2. Let $\mathbf{R}$ be the set of real numbers.

   (i) The space $\mathbf{R}$ endowed with upper-limit topology is completely normal.
   (ii) The space $\mathbf{R}$ endowed with finite-complement topology is Hausdorff.

(iii) The space **R** endowed with the topology generated by the base of all rays $a < x < \infty$ is $T_1$.

3. $(X, \tau)$ be a first countable space. Then

(i) X is Hausdorff if every convergent sequence in $X$ has a unique limit point.
(ii) Every convergent sequence in $X$ has a unique limit point if X is Hausdorff.
(iii) X is Hausdorff iff every convergent sequence in $X$ has a unique limit point.

4. (i) Let $X$ be a nonempty set and $(Y, \sigma)$ be a Hausdorff space. Given a map $f : X \to Y$, declare a subset $U \subset X$ to be open in $X$, iff $U = f^{-1}(V)$ for some $V \in \sigma$. Then $\{U\}$ defines a Hausdorff topology on $X$.

(ii) Let $(X, \tau)$ be a topological space and $X \times X$ have the product topology. If the subset

$$\Delta = \{(x, x) : x \in X\} \subset X \times X$$

is closed in $X \times X$, then the topological space $(X, \tau)$ is Hausdorff.

(iii) Zariski topology on **C** is Hausdorff.

5. (i) Every metrizable space is normal.
(ii) Every subspace of a normal space is normal.
(iii) Every closed subspace of a normal space is normal.

6. Let $(\mathbf{R}, \tau_l)$ be the topological space with the left-hand topology $\tau_l$ on **R**.

(i) The space $(\mathbf{R}, \tau_l)$ is Hausdorff.
(ii) The space $(\mathbf{R}, \tau_l)$ is metrizable.
(iii) There exists a metric $d$ on $X$ such that its induced topology coincides with $\tau_l$.

7. (i) The Euclidean 3-space $\mathbf{R}^3$ is locally Euclidean.
(ii) The 2-sphere $S^2$ in $\mathbf{R}^3$ is locally Euclidean.
(iii) The space $GL(3.\mathbf{R})$ is locally Euclidean.

# References

Adhikari, M.R.: Basic Algebraic Topology and its Applications. Springer, India (2016)
Adhikari, M.R.: Basic Topology, Volume 3: Algebraic Topology and Topology of Fiber Bundles. Springer, India (2022)
Adhikari, M.R., Adhikari, A.: Basic Modern Algebra with Applications. Springer, New Delhi, New York, Heidelberg (2014)
Adhikari, A., Adhikari, M.R.: Basic Topology, Volume 2: Topological Groups, Topology of Manifolds and Lie Groups. Springer, India (2022)
Bredon, G.E.: Topology and Geometry. Springer-Verlag, New York (1983)
Borisovich, Y.C.U., Blznyakov, N., Formenko, T.: Introduction to Topology. Mir Publishers, Moscow (1985). Translated from the Russia by Oleg Efimov

Brown, R.: Topology: A Geometric Account of General Topology, Homotopy Types, and the Fundamental Groupoid. Wiley, New York (1988)

Chatterjee, B.C., Ganguly, S., Adhikari, M.R.: Introduction to Topology. Asian Books, New Delhi (2002)

Conway, J.B.: A Course in Point Set Topology. Springer, Switzerland (2014)

Dugundji, J.: Topology. Allyn & Bacon, Newton, MA (1966)

Fuks, D.B., Rokhlin, V.A.: Beginner's Course in Topology. Springer-Verlag, New York (1984)

Hu, S.T.: Introduction to General Topology. Holden-Day, San Francisco (1966)

Kelly, J.L.: General Topology, Van Nostrand, New York, 1955. Springer-Verlag, New York (1975)

Munkres, J.R.: Topology. Prentice-Hall, New Jersey (2000)

Patterson, E.M.: Topology. Oliver and Boyd (1959)

Singer, I.M., Thorpe, J.A.: Lecture Notes on Elementary Topology and Geometry. Springer-Verlag, New York (1967)

Stephen, W.: General Topology. Addison-Wesley (1970)

# Chapter 5
# Compactness and Connectedness

This chapter is devoted to address **the concepts of compactness and connectedness** in topological settings, which first arose through the study of subsets of the Euclidean line **R** in calculus and mathematical analysis. The motivation of these two concepts has come through the three basic theorems in calculus for a continuous function defined on the closed interval $[a, b] \subset \mathbf{R}$ such as

  (i) **Intermediate value theorem,**
 (ii) **Maximum value theorem** and
(iii) **Uniform continuity theorem.**

The first one leads to the concept of connectedness, on the other hand, both of the second and third theorems lead to the concept of compactness in topological settings. These generalized concepts are fundamental in the study of modern analysis, geometry, topology, algebra and many other areas. **Compactification** of a topological space is a topological method or a result of making a noncompact topological space into a compact space.

Various interesting applications of compactness and connectedness properties of topological spaces are also available in Sects. 5.28 and 5.27. **Ascoli's theorem** on function spaces is available in Sect. 5.24.6. On the other hand, **Gelfand–Kolmogoroff theorem** on the rings $\mathcal{C}(X, \mathbf{R})$ of real-valued continuous functions from compact Hausdorff spaces $X$ says that two compact Hausdorff spaces $X$ and $Y$ are homeomorphic iff the corresponding rings $\mathcal{C}(X, \mathbf{R})$ and $\mathcal{C}(Y, \mathbf{R})$ are isomorphic. This deep result **recovers the topology** of $X$ from the ring structure of $\mathcal{C}(X, \mathbf{R})$. The study of Gelfand–Kolmogoroff theorem is given in Chap. 6. Different concepts such of compactness, locally compactness, compactly generated, countably compactness, sequentially compactness, BW-compactness paracompactness, compactification of noncompact spaces, connectedness, local connectedness and path connectedness, etc. introduced in this section in topological settings are used throughout all the three volumes of the present book series.

A. Adhikari and M. R. Adhikari, *Basic Topology 1*,
https://doi.org/10.1007/978-981-16-6509-7_5

**Historically**, Heinrich Heine (1821–1881) introduced the concept of finite subcovering in 1872 in his work on uniformly continuous functions. Emile Borel (1871–1956) proved a result in 1894 asserting that every open covering of a closed interval has a finite subcovering. The concept of **compactness** was born through the **Heine–Borel theorem** 5.2.8 for any closed and bounded set of the real line **R**, described in real analysis. A more general result was published by Henri Lebesgue (1875–1941) in 1898 and by Arthur Schnflies (1853–1928) in 1900. A characterization of compactness in terms of closed sets having finite intersection property was given by Leopold Vietoris (1891–2002) in 1921 (see Theorem 5.9.6). On the other hand, the concept of connectedness of some subsets of the Euclidean line **R**$^2$ was given by Camille Jordan (1838–1922) in 1914. A systematic study of connected topological spaces was inaugurated by Felix Hausdorff in his book **Grundzüge der Mengenlehre** of 1914, which stemmed from analysis.

The **intermediate value theorem** 5.18.4 asserts that if $f : [a, b] \to$ **R** is continuous and $r \in$ **R** lies between $f(a)$ and $f(b)$, then there is a point $\alpha \in [a, b]$ such that $f(\alpha) = r$. What is the generalization of this theorem in topology ? This theorem depends not only on the property of continuity of $f$ but also depends on a special property of $[a, b]$, called **connectedness**, which is also a topological property different from compactness property. On the other hand, motivation of the different types of compactness is given in Sect. 5.1.1.

For this chapter, the books (Armstrong 1983; Bredon 1993; Chatterjee et al. 2002; Borisovich et al. 1985; Munkres 2000; Patterson 1959) and papers and some other books are referred in Bibliography.

## 5.1 Different Types of Compactness and Compact Subsets of R$^n$

This section introduces the concepts of six different types of compactness in Sect. 5.1.1 and studies compact subsets of **R**$^n$ in Sect. 5.1.2. Their generalization in arbitrary topological spaces is given in Sects. 5.1.3 and 5.1.4. In classical analysis, bounded and closed subsets of Euclidean spaces have several basic interesting properties. The properties of the set **R** of real numbers given in Theorem 5.1.1 are well known.

**Theorem 5.1.1** *Let $X$ be a nonempty subset of* **R**. *Then $X$ and* **R** *have interesting properties:*

(i) *Every infinite bounded subset of $X$ has a limit point in $X$ (**Bolzano–Weierstrass theorem**).*

(ii) *Every open cover of a closed and bounded set $X$ has a finite subcover (**Heine–Borel–Lebesgue theorem**).*

(iii) *Every sequence in a closed and bounded set $X$ has a convergent subsequence.*

(iv) *Cantor's theorem on nested closed sets in* **R**.

(v) *Every continuous function takes a closed and bounded set onto a closed bounded set in* $\mathbf{R}$.

(vi) *Every continuous function on a closed and bounded set is uniformly continuous in* $\mathbf{R}$.

**Proof** Left as an exercise. ☐

## 5.1.1 Motivation of Six Different Types of Compactness

This subsection conveys motivation for defining different types of compactness. The properties of $\mathbf{R}$ given in Theorem 5.1.1 can be utilized to introduce various classes of compactness in a topological space, as they are formulated in terms of topological properties. But the results prescribed in (*i*), (*ii*) and (*iv*) of Theorem 5.1.1, which are equivalent in Euclidean space, are not necessarily true in an arbitrary topological space. This problem motivates to define different types of compactness in an arbitrary topological space.

**Definition 5.1.2** Let $(X, \tau)$ be a topological space. Then $X$ is said to be

(i) **compact**, if every open covering of $X$ has a finite subcovering;

(ii) **locally compact** if each point of $X$ has a compact nbd ;

(iii) **paracompact** if every open covering of $X$ has a locally finite subcovering of $X$;

(iv) **countably compact**, if every countable covering of $X$ has a finite subcovering;

(v) **sequentially compact**, if every infinite sequence on $X$ has a convergent subsequence;

(vi) **Bolzano–Weierstrass compact** (B–W compact), if every infinite subset of $X$ has a point of accumulation in $X$.

**Remark 5.1.3** Six types of compactness may not be different for a certain class of topological spaces. For example, the concepts of compactness, countably compactness and sequentially compactness coincide for a subset of a metrizable space (see Exercise 6 of Sect. 5.28). The concepts of compactness and B–W-compactness are different in an arbitrary topological space (see Example 5.3.4) but Theorem 5.3.11 asserts that these two concepts coincide in metric spaces. Let $(X, d)$ be a metric space. Then its associated topological space $(X, \tau_d)$ is compact iff it satisfies B–W property.

## 5.1.2 Compactness of Subsets of $\mathbf{R}$ and $\mathbf{R}^n$

This subsection studies the concept of compactness of subsets of $\mathbf{R}$ and $\mathbf{R}^n$. A closed and bounded subset $X$ of the real line space $\mathbf{R}$ can be characterized by the property of $\mathbf{R}$ that every open covering of $X$ has a finite covering, which is the Heine–Borel theorem. This property can be extended to a certain class of topological spaces, called compact spaces. So while studying a compact topological space, it is sufficient to study only a finite number of its open subsets. This is the advantage of the study of compact topological spaces. In metrizable spaces $X$, compactness may be examined by the property that every infinite subset of $X$ has an accumulation point.

Some results of classical analysis on compactness of a subset of the Euclidean space $\mathbf{R}^n$ are stated with an eye to describe the motivation of compactness.

**Definition 5.1.4** A subset $X$ of $\mathbf{R}^n$ is said to be **bounded** if $X \subset B_{x_0}(r)$ for some open ball

$$B_{x_0}(r) = \{x \in \mathbf{R}^n : d(x, x_0) < r\} \subset \mathbf{R}^n$$

with $x_0$ as its center and $r > 0$ its radius, where $d$ is the Euclidean metric. Equivalently, a subset $X$ of $\mathbf{R}^n$ is bounded, if there is a real number $r > 0$ such that

$$d(x, y) < r \text{ for all pairs of points } x, y \in X.$$

*Example 5.1.5* All segments, circles, triangles, etc. are bounded; on the other hand, lines, half-lines, planes, exteriors of circles, Euclidean space $\mathbf{R}^n$ and the set of rational numbers are unbounded.

*Remark 5.1.6* We state two fundamental Theorems 5.1.7 and 5.1.9 of classical analysis which historically motivated the concept of compactness in a topological setting.

**Theorem 5.1.7  (Borel)** *Let $X$ be a closed and bounded subspace of the n-dimensional Euclidean space $\mathbf{R}^n$. Then for every open covering $\{U\}$ of $X$, there is a finite subcovering in the sense that from the open sets $\{U\}$ which cover $X$, a finite number of open sets can be chosen which still covers $X$.*

**Proof** Left as an exercise or see Theorem 5.2.4                                         □

*Remark 5.1.8* **Borel Covering theorem** 5.1.7 leads to the concept of compactness for a topological space $X$ by calling $X$ to be compact if every open covering of $X$ has a finite subcovering. Hence Theorem 5.1.7 shows that the closed and bounded subspaces of Euclidean spaces $\mathbf{R}^n$ are compact. Its converse is also true by Heine–Borel theorem 5.2.4 in $\mathbf{R}^n$, which asserts that a subset of $\mathbf{R}^n$ is compact iff it closed and bounded.

**Theorem 5.1.9** *Let $X$ be a closed and bounded subspace of the n-dimensional Euclidean space $\mathbf{R}^n$. Then for every continuous map $f : X \to \mathbf{R}^n$, its image $f(X)$ is also closed and bounded in $\mathbf{R}^n$.*

**Proof** Left as an exercise.                                                           □

We start with a concrete result that a closed and bounded subset $X$ of the real line $\mathbf{R}$ is characterized by the property that for any open covering of $X$, there is a finite subcovering of $X$. If this property holds for a topological space, this space is called a compact space.

**Theorem 5.1.10**    *(i)  The real line space* $\mathbf{R}$, *with the usual topology, is not compact but a closed interval* $[a, b]$ *in* $\mathbf{R}$ *is compact with respect to its subspace topology. But* $[a, b]$ *endowed with discrete topology is not compact. This example shows that the compactness property of a topological space depends on its topology.*
*(ii)  An open interval* $(a, b)$ *in* $\mathbf{R}$, *with the usual topology, is not compact.*
*(iii)  A subset* $X$ *of* $\mathbf{R}$, *with usual topology, is compact iff* $X$ *is closed and bounded.*
*(iv)  Let* $\tau$ *be the finite complement topology on the set of real numbers* $\mathbf{R}$ *and* $X = \mathbf{R} - \{0\}$. *Then the subspace* $X$ *of the topological space* $(\mathbf{R}, \tau)$ *is compact but it is not closed in the topological space* $(\mathbf{R}, \tau)$.

**Proof**  Left as an exercise.                                                            □

*Example 5.1.11*  Compactness property depends on the topology. For example, the closed interval $[a, b]$ is compact in the subspace topology inherited from the natural topology on $\mathbf{R}$ but it is not compact as a subspace of $\mathbf{R}$ endowed with discrete topology. Because, $[a, b] = \bigcup\{x : x \in [a, b]\}$ and hence $\{x : x \in [a, b]\}$ forms an open covering of $[a, b]$ with respect to the discrete topology but it has no finite subcovering.

**Theorem 5.1.12**  *Let* $X \subset \mathbf{R}$ *be a nonempty subset. Then following statements on* $X$ *are equivalent:*

*(i)  Let* $\{x_1, x_2, \ldots, x_n, \ldots\}$ *be a sequence of points in* $X$. *Then there is a subsequence* $\{x_{n1}, x_{n2} \ldots, x_{nk}, \ldots\}$, *which converges to a point* $x \in X$.
*(ii)  Every infinite subset of* $X$ *has at least one limit point in* $X$.
*(iii)  Let* $\{U_k : k \in \mathbf{K}\}$ *be an open covering of* $X$ *by a family of open intervals of* $\mathbf{R}$. *Then it has a finite subcovering.*

**Proof**  Left as an Exercise.                                                            □

*Remark 5.1.13*  The equivalence of the properties of a subset $X \subset \mathbf{R}$ stated in Theorem 5.1.12 is given in (Heritt 1960).

**Definition 5.1.14**  A subset $X \subset \mathbf{R}$ having any one of equivalent properties of Theorem 5.1.12 is called a **compact set** in $\mathbf{R}$.

*Example 5.1.15*  There are infinitely many compact and noncompact sets in $\mathbf{R}$. For any two real numbers $a, b$ with $a < b$, the closed interval $[a, b] \subset \mathbf{R}$ is a compact subset of the real line space $\mathbf{R}$. On the other hand, its every open interval $(a, b) \subset \mathbf{R}$ is noncompact, because the family of open intervals $\mathcal{U} = \{U_n = (a, b - \frac{1}{n}), \forall n \in \mathbf{N}\}$ forms an open covering of $(a, b)$ but it has no finite subcovering.

**Proposition 5.1.16**  *Every real-valued function defined on a nonempty finite set is bounded and attains its bounds.*

**Proof**  Let $X = \{x_1, x_2, \ldots, x_n\}$ be a finite set (may or may not be a subset of $\mathbf{R}$) and $f : X \to \mathbf{R}$ be a function. Then $max\{f(x_1), f(x_2), \ldots, f(x_n)\}$ is the lub and $min\{f(x_1), f(x_2), \ldots, f(x_n)\}$ is the glb of the values of $f$, which are attained by $f$ at points of its domain of definition $X$.                                                            □

***Remark 5.1.17*** For an arbitrary nonempty compact subset of $\mathbf{R}$, every real-valued continuous function $f: X \to \mathbf{R}$ is bounded and attains its lub and glb. This gives a generalization of Proposition 5.1.16. See proof of Theorem 5.27.2.

***Remark 5.1.18*** Remark 5.1.17 needs the condition of continuity of the function $f$ but for Proposition 5.1.16 no assumption on continuity is needed. It is clear of course that every real-valued continuous function defined on a finite subset of $\mathbf{R}$ is continuous.

### 5.1.3   Compact Sets in Arbitrary Topological Spaces

This subsection conveys the concept of compactness in the language of open sets. Compactness is an important topological property in the sense that this property is shared by homeomorphic spaces. Compact sets are almost finite for certain technical purposes. The properties described in Theorem 5.1.12 can be utilized to introduce the concept of "compactness" in a topological setting as these theorems are formulated in terms of open sets. This concept was motivated by Heine–Borel theorem of analysis on closed and bounded sets of $\mathbf{R}$. In an arbitrary topological space, as boundedness carries neither any significance nor it is a topological property, it has no direct impact in defining compactness. But Borel Covering Theorem 5.1.7 relates boundedness of closed sets in Euclidean spaces to certain families of open sets, called open coverings, which lays the foundation of Definition 5.1.22 of compactness.

In more general, for a metric space $X$, given an $\epsilon > 0$, $X$ can be expressed as an union of all its open balls $B_x(\epsilon)$. There is a natural problem: Does there exist a finite collection of these open balls $B_x(\epsilon)$ that can cover $X$? If $X$ has no metric, to solve this problem, the balls $B_x(\epsilon)$ is replaced by arbitrary open sets and this process leads to the concept of compactness.

**Definition 5.1.19** Let $(X, \tau)$ be a topological space and $\mathcal{U}$ be a certain family of subsets of $X$. Then

(i)   $\mathcal{U}$ is said to be a **cover of** $A \subset X$, if $A \subset \bigcup \{U: U \in \mathcal{U}\}$ and

(ii)   $\mathcal{U}$ is said to be **an open cover (or covering)** of $A$ if $\mathcal{U}$ is a cover of $A$ and every set $U \in \mathcal{U}$ is open,

(iii)   a **subcover of** $A$ is a subset of $\mathcal{U}$ which is also a cover of $A$.

***Example 5.1.20*** An open covering of a topological space $(X, \tau)$ is a family of $\{U_i\}$ of open sets of $(X, \tau)$, whose union is the whole set $X$. A topological space may have different open coverings. For example,

(i)   the family $\mathcal{U} = \{(-n, n): n \in \mathbf{N}\}$ is an open covering of $\mathbf{R}$ in the natural topology and the family $\mathcal{U} = \{\{x\}: x \in \mathbf{R}\}$ is also an open covering of $\mathbf{R}$ in the discrete topology.

(ii)   both $\mathcal{U} = \{U_n = (1/n, 1): n = 2, 3, \ldots\}$ and $\mathcal{V} = \{V_n = (\frac{1}{n}, \frac{n}{n+1}): n = 2, 3, \ldots$ form two different open coverings of the open interval $(0, 1)$ in the subspace topology inherited from $\mathbf{R}$ with the natural topology.

***Example 5.1.21***    (i) Let $(X, \tau)$ be a discrete topological space. Then the family $\mathcal{U} = \{\{x\}: x \in X\}$ forms an open covering of $X$.

(ii) The family $\mathcal{U} = \{(-n, \ n): n \in \mathbf{N}\}$ forms an open covering of $\mathbf{R}$ both in the natural topology and in the lower-limit topology of $\mathbf{R}$.

(iii) Let $(X, d)$ be a metric space with its induced topology $\tau_d$. Then given $r > 0$, the family $\mathcal{B} = \{B_x(r): x \in X\}$ of all open balls in $X$ of radius $r$ centered at $x$ are open sets in $(X, \tau_d)$ and this family forms an open covering of $X$.

**Definition 5.1.22** A topological space $(X, \tau)$ is said to be **compact** if every open covering of $X$ has a finite subcovering. A subspace $A \subset X$ of a topological space $X$ is said to be compact if $A$ is compact with respect to its subspace topology.

***Remark 5.1.23*** Definition 5.1.22 of compactness asserts that from any open covering $\{U_i\}$ of a compact space $X$, we can choose finitely many indices $i_j$ with $j = 1, 2, \ldots, n$ such that $\bigcup_{j=1}^{n} U_{i_j} = X$. If $X$ is a compact space, every sequence of points $x_n$ of $X$ has a convergent subsequence, which means, every subsequence $x_{n_1}, x_{n_2}, \ldots, x_{n_t}, \ldots$ converges to a point of $X$. For metric spaces, this condition is equivalent to compactness (see Exercise 27 of Sect. 5.28).

### 5.1.4   Subspaces of Compact Spaces

This subsection studies subspaces of compact spaces. A subset $X$ of the real line space $\mathbf{R}$ with usual topology is compact iff $X$ is closed and bounded (see Corollary 5.14.4). The following natural problems arise :

(i) is every subspace of an arbitrary compact space compact?
(ii) is every closed subspace of an arbitrary compact space compact?
(iii) is every finite union of compact subspaces of an arbitrary topological space compact?

***Remark 5.1.24*** The answers of the problems (ii) and (iii) are affirmative by Theorem 5.1.26. On the other hand, the answer of the problems (i) is negative by Example 5.1.25.

***Example 5.1.25*** The answer of the problem (i) is not affirmative: every subspace of an arbitrary compact space is not compact. For example, consider the subspace $(0, 1)$ of the closed interval $[0, 1]$ in the real line space $\mathbf{R}$. The subspace $(0, 1) \subset [0, 1]$ is not compact. Because, $\{(\frac{1}{n}, \frac{n}{n+1}): n = 2, 3, \ldots\}$ forms an open covering of the open interval $(0, 1)$, but it has no finite subcovering. Hence the open interval $(0, 1)$ is not compact but the closed interval $[0, 1]$, is compact.

**Theorem 5.1.26**    (i) *Every closed subspace of a compact space is compact.*
(ii) *Every finite union of compact subspaces of a topological space is compact.*

**Proof** (i) Let $(X, \tau)$ be a compact space, $Y$ be a closed subset in $(X, \tau)$ and $\mathcal{A} = \{A_i : i \in \mathbf{A}\}$ be a family of closed sets in $(Y, \tau_Y)$ having the finite intersection property. Since $Y$ is closed in $(X, \tau)$, each set $A_i$ is also closed in $(X, \tau)$. This asserts that $\{A_i : i \in \mathbf{A}\}$ is also a family of closed sets in $(X, \tau)$ having the finite intersection property. Since $(X, \tau)$ is compact by hypothesis, it follows by Theorem 5.9.6 that $\bigcap \{A_i : i \in \mathbf{A}\} \neq \emptyset$. Consequently, the subspace $Y$ is also compact by Theorem 5.9.6.

(ii) Let $(X, \tau)$ be a topological space and $\{X_i : i = 1, 2, \ldots, n\}$ be a finite collection of closed subspaces of $(X, \tau)$. Suppose that $Y = X_1 \cup X_2 \cup \cdots \cup X_n \subset X$ and $\mathbf{F}$ is an open covering of $Y$. Then this $\mathbf{F}$ forms also an open covering of each compact subspace $X_i$. Hence there exists a finite subcovering $\mathbf{F_i} = \{U_{i_1}, U_{i_2}, \ldots, U_{i_k}\}$ of $\mathbf{F}$ for each $X_i$: $i = 1, 2, \ldots, n$. This asserts that the sets in $\mathbf{F_1}, \mathbf{F_2}, \ldots, \mathbf{F_n}$ form a finite open covering of $(Y, \tau_Y)$. Hence it follows that the subspace $Y$ is also compact.

$\square$

**Cantor space** defined by Cantor (1845–1918) is the Cantor set (defined in Chap. 1) endowed with the subspace topology inherited from the topological space $\mathbf{I} = [0, 1]$.

**Corollary 5.1.27** *Cantor space $C$ is compact.*

**Proof** Since Cantor space $C$ is closed in the compact set $[0, 1]$, the corollary follows immediately by the first part of Theorem 5.1.26. $\square$

**Theorem 5.1.28** *Every compact subspace of a Hausdorff space is closed.*

**Proof** Let $A$ be a compact subspace of a Hausdorff space $(X, \tau)$. We claim that $A$ is closed in $(X, \tau)$. Let $y$ be an arbitrary point of $A$ and $x \in X - A$. Since $(X, \tau)$ is Hausdorff by hypothesis, there exist disjoint open sets $U_y$ and $V_y$ such that $x \in U_y$ and $y \in V_y$. Keeping $x$ fixed and varying $y \in A$, it follows that the family $\{V_y : y \in A\}$ of open sets forms an open covering of the compact subspace $A$. Hence there exists a finite subcovering of $A$ such that

$$A \subset V_{y_1} \cup V_{y_2} \cup \cdots \cup V_{y_n},$$

for a finite number of points $y_1, y_2, \ldots, y_n \in A$. Construct two sets

$$V(A) = V_{y_1} \cup V_{y_2} \cup \cdots \cup V_{y_n},$$

and

$$U(x) = U_{y_1} \cap U_{y_2} \cap \cdots \cap U_{y_n},$$

which are disjoint and open nbds of $x$ and $A$. This asserts that the compact set $A$ and a point not in $A$ can be separated in a Hausdorff space by disjoint open nbds $U(x)$ and $V(A)$. This shows that the complement $X - A$ is open and hence $A$ is closed in $(X, \tau)$. $\square$

## 5.2 Compactness in Metric Spaces

This section generalizes the concept of compactness in $\mathbf{R}^n$ for metric spaces and discusses compact sets in metric spaces. It proves **Heine–Borel theorem in $\mathbf{R}^n$** and **Lebesgue Lemma**. It is shown that **for a metric space, the concepts of compactness and B–W properties coincide.**

### 5.2.1 Compact Sets in Metric Spaces

This subsection studies compact sets in an arbitrary metric space with an emphasis on Euclidean spaces. The concept of compact space is a generalization of closed and bounded sets in the Euclidean space but a closed and bounded set in an arbitrary metric space may not be compact (see Example 5.2.10).

**Definition 5.2.1** A subset $A$ of a metric space $X$ is said to be compact if every open covering of $A$ permits a finite subcovering.

**Example 5.2.2** (i) Every finite subset of a metric space is compact.
(ii) Every closed interval $[a, b]$ in the real line space $\mathbf{R}$ is compact.
(iii) The open interval $(0, 1)$ in the real line space $\mathbf{R}$ is not compact, because there exists an open covering such as $\mathcal{U} = \{U_n = (1/n, 1): n \, (> 1) \in \mathbf{N}\}$, of the open interval $(0, 1)$, but $\mathcal{U}$ has no subcovering.

**Remark 5.2.3** Example 5.2.2 raises the problem: Exactly which subspaces of $\mathbf{R}^n$ are compact? The answer is that the compact subsets of the Euclidean space $\mathbf{R}^n$ are precisely its closed and bounded subsets, which is the Heine–Borel theorem 5.2.4 in $\mathbf{R}^n$ named after H. E. Heine (1821–1881) and Emile Borel (1871–1956) and hence it characterizes completely a compact subspace of $\mathbf{R}^n$ by its bounded and closed subsets. Some alternative proofs of this theorem are given in Theorem 5.25.29.

**Theorem 5.2.4** *(Heine–Borel theorem in $\mathbf{R}^n$) A subset of $\mathbf{R}^n$ is compact iff it closed and bounded.*

**Proof** Let $A$ be a compact subspace of $\mathbf{R}^n$. Since $A$ is a compact subspace of the Hausdorff space $\mathbf{R}^n$, it is closed by Theorem 5.1.26. Consider the open balls $B_0(n)$ in $\mathbf{R}^n$ with center at the origin $0 \in \mathbf{R}^n$ and integral radius $n$. Then the family of open balls $\{B_0(n): n \in \mathbf{N}\}$ forms an open covering of the compact subspace $A$ of $\mathbf{R}^n$. As $A$ is compact by hypothesis, the family $\{B_0(n): n \in \mathbf{N}\}$ has a finite subcovering of $A$. Hence $A$ must be inside the union of finitely many balls of $\{B_0(n)\}$. This shows that there is an integer $n_0$ such that $A \subset B_0(n_0)$. This asserts that $A$ is bounded. Consequently, $A$ is closed and bounded. For the converse part, suppose that $A$ is closed and bounded. Then there are bounded intervals $[a_1, b_1], [a_2, b_2], \ldots, [a_n, b_n]$ in $\mathbf{R}$ such that $A \subset [a_1, b_1] \times [a_2, b_2] \times \cdots \times [a_n, b_n]$. If $\{x_k\}$ is a sequence in $A$ such that $x_k = (x_{k_1}, x_{k_2}, \ldots x_{k_n})$, where, $x_{k_i} \in [a_i, \, b_i]: i = 1, 2, \ldots, n, \, \forall \, k \in \mathbf{N}$.

Then $\{x_{k_1}: k \in \mathbf{N}\}$ is a sequence in $[a_1, b_1]$ such that it has a convergent subsequence say, $\{x_{k_1}: k \in \mathbf{N}_1 \subset \mathbf{N}\}$. Then $\mathbf{N}_1$ has also its natural ordering. Let $x_1 = lim_{k \in \mathbf{N}_1} x_{k_1}$. Consider the sequence $\{x_{k_2: k \in \mathbf{N}_1}\}$ in $[a_2, b_2]$. It has a convergent subsequence $\{x_{k_2}: k \in \mathbf{N}_2 \subset \mathbf{N}_1\}$ such that $x_2 = lim_{k \in \mathbf{N}_2} x_{k_2}$. Proceeding as above, there exist subsets: $\mathbf{N}_p \subset \cdots \subset \mathbf{N}_1 \subset \mathbf{N}$ such that

$$lim_{k \in \mathbf{N}_p} x_{k_m} = x_m, \text{ for } 1 \le m \le p.$$

This shows that $\{x_k: k \in \mathbf{N}_p\}$ is a subsequence of the original sequence $\{x_k: k \in \mathbf{N}\}$ and it converges to the point $x = (x_1, x_2, \ldots, x_p) \in A$, since $A$ is closed. This asserts that $A$ is compact by Exercise 27 of Sect. 5.28. $\qquad \square$

**Corollary 5.2.5** *A subset $A$ of $\mathbf{R}$ is compact iff it is closed and bounded.*

**Proof** It follows from Theorem 5.2.4 for $n = 1$, in particular. $\qquad \square$

**Corollary 5.2.6** *A subspace of the real line space $\mathbf{R}$ is compact iff it is closed and bounded.*

**Proof** It follows from Theorem 5.2.4. $\qquad \square$

**Corollary 5.2.7** *The closed interval $[a, b]$ is compact in the real line space $\mathbf{R}$.*

**Proof** It follows from Theorem 5.2.4. $\qquad \square$

**Theorem 5.2.8** *(Heine–Borel theorem in $\mathbf{R}$). Every closed interval in the real number space $\mathbf{R}$, with usual topology, is compact.*

**Proof** **Proof I**: It follows from Theorem 5.2.4.

**Proof II**: See proof of Theorem 5.14.1.

**Proof III**: Let $[a, b]$ be a given closed interval in $\mathbf{R}$ and $\mathcal{C}$ be an open covering of $[a, b]$ in $\mathbf{R}$. Let $\mathcal{E}$ be the set of all those elements $x \in [a, b]$ such that there is a finite subcovering of $\mathcal{C}$ for the closed interval $[a, x]$. Then $a \in \mathcal{E}$ and the point $b$ is an upper bound of the set $\mathcal{E}$. This shows that $\mathcal{E}$ has an lub $e$, say. Then it follows that $a < e$. If $U \in \mathcal{C}$ is such that $e \in U$ and $d \in (a, e) \cap U$, then $[a, d]$ has an open finite subcovering covering $\mathcal{C}'$, which is finite subcollection of $\mathcal{C}$. Hence $\mathcal{C}' \cup U$ is a finite subcollection of $\mathcal{C}$, which forms a finite open subcovering of $[a, e]$. If $e \ne b$, there is a number $f$ such that $f \in U, (e, f) \subset U$ and hence $\mathcal{C}' \cup U$ is a finite subcollection of $\mathcal{C}$ and forms an open covering of $[a, f]$ such that $f \in \mathcal{E}$, which is not possible as $f > e$ and $e$ is the lub of $\mathcal{E}$. This asserts that $e = b$ and hence $\mathcal{C}' \cup U$ is a finite subcollection of $\mathcal{C}$, which forms a finite covering of $[a, b]$. Consequently, $[a, b]$ is compact in the real number space $\mathbf{R}$.

$\qquad \square$

**Corollary 5.2.9** *(Another form Heine–Borel theorem in $\mathbf{R}$) Every closed and bounded subspace of the real number space, with usual topology, is compact.*

***Proof*** Since a closed and bounded subspace of the real line space **R** is a closed subspace of some closed interval $[a, b]$, the corollary follows from Theorem 5.2.8. □

***Example 5.2.10*** Heine–Borel theorem is true in Euclidean space but it is not true in an arbitrary metric space, because, every closed and bounded set in an arbitrary metric space (even in $(\mathbf{R}, d)$ having a metric $d$ equivalent to the Euclidean metric on **R**) is not compact. In support, consider

(i) the example, where the metric $d$ is defined

$$d: \mathbf{R} \times \mathbf{R} \to \mathbf{R}, (x, y) \mapsto \frac{|x - y|}{1 + |x - y|}.$$

This metric $d$ is different from the standard (Euclidean) metric on **R** but these two metrics are equivalent. With this metric $d$, the space **R** is closed and bounded but not compact.

(ii) another example: given two rational numbers $a, b$ with $a < b$, define the set

$$\mathcal{S} = \{x \in \mathbf{Q}: a \leq x \leq b\} = [a, b] \cap \mathbf{Q}.$$

This set $\mathcal{S}$ is closed and bounded. It is not compact, since there is a sequence $\{x_n\}$ in $\mathcal{S}$ converging to an irrational number and hence this sequence has no subsequence converging to a point of $\mathcal{S}$.

***Example 5.2.11*** A subspace of a compact space may not be compact. For example, in the real number space **R** with usual topology, an open interval is not compact (see Example 5.2.2). The open interval $(0, 1)$ is a proper subspace of $[0, 1]$. The open interval $(0, 1)$ is not compact but the closed interval $[0, 1]$ is compact by Heine–Borel theorem.

### 5.2.2 Lebesgue Lemma and Lebesgue Number

This subsection studies Lebesgue Lemma and Lebesgue number named after H. Lebesgue (1875–1941) for an open covering of a compact metric space establishing a relation between such a space and Lebesgue number and proves Lebesgue Lemma providing a technical result on open covering of a compact metric space.

Given an open covering $\mathcal{F} = \{U_\alpha: \alpha \in \mathbf{A}\}$ of a compact metric space $X$, there exists a real number $\delta > 0$ (called **Lebesgue number** of $\mathcal{F} = \{U_\alpha\}$) such that every open ball $B_x(\epsilon)$ *in* $X$ for some $\epsilon > 0$, is contained in at least one open set $\{U_\alpha\} \in \mathcal{F}$. The concept of a Lebesgue number stems from Lebesgue's work on measure theory, starting with his thesis in 1902. The existence of Lebesgue number is proved in Lemma 5.2.12.

**Lemma 5.2.12** *(Lebesgue) Let $X$ be a compact metric space. Given an open covering $\mathcal{F} = \{U_\alpha : \alpha \in \mathbf{A}\}$ of $X$, there exists a real number $\delta > 0$ (called **Lebesgue number** of $\mathcal{F}$) such that any subset $Y$ of $X$ of diameter $diam(Y) < \delta$ is contained in some member of $\mathcal{F}$, i.e., whenever $Y \subset X$ and $diam(Y) < \delta$, then $Y \subset U_\alpha$ for some $\alpha \in \mathbf{A}$.*

**Proof I**: Let $X$ be a compact metric space with metric $d$. Then for an arbitrary point $x \in X$, there is an $\epsilon(x) > 0$ depending on $x$ such that the open ball $B_x(2\epsilon(x)) \subset U_\alpha$ for some $\alpha \in \mathbf{A}$. Since $X$ is compact, there is a finite number of the open balls $B_x(\epsilon(x))$, suppose for $x = x_1, x_2, \ldots, x_m$. Let $\delta = min\{\epsilon(x_j) : j = 1, 2, \ldots, m\}$. If $dim(Y) < \delta$ and $y_0 \in Y$, there exists an index $j$ such that $1 \leq j \leq m$ with the property that $d(y_0, x_j) < \epsilon(x_j)$. Again, for $y \in Y$, $d(y, y_0) < \delta \leq \epsilon(x_j)$. Hence by triangle inequality for the metric space $X$, it follows that

$$d(y, x_j) \leq d(y, y_0) + d(y_0, x_j) < 2\epsilon(x_j).$$

This asserts that $Y \subset B_x(2\epsilon(x)) \subset U_\alpha$ for some $\alpha \in \mathbf{A}$.

**Proof II**: If the result is not true, there exists a sequence $X_1, X_2, \ldots, X_n, \ldots$ of subsets of $X$ such that none of them is contained in a member $\mathcal{F} = \{U_\alpha\}$, and whose diameter tends to 0 as we move along the sequence. For every $n$, select a point $x_n \in X_n$. Consider the two cases:

Case I: The sequence $\{x_n\}$ contains only finitely many distinct points. In this case, some point will be repeated infinitely many times.

Case II: The sequence $\{x_n\}$ contains infinitely many distinct points. In this case, it has a limit point, since by hypothesis $X$ is compact.

Let $x$ denote the repeated point for case I or limit point for case II. If $U$ is a member of $\mathcal{F}$, which contains the point $x$, then choose a real number $\epsilon > 0$ such that the open ball $B_x(\epsilon) \subset U$ and a positive integer $n_0$ sufficient large so that

(i) $diam X_{n_0} < \epsilon/2$ and
(ii) $x_{n_0} \in B_x(\epsilon/2)$.

Hence, $d(x_{n_0}, x) < \epsilon/2$ and $d(x', x_{n_0}) < \epsilon/2$ for any point $x' \in X_{n_0}$. This shows by triangle inequality that $d(x, x') < \epsilon$, $\forall x' \in X_{n_0}$, and hence $X_{n_0} \subset U$. But it contradicts our assumption that $U \notin \mathcal{F}$. This contradiction proves the theorem. $\square$

**Corollary 5.2.13** *(Lebesgue Lemma: Alternative Form ): Let $X$ be a compact metric space. Given an open covering $\{U_\alpha : \alpha \in \mathbf{A}\}$ of $X$, there exists a real number $\delta > 0$ (called **Lebesgue number of** $\{U_\alpha\}$), such that every open ball of radius less than $\delta$ lies in some element of $\{U_\alpha\}$.*

**Proof** It follows from Lemma 5.2.12. $\square$

**Remark 5.2.14** For another proof of Lebesgue Lemma, see Basic Topology, Volume III of the present series of books.

**Corollary 5.2.15**  *Let $X$ and $Y$ be two compact metric spaces. If $f : X \to Y$ is continuous, then it is uniformly continuous.*

**Proof**  Given an $\epsilon > 0$, the collection $\mathcal{C} = \{f^{-1}(B_y(\epsilon/2)) : y \in Y\}$ of open sets forms an open covering of $X$. Let $\delta > 0$ be a Lebesgue number of $\mathcal{C}$. Then $f$ sends every ball $B_x(\delta)$ in $X$ to some ball $B_y(\epsilon/2)$ in $Y$. Hence, if $d(x, z) < \delta$ then

$$d(f(x), f(z)) \leq d(f(x), y) + d(y, f(z)) < \epsilon/2 + \epsilon/2 = \epsilon$$

asserts that $f$ is uniformly continuous.                                          □

# 5.3  Bolzano–Weierstrass (B–W) Compactness

This section is devoted to the study of the concept of (B–W)-compactness in a topological setting, which is motivated by Bolzano–Weierstrass theorem of analysis asserting that every bounded infinite subset of real numbers has an accumulation point. A key link between compactness and Bolzano–Weierstrass (B–W) property is established in Theorem 5.3.11. The (B–W) property of a metric space $X$ asserts that every infinite subset of $X$ has at least one accumulation point. The main result of this section is the Theorem 5.3.9 and its Corollary 5.3.10 saying every metric space having the (B–W)-property is compact and conversely, every compact metric space has the (B–W) property.

Definition 5.3.1 asserts that a topological space $(X, \tau)$ is (B–W)-compact or limit point compact if every infinite subset of $X$ has a limit point.

**Definition 5.3.1**  Let $(X, \tau)$ be a topological space.

(i) The space $(X, \tau)$ is said to be **Bolzano–Weierstrass compact or (B–W)-compact or limit point compact** if every infinite subset of $X$ has an accumulation point.

(ii) A subset $A$ of $X$ with relative topology is said to be (B–W)-compact, if every infinite subset of $A$ has an accumulation point.

*Example 5.3.2*  Let $\mathbf{R}$ be the Euclidean line.

(i) Every bounded closed interval $X = [a, b]$, with subspace topology inherited from $\mathbf{R}$, is (B–W)-compact. Because, if $A$ is an infinite subset of $X$, then $A$ is a bounded infinite subset of real numbers and hence it has an accumulation point $a$ say, by Bolzano–Weierstrass theorem. Since, $X$ is closed, it follows that $a \in X$ and hence $X$ is (B–W)-compact.

(ii) The open interval $X = (0, 1)$ with subspace topology inherited from $\mathbf{R}$, is not (B–W) is compact. Because, the infinite subset $A = \{\frac{1}{2}, \frac{1}{3}, \cdots, \frac{1}{n}, \cdots\}$ of $X$ has accumulation point 0 but $0 \notin X$ shows that $X = (0, 1)$ is not (B–W)-compact.

**Theorem 5.3.3**  *Every compact space is a (B–W)-compact space.*

***Proof*** Let $(X, \tau)$ be a compact space and $Y$ be an infinite subset of $X$. To prove the theorem, it is sufficient to show that $Y$ has a limit point. If $Y$ has no limit point outside $Y$, then $Y$ contains all its limit points and hence $Y$ is closed in $X$. This shows that $Y$ is compact as a closed subset of the compact space $X$. Again, for each point $y \in Y$, there exists a nbd $U_y$ of $y$ such that $U_y \cap (Y - \{y\}) = \emptyset$, since $y$ is not a limit point of $Y$. Moreover, as $\{U_y\}$ forms an open covering of the compact set $Y$, there exists a finite subcovering whose number is say $n$. Again, since every nbd $U_y$ contains exactly one point of $Y$ and hence this set $Y$ contains precisely these $n$ elements. This contradicts our hypothesis that $Y$ is an infinite subset of $X$.

$\square$

***Example 5.3.4*** Compactness property of a topological space implies its (B–W)-compactness property by Theorem 5.3.3 but its converse is not true. For example, let $\mathbf{N}_\tau$ be the topology on $\mathbf{N}$ defined by

$$\mathbf{N}_\tau = \{(n, n+1) : n \in \mathbf{N}\}.$$

Then the topological space $(\mathbf{N}, \mathbf{N}_\tau)$ is (B–W)-compact. But it not compact. Because $\{(n, n+1) : n \in \mathbf{N}\}$ forms an open covering of $\mathbf{N}$ but it has no finite subcovering.

***Remark 5.3.5*** Example 5.3.4 implies that the concepts of Bolzano–Weierstrass (B–W) property and compactness property are different for an arbitrary topological space. But these two concepts coincide for a metric space by Theorem 5.3.11.

## Compactness and Bolzano–Weierstrass Properties in Metric Spaces

This subsection continues the study of compactness property in metric spaces and establishes a key link between compactness property and Bolzano–Weierstrass (B–W) property for metric spaces as stated in Remark 5.3.5. Another form of Bolzano–Weierstrass theorem is given in Theorem 5.27.1. For more connection between compactness and (B–W) properties in a topological space, see Chap. 7.

**Theorem 5.3.6** *(Bolzano–Weierstrass) Let $(X, d)$ be a compact metric space and $A$ be an infinite subset of $X$. Then $A$ has at least one accumulation (limit) point.*

***Proof*** Let $(X, d)$ be a compact metric space and $A$ be an infinite subset of $X$. Suppose $A$ has no accumulation point. This supposition implies that every point of $X$ is not an accumulation point of $A$ and hence for each point $x \in X$, there is an open ball $B_x$ with center at $x$ such that

$$B_x \cap A = \begin{cases} \emptyset, & \text{if } x \notin A \\ \{x\}, & \text{if } x \in A \end{cases}$$

The family $\mathcal{B}$ of all these open $B_x$ forms an open covering of $X$. Since by hypothesis, $X$ is compact, there exists a finite subcovering $\mathcal{S} \subset B$ of $X$. Since the set $A$ is contained in the set of centers of the open balls in this finite subcovering $\mathcal{S}$, there are finite number of points $x_1, x_2, \ldots, x_n$ such that $A = \{x_1, x_2, \ldots, x_n\}$ is finite, which contradicts the assumption that $A$ is infinite. This contradiction proves the theorem. $\quad\square$

**Theorem 5.3.7** *Let $(X, d)$ be a metric space. If every infinite subset of $X$ has at least one accumulation point, then $X$ is compact.*

**Proof** Let $\mathcal{C} = \{U_i : i \in \mathbf{A}\}$ be an open covering of $X$ and $\epsilon_L$ be its Lebesque number. Select a positive integer $n$ such that $\frac{1}{n} < \epsilon_L$. Then there exists a finite set of points $\{x_1, x_2, \ldots, x_m\}$ of $X$ such that the family of open balls

$$\left\{ B_{x_1}\left(\frac{1}{n}\right), B_{x_2}\left(\frac{1}{n}\right), \ldots, B_{x_m}\left(\frac{1}{n}\right) \right\}$$

forms an open covering of $X$. Again, for each $k = 1, 2, \ldots, m$, there is an $\alpha_i \in \mathbf{I}$ such that $B_{x_k}(\frac{1}{n}) \subset U_{\alpha_i}$. This asserts that the family

$$\{U_{\alpha_1}, U_{\alpha_2}, \ldots, U_{\alpha_m}\}$$

forms a finite subcovering of $\mathcal{C} = \{U_i : i \in \mathbf{A}\}$ of $X$. This implies that $X$ is compact. $\quad\square$

**Definition 5.3.8** Every metric space $(X, d)$ induces a topology $\tau_d$ on $X$. The resulting topological space $(X, \tau_d)$ is called the **associated topological space of** $(X, d)$.

**Theorem 5.3.9** *Let $(X, d)$ be a metric space. Then every infinite subset of $X$ has at least one accumulation point iff its associated topological space $(X, \tau_d)$ is compact.*

**Proof** Let $(X, d)$ be a metric space such that every infinite subset of $X$ has at least one accumulation point. Then its associated topological space $(X, \tau_d)$ is compact by Theorem 5.3.7. Conversely, let $X$ be a compact metric space. Then using Theorem 5.3.6, it follows that every infinite subset of $X$ has at least one accumulation point. $\quad\square$

**Corollary 5.3.10** *Let $(X, d)$ be a metric space. Then it is compact iff its associated topological space $(X, \tau_d)$ is compact.*

The above discussion is summarized in Theorem 5.3.11, which is a basic and important result characterizing compactness of topological spaces with the help of Bolzano–Weierstrass (B–W)-property.

**Theorem 5.3.11** *Let $(X, d)$ be a metric space. Then its associated topological space $(X, \tau_d)$ is compact iff it satisfies (B–W) property.*

**Remark 5.3.12** In a metric space, the three concepts: Bolzano–Weierstrass (B–W)-property, compactness property and sequentially compactness property coincide (see Exercise 16 of Sect. 5.28).

## 5.4  Basic Link between Compactness and Hausdorff Properties of Topological Spaces

This section describes some interesting properties obtained by relating compactness property with Hausdorff property of topological spaces. Theorem 5.4.5 is one of the useful results providing a simple tool to invade certain type of homeomorphism problems.

**Proposition 5.4.1** *Let $(X, \tau)$ be a compact Hausdorff space. Then it is normal.*

**Proof** Let $(X, \tau)$ be a compact Hausdorff space. Suppose $A$ and $B$ are two disjoint closed sets in $X$. Then they are also compact sets of $(X, \tau)$. For each $a \in A$, there exist disjoint open sets $U_a$ and $V_a$ such that $a \in U_a, B \subset V_a$. Running $a$ over $A$, we have an open covering $\{U_a\}$ of $A$. Then it has a finite subcovering whose union is $U$, say, and the intersection of the corresponding sets $V_a$ is $V$, say. These $U$ and $V$ are disjoint open nbds of $A$ and $B$. This proves that $(X, \tau)$ is normal.                               $\square$

**Remark 5.4.2** The imposition of the condition of Hausdorff property of $(X, \tau)$ in Theorem 5.4.1 is necessary so that one-pointic sets are closed and the condition of disjoint closed sets having open nbds includes the condition that distinct points have disjoint open nbds. Again, the restriction that $A$ and $B$ are to be closed is necessary: since arbitrary disjoint subsets may not have disjoint open nbds: for example, the subsets $\{0\}$ and $(0, 1]$ in the real line space **R** have no disjoint open nbds.

**Proposition 5.4.3** *Let $(X, \tau)$ be a compact space and $(Y, \sigma)$ be a Hausdorff space. If a map $f : (X, \tau) \to (Y, \sigma)$ is continuous and surjective, then a set $V \subset Y$ is open iff $f^{-1}(V)$ is open in $X$.*

**Proof** Let $(X, \tau)$ be a compact space and $(Y, \sigma)$ be a Hausdorff space. By hypothesis, $f$ is continuous. Then for every open set $V$ in $Y$, $f^{-1}(V)$ is open in $X$. For the converse, suppose $X$ is compact and $Y$ is Hausdorff. If $f^{-1}(V)$ is open in $X$, then its complement $X - f^{-1}(V)$ is closed in the compact space $X$, and hence it is compact. Consequently, its continuous image $f(X - f^{-1}(V))$ is compact in $Y$. As $Y$ is Hausdorff, $f(X - f^{-1}(V))$ is closed in $Y$. Again as $f$ is bijective, $f(X - f^{-1}(V)) = Y - V$. This asserts that $Y - V$ is closed and hence $V$ is open in $Y$.                               $\square$

**Proposition 5.4.4** *Let $(X, \tau)$ be a Hausdorff space and $Y \subset X$ be compact. Then $Y$ is closed.*

**Proof** Let $(X, \tau)$ be a Hausdorff space and $Y \subset X$ be compact. It is sufficient to show that $X - Y$ is open. Let $x \in X - Y$ be an arbitrary point. Then $x$ is not an element of $Y$. Let $y \in Y$. Since $X$ is a Hausdorff space by hypothesis, there exist two disjoint open sets $U_y$ and $V_y$ in $X$ such $x \in V_y$ and $y \in U_y$. Now, $Y = \bigcup(Y_y \cap A)$ asserts by compactness of $Y$ that there exist points $y_1, y_2, \ldots, y_n \in Y$ such that

$$Y \subset U_{y_1} \cup U_{y_2} \cup \cdots \cup U_{y_n} = U, say$$

and correspondingly,

$$x \in V_{y_1} \cap V_{y_2} \cap \cdots \cap V_{y_n} = V, say$$

Then $U$ and $V$ are two disjoint open sets such that

$$x \in V \subset X - U \subset X - Y.$$

Since $V$ is open and $x$ is an arbitrary point of $X - Y$, it follows that $X - Y$ is open and hence it is proved that $Y$ is closed.     □

Theorem 5.4.5 asserts that any continuous bijective map from a compact space to a Hausdorff space is a homeomorphism. It is an important theorem which is applied to solve some homeomorphism problems. For example, see Sect. 5.4.

**Theorem 5.4.5** *Let* $(X, \tau)$ *be a compact space and* $(Y, \sigma)$ *be a Hausdorff space. If a map* $f : (X, \tau) \to (Y, \sigma)$ *is continuous and bijective, then* $f$ *is a homeomorphism.*

**Proof** Let $(X, \tau)$ be a compact space and $(Y, \sigma)$ be a Hausdorff space. Let $f : X \to Y$ be a continuous bijective map. To prove that $f^{-1}$ is continuous it is sufficient to prove that $f$ sends closed sets of $X$ to closed sets in $Y$, which implies that if $A \subset X$ is closed in $X$, then $f(A) \subset Y$ is closed in $Y$. As $A$ is a closed subset of the compact subset $X$, it is compact. This asserts that its continuous image $f(A)$ is a compact subspace of the Hausdorff space $Y$. Then it follows by Proposition 5.4.4 that $f(A) \subset Y$ is closed in $Y$. It concludes that $f$ is a homeomorphism.     □

**Remark 5.4.6** The topology $\tau$ on a compact Hausdorff space $(X, \tau)$ has an optimal property in the sense that any topology finer than a Hausdorff topology is also Hausdorff and any topology coarser than a compact topology is also compact. Hence, any topology $\sigma$ strictly stronger (finer) than the given topology $\tau$ is Hausdorff; on the other hand, any topology $\sigma$ strictly weaker (coarser) than the given topology $\tau$ is compact but it is not Hausdorff, otherwise the identity map $1_X : X \to X$ would be a homeomorphism by Theorem 5.4.5 asserting that $\sigma = \tau$, which is a contradiction.

## Identification Maps from Compact Spaces

This subsection studies identification maps and relates such maps with maps from compact spaces to Hausdorff spaces to construct new geometric objects.

Recall that given topological spaces $(X, \tau)$ and $(Y, \sigma)$, a continuous surjective map

$$f : (X, \tau) \to (Y, \sigma)$$

is an identification map if $\sigma = \tau_f$, where $\tau_f$ is the quotient topology on $Y$ induced by $f$.

Theorem 5.4.7 is an important result in topology which is used to identify some quotient spaces with well-known spaces (see Examples 5.4.9 and 5.4.11).

**Theorem 5.4.7** *Let $(X, \tau)$ be a compact space, $(Y, \sigma)$ be a Hausdorff space and $f : (X, \tau) \to (Y, \sigma)$ be a surjective map. If $f$ maps closed (open) sets of $X$ to closed (open) sets of $Y$, then $f$ is an identification map.*

**Proof** To prove this theorem we first prove that if $X$ and $Y$ are topological spaces and if $f : X \to Y$ is a surjective map such that $f$ maps closed sets of $X$ to closed sets of $Y$, then $f$ is an identification map. To prove this let $A$ be a subset of $B$ such that $f^{-1}(A)$ is closed in $X$. As $f$ is onto by hypothesis, $f(f^{-1}(A)) = A$. This asserts that $A$ must be closed in the given topology on $Y$. This implies that this topology is the largest topology for which $f$ is continuous and hence $f$ is an identification map. The proof is similar for open sets.                                                                             $\square$

**Corollary 5.4.8** *Let $(X, \tau)$ be a compact space and $(Y, \sigma)$ be a Hausdorff space. If $f : X \to Y$ is continuous and surjective, then $f$ is an identification map.*

**Proof** Let $f : X \to Y$ be a continuous onto map. As $X$ is compact by hypothesis, any closed subset $A$ of $X$ is compact and its image $f(A)$ is compact in $Y$, since $f$ is continuous. Hence $f(A)$ is closed in $Y$. This shows that $f$ sends closed sets in $X$ to the closed sets in $Y$. This implies by Theorem 5.4.7 that $f$ is an identification map.                                                                                               $\square$

**Example 5.4.9** Let $\mathbf{I}$ be the closed unit interval and $\sim$ be an equivalence relation on $\mathbf{I}$ such that $[0] = [1] = \{0, 1\}$ and $[x] = \{x\}$ for $0 < x < 1$. Then $\mathbf{I}/ \sim$ is the quotient space homeomorphic to the circle $S^1$ by Theorem 5.4.7. In other words, $S^1 = \mathbf{I}/ \sim$, which is obtained from $\mathbf{I}$ by identifying the end points 0 and 1 of $\mathbf{I}$.

**Example 5.4.10** Quotient space of a compact Hausdorff space may not be Hausdorff. For example, consider the quotient space $X/A$, where $X = \mathbf{I}$ and $A = [0, 1)$ are subspaces of $\mathbf{R}$ with subspace topology of $\mathbf{R}$ and the canonical map $p : X \to X/A, x \mapsto [x]$, where $X/A$ is the quotient space corresponding to the equivalence relation $\sim$, which identifies every pair of elements in $A$ and no other pair of points is continuous and surjective. Since $p^{-1}([0]) = [0, 1)$ is open. the point $[0] \in X/A$ is open. On the other hand, since $p^{-1}([1]) = \{1\}$ is not open, the point $[1] \in X/A$ is not open. This implies that the quotient space $X/A$ is Sierpinski and hence is not Hausdorff.

**Example 5.4.11** If we identify all the points of the circumference of a disk $\mathbf{D}^2$, then the resulting quotient space is homeomorphic to the sphere $S^2$. To show it, let $N$ be the north pole of $S^2$. Then there exists a homeomorphism

$$f : \mathbf{D}^2 - S^1 \to S^2 - N.$$

Extend the map $f$ to a continuous map

$$\tilde{f} : \mathbf{D}^2 \to S^2$$

mapping the entire $S^1$ to the point $N$. This produces a continuous surjective map

$$\psi : \mathbf{D}^2 / S^1 \to S^2.$$

Hence it follows by Corollary 5.4.8 that $\psi$ is a homeomorphism.

## 5.5 Sequentially Compact Spaces

This section studies sequentially compact spaces. The concept of sequentially compactness in a metric space abstracts the Bolzano–Weierstrass property which asserts that the closed and bounded subsets of the real line $\mathbf{R}$ are sequentially compact.

**Definition 5.5.1** A topological space $(X, \tau)$ is said to be **sequentially compact**, if every infinite sequence in $X$ has a convergent subsequence.

**Example 5.5.2** Let $\tau$ be the collection of subsets of the set $\mathbf{R}$ of real numbers consisting of

  (i) all those subsets of $\mathbf{R}$, which do not contain 0; and
 (ii) the other four subsets $\mathbf{R}$, which are $\mathbf{R}$, $\mathbf{R} - \{1, 2\}$, $\mathbf{R} - \{1\}$, and $\mathbf{R} - \{2\}$.

Then $\tau$ forms a topology on $\mathbf{R}$. The topological space $(\mathbf{R}, \tau)$ is sequentially compact. Because, if $\mathbf{U}$ is any open covering of $(\mathbf{R}, \tau)$, then $\mathbf{U}$ must contain at least one of the sets prescribed in (ii) to include 0. Let $V$ be a set of this type taken from the open covering $\mathbf{U}$, then the subset $X - V$ consists of at most two points which are 1 and 2. If $V_1$ and $V_2$ are two open sets which are in $\mathbf{U}$ and contain the points 1 and 2 respectively, then $\{V, V_1, V_2\}$ forms a finite subcovering of $\mathbf{R}$. This shows that the space $(\mathbf{R}, \tau)$ is compact. It is also sequentially compact. Because, any infinite sequence $\{x_n\}$ in $\mathbf{R}$ is either of the two types, which are

(a) $x_n \neq 1$ and 2 for all $n$, excepting for a finitely many values of $n$; and the sequence $\{x_n\}$ is itself convergent and converges to the limit 0;
(b) $x_n = 1$ or 2 for infinitely many values of $n$ and then there exists an infinite subsequence of $\{x_n\}$, which converges to the limit 1 or 2.

**Proposition 5.5.3** *Every closed subspace of a sequentially compact space is also sequentially compact.*

**Proof** Left as an exercise. □

**Remark 5.5.4** The remaining part of this section studies (B–W)-compact spaces (i.e., limit point compact spaces) and relates them to sequentially compact spaces.

**Definition 5.5.5** A metric space $X$ is said to be sequentially compact if every sequence $\{x_n\}_{n=1}^{\infty}$ of points in $X$ has a convergent subsequence.

**Proposition 5.5.6** *In a metrizable space, every (B–W)-compact space is sequentially compact.*

**Proof** Let $X$ be a (B–W)-compact space. To prove the proposition, it is sufficient to show that every infinite sequence $\{x_n\}$ in $X$ has a convergent subsequence. Corresponding to the infinite sequence $\{x_n\}$ in $X$, construct the set

$$Y = \{x_n : n \in \mathbf{R}\} \subset X.$$

Case I: If the set $Y$ is finite, then there is a point $x \in Y$ such that $x_n = x$ for infinitely many values of $n$ and hence the sequence $\{x_n\}$ has a subsequence which is constant and converges automatically.

Case II: If the set $Y$ is infinite, then by definition of (B–W)-compact space, the set $Y$ has a limit point $x \in Y$. Construct a subsequence $\{x_{n_1}, x_{n_2}, \ldots x_{n_k}, \ldots\}$ of $\{x_n\}$ in $X$ that converges to $x$; take $x_{n_1}$ so that $x_{n_1} \in B_x(1)$. Let $n_{k-1}$ be a given positive integer. Since the open ball $B_x(\frac{1}{n}))$ intersects $Y$ at infinitely many points, we can take an index $n_k > n_{k-1}$ such that $x_{n_k} \in B_{x_{n_k}}(\frac{1}{n})$. This completes the construction of subsequence $\{x_{n_1}, x_{n_2}, \ldots x_{n_k}, \ldots\}$ of $\{x_n\}$ in $X$ that converges to $x$. **Alternatively** for each $n \in \mathbf{N}$, there exists a finite set of points $x_{n_1}, x_{n_2}, \ldots x_{n_k}$ such that the family of open balls

$$B_{x_{n_1}}\left(\frac{1}{n}\right), B_{x_{n_2}}\left(\frac{1}{n}\right), \ldots, B_{x_{n_k}}\left(\frac{1}{n}\right)$$

forms an open covering of $Y$, otherwise, there exists an integer $n$ such that no finite family of open balls of radius $\frac{1}{n}$ would cover $Y$. Construct a sequence $\{x_1, x_2, \ldots, x_k, \ldots\}$ of points of $Y$ such that for $k > 1$,

$$x_k \notin \bigcup_{i=1}^{i=k-1} B_{x_i}\left(\frac{1}{n}\right).$$

This implies that $d(x_k, x_j) \geq \frac{1}{n}$ if $k \neq j$. This asserts that the set of points $x_{n_1}, x_{n_2}, \ldots$ $x_{n_k}$ is infinite and hence this set has a limit point $x \in Y$. As the points $x_k, x_j \in B_x(\frac{1}{2n})$, this gives the contradiction that $d(x_k, x_j) < \frac{1}{n}$. $\qquad\square$

**Remark 5.5.7** In an arbitrary topological space, the concepts of compactness, (B–W)-compactness and sequentially compactness are different but they are closely related by Theorem 5.3.3 and Proposition 5.5.6. On the other hand, all these three concepts coincide in a metrizable space (see Exercise 103 of Sect. 5.28).

# 5.6 Locally Compact and Compactly Generated Hausdorff Spaces

This section studies locally compact and compactly generated Hausdorff spaces. There exist many topological spaces which are not compact but they contain very important compact subspaces. The Euclidean spaces is an important example. This type of spaces called locally compact spaces forms an important class of topological spaces, specially for the study of function spaces (see Sect. 5.24) and one-point compactification (see Sect. 5.13.3). A characterization of locally compact Hausdorff spaces is given in Exercise 51 of Sect. 5.28. On the other hand, the category of compactly generated Hausdorff spaces is very important, since it contains all locally compact spaces and almost all important spaces in topology. For this category, the paper (Steenrod 1967) is referred.

## 5.6.1 Locally Compact Spaces

This subsection studies locally compact spaces. Euclidean spaces $\mathbf{R}^n$ are not compact but they form an important class of locally compact spaces for every integer $n \geq 1$ (see Example 5.6.4). This example motivates to study locally compact spaces.

**Definition 5.6.1** A topological space $(X, \tau)$ is said to be locally compact at a point $a \in X$, if the point $a$ has at least one compact nbd in $X$. It is said to be **locally compact** if it is locally compact at every point $x \in X$, i.e., a topological space $(X, \tau)$ is locally compact if its every point $x$ lies in an open set $U_x$ such that its closure $\overline{U_x}$ is compact.

***Example 5.6.2*** (i) The real line space $\mathbf{R}$ is locally compact, because, the open intervals $(a, b)$ are basis elements for the natural topology on $\mathbf{R}$ and any point $x \in (a, b)$ lies in its closure $[a, b]$, which is compact in $\mathbf{R}$. On the other hand, the space $\mathbf{Q}$ of rational numbers with topology induced from the natural topology on $\mathbf{R}$ is not locally compact.

(ii) Euclidean $n$-space $\mathbf{R}^n$ is locally compact, because $x \in (a_1, b_1) \times (a_2, b_2) \times \cdots \times (a_n, b_n)$ asserts that $x \in [a_1, b_1] \times [a_2, b_2] \times \cdots \times [a_n, b_n]$, which is compact under product topology by Tychonoff product theorem 5.11.4. The subspace $X = \{(x, y) \in \mathbf{R}^2 : x, y \in \mathbf{Z}\}$ of $\mathbf{R}^2$ is also locally compact. Because, it is a discrete set and hence every point $x \in X$ has a compact nbd $\{x\}$.

***Example 5.6.3*** (i) Every discrete topological space is locally compact.
(ii) Any compact space is a locally compact space ;
(iii) Any closed subset of a locally compact space is a locally compact space;
(iv) $\mathcal{C}[0, 1]$ with compact open topology (see Sect. 5.24.1) is not locally compact.

***Example 5.6.4*** Every compact space is locally compact but its converse is not necessarily true. For example, the real line space **R** with the usual topology $\sigma$ is not compact but it is locally compact. Because, the family of open intervals $\{ \cdots, (-3, -1), (-2, 0), (-1, 1), (0, 2), \ldots, \}$ forms an open covering of **R** but it has no finite subcovering. Again, every point $x \in \mathbf{R}$ is an interior point of the closed interval $[x - \epsilon, x + \epsilon]$ for every $\epsilon > 0$, which is compact by Heine–Borel theorem.

**Definition 5.6.5** Let $A$ be a subspace of a topological space $(X, \tau)$. Then $A$ is said to be **locally closed** if every point $x \in A$ has an open nbd $U_x$ such that $A \cap U_x$ is closed in $U_x$.

Proposition 5.6.6 characterizes locally closed sets in terms of intersection of a closed set and an open set.

**Proposition 5.6.6** *Let $(X, \tau)$ be a topological space and $Y$ be a subspace of $X$. Then $Y$ is locally closed iff $Y$ can be expressed as $Y = U \cap A$, where $A$ is closed in $X$ and $U$ is open in $X$.*

***Proof*** Let $(X, \tau)$ be a topological space and $Y$ be a subspace of $X$. Let $Y$ be locally closed. If $U = \bigcup \{U_y : y \in Y\}$, where $U_y$ has the property of the Definition 5.6.5, then $U$ is an open set. Again if $A = \bar{Y}$, then $A$ is closed. Consequently,

$$U \cap A = \bigcup_{y \in Y} \{U_y : y \in Y\} \cap \bar{Y} = \bigcup_{y \in Y} (U_y \cap \bar{Y}) = \bigcup_{y \in Y} (U_y \cap Y) = U \cap Y = Y.$$

The converse part follows from Definition 5.6.5.                                      □

**Proposition 5.6.7** *Let $(X, \tau)$ be a locally compact space and $Y$ be a closed subspace of $X$. Then $Y$ is also locally compact.*

***Proof*** Let $Y$ be a closed subspace of $X$ and $x \in Y$ be an arbitrary point. As $X$ is locally compact at the point $x$, there is a compact nbd $N_x$ of $x$ in $X$. Let $Z$ be the space defined by $Z = N_x \cap Y$. As a closed nbd of $N_x$, the space $Z$ is compact. Since $N_x$ is a nbd of $x$ in $X$, the space $Z$ is also a nbd of $x$ in $Y$. This asserts that $Y$ is locally compact at $x$. As $x \in Y$ is an arbitrary point, it follows that $Y$ is also locally compact.                                      □

### 5.6.2  Locally Compact Hausdorff Space

This subsection studies locally compact Hausdorff spaces, (imposing Hausdorff condition on locally compact spaces), which play an important role in the study of function spaces (see Exponential Correspondence Theorem 5.24.9). For more properties of locally compact and locally compact Hausdorff spaces, see Sect. 5.28.

The concept of a locally compact space given in Definition 5.6.1 is reformulated in Definition 5.6.8 for a Hausdorff space.

**Definition 5.6.8** Let $(X, \tau)$ be a Hausdorff space. It is said to be a **locally compact Hausdorff** space if for every point $x \in X$ and every open set $U$ in $(X, \tau)$ containing the point $x$, there exists an open set $V$ in $(X, \tau)$ such that

(i) $\overline{V}$ is compact and
(ii) $x \in V \subset \overline{V} \subset U$.

**Theorem 5.6.9** *Let $(X, \tau)$ be a locally compact space and $x \in X$. Then the family of closed nbds of $x$ in $X$ forms a local base at $x$ if $(X, \tau)$ is either Hausdorff or regular.*

**Proof** By hypothesis, $(X, \tau)$ is a locally compact space and $x \in X$. Then there exists a compact nbd $N$ of $x$ in $X$. Let $T$ be an arbitrary nbd of $x$ in $X$. First suppose that $(X, \tau)$ is regular. Then there exists a closed nbd $W$ of $x$ in $X$ such that $W \subset N \cap T$. Since $N$ is compact and $W$ is closed in $N$, it follows that $W$ is also compact. Next suppose that $(X, \tau)$ is Hausdorff. Then $N$ is a normal Hausdorff space by Proposition 5.4.1 and hence $N$ is also regular. So, there exists a closed nbd $W$ of $x$ in $N$ such that $W \subset N \cap T$, because $N \cap T$ is a nbd of $x$ in the regular space $N$. Since every closed set in a compact space is compact, it follows that $N$ is closed. This asserts that $W$ is both closed and compact in $X$. Hence $W$ is also a nbd of $x$, because $N$ is itself a nbd of $x$ in $X$. This asserts that $W$ is a closed and compact nbd of $x$ if $(X, \tau)$ is either Hausdorff or regular. This proves the existence of a closed compact nbd $W$ of $x$ in $X$. $\qquad\square$

**Corollary 5.6.10** *Every locally compact Hausdorff space is regular.*

**Proof** Let $(X, \tau)$ be a locally compact Hausdorff space and $x \in X$ be an arbitrary point. Then the family of closed nbds of $x$ in $X$ forms a local base at $x$ by Theorem 5.6.9. This implies that $X$ is regular at the point $x$. Since $x \in X$ is an arbitrary point, it follows that $(X, \tau)$ is regular. $\qquad\square$

## 5.6.3 Category of Compactly Generated Hausdorff Spaces

This subsection is devoted to describe the category of compactly generated Hausdorff spaces, which contains all locally compact spaces, metrizable spaces, all topological spaces satisfying the first axiom of countability and almost all important spaces in topology. So, it is necessary to convey the concept of compactly generated spaces.

**Definition 5.6.11** A topological space $(X, \tau)$ is said be **compactly generated** if

(i) it is a Hausdorff space and
(ii) each subset $A$ of $X$ satisfying the property that $A \cap C$ is closed for every compact subset $C$ of $X$, is itself closed.

**Remark 5.6.12** A compactly generated space is a Hausdorff space such that it has the weak topology determined by its compact subsets. If $X$ and $Y$ are two topological spaces such that $X$ is locally compact and $Y$ is compactly generated, then their Cartesian product is compactly generated.

**Definition 5.6.13** Compactly generated Hausdorff spaces and their continuous maps form a category, called the **category of compactly generated Hausdorff spaces**, denoted by $\mathcal{CG}$.

***Example 5.6.14*** (i) Every metrizable space is in $\mathcal{CG}$. In more general, every first countable space is in $\mathcal{CG}$ by Proposition 5.6.17.

(ii) Every locally compact space is in $\mathcal{CG}$ by Proposition 5.6.17.

**Proposition 5.6.15** *Let $X$ be a Hausdorff space such that for every subset $B$ and each limit point $b$ of $B$, there exists a compact set $C$ in $X$ with the property that if $b$ is a limit point of $B \cap C$, then $X \in \mathcal{CG}$.*

***Proof*** Suppose $B$ intersects every compact set in a closed set and $b$ is a limit point of $B$. Then by hypothesis, there exists a compact set $C$ such that $b$ is a limit point of $B \cap C$. Since the set $B \cap C$ is closed, the point $b \in B \cap C$. This shows that $b \in B$ and hence $B$ is closed. Consequently, $X \in \mathcal{CG}$.                                          $\square$

**Corollary 5.6.16** *Let $X$ be a Hausdorff space such that every limit relation as stated in Proposition 5.6.15 holds in $X$, then $X \in \mathcal{CG}$.*

***Proof*** It follows from Proposition 5.6.15.                                          $\square$

Proposition 5.6.17 asserts that the category $\mathcal{CG}$ is large in the sense that it contains several important families of topological spaces.

**Proposition 5.6.17** *The category $\mathcal{CG}$ contains*

(i) *all locally compact spaces;*
(ii) *all first countable spaces;*
(iii) *all metrizable spaces.*

***Proof*** (i) Let $X$ be a locally compact space and $x \in X$. If $U_x$ is a nbd of the point $x \in X$, we take $C$ to be the compact closure of $U_x$. Hence it follows that $X \in \mathcal{CG}$.

(ii) Let $X$ be a first countable space and $x \in X$. Then there exists a compact set $C$, which consists of $x$ together with a sequence in $X$ converging to the point $x$. Hence it follows that $X \in \mathcal{CG}$.

(iii) Let $X$ be a metrizable space. Since every metrizable space is first countable, it follows by (ii) that $X \in \mathcal{CG}$.

                                                                                        $\square$

**Theorem 5.6.18** *Let $X$ be an object in the category $\mathcal{CG}$ and $Y$ be a Hausdorff space. If a function $f : X \to Y$ is continuous on each compact subset of $X$, then $f$ is continuous.*

***Proof*** Let $B$ be an arbitrary closed subset in $Y$ and $C$ be a compact subset in $X$. Since by hypothesis, $Y$ is a Hausdorff space and $f|_C$ is continuous, it follows that $f(C)$ is compact and hence it is closed in $Y$. This asserts that $B \cap f(C)$ is closed in $Y$ and hence $(f|_C)^{-1}(B \cap f(C)) = f^{-1}(B) \cap C$ is closed in $X$. Since $X$ be an object in the category $\mathcal{CG}$, the set $f^{-1}(B)$ is closed in $X$. Moreover, since $B$ is an arbitrary closed subset of $Y$, it follows that the map $f$ is continuous.                                          $\square$

**Definition 5.6.19**   Given a Hausdorff space $X$, its **associated compactly generated space** $\mathcal{A}(X)$ is the set $X$ with the topology obtained by declaring a set to be a closed set of $\mathcal{A}(X)$ if it intersects each compact set of $X$ in a closed set. If

$$f : X \to Y$$

is a mapping of Hausdorff spaces, then

$$\mathcal{A}(f) : \mathcal{A}(X) \to \mathcal{A}(Y)$$

defines the same function.

Theorem 5.6.20 proves some interesting properties of $\mathcal{A}(X)$.

**Theorem 5.6.20**   *Let $X$ be a Hausdorff and $\mathcal{A}(X)$ be its associated compactly generated space, Then*

  (i)  *the identity function $1_d : \mathcal{A}(X) \to X$ is continuous ;*
 (ii)  *$\mathcal{A}(X)$ is a Hausdorff space;*
(iii)  *$\mathcal{A}(X)$ and $X$ have the same compact subsets.*
 (iv)  *$\mathcal{A}(X)$ is an object of the category $\mathcal{CG}$;*
  (v)  *If $X$ is an object in the category $\mathcal{CG}$, then the identity map $1_d : \mathcal{A}(X) \to X$ is a homeomorphism;*
 (vi)  *If $f : X \to Y$ is continuous on compact sets, then the function $\mathcal{A}(f)$ is also continuous.*

**Proof**   (i) Let $B$ be a closed set in $X$ and $C$ be a compact subset in $X$. Then $C$ is closed in $X$ and consequently, $B \cap C$ is also closed in $X$. Hence $B$ is also closed in $\mathcal{A}(X)$. This proves (i).

(ii) Since $X$ is Hausdorff by hypothesis, (ii) follows from (i).

(iii) Let $B$ be a compact set in $\mathcal{A}(X)$. Then (i) shows that $B$ is compact in $X$. Again, if $C$ is compact in $X$ and $C_{\mathcal{A}(X)}$ denotes the set $C$ with its relative topology induced from $\mathcal{A}(X)$. Then by (i), the identity map $1_d : C_{\mathcal{A}(X)} \to C$ is continuous. To prove the continuity of its inverse, let $A$ be a closed set of $C_{\mathcal{A}(X)}$. Then $A$ intersects every compact set of $X$ in a closed set. Hence $A \cap C = A$ is closed in $C$. This implies that the identity map $C \to C_{\mathcal{A}(X)}$ is continuous. Hence it follows that $C_{\mathcal{A}(X)}$ is compact. It proves (iii). (iv) If a set $B$ intersects each compact set of $\mathcal{A}(X)$ in a closed set, then by (iii), it intersects each compact set of $X$ in a closed set. Then by (iii), it intersects each compact set of $X$ in a compact set and hence it is closed. This implies that $B$ is closed in $\mathcal{A}(X)$. It proves (iv).

(iv) It follows from (iv).

(v) To prove that $\mathcal{A}(f)$ is continuous on each compact set of $\mathcal{A}(X)$, it is sufficient by using Theorem 5.6.18 that $\mathcal{A}(f)$ is continuous on each compact set of $\mathcal{A}(X)$, which follows from the above results.

$\square$

## 5.7  Baire Space

This section studies Baire spaces named after René -Louis Baire (1874–1932), which form an important class of topological spaces containing complete metric spaces by Corollary 5.7.7 and compact Hausdorff spaces and also locally compact Hausdorff spaces (see Exercise 99 of Sect. 5.28). A Baire space $X$ given in Definition 5.7.1 cannot be expressed as a countable union of closed sets with empty interior in $X$. The nomenclature originally used by René -Louis Baire for defining a Baire space involved the word **category**.

**Definition 5.7.1** A topological space $(X, \tau)$ is said to be a **Baire space** if given any countable family $\{X_n\}$ of closed sets of $X$ with each $X_i$ having empty interior in $X$, their union $\bigcup X_n$ has also empty interior in $X$. **Equivalently**, a topological space $(X, \tau)$ is said to be a Baire space if intersection of every countable family of open dense sets in $X$ is dense.

**Example 5.7.2** Every locally compact space is a Baire space.

**Example 5.7.3** In the real number space **R**,

  (i) the subspace **N** of positive integers is a Baire space, because it satisfies the condition of a Baire space vacuously. Because every subset of $N$ is open, and hence there exist no subsets of **N** with empty interior, except the empty set $\emptyset$;
  (ii) the subspace **Q** of the rational number space is not a Baire space, since every one-pointic set in **Q** is closed and hence it has empty interior in **Q**. Moreover, **Q** is a countable union of its one-pointic subsets;
  (iii) every closed subset of **R** is a Baire space by Exercise 99 of Sect. 5.28, because it is a complete metric space;
  (iv) the subspace of irrational numbers in **R** is also a Baire space.

**Definition 5.7.4** A topological space $(X, \tau)$ is said to be of the **first category** if it is a countable union of nowhere dense subsets of $X$. Otherwise, $(X, \tau)$ is said to be of the **second category**.

**Example 5.7.5** In the real number space **R**, the set **Q** is of first category, since the one-pointic subsets $\{x\}$ of **Q** are nowhere dense in **Q** whose union is **Q**. Hence **Q** is the countable union of nowhere dense subsets subset of **Q**. On the other hand, the real number space **R** is of second category by Baire Category theorem 5.7.8.

**Theorem 5.7.6** *Let $(X, d)$ be a complete metric space and $\{X_n\}$ be a countable family of open sets such that each of them is dense in $X$. Then their intersection*

$$\bigcap_{n=1}^{\infty} X_n \neq \emptyset.$$

**Proof** By hypothesis, $(X, d)$ is a complete metric space and $\{X_n\}$ is a countable family of open sets such that each of them is dense in $X$. Then

$$\overline{X_n} = X, \quad \forall n = 1, 2, \ldots.$$

Construct a sequence in $X$ converging to a point in $\bigcap_{n=1}^{\infty} X_n$. For this construction, take a point $x_1 \in X_1$. By hypothesis, $X_1$ is an open set in $X$. Hence, there is an open ball $B_{x_1}(r_1) \subset X_1$ for some $r_1 > 0$. Now, $x_1 \in X = \bar{X}_2$ implies that $B_{x_1}(r_1) \cap X_2 \neq \emptyset$. Again, take another point $x_2 \in B_{x_1}(r_1) \cap X_2$. Choose $r_2 > 0$ such that $B_{x_2}(r_2) \subset B_{x_1}(r_1) \cap X_2$ and $r_2 < \min\{r_1/2, r_1 - d(x_1, x_2)\}$. For $x \in \overline{B_{x_2}(r_2)}$, it follows that

$$d(x, x_1) \leq d(x, x_2) + d(x_2.x_1) \leq r_2 + d(x_1, x_2) < r_1$$

and hence it asserts that $\overline{B_{x_2}(r_2)} \subset B_{x_1}(r_1)$. Go on continuing induction process to construct a sequence $\{x_n\}$ of points in $X$ and a sequence of balls $B_{x_n}(r_n)$ with center at $x_n$ and radius $r_n$ such that)

$$r_n < \frac{r_1}{2^{n-1}} \text{ and } \overline{B_{x_n+1}(r_{n+1})} \subset B_{x_n}(r_n) \subset X_n.$$

The sequence $\{x_n\}$ is clearly a Cauchy sequence in $X$ by its construction. Since by hypothesis, $X$ is a complete metric space, it follows that the sequence $\{x_n\}$ converges to some point $x_0 \in X_n$ for all $n = 1, 2, 3, \ldots$. This asserts that

$$\bigcap_{n=1}^{\infty} X_n \neq \emptyset.$$

$\square$

**Corollary 5.7.7** *Every complete metric space is a Baire space.*

**Proof** It follows from Theorem 5.7.6. $\square$

**Theorem 5.7.8 (Baire Category theorem)** *Every complete metric space is of the second category in the sense that it is not expressible as the union of a countable number of nowhere dense sets.*

**Proof** Let $(X, d)$ be a complete metric space. If

$$X = \bigcup_{n=1}^{\infty} X_n,$$

where every $X_n$ is nowhere dense, then $X = \bigcup_{n=1}^{\infty} \bar{X}_n$. By taking complements, it follows that

$$\emptyset = \bigcap_{n=1}^{\infty} (\bar{X}_n)^c.$$

But each $(\bar{X}_n)^c$ is open and also dense in $X$. Hence $\emptyset = \bigcap_{n=1}^{\infty} (\bar{X}_n)^c$ contradicts Theorem 5.7.6. $\qquad\qquad\square$

**Remark 5.7.9** Theorem 5.7.6 and Baire Category theorem 5.7.8 are also valid if the condition "complete metric space" in these theorems are replaced by the condition "compact Hausdorff space." For other forms of Baire Category theorem see Exercise 100 of Sect. 5.28.

## 5.8   Compactness Is a Topological Property

This section proves that compactness is a topological property, i.e., it is preserved under every homeomorphism by Corollary 5.8.3. Even compactness is preserved by every continuous surjective map by Theorem 5.8.1. This topological property is very important to solve many problems of topology such as classification of topological spaces up to homeomorphism. Applications of this topological property to homeomorphism problems are available in Sect. 5.25.4. Compactness property is also used to study maximal ideals in ring theory offering an interplay between topology and algebra (see Chap. 6).

**Theorem 5.8.1** *Let $(X, \tau)$ and $(Y, \sigma)$ be topological spaces and $f : (X, \tau) \to (Y, \sigma)$ be a continuous onto map. If $(X, \tau)$ is compact, then $f(X) = Y$ is also compact in $(Y, \sigma)$.*

**Proof** Let $(X, \tau)$ be compact and $\mathcal{C}$ be an open covering of $Y$. If $U \in \mathcal{C}$, then by continuity of $f$, it follows that $f^{-1}(U) \in \tau$. Hence the family of open sets

$$\mathcal{D} = \{f^{-1}(U) : U \in \mathcal{C}\}$$

forms an open covering of $X$. As $X$ is compact, the open covering $\mathcal{D}$ of $X$ has a finite subcovering such as

$$X = f^{-1}(U_1) \cup f^{-1}(U_2) \cup \cdots \cup f^{-1}(U_n).$$

Since $f$ is onto by hypothesis, it follows that

$$f(f^{-1}(U_i)) = U_i, \ \forall i = 1, 2, \ldots, n \text{ and hence } Y = U_1 \cup U_2 \cup \cdots \cup U_n.$$

This proves that $(Y, \sigma)$ is also compact.

$\qquad\qquad\square$

**Corollary 5.8.2** *Every continuous image of a compact space is compact.*

***Proof*** Let $(X, \tau)$ be a compact space and $f: (X, \tau) \to (Y, \sigma)$ be a continuous map. Then $f(X)$ is compact by Theorem 5.8.1.                                                                                            □

Corollary 5.8.3 asserts that compactness of a topological space is a topological property.

**Corollary 5.8.3** *Let $(X, \tau)$ and $(Y, \sigma)$ be two homeomorphic spaces. Then $(Y, \sigma)$ is compact iff $(X, \tau)$ is compact.*

***Proof*** Let $f: (X, \tau) \to (Y, \sigma)$ be a homeomorphism. If $(X, \tau)$ is compact, then $(Y, \sigma)$ is its homeomorphic image and hence it is compact by Theorem 5.8.1. Its converse part is similar.                                                            □

**Theorem 5.8.4** *Let $f: X \to \mathbf{R}$ be a continuous function from a compact space $X$. Then $f$ assumes its maximum and minimum.*

***Proof*** Since $X$ is compact, it follows by Theorem 5.8.1 that $f(X)$ is a compact subset of $\mathbf{R}$. This asserts that $f(X)$ is closed and bounded in $\mathbf{R}$. Consequently, the maximum of $f$ exists and it is finite and is the lub of $f(X)$, which is a limit point of $f(X)$. Since $f(X)$ is closed in $\mathbf{R}$, the maximum $M$ of $f$ is in $f(X)$. Similarly, its minimum $m$ is in $f(X)$.                                                                                               □

**Corollary 5.8.5** *If $f: X \to \mathbf{R}$ is a continuous function and $X$ is compact, then $\sup f(X)$ and $\inf f(X)$ exist in $f(X)$ and they are finite.*

***Proof*** Since $X$ is compact and $f$ is continuous, the set $f(X)$ is compact in $\mathbf{R}$. Hence $f(X)$ is a closed and bounded subset of $\mathbf{R}$. This shows that $\sup f(X)$ exists and it is finite. Moreover, since, $f(X)$ is closed, $\sup f(X) \in f(X)$. Similarly, it is proved that $\inf f(X)$ is in $f(X)$ and it is finite.                                                        □

**Corollary 5.8.6** *Every continuous real-valued function on a compact set is bounded.*

***Proof*** It follows from Corollary 5.8.5.                                                    □

The converse of Corollary 5.8.6 is given in the following form.

**Proposition 5.8.7** *Let $X \subset \mathbf{R}^n$ be a subspace such that every continuous real-valued function on $X$ is bounded. Then $X$ is compact.*

***Proof*** Suppose $X$ is not bounded. Then the function

$$f: X \to \mathbf{R}, x \mapsto ||x||$$

is not bounded on $X$. Because, if possible, $X$ is bounded but it is not compact. Then it is not closed by Heine–Borel theorem and there exists a point $\alpha$ in $\overline{X} - X$. Under this situation, the function

$$g: X \to \mathbf{R}, x \mapsto ||x - \alpha||^{-1}$$

cannot be bounded on $X$. This contradiction proves that $X$ is compact.                    □

## 5.9   Characterization of Compactness by Finite Intersection Property with Motivation

This section addresses the concept of compactness in terms of closed sets and characterizes compact spaces by finite intersection property. The motivation of this characterization comes from the observation: if $\mathcal{O}$ is an open covering of topological space $X$, then the collection $\mathcal{F}$ of complements of sets in $\mathcal{O}$ is a collection of closed sets, where their intersection is $\emptyset$ and conversely, if $\mathcal{F}$ is a collection of closed sets, where their intersection is $\emptyset$, the collection $\mathcal{O}$ of complements of sets of $\mathcal{F}$ is an open covering. This asserts that the space $X$ is compact iff every collection of closed sets with an empty intersection has a finite subcollection, where intersection is also $\emptyset$. The compactness property of a topological space $X$ is characterized in Theorem 5.9.6 with the help of "finite intersection property of closed sets of $X$."

**Definition 5.9.1**  A collection of subsets $\{X_i : i \in \mathbf{A}\}$ of a given nonempty set $X$ is said to have the **finite intersection property** if every finite subcollection of $\{X_i\}$ has a nonempty intersection.

**Example 5.9.2**  The collection of open intervals

$$\mathcal{C} = \{(0, 1), (0, 1/2), (0, 1/3), \ldots\}$$

has the finite intersection property, because, if

$$x_1 \in (0, 1), x_2 \in (0, 1/2), \ldots, x_n \in (0, 1/n),$$

and

$$x = \min\{x_1, x_2, \ldots, x_n\} > 0,$$

then $\mathcal{C}$ has the finite intersection property:

$$(0, x_1) \cap (0, x_2) \cap \cdots \cap (0, x_n) = (0, x) \neq \emptyset$$

but $\mathcal{C}$ has itself an empty intersection.

**Remark 5.9.3**  It is well known in analysis that

   (i)  the compact subsets of $\mathbf{R}^n$ are closed and bounded subsets of $\mathbf{R}^n$ (Heine–Borel theorem);
   (ii) an open interval $(a, b)$ of the real line space $\mathbf{R}$ is not compact.

**Definition 5.9.4**  A topological space $(X, \tau)$ is said to have **finite intersection property (FIP)** with respect to closed sets, if every collection $\mathcal{F} = \{X_i : i \in \mathbf{A}\}$ of closed sets in $(X, \tau)$ has the property

$$X_1 \cap X_2 \cap \cdots \cap X_n \neq \emptyset$$

for every finite collection of closed sets

$$\{X_1, X_2, \ldots, X_n\} \subset \mathcal{F},$$

then $(X, \tau)$ has also the property

$$\bigcap \{X_i : i \in \mathbf{A}\} \neq \emptyset.$$

**Remark 5.9.5** Compactness is now characterized with the help of finite intersection property of closed subsets of a topological space in Theorem 5.9.6 given by Leopold Vietoris. On the other hand, countably compactness is characterized by Cantor intersection theorem 5.16.4. Theorem 5.9.6 just translates the Definition 5.1.22 of compactness in terms of open sets to a statement in terms of closed sets. The motivation of this result comes from the bijective correspondence between the family $\mathcal{F}$ of closed sets in a topological space $(X, \tau)$ and the family $\mathcal{O}$ of open sets in $(X, \tau)$ obtained by complementation.

**Theorem 5.9.6** *(Vietoris) A topological space $(X, \tau)$ is compact iff $(X, \tau)$ has the finite intersection property (FIP) with respect to closed sets.*

**Proof** Let $(X, \tau)$ be a compact space and $\mathcal{F} = \{X_i : i \in \mathbf{A}\}$ be a collection of closed sets in $(X, \tau)$ having the finite intersection property. If possible, their intersection $\bigcap \{\{X_i : i \in \mathbf{A}\} = \emptyset$. It follows by **De Morgan rule** that $\bigcup \{X - X_i : i \in \mathbf{A}\} = X$, which shows that the collection $\{X - X_i : i \in \mathbf{A}\}$ forms an open covering of the compact space $(X, \tau)$. As by assumption, $(X, \tau)$ is compact, there exists a finite subcovering of $X$ formed by the open sets say, $X - X_1, X - X_2, \ldots, X - X_n$. Hence

$$X = (X - X_1) \cup (X - X_2) \cup \cdots \cup (X - X_n) = X - (X_1 \cap X_2 \cap \cdots \cap X_n)$$

asserts that $X_1 \cap X_2 \cap \cdots \cap X_n = \emptyset$, which contradicts the assumption that $\mathcal{F}$ has the finite intersection property. Conversely, let $(X, \tau)$ be a topological space satisfying the given conditions. To prove that $(X, \tau)$ is compact, let $\mathcal{C} = \{U_i : i \in \mathbf{A}\}$ be an open covering of $X$. Then $X = \bigcup \{U_i : i \in \mathbf{A}\}$. Taking complements we have

$$\emptyset = X - \bigcup \{U_i : i \in \mathbf{A}\} = \bigcap \{X - U_i : i \in \mathbf{A}\}.$$

Hence $\mathcal{F} = \{X - U_i : i \in \mathbf{A}\}$ forms a family of closed sets in $X$ such that $\emptyset = \bigcap \{X - U_i : i \in \mathbf{A}\}$. This asserts that the family $\mathcal{F}$ has not finite intersection property. Hence there exists a certain finite subcollection $\{X - U_1, X - U_2, \ldots, X - U_n\}$, say, must have an empty intersection. This shows that

$$\emptyset = (X - U_1) \cap (X - U_2) \cap \cdots \cap (X - U_n) = X - (U_1 \cup U_2 \cap \cdots \cup U_n),$$

which implies that
$$X = U_1 \cup U_2 \cup \cdots \cup U_n.$$

Consequently,
$$\{U_1, U_2, \ldots, U_n\}$$

forms a finite subcovering of the open covering $\mathcal{C}$ of $(X, \tau)$. This proves that $(X, \tau)$ is compact.  □

**Corollary 5.9.7** *Let $(X, \tau)$ be a compact set and $X_1 \supset X_2 \supset \cdots$ be a nested sequence of closed sets in $(X, \tau)$. If every member $X_n$ of this sequence is nonempty, then*
$$\bigcap \{X_n : n \in \mathbf{N}\} \neq \emptyset.$$

**Proof** It follows from Theorem 5.9.6 as a particular case.  □

*Example 5.9.8* All topological spaces are not compact. For example, the real number space $\mathbf{R}$ with usual topology is not compact. Because, the open covering $\{(-n, n) : n \in \mathbf{N}\}$ of $\mathbf{R}$ has no finite subcovering. Even, in the real number space $\mathbf{R}$ with usual topology an open interval is not compact. For example, the open covering $\{(1/n, n/n + 1) : n \geq 2\}$ of the open interval $(0, 1)$ has no finite subcovering. On the other hand, by Heine–Borel theorem every closed interval of the real number space $\mathbf{R}$ with usual topology is compact. Theorem 5.9.9 gives a generalization of Heine-Borel theorem for an arbitrary compact space and Theorem 5.14.1 gives an independent proof of Heine–Borel theorem.

Every closed interval $[a, b] \subset \mathbf{R}$ is compact. Theorem 5.9.9 gives a generalization of Heine–Borel theorem in a topological setting.

**Theorem 5.9.9** *Every closed subset $A$ of a compact space $(X, \tau)$ is compact in the relative topology induced from $X$.*

**Proof** By hypothesis, $A$ is a closed subset of a compact space $(X, \tau)$. We claim that $A$ has the finite intersection property. Let $\mathcal{F}$ be a collection of closed subsets of $A$ satisfying the finite intersection property given in Definition 5.9.4. Every $F \in \mathcal{F}$ is closed in $X$, because, $A$ is closed by hypothesis and $F$ is closed in $A$. This asserts that $\mathcal{F}$ is a collection of closed sets in $X$ and since $X$ is compact, the intersection $\bigcap_{F \in \mathcal{F}} F \neq \emptyset$. But $\bigcap_{F \in \mathcal{F}} F \subset A$ asserts that $A$ has the finite intersection property. This proves that $A$ is compact by Theorem 5.9.6.  □

**Corollary 5.9.10** *(Heine–Borel theorem) Every closed interval of the real number space $\mathbf{R}$ with usual topology is compact.*

**Proof** It follows from Theorem 5.9.9.  □

## 5.10 Paracompact Spaces

This section addresses the concept of paracompactness of topological spaces introduced by Jean Dieudonné in 1944. Paracompact spaces include regular and normal spaces by Proposition 5.10.11. There are many important topological spaces which are not compact but they are paracompact. For example, the real number space **R** is paracompact but it is not compact. Paracompact spaces represent a special class of topological spaces having localization of its compactness. The significance of paracompactness is the assertion of the existence of partition of unity in Theorem 5.10.15.

The concept of paracompactness is closely related to that of metrizability of topological spaces. The former concept is sometimes applied in a easier way to study metrizable spaces. Paracompactness is an important tool to study some problems in algebraic topology and geometry such as homotopy classification of vector bundles over paracompact spaces (see **Basic Topology: Volume 3** of the present series of books). Its other importance lies in the results that the class of paracompact spaces contains the compact Hausdorff spaces and metrizable spaces. By Stone's theorem, every metrizable space is paracompact. Its converse is true in the sense that every paracompact locally metrizable space is metrizable (see Nagata–Smirnov theorem). For a paracompact topological space $(X, \tau)$, a locally finite covering of $(X, \tau)$ always exists.

**Definition 5.10.1** Let $C$ and $D$ be two open coverings of a topological space $(X, \tau)$. Then $C$ is said to be a **refinement of** $D$ if every element of $C$ is a subset of some element of $D$.

**Definition 5.10.2** Let $(X, \tau)$ be a topological space. A family $\mathcal{F}$ of open covering of $X$ is said to be **locally finite** if every point of $X$ has a nbd which interests, nontrivially, only a finite number of members belonging to $\mathcal{F}$ i.e., if for every point $x \in X$, there is an open set $U_x \in \tau$ containing $x$ such that the set

$$\{U \in \mathcal{F} : U \cap U_x \neq \emptyset\} \text{ is finite.}$$

**Definition 5.10.3** A Hausdorff space $(X, \tau)$ is said to be **paracompact** if every open covering $C$ of $X$ has an open, locally finite refinement, i.e., for every open covering $C$ of $X$, there is a locally finite open covering $D$ such that for every open set $V \in D$, there is an open set $U \in C$ with the property that $V \subset U$.

**Remark 5.10.4** Hausdorff axiom is not assumed for defining a paracompact space by some authors.

**Example 5.10.5** (i) $\mathbf{R}^n$ is paracompact by Proposition 5.10.6.
(ii) Every regular space with topology having a countable basis is paracompact.
(iii) Every compact space is paracompact but its converse is not necessarily true. For example, the Euclidean space $\mathbf{R}^n$ is paracompact but it is not compact.

(iv) Every metric space is paracompact by Theorem 5.10.9.

(v) Every closed subspace of a paracompact space is paracompact by Proposition 5.10.7; but an arbitrary subspace of a paracompact space is not necessarily paracompact (see Exercise 86) of Sect. 5.28.

**Proposition 5.10.6** *The Euclidean n-space* $\mathbf{R}^n$ *is paracompact.*

**Proof** Let $C$ be an open covering of $\mathbf{R}^n$. Let $\{B_n = B_0(n) : n \in \mathbf{N}\}$ be a countable family of open balls centered at the origin $0 \in \mathbf{R}^n$ and radius $n$. Take $B_0(0) = \emptyset$. For an arbitrary $n \in \mathbf{N}$, select finitely many members of $C$ that cover $\overline{B}_0(n)$ and meet every open set $\mathbf{R}^n - \overline{B}_0(n-1) = \mathbf{R}^n - \overline{B}_{n-1}$. Denote this finite collection of open sets by $\mathcal{A}_n$. Consider the family $\mathcal{A} = \bigcup \mathcal{A}_n$. It is a refinement of $C$ and is locally finite, since the open set $B_n$ meets only finitely many members of $\mathcal{A}$ such as those members which are expressible as $\mathcal{A}_1 \cup \mathcal{A}_2 \cup \cdots \cup \mathcal{A}_n$. Moreover, $\mathcal{A}$ is a covering of $\mathbf{R}^n$, since for an arbitrary element $x \in \mathbf{R}^n$, let $n$ be the smallest positive integer such that $x \in \overline{B}_n$. This asserts that $x \in \mathcal{A}_n$ for some $n$. $\qquad\square$

**Proposition 5.10.7** *Every closed subspace of a paracompact space is paracompact.*

**Proof** Let $(X, \tau)$ be a paracompact space and $A$ be a closed subspace of $(X, \tau)$. Consider an open covering $\mathcal{A}$ of $A$. Corresponding to every $A \in \mathcal{A}$, take an open set $U_A \in \tau$ such that $U_A \cap A = A$. Let $C$ be an open covering of $X$ consisting of open sets $U_A$ and open sets $X - A$ and $\mathcal{L}$ be a locally finite refinement of $C$. Then the family $\mathcal{F} = \{Y \cap A : Y \in \mathcal{L}\}$ forms a locally finite open refinement of the original covering $\mathcal{A}$ of $A$.

$\qquad\square$

**Remark 5.10.8** Paracompact space is close to being a metric space (in the sense of Theorem 5.10.10 and Proposition 5.10.11). Theorem 5.10.9 asserts that every metric space is paracompact. It was first proved by Stone (1948a) by using the axiom of choice. But M.E. Rudin proved this result in an alternative but short-end method in 1969 (Rudin 1969) by using well-ordering principle of an arbitrary open covering of the given metric space. Theorem 5.10.9 is now proved following the technique used by Rudin.

**Theorem 5.10.9** *(Stone) Every metric space is paracompact.*

**Proof** The proof of the theorem consists of several steps. Let $(X, d)$ be a metric space. Then it is Hausdorff and every open covering of $X$ can be well-ordered. Let $\mathcal{U} = \{U_\alpha\}$ be an open covering of $X$ indexed by ordinals $\alpha$ and $S_x(r)$ be the open ball in $X$ with center $x$ and radius $r$. For every positive integer $n$, define $S_{\alpha n}$

$$S_{\alpha n} = \bigcup \{S_x(\frac{1}{2^n})\}$$

such that

(i) $\alpha$ is the smallest ordinal with $x \in U_\alpha$;

(ii) $x \notin S_{\beta m}$, *if* $m < n$;

(iii) $S_x \left( \frac{3}{2^n} \right) \subset U_\alpha$.

We claim that $\{S_{\alpha n}\}$ forms an open locally finite refinement of the open covering $\mathcal{U} = \{U_\alpha\}$ of $X$ indexed by ordinals $\alpha$. It is clearly open. To show that it is locally finite, let $x \in X$ and $\alpha$ be the smallest cardinal such that $x \in S_{\alpha n}$ for some $n \in \mathbf{N}$. Select an $m \in \mathbf{N}$ such that

$$S_x \left( \frac{1}{2^m} \right) \subset S_{\alpha n}.$$

We now prove the following two results:

(**A**): if $k \geq n + m$, then $S_x \left( \frac{1}{2^{n+m}} \right) \cap S_{\beta k} = \emptyset$ for any ordinal $\beta$;

(**B**) : if $k < n + m$, then $S_x \left( \frac{1}{2^{n+m}} \right) \cap S_{\beta k} \neq \emptyset$ for at least one ordinal $\beta$.

For (**A**), since $k > n$, it follows from (*ii*) that the ball $S_y \left( \frac{1}{2^k} \right)$ has its center $y$ not in $S_{\alpha n}$. Again since, $S_x \left( \frac{1}{2^m} \right) \subset S_{\alpha n}$, it follows that

$$d(x, y) \geq \frac{1}{2^m}.$$

Since $k \geq m + 1$ and $n + m \geq m + 1$, it follows that

$$S_x \left( \frac{1}{2^{n+m}} \right) \cap S_y \left( \frac{1}{2^k} \right) = \emptyset.$$

For (**B**), let $z \in S_{\beta k}$, and $t \in S_{\gamma k}$ with ordinals $\beta < \gamma$. Then

$$d(z, t) > \frac{1}{2^{n+m+k}},$$

because, there are points $w, p \in X$ such that

$$z \in S_w \left( \frac{1}{2^k} \right) \subset S_{\beta k} \text{ and } t \in S_p \left( \frac{1}{2^k} \right) \subset S_{\gamma k}.$$

Hence it follows by using the condition (*iii*) that

$$S_w \left( \frac{3}{2^k} \right) \subset U_\beta.$$

To show that the family $\{S_{\alpha n}\}$ covers $X$, take an arbitrary point $x \in X$. Let $\alpha$ be the smallest ordinal number such that $x \in U_\alpha$ and $n \in \mathbf{N}$ be taken so large such that the condition (*iii*) holds. Then it follows by using the condition (*ii*) that $x \in U_{\beta m}$ for some $m \leq n$. This asserts that every element of $X$ is in some $U_{\beta m}$. This implies that $\{S_{\alpha n}\}$ covers $X$. Consequently, the family $\{S_{\alpha n}\}$ of open sets refines the open covering

$\mathcal{U} = \{U_\alpha\}$ of $X$ indexed by ordinals $\alpha$. This concludes that the metric space $(X, d)$ is paracompact. $\qquad\square$

**Theorem 5.10.10**  *Let $(X, \tau)$ be a metrizable space. Then it is paracompact.*

**Proof**  Let $\mathcal{C}$ be an open covering of $X$. Then by Exercise 61 of Sect. 5.28, the covering $\mathcal{C}$ has an open refinement which also covers $X$ and is countably locally finite. Finally, Exercise 62 of Sect. 5.28 asserts that $\mathcal{C}$ has an open refinement that also covers $X$ and is locally finite. Hence it follows by Definition 5.10.3 that the metrizable space $(X, \tau)$ is paracompact.

$\qquad\square$

**Proposition 5.10.11**  *Let $(X, \tau)$ be a paracompact space. Then*

*(i)  $(X, \tau)$ is a regular space;*
*(ii)  $(X, \tau)$ is a normal space.*

**Proof**  Let $(X, \tau)$ be a paracompact space.

(i)  Given a point $x \in X$, let $A$ be a closed subset of $X$ such that $x \notin A$. For every point $a \in A$, there are open sets $U_a, V_a$ such that $x \in U_a$ and $a \in V_a$. Consider a covering of $X$ by the open set $X - A$ together with the open sets $\{V_a : a \in A\}$. Then there exists an open locally finite refinement by the sets $U_i$, say. Consider the set $U = \bigcup\{U_i : U_i \subset V_a$ for some $a \in A\} \supset A$. As this is a locally finite collection, $\bar{U}$ is the union of $\bar{U}_i's$. Since $x \notin \bar{U}_i$ for any $i$, it follows that $x \notin \bar{U}$, and hence the pair $U$ and $X - \bar{U}$ of open sets form a separation, implying that $X$ is a regular space.

(ii)  To show that $(X, \tau)$ is normal, take $A$ and $B$ be any two disjoint closed sets in $(X, \tau)$ and proceed as in (i) with $A$ playing the role of $x$ and the other closed set $B$ playing the role of $A$.

$\qquad\square$

## *A Characterization of Paracompact Spaces by Partition of Unity*

This subsection conveys the concept of partition of unity subordinate to a given open covering which is important in mathematics and is used to characterize paracompact spaces in Theorem 5.10.15. Such a characterization is very important in the study of paracompact spaces, because, by using just its definition is difficult to examine that the given space is compact. Another characterization of compact spaces by its $\sigma$-subsets is given by its disjoint union of open $\sigma$-compact subsets in Exercise 104 of Sect. 5.28. Many problems arising in the study of differential manifold theory are easy to solve locally by using the concept of local coordinate system. Sometimes, global solutions are obtained from such local solutions by using a partition of unity (see **Basic Topology, Volume 2**) of the present series of books.

**Definition 5.10.12** Let $f : X \to \mathbf{R}$ be continuous function. Then the support of $f$, denoted by *supp* $(f)$ is defined by

$$supp\ (f) = \overline{\{x \in X : f(x) \neq 0\}} \subset X$$

i.e., *supp* $(f)$ is the closure of the set

$$S_f = \{x \in X : f(x) \neq 0\}.$$

Thus if $x \notin supp\ (f)$, then there exists a nbd $N_x$ in $X$ on which $f$ vanishes.

**Definition 5.10.13** Let $(X, \tau)$ be topological space and $\mathcal{C} = \{U_a : a \in \mathbf{A}\}$ be an open covering of $X$. A **partition of unity** subordinate to $\mathcal{C}$ consists of a family of continuous functions

$$\mathcal{F} = \{f_b : X \to \mathbf{I} : b \in \mathbf{B}\}$$

such that

(i) there is a locally finite open refinement $\{V_b : b \in \mathbf{B}\}$ with the property that $supp(f_b) \subset V_b$, $\forall\, b \in \mathbf{B}$; and
(ii) $\sum_b f_b(x) = 1$, $\forall\, x \in X$.

$\sum_b f_b(x)$ is always a finite sum for every $x \in X$, since at each point $x \in X$, only a finite number of the functions $f_b(x)$ is different from 0 and hence the condition $(ii)$ caries sense.

**Example 5.10.14** Let $(X, d)$ be a metric space and $\mathcal{C}$ be an open covering of $X$. Define a map

$$\sigma : X \to \mathbf{R}, x \mapsto \Sigma_{U \in \mathbf{C}}\, d(x, X - U).$$

Since $\mathcal{C}$ is an open covering of $X$ and $d(x, X - U) > 0$ for every $x \in U$, it follows that $\sigma(x) > 0$ for every $x \in X$. This defines for every $U \in \mathcal{U}$ a continuous map

$$f_U : X \to \mathbf{R}. x \mapsto d(x, X - U)/\sigma(x).$$

Then the family of continuous functions

$$\mathcal{F} = \{f_U : X \to \mathbf{I} : U \in \mathcal{C}\}$$

forms a partition of unity in $X$ such that the support $supp(f_U)$ of the map $f_U$ is precisely the open set $U \in \mathcal{C}$.

Theorem 5.10.15 characterizes a paracompact space in term of partition of unity subordinate to its open covering.

**Theorem 5.10.15** *Let $(X, \tau)$ be a topological space. It is paracompact iff every open covering $\mathcal{C}$ of $X$ has a partition of unity subordinate to the covering $\mathcal{C}$.*

*Proof* First suppose that $C$ is an open covering of $X$ and $\{\psi_a : a \in \mathbf{A}\}$ is a partition of unity subordinate to $C$. Then the collection $\{x \in X : \psi_a(x) > 0, \ a \in \mathbf{A}\}$ forms a locally finite open covering of $X$, which is a refinement of $C$. This asserts that $(X, \tau)$ is paracompact. Conversely, suppose that $(X, \tau)$ is paracompact. Then by Proposition 5.10.11, it is normal. Let $C$ be an open covering of $X$ and $\mathcal{L}$ be a locally finite refinement of $C$. Then by using Exercise 63 of Sect. 5.28, there is a locally finite over covering $\mathcal{V} = \{V_U : U \in \mathcal{L}\}$ of $X$. Since $(X, \tau)$ is normal, finally, by using Exercise 64 of Sect. 5.28, it follows that there exists a partition of unity on $X$ subordinate to the covering $C$.

$\square$

## 5.11 Alexander's Subbase Theorem and Tychonoff Product Theorem

This section proves Alexander's Subbase Theorem 5.11.2, which is a powerful theorem in topology. This theorem is applied to prove the celebrated Tychonoff Product Theorem 5.11.4 for the product space of compact spaces asserting that compactness is a product invariant property in the sense that the product of compact spaces is also compact. As its consequence, this section also proves the compactness of $n$-cube $\mathbf{I}^n$ for every integer $n \geq 1$ and the cube $\mathbf{I}^\beta$, where $\beta$ is any transfinite cardinal number.

Finally, Tychonoff Embedding Theorem 5.11.8, which is a basic theorem in topology is proved. An important application of Tychonoff theorem is the generalization of the classical Heine–Borel theorem (see Proof II of Theorem 5.25.29).

**Definition 5.11.1** A family $\mathcal{F}$ of sets is said to have **finite character**, if $\emptyset \in \mathcal{F}$, and a nonempty set $A$ is a member of $\mathcal{F}$ iff every finite subset of $A$ is in $\mathcal{F}$.

To prove Alexander's subbase theorem, we use **Tukey's lemma** (see Chap. 1), which asserts that every nonempty family of sets of finite character has a maximal member.

**Theorem 5.11.2** *(Alexander's subbase theorem) Let $(X, \tau)$ be a topological space and $\mathcal{B}$ be a subbase for the topology $\tau$. If for any open covering of $X$ by a subcollection of members of $\mathcal{B}$, there is a finite subcovering of $X$, then the topological space $(X, \tau)$ is compact.*

*Proof* Let $\mathcal{F}$ be a given family of open sets in $(X, \tau)$ having the property that no finite subcollection of $\mathcal{F}$ forms an open covering of $X$. Then $\mathcal{F}$ is of finite character. Applying Tukey's lemma, it follows that there is a maximal family $\mathcal{A}$ having this property. To prove the theorem it is sufficient to prove that $\mathcal{A}$ does not form an open covering of the topological space $X$. Let $\mathcal{S}$ be the collection of all those members of $\mathcal{A}$ which are in the subbase $\mathcal{B}$ for the topology $\tau$. Then there exists no finite subcollection of $\mathcal{S}$, which forms a covering of the topological space $X$. This asserts by hypothesis that $\mathcal{S}$ does not form a covering of $X$. Again, as $\mathcal{B}$ forms a subbase for the topology

$\tau$, given any point $x \in F \in \mathcal{F}$, there is a finite subfamily $\{B_1, B_2, \ldots B_n\}$, say of the family $\mathcal{B}$ such that

$$x \in B_1 \cap B_2 \cap \cdots \cap B_n \subset F \tag{5.1}$$

If no set $B_i \in \{B_1, B_2 \ldots B_n\}$, is in $\mathcal{A}$, then by using its maximal property of $\mathcal{A}$, it follows that for the family $\mathcal{A} \cup \{B_i\}$, there is a finite subfamily forming a covering of $X$, say $X = B_i \cup A_i$, where $A_i$ is a finite union of members belonging to $\mathcal{A}$, say.

$$X = (B_1 \cup B_2 \cup \cdots \cup B_n) \cup (A_{11} \cup A_{12} \cup \cdots \cup A_{1r_1}) \cup \cdots$$
$$\cup (A_{n1} \cup A_{n2} \cup \cdots \cup A_{nr_n} : A_{ij} \in \mathcal{A}).$$

Hence, it follows that

$$X = F \cup (A_{11} \cup A_{12} \cup \cdots \cup A_{1r_1}) \cup \cdots \cup (A_{n1} \cup A_{n2} \cup \cdots \cup A_{nr_n}).$$

But it is not possible, since no finite subfamily of $\mathcal{A}$ forms an open covering of $X$. This asserts that our supposition that no set $B_i \in \{B_1, B_2 \ldots B_n\}$, is in $\mathcal{A}$ is not tenable. Hence there exists at least one of the sets $B_i \in \{B_1, B_2, \ldots B_n\}$ is in $\mathcal{A}$. Let $B_k$ be this set in $\mathcal{A}$ and hence $B_k \in \mathcal{S}$. This implies that $x \in \bigcup_{S \in \mathcal{S}} \{S\}$. It asserts that

$$\bigcup_{A \in \mathcal{A}} \{A\} = \bigcup_{S \in \mathcal{S}} \{S\}.$$

This proves that $\mathcal{A}$ does not form an open covering of $X$, because $\mathcal{S}$ does not form an open covering of $X$. Hence it follows that the topological space $(X, \tau)$ is compact. **Alternative proof**: Suppose $(X, \tau)$ is not compact. Let $\mathcal{B}$ be a subbase for the topology $\tau$. Then the family $\mathcal{F}$ of all open coverings of $X$ with no finite subcovering of $X$ is empty. It is a partially ordered set by set inclusion. Let $\{F_\alpha\}$ be a totally ordered subset in $\mathcal{F}$. Then $F = \bigcup_\alpha F_\alpha$ is an upper bound and hence by Zorn's lemma, $\mathcal{F}$ has a maximal element $\mathcal{A}$. Consider the family $\mathcal{S} = \mathcal{A} \cap \mathcal{B}$. Then $\mathcal{S}$ forms a covering of $X$. Since $\mathcal{S} \subset \mathcal{B}$, there is a finite subcover by assumption. But it is not possible, since $\mathcal{S} \subset \mathcal{A}$. This contradiction shows that the original family $\mathcal{F} = \emptyset$. This proves that $(X, \tau)$ is compact.

$\square$

**Definition 5.11.3**  The topological product of compact spaces is called the **Tychonoff product space**

**Theorem 5.11.4**  (*Tychonoff product theorem*)  *The product space of any nonempty family of compact spaces is compact.*

**Proof**  **Proof I**: It is proved by using Alexander's subbase theorem. Let $\mathcal{F} = \{(X_k, \tau_k) : k \in \mathbf{K}\}$ be a given family of compact spaces and $(X, \tau)$ be their topological product space. Let

$$p_k : X \to X_k$$

be the natural projection maps for all $k \in \mathbf{K}$. Consider the subbase

$$\mathcal{B} = \{p_k^{-1}(U_k) : U_k \in \tau_k, k \in \mathbf{K}\}$$

for the product topology $\tau$. Then by Alexander's Subbase Theorem 5.11.2, it follows that $X$ is to be compact if each subfamily $\mathcal{A}$ of $\mathcal{B}$ with the property that no finite subcollection of $\mathcal{A}$ forming a covering of $X$ does not form a covering of $X$. Let $\mathcal{S}_k$ be the collection of all those open sets $U_k \in \tau_k$ for which $p_k^{-1}(U_k) \in \mathcal{A}$ for each index $k \in \mathbf{K}$. Then no finite subcollection of $\mathcal{S}_k$ forms a covering of the compact space $X_k$. This asserts that there is a point $x_k \in X_k$, which is not in any open set $U_k \in \mathcal{S}_k$. This shows that the point $x \in X$ having the $k$-th coordinate $x_k$ is not in any member of $\mathcal{A}$. This implies that $\mathcal{A}$ fails to form a covering of the topological space $X$. Hence it follows by Theorem 5.11.2 that the product space $(X, \tau)$ is compact.

**Proof II**: It is proved by finite intersection property (FIP) of compactness. To prove the compactness of the product space $(X, \tau)$, it is sufficient to prove that given any family $\mathcal{C}$ of subsets of $X$ having FIP, their intersection

$$\bigcap\{\bar{C} : C \in \mathcal{C}\} \neq \emptyset.$$

Let $\mathcal{B}$ be the class of all families of subsets of $X$ having FIP. Then $\mathcal{B}$ is of finite character and hence by Tukey's lemma it has a maximal element $\mathcal{M}$, which contains the given family $\mathcal{C}$. Hence it follows that the intersection of the elements of every finite subfamily of $\mathcal{M}$ is also in $\mathcal{M}$. Moreover, if a subset $A \subset X$ intersects every member of $\mathcal{M}$, then $A \in \mathcal{M}$. Consider the family of natural projections

$$p_k : X \to X_k, \quad \forall k \in \mathbf{K}$$

and the family of sets

$$\mathcal{F}_k = \{p_k(M) : M \in \mathcal{M}\} \subset X_k, \forall k \in \mathbf{K}.$$

Since, each $X_k$ is compact by hypothesis, it follows by its FIP that

$$D_k = \bigcap\{\overline{p_k(M)} : M \in \mathcal{M}\} \neq \emptyset.$$

Take a point $x_k \in D_k$ for each $k \in \mathbf{K}$. If $x \in X$ is the point whose $k$-th coordinate $x(k)$ is the point $x_k \in X_k$, then $x \in \bar{M}$ for every $M \in \mathcal{M}$. Hence it follows that the product space $(X, \tau)$ is compact.

$$\square$$

**Corollary 5.11.5** *(Tychonoff theorem for finite products) Let* $(X_1, \ \tau_1)$, $(X_2, \ \tau_2)$, ..., $(X_n, \ \tau_n)$ *be compact spaces. Then their product space*

$$(X_1, \ \tau_1) \times (X_2, \ \tau_2) \times \cdots \times (X_n, \ \tau_n)$$

*is also compact.*

**Corollary 5.11.6**  *Let* $\mathbf{I} = [0, 1]$ *be the closed interval in the real line space* $\mathbf{R}$ *with usual topology* $\sigma$. *Then*

*(i)   the n-cube* $\mathbf{I}^n$ *is compact for every integer* $n \geq 1$;
*(ii)  the cube* $\mathbf{I}^\beta$ *is compact, where* $\beta$ *is any transfinite cardinal number.*

**Proof**  Let $\mathbf{R}$ be the real line space with natural topology $\sigma$ and $\sigma_\mathbf{I}$ be the relative topology on $\mathbf{I}$ induced from $\sigma$. Then $(\mathbf{I}, \ \sigma_\mathbf{I})$ is compact by Heine–Borel theorem. Let $B$ be a set such that card $B = \beta$. Then $\mathbf{I}^\beta$ is the topological product of the family of compact spaces $\{(X_b, \ \sigma_b) : X_b = \mathbf{I}, \ \sigma_b = \sigma_\mathbf{I}, \ \forall \, b \in B\}$. Then $\mathbf{I}^\beta$ is compact by Tychonoff Product Theorem 5.11.4.

(i)  If $B = \{1, 2, \ldots, n\}$ is a finite set, then $\beta = n$ and hence (i) follows.
(ii) If $B$ is an infinite set such that card $B = \beta$, an infinite cardinal. Then (ii) follows.

$\square$

## Tychonoff Embedding Theorem

This subsection deals with the problem of embeddings of a given topological space in a cube and proves Tychonoff embedding theorem which asserts that every Tychonoff space, i.e., a completely regular $T_1$ space (such as $\mathbf{I}$) can be embedded as a subspace of a cube. This gives an equivalent definition of a Tychonoff space in Corollary 5.11.9.

**Definition 5.11.7**  Let $\mathcal{F} = \{f_k : X \rightarrow Y_k : k \in \mathbf{K}\}$ be a given family of maps from a topological space $X$ into a topological space $Y_k$. Then this family is said to **distinguish points** of $X$ if for any two distinct points $p, q \in X$, there exists a map $f_k \in \mathcal{F}$ such that $f_k(p) \neq f_k(q)$. Again, this family is said to **distinguish points from closed sets** of $X$ if for any closed set $A$ in $X$ and any point $p \in X - A$, there exists a map $f_k \in \mathcal{F}$, such that $f_k(p)$ is not an element of $\overline{f_k(A)}$. A a completely regular $T_1$ space is called a **Tychonoff space**.

**Theorem 5.11.8**  *(Tychonoff embedding theorem)* *Let* $(X, \tau)$ *be a Tychonoff space. Then it can be embedded as a subspace of a cube.*

**Proof**  Let $(X, \tau)$ be a Tychonoff space. Then it is a completely regular $T_1$ space. Let $\mathcal{F} = \{f_k : X \rightarrow \mathbf{I} : k \in \mathbf{K}\}$ be the family of all continuous real functions $f_k : X \rightarrow \mathbf{I}$. Then by completely regularity property of $X$, the family $\mathcal{F}$ distinguishes points from the closed sets of $X$. Again, since, every one-pointic set of $X$ is closed, $\mathcal{F}$ also distinguishes points of $X$. This asserts that the map

$\psi: X \to \mathbf{I}^{\mathcal{F}}, x \mapsto \psi_x$ where $\psi_x: \mathcal{F} \to \mathbf{I}$, $f_k \mapsto f_k(x)$, equivalently. $\psi(x)(f_k) = f_k(x)$ is an embedding, where $\mathbf{I}^{\mathcal{F}}$ is a topological power of $\mathbf{I}$. $\qquad \square$

Corollary 5.11.9 characterizes a Tychonoff space in terms of an embedding as a subspace of a cube.

**Corollary 5.11.9** *A topological space* $(X, \tau)$ *is a Tychonoff space iff* $X$ *is homeomorphic to a subspace of a cube.*

***Proof*** Let $\mathcal{F} = \{f_k: X \to \mathbf{I}: k \in \mathbf{K}\}$ be the family of all continuous real functions $f_k: X \to \mathbf{I}$. Since $\mathbf{I}^{\mathcal{F}}$ is a compact Hausdorff space, it follows that $\mathbf{I}^{\mathcal{F}}$ is Tychonoff and its every subspace is also Tychonoff. Hence the Corollary follows from Theorem 5.11.8. $\qquad \square$

***Remark 5.11.10*** Every subspace of a normal space may not be normal. In addition to its supporting examples in Chap. 4, it is observed that under the notation of Theorem 5.11.8, every cube $\mathbf{I}^{\mathcal{F}}$ is normal but this theorem shows that a subspace of $\mathbf{I}^{\mathcal{F}}$ may not be normal, since every Tychonoff space is a completely regular $T_1$ space by its definition (not necessarily normal) and it can be embedded as a subspace (not necessarily normal) of a cube. For more properties of completely regular and Tychonoff spaces see Chap. 6.

## 5.12   Net and Convergence

This section starts with the concept of **Net** by generalizing the concept of a sequence in a metric space (studied earlier) for an arbitrary topological space and studies convergence problems by using this concept. This concept has wide applications in topology such as it is used in the study of compactness and continuity of functions and others (see Exercises 54–57 of Sect. 5.28). The convergence of sequences in metric spaces plays a key role in the study of continuity of functions. But it is not so for arbitrary topological spaces in general. The concept of "Net" provides useful tools to prove some results in arbitrary topological spaces, which are analogous to sequences in metric spaces.

### 5.12.1   Net: Introductory Concepts

This subsection conveys the concept of net in an arbitrary topological space $X$ as a mapping from a directed set $\mathcal{D}$ to $X$, instead, of a mapping from $\mathbf{N}$ to $X$ for a sequence in $X$.

**Definition 5.12.1** A directed set $\mathcal{D}$ is a partially ordered set with respect to a binary relation "$\geq$" (or "$\leq$") such that if $a, b \in \mathcal{D}$, then there exists an element $c \in \mathcal{D}$ with the property that $c \geq a$ and $c \geq b$.

**Example 5.12.2** (i) The set **N** of natural numbers is a directed set by the **natural ordering** "$\geq$" of real numbers.

(ii) Let $X$ be a nonempty set and $\mathcal{P}(X)$ be its power set. Then it is a directed set by **set theoretic inclusion** "$\subset$", i.e., for $A, B \in \mathcal{P}(X)$, $A \leq B$ holds if $A \subset B$.

(iii) Let $\mathcal{C}(X)$ be the set of all open coverings of the topological space $X$ is a directed set by **refinement relation** "$\leq$" given by $\{V\} \leq \{U\}$ if $\{U\}$ refines $\{V\}$, because any two coverings have a common refinement.

**Definition 5.12.3** (**Net**) Let $X$ be a topological space. A net in $X$ is a function $\psi : \mathcal{D} \to X$ from a directed set $\mathcal{D}$ into $X$. In particular, if $\mathcal{D} = \mathbf{N}$ with its natural ordering, then $\psi$ is said to be a sequence in $X$. A net $\psi : \mathcal{D} \to X$ in $X$ is usually expressed by $\{x_d : d \in \mathcal{D}\}$, where $x_d$ represents the element $\psi(d) \in X$.

**Example 5.12.4** (i) Given a topological space $X$, every sequence $\psi : \mathbf{N} \to X$ in $X$ is a net.

(ii) Let $\mathcal{P}(X)$ be the power set of $X$. It is a directed set by set theoretic inclusion "$\subset$". Take a point $x_A \in A$ for each $A \in \mathcal{P}(X)$. Clearly $\{x_A : A \in \mathcal{P}(X)\}$ is a net.

(iii) Let $X$ be a topological space and $x_0 \in X$ be a given point. If $\mathcal{N}$ be the collection of all nbds of $x_0$, define "$\geq$" on $\mathcal{N}$ by the rule $A \geq B$ holds iff $A \subset B$. Take a point $x_A \in A$ for each $A \in \mathcal{N}$. Clearly $\{x_A : A \in \mathcal{N}\}$ gives an example of a net.

**Definition 5.12.5** Let $\mathcal{D}$ be a directed set and $L \subset \mathcal{D}$ be a given subset. Then $L$ is said to be a **residual subset** of $\mathcal{D}$ if there is an element $\alpha_0 \in \mathcal{D}$ such that $\alpha \geq \alpha_0$ implies $\alpha \in L$ for every $\alpha \in \mathcal{D}$. If for each $\alpha \in \mathcal{D}$, there is an element $\beta \in L$ with the property that $\beta \geq \alpha$, then $L$ is said to be a **cofinal subset** of $\mathcal{D}$.

**Example 5.12.6** (i) Every residual subset of a directed set $\mathcal{D}$ is cofinal.

(ii) Every cofinal subset of $\mathcal{D}$ is also directed by "$\geq$".

**Definition 5.12.7** Let $X$ be a topological space and $\psi : \mathcal{D} \to X$ be a net. Given a subset $V \subset X$. the net $\psi$ is said to be in $V$ if $\psi(\mathcal{D}) \subset V$.

(i) The net $\psi$ is said to be **eventually in** $V$ if there is a residual subset $K \subset \mathcal{D}$ such that $\psi(K) \subset V$.

(ii) The net $\psi$ is said to be **frequently in** $V$ if there is a cofinal subset $L \subset \mathcal{D}$ such that $\psi(L) \subset V$.

## 5.12.2 Convergence

This subsection addresses the convergence of nets in a topological space $X$ as a generalization of convergence of sequences in metric spaces and it characterizes completely the topology on $X$ in terms of open sets involving the concept of net in Theorem 5.12.10 and hence it determines the topology on $X$ by using the concept of convergence of nets, which is similar in case of convergence of sequences in metric spaces.

**Definition 5.12.8** (**Convergence of Net**) Let $X$ be a topological space and $\psi : \mathcal{D} \to X$ be a net. Then $\psi$ is said to converge to a point $a \in X$, if for every nbd $U$ of $a$ in $X$, the net $\psi$ is eventually in $U$. Equivalently, $\psi$ is said to converge to a point $a \in X$, if for every nbd $U$ of $a$ in $X$, there exists point $d_0 \in D$ such that $x_d = \psi(d) \in U$, $\forall d \geq d_0$. It is symbolized, $x_d \to a$.

This definition of convergence of a net is analogous to the classical definition of convergence of a sequence in a metric space.

***Example 5.12.9***    (i)  If $X$ is a discrete space, then every net $\psi : \mathcal{D} \to X$ converges to a point $a \in X$ iff there is some $\alpha_0 \in \mathcal{D}$ such that $\psi(\alpha) = a \, \forall \alpha \geq \alpha_0$.
(ii)  If $X$ is an indiscrete space, then every net $\psi : \mathcal{D} \to X$ converges to every point of $X$.

Theorem 5.12.10 determines completely the topology of a topological space with the help of convergence of nets.

**Theorem 5.12.10**  *Let $X$ be a topological space. A subset $V \subset X$ is open in $X$ iff no net in $X - V$ can converge to a point in $V$.*

***Proof*** First suppose that $V$ is a subset of $X$ such that there is no net in $X - V$. We claim that $V$ is open in $X$. If $V$ is not open set in $X$, then there is a point $a \in V$ such that every nbd of $a$ intersects $X - V$. If $\mathcal{D}$ is a local base at the point $a$ in $X$, then $\mathcal{D}$ becomes a directed set by the set-theoretic inclusion "$\subset$." Take a point $\psi(A)$ in $A \cap (X - V)$ for each nbd $A \in \mathcal{D}$. Then the function $\psi : \mathcal{D} \to X, A \mapsto \psi(A)$ gives a net which converges to the point $a \in X - V$. But it contradicts the hypothesis. This forces to conclude that $V$ is open in $X$.

Conversely, suppose that $V \subset X$ is open in $X$ and $\psi : \mathcal{D} \to X$ is a net such that it converges to a point $a \in V$. As $V$ is nbd of $a$, and $\psi : \mathcal{D} \to X$ is a net such that it converges to a point $a \in V$, it follows that the net $\psi$ is eventually in $V$. But it asserts that $\psi$ is not in $X - V$.                                                                                        □

**Corollary 5.12.11**  *Let $X$ be topological space such that it has a countable local base at each point $x \in X$ (i.e., it satisfies the first axiom of countability). Then a subset $V \subset X$ is open iff no sequence in $X - V$ converges to a point of $V$.*

***Proof*** Let $\mathcal{B} = \{V_1, V_2, V_3, \ldots\}$ be a countable local base at the point $a$ of $X$. Set $U_n = V_1 \cap V_2 \cap \cdots \cap V_n$ for each $n = 1, 2, \ldots$. Then the sequence $\{U_1, U_2, U_3, \ldots\}$ also forms a local base of $X$ at the point $a \in X$. As $U_n \supset U_{n+1}$, $\forall n = 1, 2, \ldots$, the sequence $\{U_1, U_2, U_3, \ldots\}$ is a decreasing sequence of open nbds of the point $a$ that forms a local base. Then the corollary follows from the proof of Theorem 5.12.10.                                                                               □

**Definition 5.12.12** (**Cofinal map**) Let $\mathcal{D}$ and $\mathcal{F}$ be two directed sets. A function $\alpha: \mathcal{F} \to \mathcal{D}$ is said to be cofinal if for every residual subset $\mathcal{C} \subset \mathcal{D}$, there is a residual subset $\mathcal{G} \subset D$ with the property that $f(\mathcal{G}) \subset \mathcal{C}$. The set $\mathcal{F}$ with a cofinal function is called a cofinal set. If a function $\alpha: \mathcal{F} \to \mathcal{D}$ is cofinal, then for every element $d$ of $\mathcal{D}$, there is an element $f$ of $\mathcal{F}$ with the property that $\alpha(x) \geq d$, $\forall x \geq f$ in $\mathcal{F}$.

**Definition 5.12.13** (**Subnet of Net**) Let $\psi: \mathcal{D} \to X$ be a net. A net $\beta: \mathcal{F} \to \mathcal{X}$ is said to be a subnet of $\psi$ if there is a cofinal function $\alpha: \mathcal{F} \to \mathcal{D}$ such that $\beta = \psi \circ \alpha$.

**Example 5.12.14** Let $\psi: \mathcal{D} \to X$ be a net and $\mathcal{F}$ be a cofinal subset of $\mathcal{D}$, which is directed by the ordering induced by the ordering in $\mathcal{D}$. Then the inclusion function $i: \mathcal{F} \hookrightarrow \mathcal{D}$ is cofinal and the restriction map $\psi_{\mathcal{F}} = \psi \circ i: \mathcal{F} \to X$ is a subnet of $\psi$.

**Definition 5.12.15** (**Cluster point of Net**) Let $\psi: \mathcal{D} \to X$ be a net. A point $a \in X$ is said to be a cluster point of the net $\psi$ if the net $\psi$ is frequently in every nbd of $a$.

***Remark 5.12.16*** Theorem 5.12.10 characterizing the topology of a space with the help of convergence of nets facilitates a study of cluster point of a subset of a topological space and characterization of Hausdorff spaces by using the concept of net. Let $\psi: \mathcal{D} \to X$ be a net. In particular, if $X$ is a first countable space, then all the results regarding "net," excepting the characterization of compactness (Exercise 56 of Sect. 5.28) hold if one replaces " net" by "sequence." For more study on net, see Exercises 54–57 of Sect. 5.28.

# 5.13 Compactification Problems: Stone-Čech Compactification and Alexandroff One-point Compactification

This section is devoted to the study of the concept of compactification of a noncompact space $X$ by adjoining one or more points to $X$ and by declaring a topology on the enlarged space $X^+$ such that $X^+$ is a compact space containing $X$ as a dense subspace. There are many noncompact spaces which need compactification for their deep study. For example, the Euclidean $n$-space $\mathbf{R}^n$ is not compact but its one-point compactification is the $n$-sphere $S^n$ (see Corollary 5.13.16).

Compactification is a process or result of making a topological space into a compact space. Considering the importance of compactness in mathematics, it is a natural problem: given a noncompact space $X$, how to construct a compact space $Y$ which contains $X$ as a dense subspace? This method of construction is called compactification. This section describes Stone-Čech compactification in Sect. 5.13.2 and Alexandroff one-point compactification in Sect. 5.13.3, which solve the above problem.

## 5.13.1  Motivation of Compactification

This subsection conveys the basic motivation of compactification of noncompact topological spaces. It gives a solution of the problem: given a noncompact space $X$, how to construct a compact space $Y$ which contains $X$ as a dense subspace ? A compactification of a noncompact space $X$ to a compact space $Y$ is an embedding $i: X \hookrightarrow Y$ such that $i(X)$ is dense in $Y$.

**Definition 5.13.1**  A **compactification** of a topological space $(X, \tau)$ is any compact space $(X^+, \sigma)$ containing $(X, \tau)$ as a dense subspace. In the language of mapping, a compactification $X^+$ of $X$, is an embedding

$$f: X \to X^+$$

of $X$ into a compact space $X^+$ such that $f(X)$ is dense in $X^+$.

***Example 5.13.2***  A compactification of the real line space $\mathbf{R}$ is the compact space obtained by adjoining two new points, abbreviated, $+\infty$ and $-\infty$. The new space thus obtained is called the **extended real line**.

***Example 5.13.3***  One-point compactification of the Euclidean $n$-space $\mathbf{R}^n$ is the $n$-sphere $S^n$ proved in Corollary 5.13.16, which is a very useful result in topology.

***Example 5.13.4***  (**Construction**)  Let $(X, \tau)$ be $T_1$-space and $\infty$ be a point not in $X$. The simplest compactification of a noncompact space $X$ is the one-point compactification $X^+$ of $X$ is a compact topological space constructed from the original space $X$ as follows:

  (i)  The underlying sets of $X^+$ and $X \cup \{\infty\}$ are the same;
  (ii)  The topology $\sigma$ on $X^+$ are precisely the open sets in $\tau$ together with all subsets $U$ of $X^+$ such that $X^+ - U$ is a closed compact subset of $X$ , i.e., the topology $\tau$ on $X$ is the same as the relative topology on $X$ induced by the topology on $X^+$. This topological space $(X^+, \sigma)$ is a one-point compactification of $X$.

***Example 5.13.5***   (i)  Let $X = \mathbf{R}$ be the Euclidean line space. Hence its one-point compactification $X^+ = X \cup \{\infty\}$, where an open set containing the point $\infty$ is a complement of a compact subset of $X$. Then $X^+$ is the topological space homeomorphic to the circle $S^1$.

  (ii)  Let $X = \mathbf{R}^2$ be the Euclidean plane. Then the one-point compactification $X^+ = X \cup \{\infty\}$ is the topological space homeomorphic to the 2-sphere $S^2$ by a homeomorphism

$$\psi: S^2 \to X^+,$$

which is given by the stereographic projection

$$p: S^2 - N \to \mathbf{R}^2.$$

Given a point $x \in S^2$, the point $\psi(x)$ is the point of intersection of the line through the north pole N of $S^2$ and $x$ with the plane $\mathbf{R}^2$, which is the point $p(x)$.

**Theorem 5.13.6** *Let $(X, \tau)$ be a noncompact $T_1$-space and $X^+$ be the space constructed in Example 5.13.4.*

(i) *If $i: X \hookrightarrow X^+$ is the inclusion map, then $i$ is an embedding.*
(ii) *The space $X^+$ is compact.*
(iii) *$X$ is dense in $X^+$.*
(iv) *The map $i$ is a compactification of $X$.*

**Proof** Consider the topological space $(X^+, \sigma)$ constructed in Example 5.13.4.

(i) Since the topology $\sigma$ of $X^+$ consists of precisely the open sets in $\tau$ together with all subsets $U$ of $X^+$ such that $X^+ - U$ is a closed compact subset of $X$, the inclusion map

$$i: X \hookrightarrow X^+$$

is an embedding.
(ii) Assume that the space $X$ is noncompact. Let $\Omega$ be an arbitrary open covering of $X^+$. Then there is a member $U_\infty \in \Omega$ which contains the point $\infty$. Hence $X^+ - U_\infty$ is a compact subset of $X$. This asserts that there exists a finite subcovering $\{U_1, U_2, \ldots, U_k\}$, say, of $\Omega$. such that

$$X^+ - U_\infty \subset U_1 \cup U_2 \cup \cdots \cup U_k.$$

This shows that $\{U_\infty, U_1, U_2, \ldots U_k\}$ forms a finite subcover of $X^+$ and hence $X^+$ is compact.
(iii) To show that $X$ is dense in $X^+$, let $U$ be an arbitrary nonempty open set of $X^+$. If the point $\infty$ is not in the open set $U$, then $U \subset X$. On the other hand, if the point $\infty$ is in the open set $U$, then $X^+ - U$ is a compact subset of $X$. Since by assumption $X$ is noncompact, $U$ contains at least one point of $X$. This asserts that $X$ is dense in $X^+$.
(iv) It follows from the above discussion in the language of a mapping.

$\square$

**Remark 5.13.7** One-point compactification process of a topological space $X$ is simple and elegant but its deep study needs additional conditions such as locally compactness of $X$. More precisely, the one-point compactification of a noncompact space $X$ is closely related to locally compactness of property of $X$ (if it exists). The Hausdorff property of one-point compactification of a noncompact space $X$ is characterized in Proposition 5.13.8 by its locally compact Hausdorff property.

**Proposition 5.13.8** *Let $X^+$ be the one-point compactification of a noncompact space $X$. Then the space $X^+$ is Hausdorff iff $X$ is locally compact Hausdorff*

*Proof* First assume that the space $X$ is locally compact Hausdorff. To show that $X^+$ is Hausdorff, let $x, y \in X^+$ be two distinct points. If both of them are points different from $\infty$, then there exist disjoint open sets $U$ and $V$ of $X$ such that

$$x \in U, \ y \in V, \ U \cap V = \emptyset.$$

Again, if one of the points $x$ and $y$ is the point $\infty$, say, $x = \infty$, then $y \in X$. Since, by hypothesis, $X$ is locally compact, the point $y$ has a compact nbd $U_y$ in $X$. Suppose that $U = X^+ - U_y$ and $V = Int(U_y)$. Since $U_y$ is a compact set in a Hausdorff space $X$, it follows that $U_y$ is closed. This asserts that $U$ and $V$ are open sets of $X^+$ such that

$$x \in U, \ y \in V, \ U \cap V = \emptyset.$$

This proves that the space $X^+$ is Hausdorff.

Conversely, let the space $X^+$ be Hausdorff. Then the open subspace $X = X^+ - \{\infty\}$ of the compact Hausdorff space $X^+$ is a locally compact Hausdorff space. $\quad\square$

### 5.13.2  Stone-Čech Compactification

This subsection proves Stone–Cech compactification in Theorem 5.13.9, which is a basic result in topology. This theorem named after M. H. Stone (1902–1989) and E. Čech (1893–1960) asserts that every completely regular space $X$ is embeddable as a dense subspace in a specified compact Hausdorff space $\beta(X)$, which has an important property that every bounded real-valued function continuous on $X$ has a unique extension to a bounded continuous real-valued function on $\beta(X)$. Stone published a paper (Stone 1948a, b) on compactification of topological spaces.

**Theorem 5.13.9** *(Stone-Čech Compactification) If $X$ is a completely regular space, then there exists a compact Hausdorff space $\beta(X)$ such that*

*(i)  $X$ is a dense subspace of $\beta(X)$ and*
*(ii)  every bounded continuous function*

$$f : X \to \mathbf{R}$$

*has a unique extension to a bounded continuous function*

$$\tilde{f} : \beta(X) \to \mathbf{R}.$$

*Proof* **Existence of** $\beta(X)$: Let $\mathbf{I} = [0, 1]$ be the closed unit interval which is a subspace of the real line space $\mathbf{R}$ and $\mathcal{C} = \mathcal{C}(X, \mathbf{I})$ be the set of all continuous functions

$\{f_\alpha : X \to \mathbf{I} : \alpha \in \mathbf{B}\}$. Then by using Tychonoff Embedding Theorem 5.11.8, there exists an embedding

$$\psi : X \to \mathbf{I}^{\mathcal{C}} : x \mapsto \psi_x \text{ where } \psi_x : \mathcal{C} \to \mathbf{I}, \; f_\alpha \mapsto f_\alpha(x),$$

which asserts that $\psi(x)(f_\alpha) = f_\alpha(x)$. Since $\mathbf{I}^{\mathcal{C}}$ is a compact Hausdorff space by Corollary 5.11.6, it follows that

$$\psi : X \to \psi(X)$$

is a homeomorphism and hence $\beta(X) = \overline{\psi(X)}$ proves the existence of $\beta(X)$.

(i) It follows from the construction of $\beta(X)$.

(ii) For the proof of this part, it is sufficient to consider the image of $f$ in $\mathbf{I}$. Let $\{\mu_\beta\}$ be a net in $\mathbf{I}^{\mathcal{C}}$ converging to $\mu$. Define a function

$$\tilde{f} : \mathbf{I}^{\mathcal{C}} \to \mathbf{R}, \; \mu \mapsto \mu(f).$$

Then

$$lim\{\tilde{f}(\mu_\alpha)\} = lim\{f(\mu_\alpha)\} = \mu(f) = \tilde{f}(\mu)$$

asserts that $\tilde{f}$ is continuous by Exercise 57 of Sect. 5.28. Since

$$\tilde{f}(\psi(x)) = (\psi(x))f) = f(x), \; \forall \, x \in X,$$

it implies that $\tilde{f}$ is a continuous extension of $f$ over $\beta(X)$. The uniqueness of $\tilde{f}$ follows from its construction.

$\square$

**Definition 5.13.10** Given a completely regular space $X$, the compact Hausdorff space $\beta(X)$ defined in Theorem 5.13.9 is called the **Stone–Cech compactification** of $X$.

**Example 5.13.11** In the real line space $\mathbf{R}$, consider its subspaces $X = (0, 1]$ and $\mathbf{I} = [0, 1]$. Then $X$ is a completely regular space and it is a dense subspace of the compact Hausdorff space $\mathbf{I}$. But its Stone–Cech compactification $\beta(X) \neq [0, 1] = \mathbf{I}$. Because, the map $f : X \to \mathbf{R}$, $t \mapsto \sin t^{-1}$, is a bounded continuous real function, but it cannot be continuously extended over $\mathbf{I}$. On the contrary, the space $\mathbf{I}$ is a compact Hausdorff space containing $X$ as a dense subspace.

## 5.13.3 Alexandroff One-point Compactification

This subsection proves Alexandroff one-point compactification in Theorem 5.13.14, which is the simplest important one-point compactification. Given a noncompact space $X$, this subsection presents a construction process of a compact space $X^+$ containing $X$ as a dense subspace by adjoining one extra point, known as a one-point compactification of Alexandroff by imposing certain conditions on $X$ prescribed in Theorem 5.13.14 and also establishes some relations between the topologies of $X$ and $X^+$ as well as the properties of functions defined on these topological spaces.

Locally compactness of a topological space is characterized by one-point compactification of the topological space in Theorem 5.13.14. An important consequence of this theorem asserts that one-point compactification of $\mathbf{R}^n$ is $S^n$, which has wide applications in **geometric topology (see Basic Topology, Volume 3** of the present series of books).

**Definition 5.13.12 (Alexandroff one-point compactification)** Given a noncompact space $(X, \sigma)$, a compact Hausdorff space $(X^+, \sigma^+)$ is said to a one-point compactification of $X$. where $X^+ = X \cup \{\infty\}$, where $\infty$ is a point not in $X$ and a topology $\sigma^+$ is defined on $X^+$ making $(X^+, \sigma^+)$ it a compact Hausdorff.

***Remark 5.13.13*** The one-point compactification $X^+$ of a noncompact space $X$ plays a key role to obtain interesting results when $X$ is taken to be locally compact. The concept of Alexandroff one-point compactification is now applied to characterize locally compactness of a topological space.

**Theorem 5.13.14** *(Alexandroff) Let $(X, \sigma)$ be a topological space.*

  (i) *There exists a one-point compactification $X^+ = X \cup \{\infty\}$ iff the space $(X, \sigma)$ is locally compact.*
  (ii) *Moreover, the topology on $X^+$ is uniquely determined and coincides with the topology on $X$ as a subspace in $X^+$.*

***Proof*** Let $(X, \sigma)$ be a locally compact space and $X^+ = X \cup \{\infty\}$. A subset $U$ in $X^+$ is declared to be open if either $U$ is an open set in $X \subset X^+$ or it is of the form $V \cup \{\infty\}$, where $V$ is an open set in $X$ such that $X - V$ is compact. The existence of $V$ follows from the local compactness of $X$. The family of all open sets thus described form a topology $\sigma^+$ on $X^+$. We claim that $X^+$ is compact with respect to this topology $\sigma^+$.

**Hausdorff property of $X^+$:** Under this topology $\sigma^+$, the space $X^*$ is Hausdorff. Because, given a point $x \in X$ and the point $\infty$, they have disjoint nbds: take a nbd $U_x$ of $x$ such that $\bar{U}_x$ is compact in $X$ and a nbd $V_\infty = (X - \bar{U}_x) \cup \{\infty\}$, which are clearly disjoint.

**Compactness property of $X^+$:** Next we show that the space $X^+$ is compact. For this, let $\mathbf{C}^+ = \{U_\beta\}$ be an open covering of $X^+$. Then there is a member $U_{\beta_0}$ in $\mathbf{C}^+$, which contains the point $\infty$ and hence $V \cup \{\infty\} = U_{\beta_0}$, where $X - V = A$ is compact in $X^*$. The subcovering $\mathbf{C}_1^+ = \{U_\beta\}_{\beta \neq \beta_0}$ of $A$ has a finite subcovering $\{U_{\beta_1}, \ldots, U_{\beta_m}\}$. Hence $\{U_{\beta_0}, U_{\beta_1}, \ldots, U_{\beta_m}\}$ is finite covering of the space $X^+$.

Combining the above results, it follows that the topological space $(X^+, \sigma^+)$ is a compact Hausdorff space, called **Alexandroff one-point compactification of** $X$.

Conversely, let $X^+ = X \cup \{\infty\}$ be a one-point compactification of $X$. We claim that $X$ is locally compact. As $X$ is open in $X^+$, every point $x \in X$ has an open nbd $U_x$ in the topology $\sigma^+$ on $X^+$ such that $\bar{U}_x \subset X$. Then $\bar{U}_x$ is a closed set in $X^+$ and hence it is compact in both $X^+$ and $X$. This asserts the local compactness of $X$. Any open set in $X^+$ containing a point $\infty$ is of the form $V \cup \{\infty\}$, where $V$ is an open set in $X$. On the other hand, the closed set $X^+ - (V \cup \{\infty\}) = X - V$ is compact, since $X^+$ is so.

**Uniqueness of the topology** $\sigma^+$ on $X^+$: Let $V$ be an open set in some such topology $\sigma_1^+$ on $X^+$. Then the set $X^+ - V = A$ is closed and hence $A$ is compact. If $A \subset X$, then $V$ is open in the topology $\sigma^+$. On the other hand, if $A$ is not a subset of $X$, then $V \subset X$ is open in $X$, because, $X$ is a subspace. $V$ is also open in the topology $\sigma^+$. Again, since $X$ is a subspace of $X^+$, if $V \subset X$ is open in $X$, then $V$ is of the form $V = U \cap X$ for some open set $U$ in $X^+$. Then it follows that $V = U \cap X$ is also open in $X^+$, since $X$ is open in $X^+$. Again, if $A$ is compact in $X$, then it is also compact in $X^+$, since the property of compactness does not depend on the space which it contains. This shows that $A$ is closed in $X^+$ and hence the set $X^+ - A$ is open in $X^+$. This asserts that $\sigma_1^+ \subset \sigma^+$. Proceeding likewise, it also follows that $\sigma^+ \subset \sigma_1^+$. This implies that $\sigma^+ = \sigma_1^+$.

$\square$

**Corollary 5.13.15** *If $X$ is a compact space (already), then the isolated point set $\{\infty\}$ and $X$ are both clopen in $(X^+, \sigma^+)$.*

**Proof** It follows from Theorem 5.13.14. $\square$

**Corollary 5.13.16** *The one-point compactification of the Euclidean n-space $\mathbf{R}^n$ is the n-sphere $S^n$.*

**Proof** Let $N$ be the north point of the $n$-sphere $S^n$. Then the punctured sphere $S^n - N$ is homeomorphic to $\mathbf{R}^n$ by stereographic projection. Hence it follows by Alexandroff one-point compactification that $S^n$ is a one-point compactification of $\mathbf{R}^n$, abbreviated $\mathbf{R}^n \cup \{\infty\} = S^n$. $\square$

**Corollary 5.13.17** *The 3-sphere $S^3$ is the one-point compactification $\mathbf{R}^3 \cup \{\infty\}$ of $\mathbf{R}^3$ and a homeomorphism $h: \mathbf{R}^3 \to \mathbf{R}^3$ has a unique extension to a homeomorphism $\tilde{h}: S^3 \to S^3$.*

**Proof** It follows from Corollary 5.13.16. $\square$

**Remark 5.13.18** The compactification of a topological space $X$ can also be done by adjoining one or more points to $X$ and subsequently, a suitable topology is defined on the larger set $X^+$ with an eye to make the set $X^+$ compact, which contains $X$ as an dense subspace of $X^+$.

**Example 5.13.19** (**Two-point compactification**) Let $\mathbf{R}$ be the real line space with usual topology $\sigma$ and $\mathbf{R}^+ = \mathbf{R} \cup \{-\infty, \infty\}$ be the extended real line obtained by adjoining two new points $-\infty$ and $\infty$ to $\mathbf{R}$. The usual order relation on $\mathbf{R}$ is now extended to $\mathbf{R}^+$ by defining $-\infty < x < \infty$ for any $x \in \mathbf{R}$. Then the family of subsets of $\mathbf{R}^+$ of the forms

(i) $[-\infty, a) = \{x \in \mathbf{R}^+ : x < a\}$;

(ii) $(a, b) = \{x \in \mathbf{R} : a < x < b\}$ and

(iii) $(a, \infty] = \{x \in \mathbf{R}^+ : a < x\}$

forms an open base for a topology $\sigma^+$ on $\mathbf{R}^+$. The topological space $(\mathbf{R}^+, \sigma^+)$ is compact and it contains the noncompact space $(\mathbf{R}, \sigma)$, as a dense subspace and hence by definition, $(\mathbf{R}^+, \sigma^+)$ is a two-point compactification of $(\mathbf{R}, \sigma)$.

Theorem 5.13.20 conveys an important application of Alexandroff one-point compactification and proves equivalent formulation of a locally compact Hausdorff space.

**Theorem 5.13.20** *Let $H$ be a Hausdorff space. Then the following statements are equivalent:*

*(i) The space $H$ is locally compact.*

*(ii) The space $H$ is a locally closed subspace of a compact Hausdorff space.*

*(iii) The space $H$ is a locally closed subspace of a locally compact Hausdorff space.*

**Proof** $(i) \implies (ii)$: Let the space $H$ be locally compact. Then by Alexandroff one-point compactification $H^+$ (see Theorem 5.13.14), $H$ is an open subspace of its one-point compactification $H^+$. This shows that (i) implies (ii).

$(ii) \implies (iii)$ trivially.

$(iii) \implies (i)$: Let $X$ be a locally compact space containing $H$ and $H$ be expressed as $H = A \cap U$, where $A \subset X$ is closed and $U \subset X$ is open. Then $A$ is locally compact and $H = U \cap A$ is open in $A$. This asserts that $H$ is locally compact which shows that (iii) implies (i).                                                                          □

## 5.14   Haar–Konig Theorem: Characterization of Compactness in Linearly Ordered Spaces

This section communicates a complete characterization of compactness in linearly ordered spaces in Theorem 5.14.2, from which Heine–Borel theorem in $\mathbf{R}$ follows as a particular case. We start with an independent proof of Heine–Borel theorem in $\mathbf{R}$ to facilitate the proof of Theorem 5.14.2.

**Theorem 5.14.1** *(Heine–Borel theorem in $\mathbf{R}$) Every closed interval in the real number space $(\mathbf{R}, \sigma)$ with usual topology $\sigma$, is compact.*

**Proof** Let $[a, b]$ be a given closed interval in $\mathbf{R}$ and $\mathcal{C}$ be an open covering of $[a, b]$ in $\mathbf{R}$. Let $\mathcal{E}$ be the set of all those elements $x \in [a, b]$ such that there is a finite subcovering of $\mathcal{C}$ for the closed interval $[a, x]$. Then $a \in \mathcal{E}$ and the point $b$ is an upper bound of the set $\mathcal{E}$. This shows that $\mathcal{E}$ has an lub $e$, say. Then it follows that $a < e$. If $U \in \mathcal{C}$ is such that $e \in U$ and $d \in (a, e) \cap U$, then $[a, d)$ has an open finite subcovering covering $\mathcal{C}'$, which is finite subcollection of $\mathcal{C}$. Hence $\mathcal{C}' \cup U$ is a finite subcollection of $\mathcal{C}$, which forms a finite open subcovering of $[a, e]$. If $e \neq b$, there is a number $f$ such that $f \in U$, $(e, f) \subset U$ and hence $C' \cup U$ is a finite subcollection of $C$ and it forms an open covering of $[a, f]$ such that $f \in \mathcal{E}$, which is not possible as $f > e$ and $e$ is the lub of $\mathcal{E}$. This asserts that $e = b$ and hence $C' \cup U$ is a finite subcollection of $C$, which forms a finite covering of $[a, b]$. Consequently, $[a, b]$ is compact in the real number space $(\mathbf{R}, \sigma)$.    □

Haar–Konig Theorem 5.14.2 characterizes compactness of a linearly ordered space (with interval topology) in terms of its order-completeness, from which Heine–Borel theorem in $\mathbf{R}$ follows as a corollary.

**Theorem 5.14.2** (*Haar–Konig theorem*) *Let $X$ be a linearly ordered set $X$ with interval topology. Then $X$ is compact iff $X$ is order-complete in the sense that every nonempty subset of $X$ has both a lub and a glb.*

**Proof** Let $X$ be a linearly ordered set $X$ and $\tau$ be the interval topology on $X$. First suppose that $X$ is order-complete. Let $m$ be the smallest element and $M$ be the greatest element of $X$. Proceed likewise as in proof of Theorem 5.14.1 to show that the topological space $(X, \tau)$ is compact.

For the converse part, suppose that $(X, \tau)$ is compact. If the linearly ordered set $X$ is not order-complete, then we show that the topological space $(X, \tau)$ is not compact. To show it, let $Y \subset X$ be nonempty and suppose that $Y$ has no lub in $X$. Consider the family of open intervals $\mathcal{O}$ consisting of the open intervals of the forms $(-\infty, y)$, $\forall y \in Y$ and $(v, +\infty)$ for every upper bound v of $Y$. Then $\mathcal{O}$ forms an open covering of $X$. But this covering has no finite subcovering, otherwise, the largest element $y$ or the smallest element $v$ appearing in that subcovering would be the lub of the set $Y$. Hence $(X, \tau)$ would refuse to be compact. For the other possibility, if $Y$ has no glb in $X$, similar argument also produces a contradiction. This contradiction proves that $(X, \tau)$ is compact.    □

**Corollary 5.14.3** *Every well-ordered set endowed with interval topology is compact iff it contains a maximal element.*

**Proof** It follows from Theorem 5.14.2.    □

Corollary 5.14.4 proves the Heine–Borel theorem saying that every closed interval in the real line space with usual topology is compact.

**Corollary 5.14.4** *A subset $X$ of the real number space $\mathbf{R}$ with usual topology is compact iff $X$ is closed and bounded.*

**Proof** The subset $X \subset \mathbf{R}$ is order-complete iff it is closed and bounded. Hence the corollary follows by using Theorem 5.14.2.    □

## 5.15   Compact Subsets of Metrizable Spaces

This section is devoted to address compact subsets of a metrizable space in Proposition 5.15.1 and disjoint compact subsets in metric spaces in Proposition 5.15.2. There are different formulations of compactness. For example, every closed and bounded subset of the real line space $\mathbf{R}$ is compact by Corollary 5.14.4. Here, the boundedness is with respect to the usual distance function

$$d: \mathbf{R} \times \mathbf{R} \to \mathbf{R}, (x, y) \mapsto |x - y|.$$

**Proposition 5.15.1** *Let $(X, \tau)$ be a metrizable space and $d$ be an arbitrary metric on $X$. If $Y \neq \emptyset$, is a compact subset of $X$, then*

  *(i)   $Y$ is closed;*
 *(ii)   there exists a pair of points $x$ and $y$ in $Y$ such that diam $Y = d(x, y)$;*
*(iii)   $Y$ is bounded.*
 *(iv)   diam $Y < \infty$;*

*Proof* Let $\tau_d$ be the topology on $X$ induced by the metric $d$. Then topological space $(X, \tau_d)$ is a Hausdorff space, because for any two distinct points $x, y \in X$, if we take a real number $\epsilon$ such that $0 < 2\epsilon < d(x, y)$, then $B_x(\epsilon) \cap B_y(\epsilon) = \emptyset$ and hence the points $x$ and $y$ are strongly separated by the open sets $B_x(\epsilon)$ and $B_y(\epsilon)$.

  (i) By hypothesis, $Y$ is compact and hence it is a compact subset of a Hausdorff space. This implies that $Y$ closed by Proposition 5.4.4.
 (ii) Let $P_X = X \times X$, $P_Y = Y \times Y$ be product spaces. Define a function

$$f: P_X \to \mathbf{R}, \ (x, y) \mapsto d(x, y).$$

Then $f$ is continuous. By hypothesis, $Y$ is a compact subset of $X$ and hence $P_Y$ is a compact subset of $P_X$ and $f(P_Y)$ is a compact subset of $\mathbf{R}$. Then $f(P_Y)$ is a closed and bounded set in $\mathbf{R}$. If $m = lub f(P_Y)$, then $m \in f(P_Y)$ and hence there exists a point $z \in P_Y$ such that $f(z) = m$. If we take $z = (x, y)$ *for some $x, y \in Y$,* then $d(x, y) = diam\ Y$.
(iii) It follows from (ii).
(iv) It also follows from (ii).

$\square$

**Proposition 5.15.2** *Let $(X, d)$ be a metric space and $Y, Z$ be two nonempty disjoint compact subsets of $X$. Then*

 *(i)   $d(Y, Z) > 0$;*
*(ii)   there exist points $a \in Y$ and $b \in Z$ such that $d(Y, Z) = d(a, b)$.*

*Proof* By hypothesis, $(X, d)$ is a metric space and $Y, Z$ are two nonempty disjoint compact subsets of $X$

(i) If possible, $d(Y, Z) = 0$. Then $m = \inf\{d(y, z) \colon \forall (y, z) \in Y \times Z\} = 0$. Since $d_{Y \times Z} \colon Y \times Z \to \mathbf{R}$ is continuous and $Y \times Z$ is compact in $X \times X$, it follows that there exists a point $(a, b) \in Y \times Z$ such that $d(a, b) = m = 0$. Then $a \in \bar{Y} = Y$ implies a contradiction. This proves (i).

(ii) From above discussion, it follows that there exist points $a \in Y$, $b \in B$ such that $d(a, b) = d(Y, Z)$.

$\square$

## 5.16 Countably Compactness and Its Characterization

This section introduces the concept of countably compactness and characterizes countably compact spaces by FIP in Theorem 5.16.3, by Cantor intersection theorem 5.16.4 and also by infinite sequences in Theorem 5.16.6.

**Definition 5.16.1** A topological space $(X, \tau)$ is said to be **countably compact** if every countable open covering $X$ has a finite subcovering.

**Remark 5.16.2** The concepts of compactness and countably compactness are different, because every compact space is countably compact. But its converse is true for spaces $(X, \tau)$ having the second axiom of countability property, i.e., if there exists a countable open base for $\tau$.

**Theorem 5.16.3** *A topological space $(X, \tau)$ is countably compact iff every countable collection of closed sets having the finite intersection property (FIP) has a nonempty intersection in $(X, \tau)$.*

**Proof** It follows from Theorem 5.9.6 by taking the indexing set to be a countable set. $\square$

Theorem 5.16.4 also characterizes countably compact spaces in terms of nested sequence of their closed sets.

**Theorem 5.16.4** *(Cantor intersection theorem for countably compactness) A topological space $(X, \tau)$ is countably compact iff every decreasing sequence of nonempty closed sets $A_1 \supset A_2 \supset \cdots$ in $(X, \tau)$ has a nonempty intersection, i.e., iff $\bigcap\{A_i \colon i = 1, 2, \ldots.\} \neq \emptyset$.*

**Proof** Let $(X, \tau)$ be a countably compact space and $A_1 \supset A_2 \supset \cdots$ be a decreasing sequence of nonempty closed sets in $(X, \tau)$. Hence $\{A_k \colon k = 1, 2, \ldots\}$ is a countable collection of closed sets in $(X, \tau)$ having the finite intersection property. This asserts by Theorem 5.16.3 that

$$\bigcap\{A_k \colon k = 1, 2, \ldots\} \neq \emptyset.$$

Conversely, let $\{B_i \colon i = 1, 2, \ldots\}$ be a countable collection of closed sets in $(X, \tau)$ having the finite intersection property and $A_i = B_1 \cap B_2 \cap \cdots \cap B_i$, $i = 1, 2, \ldots.$

Then $A_1 \supset A_2 \supset \cdots$ is a decreasing sequence of nonempty closed sets in $(X, \tau)$. This implies by hypothesis that

$$\bigcap \{A_i \colon i = 1, 2, \ldots.\} \neq \emptyset.$$

This asserts that

$$\bigcap \{A_i \colon i = 1, 2, \ldots.\} = \bigcap \{B_i \colon i = 1, 2, \ldots\} \neq \emptyset.$$

This implies by Theorem 5.16.3 that $(X, \tau)$ is countably compact.                    □

**Remark 5.16.5** Cantor's intersection theorem 5.16.4 gives a characterization of countably compactness of a topological space in terms of its nested sequence of closed sets. On the other hand, Theorem 5.16.6 also gives its another characterization of countably compactness of a topological space in terms of its cluster point. Recall that a point $x$ of a topological space $X$ is a cluster point of an infinite sequence $\{x_n\}$ in $X$, if given any open set $U$, containing the point $x$ and any positive integer $n_0$, there exists a positive integer $m \geq n_0$ such that $x_m \in U$.

Theorem 5.16.6 also characterizes countably compact spaces in terms of infinite sequences together with their cluster points.

**Theorem 5.16.6** *Let $(X, \tau)$ be a topological space. Then it is countably compact iff every infinite sequence $\{x_n\}$ in $X$ has a cluster point in $X$.*

**Proof** Let $(X, \tau)$ be a topological space. Suppose that every infinite sequence $\{x_n\}$ in $X$ has a cluster point in $X$ but $X$ is not countably compact. Then there exists a countable open covering $\mathcal{F} = \{U_i\}$ of $X$ such that it has no finite subcovering. Let $V_1 = U_1$ and $x_1 \in V_1$. For an integer $n > 1$, let $V_n$ be the first one of the open sets $U_2, U_3, U_4, \ldots$ which is not contained in $V_1 \cup V_2 \cup \cdots \cup V_{n-1}$. Suppose $x_n \in V_n - (V_1 \cup V_2 \cup \cdots \cup V_{n-1})$ and $x \in X$. Then there exists a positive integer $n_0$ such that $x \in V_{n_0}$. This implies that $x$ is not a cluster point of the infinite sequence $\{x_n\}$ in $X$, because, $V_{n_0} \cap \{x_{n_0+1}, x_{n_0+2}, \ldots\} = \emptyset$. This shows that the infinite sequence $\{x_n\}$ in $X$ has no cluster point in $X$. This contradiction implies that $(X, \tau)$ is countably compact.

Conversely, let $(X, \tau)$ be a countably compact space. Suppose the infinite sequence $\{x_n\}$ in $X$ has no cluster point in $X$. Then for each $x \in X$, there is an open set $U_x$ such that $x \in U_x$ and $U_x \cap \{x_{n_0+1}, x_{n_0+2}, \ldots\} = \emptyset$ for some positive integer $n_0$. For every positive integer $n$, let $V_n$ be the union of all open sets $U_x$ which do not contain any point of set $\{x_{n_0+1}, x_{n_0+2}, \ldots\}$. Then $\mathcal{F} = \{V_1, V_2, \ldots\}$ forms a countable open covering of $X$, such that there exists no finite subcovering of $\mathcal{F}$. This implies that the given topological space $(X, \tau)$ cannot be countably compact. This contradiction asserts that the infinite sequence $\{x_n\}$ in $X$ has a cluster point in $X$.                    □

The above discussion is assembled together in the basic and important Theorem 5.16.7 to provide different equivalent formulations of countable compact spaces.

**Theorem 5.16.7** *A topological space* $(X, \tau)$ *is countably compact iff it satisfies any one of the following conditions:*

(i) *Every countable collection of closed sets* $(X, \tau)$ *having the finite intersection property has a nonempty intersection in* $(X, \tau)$.

(ii) *Every decreasing sequence of nonempty closed sets in* $(X, \tau)$ *has a nonempty intersection.*

(iii) *Every infinite sequence in* $X$ *has a cluster point in* $X$.

**Example 5.16.8** A subspace of a countably compact space may not be countably compact. For example, in the real line space **R**, the closed interval $[0, 1]$ with subspace topology is compact by Heine–Borel theorem and hence for every countable open covering of $[0, 1]$ has also a finite subcovering. This implies that closed interval $[0, 1]$ is countably compact. But its subspace $(0, 1) \subset [0, 1]$ is not countably compact, because $\mathcal{F} = \{(\frac{1}{n}, \frac{n}{n+1}): n = 2, 3, \ldots\}$ forms a countable open covering of the open interval $(0, 1)$ but it has no finite subcovering.

## 5.17 Connectedness

This section conveys the concept of connectedness, which is a topological property different from compactness property. This concept generalizes the intuitive idea of the wholeness of a geometric figure and is used to solve some classification problems up to homeomorphism. A connected space is a topological space, which is one piece in the sense that it cannot be decomposed into two disjoint nonempty open sets. Connectedness property answers the natural question: how is the intermediate value theorem generalized in topology? The motivation of the concept of connectedness for a topological setting comes from the standard result of analysis given in Theorem 5.17.6 asserting that the real line space **R** is connected.

### 5.17.1 Three Different Types of Connectedness

Three different types of compactness such as

(i) connectedness;
(ii) local connectedness;
(iii) path connectedness

and the concept of disconnectedness are studied in this chapter.

## 5.17.2   Connectedness: Introductory Concepts

This subsection conveys the introductory concepts of connectedness and their basic properties. Intuitively, a space like **R** or the torus in which one can move from one point to any other point without jump, is considered as a single piece, called connected. This simple idea leads to many important applications of topology to geometry and analysis. For example, connectedness plays a vital role in topology which discusses continuous curves and their properties.

**Definition 5.17.1**   A topological space $(X, \tau)$ is said to be a **connected space** if whenever $X = A \cup B$, where $A \neq \emptyset$ and $B \neq \emptyset$, then $A \cap \overline{B} \neq \emptyset$ or $\overline{A} \cap B \neq \emptyset$.

***Remark 5.17.2***   Definition 5.17.1 asserts that whenever $X$ is decomposed as $X = A \cup B$, where $A \neq \emptyset$ and $B \neq \emptyset$, then $A$ and $B$ have a common point or some point of $A$ is a limit point of $B$ or some point of $B$ is a limit point of $A$.

***Example 5.17.3***   The closed interval $\mathbf{I} = [0, 1]$ decomposes as $[0, \frac{1}{3}) \cup [\frac{1}{3}, 1]$, then $\frac{1}{3} \in \overline{[0, \frac{1}{3})} \cap [\frac{1}{3}, 1]$. The subspace $[0, 1]$ is connected in the real line space **R**. Its connectedness also follows from Corollary 5.18.2.

**Definition 5.17.4**   : Let $(X, \tau)$ be a topological space. A pair $\{U, V\}$ of open sets $U$ and $V$ in $(X, \tau)$ is said to form a **separation** of $X$ by open sets if $U \cup V = X$ *and* $U \cap V = \emptyset$. This separation is said to be trivial if either $U = \emptyset$ or $V = \emptyset$, otherwise, it is said to be nontrivial.

***Remark 5.17.5***   For a topological space $(X, \tau)$, the condition for a separation of $X$ by the pair $\{U, V\}$ of open sets asserts that the sets $U$ and $V$ are both open and closed.

Theorem 5.17.6 proves the connectedness of the real line space, which is the motivating example of connectedness.

**Theorem 5.17.6**   *The real line space* $(\mathbf{R}, \sigma)$ *is connected.*

***Proof***   In the space $(\mathbf{R}, \sigma)$, let $\mathbf{R} = A \cup B$ for some disjoint nonempty subsets $A$ and $B$ of **R**. To prove the theorem, it is sufficient to show that there is a point of $B$, which is a limit point of $A$ or there is a point of $A$, which is a limit point of $B$, because it will prove that either $A$ intersects $\overline{B}$ or $B$ intersects $\overline{A}$. To show it, let $a \in A, b \in B,$ *with* $a < b$. Define the set

$$X = \{x \in A : x < b\} \subset A.$$

Since $a \in X$, the set $X \neq \emptyset$. Suppose $sup\ X = m$. Then $m \in A$ *or* $m \in \overline{A}$ (by definition of supremum). If $m \in A$, then $m < b$ and all the points $x$ such that $m < x < b$ are in $B$, since $m$ is an upper bound for $X$. This asserts that $m$ is a limit point of $B$. Again, if $m \notin A$, then $m \in B$, since $\mathbf{R} = A \cup B$. Hence, in this case, $m$ is a limit point of $A$. This implies that either $A \cap \overline{B} \neq \emptyset$ or $\overline{A} \cap B \neq \emptyset$ and hence **R** is connected.

$\square$

**Remark 5.17.7** An alternative proof of Theorem 5.17.6 is given in Theorem 5.25.10 by using the compactness property of the closed interval of the real line space **R**.

Theorem 5.17.8 presents different equivalent formulations of connectedness, and one of them may be taken as the definition of connectedness according to the convenience.

**Theorem 5.17.8** *Let* $(X, \tau)$ *be a topological space. Then following statements on* $(X, \tau)$ *are equivalent:*

(i) $(X, \tau)$ *is connected;*

(ii) *The only subsets of* $X$ *which are both open and closed in* $(X, \tau)$ *are precisely the whole set* $X$ *and the empty set* $\emptyset$;

(iii) $X$ *cannot be expressed as the union of two nonempty disjoint open sets in* $(X, \tau)$;

(iv) *There exists no continuous surjective map* $h: X \to D$ *for a discrete space* $D$ *having card* $D > 1$ *(i.e.,* $D$ *has more than one point).*

**Proof** (i) $\implies$ (ii): Let $(X, \tau)$ be connected and $Y \subset X$ be both open and closed in $(X, \tau)$. Then $Z = X - Y$ is also both open and closed in $(X, \tau)$. Hence $\overline{Y} = Y$ and $\overline{Z} = Z \implies Y \cap \overline{Z} = \overline{Y} \cap Z = Y \cap Z = \emptyset$. Since by hypothesis, $X$ is connected, one of the sets $Z$, $Y$ is $\emptyset$ and the other is $X$.

(ii) $\implies$ (iii): It is trivial

(iii) $\implies$ (iv): Suppose (iii) holds for $(X, \tau)$. Let $D$ be a discrete space having more than one element such that there exists a continuous surjective map $h: X \to D$. If $D = U \cup V$ for two disjoint nonempty open sets $U$ and $V$, then

$$X = h^{-1}(U) \cup h^{-1}(V)$$

implies a contradicts (iii). This contradiction proves that (iii) $\implies$ (iv).

(iv) $\implies$ (i): Let (iv) hold for $(X, \tau)$. Suppose $(X, \tau)$ is not connected. Then $X$ can be expressed as

$$X = A \cup B$$

for two nonempty sets $A$ and $B$ such that

$$A \cap \overline{B} = \overline{A} \cap B = \emptyset.$$

This shows that $A$ an $B$ are both open sets in $(X, \tau)$, since $A$ is the complement of the closed set $\overline{B}$ and $B$ is the complement of the closed set $\overline{A}$ in $X$. Given the discrete subspace $\{-1, 1\}$ of the real number space **R** construct a map

$$h: X \to \{-1, 1\}, x \mapsto \begin{cases} -1, & \text{for all } x \in A \\ 1, & \text{for all } x \in B. \end{cases}$$

Since $h$ is continuous and surjective, the existence of this map $h$ contradicts (iv). This contradiction proves that $(iv) \implies (i)$.

This proves that all the statements given in this theorem are equivalent. $\qquad \square$

**Remark 5.17.9** In view of Theorem 5.17.8, any one of its equivalent statements may be conveniently taken as a definition of connectedness according to the situation. Moreover, a continuous function cannot send a connected space onto a space which is not connected. This leads to characterize connected spaces with the help of discrete valued functions (see Theorem 5.17.19.) To the contrary, the continuous image of a connected space is always connected (see Corollary 5.17.24).

**Example 5.17.10** Consider the real line space **R**.

(i) All the intervals: $[a, b]$, $(a, b)$, $[a, b)$, $(a, b]$, $(-\infty, a]$, $[a, +\infty)$ of the real line space **R** are connected subsets of **R** (see Corollary 5.18.2).
(ii) The subspace **Q** of the real line space **R** is not connected.
(iii) The only connected subsets of the subspace **Q** in the real line space **R**, are the one-pointic sets.

**Definition 5.17.11** Let $X$ be a topological space and $A, B \subset X$ be subsets. Then $A$ and $B$ are said to be separated if

$$A \cap \bar{B} = \bar{A} \cap B = \emptyset.$$

**Example 5.17.12** The intervals $A = (1, 2), B = (2, 3)$ in the real line space **R** are separated. On the other hand, intervals $A = (1, 2], B = (2, 3)$ in the real line space **R** are not separated, because, $A \cap \bar{B} = \{2\}$.

**Remark 5.17.13** Theorem 5.17.8 shows that a topological space $(X, \tau)$ is connected if the only sets which are both open and closed are $\emptyset$ and $X$. Hence a subset $A \subset X$ is connected if $A$ is connected with respect to its relative topology. Thus a connected space cannot be represented as the union of two nonempty separated sets. A topological space $X$ is connected iff for $X = A \cup B$ with $A \cap B = \emptyset$, and both $A, B$ are open (closed), either $A = \emptyset$ or $B = \emptyset$.

**Theorem 5.17.14** *A topological space $(X, \tau)$ is connected iff it is not the union of two disjoint nonempty open subsets.*

**Proof** It follows from Theorem 5.17.8. $\qquad \square$

**Theorem 5.17.15** *Let $X$ be a convex subset of the Euclidean n-space $\mathbf{R}^n$ $(n \geq 1)$. Then $X$ is connected.*

**Proof** Let $X = A \cup B$ for two nonempty separated open sets $A$ and $B$. Take a point $a \in A$ and a point $b \in B$. Then the line segment $Y = [a, b]$ has the property

$$Y \cap A = A_Y \neq \emptyset, Y \cap B = B_Y \neq \emptyset \text{ and } Y = A_Y \cup B_Y = [a, b].$$

It contradicts the connectedness of the line segment $[a, b]$. $\qquad \square$

**Corollary 5.17.16** *Euclidean n-space* $\mathbf{R}^n$ ($n \geq 1$) *is connected.*

**Proof** It follows from Theorem 5.17.15. □

**Definition 5.17.17** Let $(X, \tau)$ be a topological space and $Y$ be a discrete space. Then a continuous map $f: X \to Y$ is called a **discrete-valued map**.

**Remark 5.17.18** A connected space is characterized in Theorem 5.17.19 in term of a discrete-valued map on it. This theorem is used to show in Theorem 5.17.21 that every continuous image of a connected space is also connected.

**Theorem 5.17.19** *Let* $(X, \tau)$ *be a topological space. It is connected iff every discrete-valued map on $X$ is a constant map.*

**Proof** Suppose every discrete-valued map on $X$ is constant. If $X$ is not connected, then there exist disjoint clopen sets $A$ and $B$ in $X$ such that $X = A \cup B$. Define the map

$$f: X \to \{0, 1\}, \ x \mapsto \begin{cases} 0, & \text{for all } x \in A \\ 1, & \text{for all } x \in B \end{cases}$$

Then $f$ is a discrete valued-map. But $f$ is not constant on $X$. This contradicts our hypothesis that every discrete-valued map on $X$ is constant. This contradiction implies that $X$ is connected. Conversely, suppose $X$ is connected and $f: X \to Y$ is a discrete-valued map. If $y \in Y$ is such that $y \in Im(f)$, then the set $\{f^{-1}(y)\}$ is nonempty and is clopen in $X$, and hence this set is the same as $X$. This implies that $f(x) = y$ for all $x \in X$. This asserts that $f$ is a constant map with $y$ as its only value. □

**Remark 5.17.20** Theorem 5.17.8 asserts that there exists no continuous map sending a connected space onto a topological space which is not connected. On the other hand, compactness is preserved by a continuous map by Theorem 5.17.21. It is proved in two different ways one by using Theorem 5.17.8 and the other by using Theorem 5.17.19. A continuous map cannot tear a piece into several pieces.

**Theorem 5.17.21** *Let* $(X, \tau)$ *and* $(Y, \sigma)$ *be two topological spaces and $f: X \to Y$ be a continuous onto map. If $X$ is connected, then $Y$ is also connected.*

**Proof** Let the space $X$ be connected and $f: X \to Y$ be a continuous onto map. Claim that $Y = f(X)$ is connected.

**Proof I by using Theorem** 5.17.8: Suppose $U$ is a subset of $Y = f(X)$, which is both open and closed in $Y$. Then $f^{-1}(U)$ is both open and closed in $X$ by continuity of $f$. Since by hypothesis $X$ is connected, it follows by Theorem 5.17.8 that $f^{-1}(U)$ is either $X$ itself or it is $\emptyset$. This implies that either $U = Y$ or $U = \emptyset$. This proves by Theorem 5.17.8 that the space $Y$ is connected.

**Proof II by using Theorem** 5.17.19: Let $X$ be connected, $f : X \to Y$ be a continuous map and $g: Y = f(X) \to D$ be a discrete-valued map on $Y$. Then

$$g \circ f: X \to D$$

is also a discrete-valued map on $X$. Since by hypothesis $X$ is connected, it follows from Theorem 5.17.19 that $g \circ f$ is a constant map. This asserts that the map $g$ is constant and hence $f(X) = Y$ is connected by Theorem 5.17.19.                     □

**Corollary 5.17.22**  *Let $(X, \tau)$ and $(Y, \sigma)$ be two topological spaces and $f: X \to Y$ be a continuous map. Then $f(X)$ is also connected.*

**Proof**  The corollary follows from Theorem 5.17.21.                                □

**Remark 5.17.23**  Corollary 5.17.24 asserts that connectedness of a topological space is a topological property in the sense that it is preserved by every homeomorphism. It is also product invariant in the sense that the topological product of any family of connected sets is also connected (see Exercise 58 of Sect. 5.28).

**Corollary 5.17.24**  *Let $(X, \tau)$ and $(Y, \sigma)$ be two topological spaces and $f: X \to Y$ be a homeomorphism. Then $X$ is connected iff $Y$ is connected.*

**Proof**  The corollary follows from Theorem 5.17.21.                                □

**Corollary 5.17.25**  *Let $f: X \to Y$ be a continuous map from a connected space $X$ to a space $Y$. Then the graph $G_f = \{(x, f(x))\}$ of $f$ is also connected.*

**Proof**  The corollary follows from Theorem 5.17.21.                                □

**Corollary 5.17.26**  *The circle $S^1$ is connected.*

**Proof**  Consider the continuous surjective map

$$f: [0, 1] \to S^1, t \mapsto e^{2\pi i t}.$$

Since $[0, 1]$ is connected, the corollary follows from Theorem 5.17.21.            □

Corollary 5.17.27 is another form of the Bolzano theorem of classical analysis (see Corollary 5.18.6).

**Corollary 5.17.27**  *Let $f: \mathbf{I} = [0, 1] \to \mathbf{R}$ be a continuous function such that $f(0) \cdot f(1) < 0$, then $f$ has a zero at a point $\alpha \in (0, 1)$, i.e., $f(\alpha) = 0$ for some $\alpha \in (0, 1)$.*

**Proof**  The hypothesis $f(0) \cdot f(1) < 0$ asserts that $f(0) \neq f(1)$. The corollary follows from Theorem 5.17.21 by using the connectedness of $[0, 1]$.          □

**Theorem 5.17.28**  *Let $(X, \tau)$ be a topological space and $Y \subset X$ be any subset dense in $X$. If $Y$ is connected, then $X$ is also connected.*

***Proof*** Let $B$ be a nonempty subset of $X$ such that $B$ is both open and closed in $X$. Then $B \cap Y \neq \emptyset$, since $Y$ being dense in $X$, it intersects every nonempty open subset of $X$. Clearly, $B \cap Y$ is both an open and closed subset of $Y$. Finally, since $Y$ is connected by hypothesis, it follows that $Y \cap B = Y$, and hence $Y \subset B$. This asserts that $X = \overline{Y} \subset \overline{B} = B$, which implies that $X = B$. Thus the only nonempty subset of $X$ which are both open and closed is $X$ itself. Hence $X$ is connected by Theorem 5.17.8. $\qquad\square$

**Corollary 5.17.29** *Let $(X, \tau)$ be a topological space and $B \subset X$ be connected. If $Y$ is a subset of $X$ such that $B \subset Y \subset \overline{B}$, then*

*(i) $Y$ is connected;*
*(ii) $\overline{B}$ is also connected (i.e., closure of a connected set is connected).*

***Proof*** If $Y$ is not connected and $Y = U \cup V$, where the pair $(U, V)$ forms a separation of $Y$ by open sets. Then either $B \subset U$ or $B \subset V$, because, $Y$ is connected by hypothesis. If $B \subset U$, then $\overline{B} \subset \overline{U}$. Then $Y$ cannot intersect $Y$, since $\overline{U} \cap V = \emptyset$. This contradicts the assumption that $V$ is a nonempty subset of $Y$. This contradiction proves part(i). By hypothesis, $B \subset Y \subset \overline{B}$ and hence part (ii) of the corollary follows from part (i). $\qquad\square$

**Theorem 5.17.30** *Let $(X, \tau)$ and $(Y, \sigma)$ be two topological spaces. If their product space $(X \times Y, \tau \times \sigma)$ is connected, then the spaces $(X, \tau)$ and $(Y, \sigma)$ are both connected. In general, if $\Pi_{a \in \mathbf{A}}(X_a, \tau_a)$ is the product space of a family $\{(X_a, \tau_a): a \in \mathbf{A}\}$ of connected spaces, then $(X_a, \tau_a)$ is connected for every $a \in \mathbf{A}$.*

***Proof*** Let $(X, \tau)$ and $(Y, \sigma)$ be two topological spaces and

$$p_X : X \times Y \to X, \ (x, y) \to x,$$

$$p_Y : X \times Y \to Y, \ (x, y) \to y$$

be usual projection maps. Since $p_X$ and $p_Y$ are both continuous onto maps and $X \times Y$ is connected by hypothesis, it follows that both $X$ and $Y$ are connected. For the general case, the proof is similar.

$\qquad\square$

***Remark 5.17.31*** The converse of Theorem 5.17.30 is also true. It says that if both the topological spaces $(X, \tau)$ and $(Y, \sigma)$ are connected, then the product space is also conceited. For general case see Exercise 44 of Section 5.28.

## 5.18 Connectedness Property of R and Its Subspaces

This section studies the connectedness property of the real line space $\mathbf{R}$ and its subspaces and establishes some consequences of classical intermediate value theorem, which is a classical result of analysis.

**Theorem 5.18.1** *A nonempty subset A of the real line space* **R**, *which consists of at least two distinct points. Then A is connected iff A is an interval.*

*Proof* Suppose $A$ is connected. If $A$ is not an interval, there exist points $a, b \in A$ with $a < b$ and another point $c \in \mathbf{R} - A$ such that $a < c < b$. Consider the sets

$$X = \{x \in A : x < c\} \subset A$$

and

$$Y = A - X.$$

Let $\bar{X}$ be the closure of $X$ in $A$ and $\bar{Y}$ be the closure of $Y$ in $A$. As the point $c$ is not in $A$, every point $\alpha$ of $\bar{X}$ in $A$ is such that $\alpha < c$, and every point $\beta$ of $\bar{Y}$ in $X$ is such that $\beta > c$. This implies that $X \cap \bar{Y}$ and $\bar{X} \cap Y$ are both $\emptyset$. This shows that $A$ is not connected. This contradiction asserts that $A$ is an interval. Since every interval in **R** is convex, the converse part follows by Theorem 5.17.15 for $n = 1$.

**An independent proof of the converse part**: Consider the interval $X = [a, b]$ as a subspace in the real line space **R**. Claim that $X$ is connected. If it is not so, then there exist nonempty disjoint open sets $U, V$ in **R** such that $(X \cap U) \cup (X \cap V) = X$. Consider the map

$$f : X \to \mathbf{R}, \; x \mapsto \begin{cases} 0, & \text{for all } x \in X \cap U \\ 1, & \text{for all } x \in X \cap V. \end{cases}$$

Since the inverse image of $f$ of any open set in **R** is either $X \cap U$, $X \cap V$, $\emptyset$ or $X$, each of which is an open set in $X$, it follows that $f$ is continuous. By definition of $f$, for any point $a \in X \cap U$, $f(a) = 0$ and for any point $b \in X \cap V$, $f(b) = 1$. Then by intermediate value theorem, there exists a point $c \in X$ such that $f(c) = 1/2$, which is different from 0 or 1. But this not possible. This asserts that $X$ is connected. $\square$

**Corollary 5.18.2** *All the intervals:* $[a, b], (a, b), [a, b), (a, b], (-\infty, a], [a, +\infty)$ *of the real line space* **R** *are connected subsets of* **R**.

*Proof* It follows from Theorem 5.18.1. $\square$

*Remark 5.18.3* Theorem 5.18.1 asserts that intervals of **R** are its only connected subsets, which may be open, closed, half open or it can be stretched to infinity in either direction and all other subsets of **R** have gaps and hence consist of several distinct pieces.

Intermediate value theorem is a classical theorem in calculus. It is proved in Theorem 5.18.4 by the concept of connectedness of a topological space. The specific property of the subspace $[a, b]$ of the real line space **R** on which intermediate value theorem is based is the connectedness property of $[a, b]$.

**Theorem 5.18.4  (*Intermediate value theorem*)** *Let* $f \colon X = [a, b] \to \mathbf{R}$ *be a continuous function such that* $f(a) \neq f(b)$. *Then for each real number* $r$ *between with* $f(a)$ *and* $f(b)$, *there is an element* $c \in [a, b]$ *such that* $f(c) = r$.

**Proof** Let $G_f = \{(x, f(x))\}$ be the graph of $f$. Then $G_f$ is also connected by Corollary 5.17.25. Hence the theorem follows from Theorem 5.17.21 and Corollary 5.17.27 by taking the nonempty intersection of the graph $G_f$ of $f$ in the Euclidean plane $\mathbf{R}^2$ with the straight line $y = r$ by using the connectedness of the graph $G_f$ and the choice of the number $r \in \mathbf{R}$.

**Alternative Proof**: As $X = [a, b]$ is connected and $f$ is continuous, it follows that $f(X) \subset \mathbf{R}$ is an interval by Theorem 5.18.1. Then $f(a) \neq f(b) \in f(X)$ and $r \in f(X)$. By hypothesis $r$ is a real number between $f(a)$ and $f(b)$. This asserts that there is a point $c \in X$ such $f(c) = r$.                                                                  $\square$

**Remark 5.18.5  Geometrical interpretation** of intermediate value theorem: For each real number $r$ between $f(a)$ and $f(b)$, the horizontal line $y = r$ intersects the graph $G_f = \{(x, f(x)) \colon x \in [a, b]\}$ of $f$ at some point $(c, r)$ for some $c \in [a, b]$.

Corollary 5.18.6 is a generalization of the classical Bolzano theorem used in analysis and is closely related to intermediate value theorem 5.18.4.

**Corollary 5.18.6  (*Bolzano theorem*)** *Let* $f \colon [a, b] \to \mathbf{R}$ *be a continuous function such that* $f(a)f(b) < 0$. *Then there is an element* $x \in [a, b]$ *such that* $f(x) = 0$.

**Proof** $f(a)f(b) < 0$ implies that $f(a)$ and $f(b)$ are of opposite signs and hence $f(a) \neq f(b)$. Moreover, the point 0 lies between them. Hence the Corollary follows from intermediate value theorem 5.18.4.                                                    $\square$

**Corollary 5.18.7  (*Brower fixed point theorem for dimension1*)** *Let* $\mathbf{I} = [0, 1]$ *be the closed unit interval and* $f \colon \mathbf{I} \to \mathbf{I}$ *be a continuous function. Then there exists a point* $x \in \mathbf{I}$ *such that* $f(x) = x$.

**Proof** Suppose $f(x) \neq x$, $\forall x \in \mathbf{I}$. Then

$$\mathbf{I} = \{x \in \mathbf{I} \colon f(x) < x\} \cup \{x \in \mathbf{I} \colon f(x) > x\}.$$

Hence $f(1) < 1$ and $f(0) > 0$ imply that both the sets $U = \{x \in \mathbf{I} \colon f(x) < x\}$ and $V = \{x \in \mathbf{I} \colon f(x) > x\}$ are nonempty disjoint open sets in $\mathbf{I}$. This produces a decomposition $\mathbf{I} = U \cup V$, asserting that $\mathbf{I}$ is not connected. This contradicts the fact that $\mathbf{I}$ is connected.

$\square$

**Remark 5.18.8** The proof of each of Bolzano theorem 5.18.6 and intermediate value theorem (IVT) 5.18.4 depends on the connectedness property of the image of $[a, b]$ under the continuous map $f$ in the real line space $\mathbf{R}$ and the property of its connected sets to contain each intermediate point together with the end points of $[a, b]$. These two theorems and Brower fixed point theorem for dimension 1 are closely related. More precisely, Bolzano theorem is an immediate consequence of IVT and Corollary 5.18.7, which is the Brower fixed point theorem 5.27.14 for dimension 1 directly follows from Bolzano theorem 5.18.6. Its proof is also given in Theorem 5.27.14.

## 5.19  Disconnected and Totally Disconnected Spaces

This section studies disconnectedness of topological spaces, which generalizes the concept of the negation of separateness of geometric objects and also studies totally disconnectedness. This property is topological property, because of the preservation of separateness of sets under a homeomorphism.

**Definition 5.19.1** $(X, \tau)$ be a topological space. Two nonempty subsets $A, B$ of $X$ are said to be separated if

$$A \cap \bar{B} = \bar{A} \cap B = \emptyset.$$

**Example 5.19.2** The intervals $A = (1, 2), B = (2, 3)$ in the real line space $\mathbf{R}$ are separated. On the other hand, intervals $A = (1, 2], B = (2, 3)$ in the real line space $\mathbf{R}$ are not separated, because, $A \cap \bar{B} = \{2\}$.

**Example 5.19.3** Consider two nonempty subsets $X_1$ and $X_2$ in the Euclidean plane $\mathbf{R}^2$.

$$X_1 = \{(x, y) \in \mathbf{R}^2 : y = \sin\left(\frac{1}{x}\right), \ 0 < x \leq 1\} \subset \mathbf{R}^2;$$

$$X_2 = \{(0, y) \in \mathbf{R}^2 : \frac{1}{2} \leq y \leq 1\} \subset \mathbf{R}^2.$$

Every point of $X_2$ is an accumulation point of $X_1$ asserts that $X_1$ and $X_2$ are not separated sets.

**Definition 5.19.4** Let $(X, \tau)$ be a topological space. It is said to be **disconnected** if it can be expressed as the union of two nonempty separated sets in $(X, \tau)$. Otherwise, $(X, \tau)$ is said to be connected.

**Remark 5.19.5** There is another formulation of disconnectedness of topological spaces given in Definition 5.19.6, which is equivalent to the formulation given in Definitions 5.19.4. Their equivalence is established in Theorem 5.19.9. The suitable one is used according to the nature of the problem.

**Definition 5.19.6** Let $(X, \tau)$ be a topological space. It is said to be **disconnected** if it decomposes into the union of two disjoint nonempty open sets in $(X, \tau)$.

**Remark 5.19.7** Two mutually complementary closed (open) sets in $(X, \tau)$ are simultaneously open (closed) in $(X, \tau)$.

**Example 5.19.8** The subspace $A = \{(x, y) \in \mathbf{R}^2 : y = mx\} - \{(0, 0)\}$ of the Euclidean plane $\mathbf{R}^2$ is disconnected. Because, geometrically, $A$ represents the family of straight lines deleted (minus) the origin in $\mathbf{R}^2$ and can be mathematically expressed as

$$A = \{(x, mx): x > 0\} \cup \{(x, mx): x < 0\},$$

which shows that $X$ decomposes into the union of disjoint nonempty open sets in $\mathbf{R}^2$.

**Theorem 5.19.9** *The two definitions 5.19.4 and 5.19.6 of disconnectedness of a topological space are equivalent.*

**Proof** To prove this theorem, we have to prove that Definition 5.19.4 $\Longrightarrow$ Definition 5.19.6 and Definition 5.19.6 $\Longrightarrow$ Definition 5.19.4. We prove this by using the Remark 5.19.7.

Definition 5.19.6 $\Longrightarrow$ Definition 5.19.4: Let $X$ be a disconnected space according to Definition 5.19.6. Then $X$ decomposes into the union of two nonempty disjoint open sets $A$ and $B$. Hence, $X = A \cup B$ and $A \cap B = \emptyset$. It shows that $A$ and $B$ are both closed sets in $X$. Since $\bar{A} = A$ and $\bar{B} = B$, it follows that $\bar{A} \cap B = \emptyset$ and $A \cap \bar{B} = \emptyset$. It asserts that $X$ is disconnected according to Definition 5.19.4.

Definition 5.19.4 $\Longrightarrow$ Definition 5.19.6: Let $X$ be a disconnected space according to Definition 5.19.4. Then $X$ decomposes into the union of two nonempty sets $A$ and $B$ such that $X = A \cup B$, where $\bar{A} \cap B = \emptyset$ and $A \cap \bar{B} = \emptyset$. Then $\bar{B} \subset X - A$, and $\bar{A} \subset X - B$ show that $\bar{A} = A$ and $\bar{B} = B$. Thus $A$ and $B$ are closed sets in $X$ and hence $X$ is disconnected according to Definition 5.19.6. $\qquad\square$

**Example 5.19.10** (i) The subspace $\mathbf{Q}$ of rational numbers in the real line space $\mathbf{R}$ is not connected. Because, given an arbitrary irrational number $k$, the sets

$$A = \{x \in \mathbf{Q}: x < k\}, \ B = \{x \in \mathbf{Q}: x > k\}$$

are disjoint nonempty open sets in $\mathbf{Q}$. This gives the decomposition $\mathbf{Q} = A \cup B$, which asserts that the space $\mathbf{Q}$ is not connected. For an alternative argument, if we take the open sets $A = \mathbf{Q} \cap (-\infty, \sqrt{2})$ and $B = \mathbf{Q} \cap (\sqrt{2}, \infty)$, then the decomposition $\mathbf{Q} = A \cup B$, shows that the space $\mathbf{Q}$ is not connected.

(ii) In general, a subspace $X$ of the real line space $\mathbf{R}$ is not connected, if $X$ is not an interval. Because, if $X$ is not an interval, then take three distinct real numbers $a, b, c$ such that
$$a, b \in X, \ a < c < b \text{ but } c \notin X.$$

If we take $A = X \cap (-\infty, c)$ and $B = X \cap (c, \infty)$, then $X$ is the union of two disjoint open sets $A$ and $B$.

(iii) On the other hand, every open interval $(a, b)$ in the real number space **R** is connected by Corollary 5.25.11 and hence every closed interval $[a, b]$ in the real number space **R** is also connected by Corollary 5.25.13.

**Definition 5.19.11** Let $(X, \tau)$ be a topological space. It is said to be **totally disconnected** if its only connected subspaces are one-pointic set. Equivalently, $(X, \tau)$ is totally disconnected if for every pair of distinct points $a, b \in X$, there exist open sets $A$ and $B$ such that

   (i) $a \in A$ and $b \in B$;
  (ii) $X \cap A$ and $X \cap B$ are disjoint open sets with $X = A \cup B$.
       Then $A \cup B$ is called a **disconnection** of $X$ with $a \in A$ and $b \in B$.

*Example 5.19.12* The space **Q** endowed with induced topology from the real line space **R** is a totally disconnected space. Because, for any pair of distinct rational numbers $a$ and $b$ with $a < b$, there exists an irrational number $c$ such that $a < c < b$. Then

$$A = \{x \in \mathbf{Q} : x < c\}$$

and

$$B = \{x \in \mathbf{Q} : x > c\}$$

form the required disconnection of **Q** with $a \in A$ and $b \in B$. This proves that **Q** is totally disconnected.

*Example 5.19.13* Cantor space defined in Chap. 3 with subspace topology inherited from the usual topology of **R** is a totally disconnected space.

*Example 5.19.14* Every topological space with discrete topology is totally disconnected.

## 5.20   Components of a Topological Space

This section describes the decomposition of disconnected spaces. Given an arbitrary topological space $X$, there is a natural way to decompose $X$ in connected and path connected pieces, called components by equivalence relations on $X$ described in Definition 5.20.1 and in Definition 5.22.4 respectively. The first one gives connected components that is studied in this subsection. On the other hand, the second one gives connected path components that is studied in Section 5.22. Let $(X, \tau)$ be a topological space and "$\sim$" be a binary relation on $X$ defined by $x \sim y$ iff there is a path in $X$ from $x$ to $y$. Then $\sim$ is an equivalence relation and each equivalence class is called a path component of $X$.

A topological space which is not connected also plays a key role in topology and geometry. Such a space $X$ can be uniquely expressed as a union of connected pieces, called components of $X$. So it is natural to decompose a disconnected space into connected spaces. We construct such a decomposition.

**Definition 5.20.1**  Let $(X, \tau)$ be a topological space. Given a point $x$ in $X$, the largest (maximal) connected set $C_x$ (from set theoretic viewpoint) containing the point $x$, is called the **component or connected component** of $x$ in $X$. The set $C_x$ is the union of all connected sets in $X$ containing the point $x$. The connected components of $X$ can be equally well defined by an equivalence relation on $X$: $x \sim y$ iff there is a connected subspace of $X$ containing both $x$ and $y$. The equivalence class containing $x$ denoted by $C_x$ is the connected component of $x$ in $X$.

**Remark 5.20.2**  Let $(X, \tau)$ be a topological space. Given a point $x$ in $X$, the set $C_x$ is the union of all connected sets in $X$ containing the point $x$.

Theorem 5.20.3 asserts that distinct connected components of a topological space are separated from one another.

**Theorem 5.20.3**  *Let $(X, \tau)$ be a topological space. Then it can be decomposed into the union of connected components which are closed and disjoint.*

**Proof**  As the connected components of $X$ are determined by an equivalence relation on $X$: $x \sim y$ iff there is a connected subspace of $X$ containing both $x$ and $y$ and the equivalence class $C_x$ containing $x$ is the connected component of $x$ in $X$, it follows that $X$ can be decomposed into the union of connected components which are disjoint. It can also be explained in an **alternative way**. Let $x$ and $y$ be two distinct points of $X$ with $C_x$ and $C_y$ as their connected components. As the sets $C_x$ and $C_y$ are both connected and maximal (i.e., largest from the set theoretic viewpoint) only two possibilities occur :

(i) either $C_x = C_y$
(ii) or $C_x \cap C_y = \emptyset$.

Since $C_x \cap \bar{C}_y = \emptyset$ and $C_y \cap \bar{C}_x = \emptyset$, it follows that $C_x$ is separated from $C_y$. Consequently, $X = \bigcup C_x$, where the union is performed over all $x \in X$. Again, since the closure of the connected set $C_x$ is $\overline{C_x}$ and hence it is connected by Corollary 5.17.29. By maximality of the connected set $C_x$ containing the point $x$, it follows that $\overline{C_x} \subset C_x \subset \overline{C_x}$ and hence $\overline{C_x} = C_x$ implies that $C_x$ is closed.  □

**Proposition 5.20.4**  *Let $(X, \tau)$ be topological space and $S$ be a connected subset of $X$. Then $S$ is contained in a component of $X$.*

**Proof**  Let $C_S$ be the union of the family of all connected subsets of $X$, which contain $S$. Then $C_S$ is also connected. It is also maximal by its construction. Hence it follows that $C_S$ is a component of $X$, which contains $S$.  □

**Example 5.20.5**  (i)  The subspace $X = \mathbf{R} - \{0\}$ of $\mathbf{R}$ has two components

$$C_1 = \{x \in X : x < 0\}, \; C_2 = \{x \in X : x > 0\}.$$

(ii)  The space $\mathbf{Q}$ endowed with topology induced from the real line space $\mathbf{R}$ has component $C_x = \{x\}$ for every point $x \in X$.

(iii)  The subspace $X = \mathbf{R}^{n+1} - S^n$ of the Euclidean space $\mathbf{R}^{n+1}$ has two components $C_1$ and $C_2$ defined by

$$C_1 = \{x \in X : ||x|| < 1\}, C_2 = \{x \in X : ||x|| > 1\}.$$

(iv)  The space $\mathbf{R}^1 - S^0$ has three components $C_1, C_2$ and $C_3$, where $C_1 = (-\infty, -1)$, $C_2 = (-1, 1)$ and $C_3 = (1, \infty)$.

(v)  A torus $T$ is connected and hence it has only one component.

(vi)  In a discrete space each point is a component.

(vii)  Every connected space has only one component which is the whole space itself.

***Example 5.20.6*** Let $A$ be the subspace of the Euclidean plane $\mathbf{R}^2$ defined by

$$A = \{(x, y) \in \mathbf{R}^2 : 3x + 5y + 1 = 0\}$$

For $A^* = \mathbf{R}^2 - A$ and $f(x, y) = 3x + 5y + 1$, define

$$C_1 = \{(x, y) \in A^* : f(x, y) > 0\} \subset A^*$$

and

$$C_2 = \{(x, y) \in A^* : f(x, y) < 0\} \subset A^*.$$

Then the straight line $3x + 5y + 1 = 0$ separates the Euclidean plane $\mathbf{R}^2$ into two disjoint connected components $C_1$ and $C_2$. The point $(0, 0) \in C_1$, since $f(0, 0) = 1 > 0$. On the other hand, the point $(1, -3) \in C_2$, since $f(1, -3) = -11 < 0$.

## 5.21  Local Connectedness and Its Characterization

This section studies the local connected spaces and characterize them in Proposition 5.21.5. Connectedness is an useful topological property. On the other hand, local connectedness is not a topological property (see Example 5.21.12) but at some situations, it is convenient and important to study spaces satisfying conditions analogous to connected spaces locally. This facilitates to study local connected spaces. Every connected component of a topological space is a closed set by Theorem 5.20.3. But it may not be open. For example, for the subspace $X = \{0\} \cup \{1/n : n \in \mathbf{N}\}$ of the real line space $\mathbf{R}$, the only connected component containing 0 is $\{0\}$, is not an open subset of $X$, since $\{0\}$ is not a nbd of 0 in $X$. This problem leads to the concept of local connectedness. A characterization of locally connected spaces in terms of open connected components is given in Proposition 5.21.8

**Definition 5.21.1** Let $(X, \tau)$ be a topological space, and $x$ be a point of $X$. Then $X$ is said to be locally connected at $x$ if for every open set $V$ containing x, there exists a connected open set $U$ with $x \in U \subset V$, i.e., any two points of $U$ are connected in $V$. The space $X$ is said to be **locally connected** if it is locally connected at each of its points.

**Example 5.21.2** The Euclidean space $\mathbf{R}^n$ is connected and locally connected for all $n \geq 1$ but its subspace $X = [-1, 0) \cup (0, 1]$ is not connected but it is locally connected.

**Example 5.21.3** The concepts of connectedness and locally connectedness are independent. Consider the following examples.

(i) Example of a locally connected space which is not connected: $X$ considered in Example 5.21.2 is not connected but it is locally connected.
(ii) Example of a connected space which is not locally connected: Consider the graph $G_f = \{(t, f(t)) : t \in \mathbf{I}\}$ of the function

$$f : \mathbf{I} \to \mathbf{R}, \ t \mapsto \begin{cases} 0, & \text{if } t = 0 \\ \sin \frac{\pi}{t}, & \text{if } t \in (0, 1] \end{cases}$$

with subspace topology inherited from the Euclidean topology on $\mathbf{R}^2$.
Since $f$ is continuous on $(0, 1]$ and $(0, 1]$ is connected the graph $G_1$ of the restriction $f|_{(0,1]}$ is connected. Again, $G_f - \{0, 0\} = G_1$ has its limit point $\{(0, 0)\}$ asserts $G_f$ is connected. On the other hand, $G_f$ is not locally connected at the point $(0, 0)$ implies that $G_f$ is not locally connected.
(iii) Example of a locally connected space which is not connected. Consider the topological space $X = \{0, 1\}$ endowed with discrete topology. Then this space is locally connected but it not connected.

**Example 5.21.4** (i) Any space which is locally Euclidean is locally connected.
(ii) Every discrete topological space $X$ is locally connected, because if $x \in X$, then $\{x\}$ is an open connected set containing $x$. But it is not connected if $X$ consists of more than one element.
(iii) A surface such as torus, sphere $S^2$ are locally connected.
(iv) The topological space $X = \{0\} \cup \{1/n : n = 1, 2, \ldots\}$ endowed with the subspace topology from the real line space $\mathbf{R}$ is not locally connected.

Proposition 5.21.5 gives a characterization of local connectedness at a point in terms of its open nbds. So, locally connected spaces can be restated in the following form:

**Proposition 5.21.5** *A topological space $(X, \tau)$ is locally connected at a point $x \in X$ iff every nbd of $x$ contains a connected nbd of $x$.*

***Proof*** Let $X$ satisfy the given condition and $V$ be a given nbd of $x$ in $X$. Then by hypothesis, $V$ contains a connected nbd $U$ of $x$. Consequently, any two points of $U$ are connected in $U$ and thus also in $V$. This asserts that $X$ is locally connected at $x$. Conversely, suppose that $X$ is locally connected at a point $x \in X$. Let $V$ be a given nbd of $x$ in $X$. Then by definition, there is a nbd $U \subset V$ of $x$ with the property that any two points of $U$ are connected in $V$. Hence for an arbitrary point $y \in U$, there is a connected set $C_y \subset V$ which contains both the points $x$ and $y$. If

$$C = \bigcup \{C_y : y \in U\} \subset V,$$

then $C$ is a connected nbd of $x$ contained in $V$.

$\square$

**Corollary 5.21.6** *Let $(X, \tau)$ be a locally connected space at a point $x \in X$. Then $x$ is an interior point of the component $C_x$ of $X$.*

***Proof*** It follows from Proposition 5.21.5 that the point $x \in X$ has a connected nbd $U_x$. Since $C_x$ is the largest connected set in $X$, which contains the point $x$, it asserts that $U_x \subset C_x$. This implies that $x$ is an interior point of $C_x$.    $\square$

**Proposition 5.21.7** *Let $(X, \tau)$ be a locally connected space. Then*

(i) *every subspace of $X$ is locally connected;*
(ii) *for every point $x \in X$, the component $C_x$ of $X$, is a open set in $(X, \tau)$.*

***Proof*** By definition of a locally connected space, $X$ is locally connected at every point of $X$. Hence it follows that every subspace of $X$ is locally connected. This proves the part (i) of the proposition. On the other hand, part (ii) follows from Corollary 5.21.6.    $\square$

Proposition 5.21.8 characterizes locally connectedness property of a topological space in terms of the components of its open sets.

**Proposition 5.21.8** *Let $(X, \tau)$ be a topological space. Then it is locally connected iff the component of an open set is also open.*

***Proof*** First suppose that component of every open set is also open in $X$. Let $N_x$ be a nbd of a point $x \in X$ and $U$ be a component of the open set $N_x$, containing the point $x$. Then by hypothesis, $U$ is open and hence for a nbd $N_x$, containing the $x$, there exists a connected nbd $U$ of $x$ such that $x \in U \subset N_x$. This asserts that $(X, \tau)$ is locally connected.

Conversely, suppose that $(X, \tau)$ is locally connected and $V$ is a component containing $x$ of the open set $U$ containing $x$. Then there exists a connected set $W$ such that $x \in Int(W) \subset W \subset U$. Since $x \in V \cap W$, both $V$ and $W$ are connected. it follows that $V \cup W$ is connected and is contained in $U$. Again, since $V$ is a component, it follows that $V \cup W = V$, $W \subset V$ and $x \in Int(W) \subset V$. It shows that $V$ is open. $\square$

**Theorem 5.21.9** *Let $(X, \tau)$ and $(Y, \sigma)$ be two topological spaces and $f: X \to Y$ be a continuous closed map. If $X$ is locally connected, then $f(X)$ is also so.*

**Proof** Let $(X, \tau)$ be locally connected, $U$ be an open set in $f(X)$ and $A$ be a component of $U$. We claim that $A$ is open in $f(X)$. By hypothesis, $f: X \to Y$ is continuous. It implies that $f: (X, \tau) \to (f(X), \sigma_{f(X)})$ is also continuous. Hence, for every open set $U$ in $f(X)$, its inverse image $f^{-1}(U)$ is also open in $X$. Any component $C$ of $f^{-1}(U)$ in $X$, is open, since $X$ is locally connected. Since $C$ is connected, $f(C)$ is also connected. Hence it follows that $f^{-1}(A)$ is the union of a family of components of $f^{-1}(U)$ in $X$. Since, $X$ is locally connected by hypothesis, it follows that each component of $f^{-1}(U)$ is open in $X$ and hence $X - f^{-1}(A)$ is closed in $X$. By hypothesis $f$ is a closed map and hence

$$f(X - f^{-1}(A)) = f(X) - A$$

is closed. This shows that $A$ is open in $f(X)$ and hence $f(X)$ is locally connected by Proposition 5.21.8. $\qquad\square$

**Corollary 5.21.10** *Local connectedness is preserved by every continuous closed map.*

**Proof** It follows from Theorem 5.21.9. $\qquad\square$

**Remark 5.21.11** Example 5.21.12 shows that local connectedness is not a topological property, because it is not preserved by an arbitrary continuous map (though it is preserved by every continuous closed map).

**Example 5.21.12** The continuous image of a locally connected space may not be locally connected. For example, consider the topological spaces $X = \{0, 1, 2, \ldots\}$ endowed with discrete topology and $Y = \{0\} \cup \{\frac{1}{n}: n = 1, 2, 3, \ldots\}$ with subspace topology inherited from the real line space **R**. Then $X$ is locally connected but $Y$ is not locally connected. Consider the continuous bijective map

$$f: X \to Y, \ n \mapsto \begin{cases} 0, & \text{if } n = 0 \\ \frac{1}{n}, & \text{if } n \in \mathbf{N}. \end{cases}$$

Then $X$ is locally connected but its continuous image $f(X) = Y$ is not locally connected. This example justifies the assumption $f: X \to Y$ to be a continuous closed map in Theorem 5.21.9 to make $f(X)$ locally connected.

**Remark 5.21.13** For more properties of local connectedness, see Exercises 47 and 48 of Section 5.28.

## 5.22   Path Connectedness, Path Component and Locally Path Connectedness

This section studies the concept of path connectedness, sometimes, written also as path connectedness. It is an important concept in topology. It is closely related to the concept of connectedness, since every path connected space is connected (see Corollary 5.22.11). The path components of a topological space $X$ are described in Definition 5.22.4 by an equivalence relation on $X$. This section studies locally path connected spaces. Roughly speaking, by a local path connected space, it is meant that its every point has an arbitrary small nbd such that it is path connected.

**Definition 5.22.1**  Let $(X, \tau)$ be a topological space. A continuous map

$$f : \mathbf{I} \to X$$

is said to be a path in $X$. If $f(0) = a$ and $f(1) = b$, then $f$ is said to be a path connecting the points $a, b \in X$. The point $a$ is said to be the **initial point** and the point $b$ is said to be the **terminal point** of the path $f$.

**Definition 5.22.2**  A topological space $(X, \tau)$ is said to be **path connected** if for every pair of points $a, b \in X$, there is a path connecting the points $a, b$ in $X$, i.e., if there exists a continuous map $f : \mathbf{I} \to X$ such that $f(0) = a$, $f(1) = b$ and $f(\mathbf{I}) \subset X$. A path connected space is sometimes written as a path connected space. It is said to be **locally path connected** at a point $x \in X$, if every open nbd $U$ of $x$, there is a path connected nbd $V$ of $x$ such that such that $V \subset U$. The space $X$ is said to be locally path connected if it is locally path connected at every point $x \in X$.

*Example 5.22.3*  Every convex set in $\mathbf{R}^n$ is path connected.

**Definition 5.22.4**  Let $(X, \tau)$ be a topological space and "$\sim$" be a binary relation on $X$ defined by $x \sim y$ iff there is a path in $X$ from $x$ to $y$. Then $\sim$ is an equivalence relation and each equivalence class is called a **path component** of $X$.

*Remark 5.22.5*  Let $(X, \tau)$ be a topological space. It follows from Definition 5.22.4 that the path components of $X$ are path connected disjoint subspaces of $(X, \tau)$ such that

 (i)   their union is $X$ and
 (ii)  they satisfy the property: every nonempty path connected subspace of $(X, \tau)$ meets only of them.

*Example 5.22.6*  In the real line space $\mathbf{R}$, each path component of its subspace $\mathbf{Q}$ consists of a single rational point and no path component of $\mathbf{Q}$ is open in the subspace $\mathbf{Q}$.

*Example 5.22.7*  (**Comb space**) Consider the subspace **Comb** of the Euclidean plane $\mathbf{R}^2$

**Fig. 5.1** Comb space

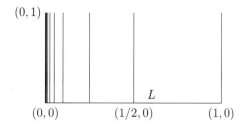

$$\textbf{Comb} = \{(x, y) \in \mathbf{R}^2 : 0 \le y \le 1, \ x = 0, \ 1/n,$$
$$\forall n \in \mathbf{N} \ or \ y = 0, \ 0 \le x \le 1, \} \subset \mathbf{R}^2.$$

The set **Comb** endowed with topology inherited from the usual topology of $\mathbf{R}^2$ is called the comb space. Every small nbd of the point $(0, 1)$ has infinite number of path components in the comb space **Comb**. Geometrically, the comb space **Comb** consists of the horizontal closed unit line segment lying on the $x$-axis, joining the point $(0, 0)$ to the point $(1, 0)$ together with the the vertical closed unit line segment standing on each of the points $(1/n, 0)$ for every $n \in \mathbf{N}$ as shown in Diagram in Fig. 5.1.

**Remark 5.22.8** The more properties of Comb space are studied in **Basic Topology, Volume 3** of the present book series.

**Example 5.22.9** (**Topologist's sine curve**): Let $X = \{(x, y) \in \mathbf{R}^2 : y = \sin(\frac{1}{x}), x > 0\} \cup \{(0, y) \in \mathbf{R}^2 : |y| \le 1\} \subset \mathbf{R}^2$. Then the space $X$ endowed with subspace topology inherited from Euclidean topology $\mathbf{R}^2$ is called topologist's sine curve (see Fig. 5.2). Let $A = \{(x, y) \in \mathbf{R}^2 : y = \sin(\frac{1}{x}), x > 0\}$ and $B = \{(0, y) \in \mathbf{R}^2 : |y| \le 1\}$. Then $X = A \cup B$. Clearly, $A$ is the image of the continuous map

$$f : (0, 1) \to \mathbf{R}^2 : x \mapsto \left(x, \sin\left(\frac{1}{x}\right)\right)$$

Hence $A$ is connected. But $X$ has two path components, one is $A$ and the other one is the vertical interval given by $B$.

**Theorem 5.22.10** *Let $(X, \tau)$ be a topological space such that every pair of points can be joined by some connected subset of $X$. Then $X$ is connected.*

**Proof** If $X$ is not connected, then there exist two disjoint open sets $U$ and $V$ such that $X = U \cup V$. Let $a \in U$ and $b \in V$ be two points and $W \subset X$ be a connected set containing the points $a$ and $b$. Then $U_1 = U \cap W$ and $V_1 = V \cap W$ are two disjoint nonempty open sets in $W$ such that $W = U_1 \cap V_1$. But it contradicts the connectedness of $W$. □

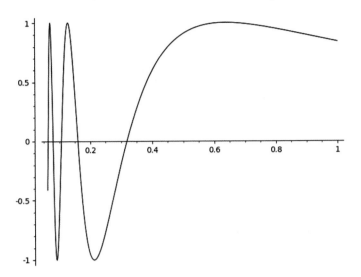

**Fig. 5.2** Topologist's sine curve

**Corollary 5.22.11** *Every path connected space is connected.*

*Proof* **Proof I** It follows from Theorem 5.22.10.

**Proof II**(*An independent proof*): Let $(X, \tau)$ be a path connected space. If it is not connected, then $X$ can be expressed as $X = A \cup B$, where $A$ and $B$ are nonempty disjoint open sets in $(X, \tau)$. Hence, there exist two distinct elements $a \in A, b \in B$. If $f : \mathbf{I} \to X$ is a path in $X$ with $f(0) = a$ and $f(1) = b$, then $f^{-1}(A)$ and $f^{-1}(B)$ are nonempty disjoint open subsets in $\mathbf{I}$, such that $\mathbf{I} = f^{-1}(A) \cup f^{-1}(B)$. But it contradicts the connectedness of $\mathbf{I}$.                                                      □

*Example 5.22.12* An immediate application of Corollary 5.22.11 in matrix algebra is now given. Let $M(n, \mathbf{R})$ be the set of all $n \times n$ matrices over $\mathbf{R}$. It is considered as a subspace of the Euclidean $\mathbf{n}^2$-space $\mathbf{R}^{n^2}$. If

$$X = \{M \in M(n, \mathbf{R}) : x^t M x \geq 0, \ \forall x \in \mathbf{R}^n\},$$

then $X$ is path connected, since for any two matrices $M, N \in X$, and any $\lambda \in [0, 1]$

$$x^t(\lambda M + (1 - \lambda)N)x \geq 0$$

asserts that $X$ is path connected and hence it is also connected by Corollary 5.22.11.

Converse of Corollary 5.22.11 is not true. In support consider the Example 5.22.13.

***Example 5.22.13*** A connected space may not be path connected.

(i) Let $[A_n, B_n]$ denote the line segment joining the points $A_n = (1/n, 0)$ and $B_n = (1/n, 1)$ in $\mathbf{R}^2$. If $[P, Q]$ represents the line segment joining the points $P = (0, 0)$ and $Q = (1, 0)$ in $\mathbf{R}^2$, then the subspace

$$X = [P, Q] \cup (0, 1) \bigcup_{n=1}^{\infty} [A_n, B_n] \subset \mathbf{R}^2$$

is connected. But it is not path connected, since the point $(0, 1)$ cannot be connected by a path to any other point in $X$.

(ii) Let $X_1$ and $X_2$ be two nonempty subsets of the Euclidean plane $\mathbf{R}^2$ defined by

$$X_1 = \{(x, y) \in \mathbf{R}^2 \colon 0 \leq x \leq 1, \ y = \frac{x}{n}, \ n \in \mathbf{N}\} \subset \mathbf{R}^2;$$

$$X_2 = \{(x, 0) \in \mathbf{R}^2 \colon \frac{1}{2} \leq x \leq 1\} \subset \mathbf{R}^2.$$

Then $X_1$ and $X_2$ are both path connected subsets of $\mathbf{R}^2$ and hence each of them is connected by Corollary 5.22.11. On the other hand, since every point of $X_2$ is a limit point of $X_1$, it follows that $X_1$ and $X_2$ are not separated and hence their union $X_1 \cup X_2$ is connected. As there exists no path from any point of $X_1$ to a point of $X_2$, $X_1 \cup X_2$ is not path connected. This example shows that a connected space may not be path connected.

***Example 5.22.14*** Consider the subspace

$$X = \bigcup_{n=1}^{\infty} \{(x, y) \in \mathbf{R}^2 : x = ny, n \in \mathbf{N}\}$$

of the Euclidean plane $\mathbf{R}^2$. To show that the subspace $X$ is path connected it is sufficient to show that any two points of $X$ can be joined by a path in $X$. Let $\alpha = (ny, y)$ and $\beta = (my', y')$ be two arbitrary points in $X$. Define the map

$$f \colon \mathbf{I} = [0, 1] \to X \colon t \mapsto \begin{cases} (1 - 2t)(ny, y), & \text{for all } t \in [0, \frac{1}{2}] \\ (2t - 1)(my', y'), & \text{for all } t \in [\frac{1}{2}, 1] \end{cases}$$

Then $f$ is well defined and is continuous by pasting lemma. Hence $f$ is a path in $X$. Moreover,

$$f(0) = \alpha, \ f(1) = \beta \implies X \text{ is path connected}$$

and hence $X$ is connected by Corollary 5.22.11.

**Definition 5.22.15** A topological space $X$ is said to be locally path connected if every nbd $N_x$ of $x \in X$, contains a path connected nbd.

**Remark 5.22.16** In a locally connected topological space the components are both open and closed. Converse of Corollary 5.22.11 is not true, i.e., a connected space may not be path connected see Example 5.22.13. But Proposition 5.22.17 and Corollary 5.22.11 taken together assert that under certain specified conditions the concepts of connectedness and path connectedness coincide.

**Proposition 5.22.17** *Every connected, locally path connected space is path connected.*

**Proof** Let $X$ be a connected, locally path connected space with base point $x_0$. Then the subset $A = \{x \in X : x, x_0$ can be joined by a path in $X \}$ is an open set in $X$. But if $x$ is not in $A$, then $x$ has a path connected nbd $N_x$ with the property that no point of $N_x$ is in $A$. This asserts that $X - A$ is an open set. Again since $X$ is path connected by hypothesis and $A$ is nonempty, the only possibility is that $A = X$. This implies that $X$ is path connected. $\qquad\square$

**Remark 5.22.18** For more study of path connectedness and local path connectedness see Sects. 5.25 and 5.28.

## 5.23  Space-Filling Curve Theorem

This section proves the existence and makes a construction of a continuous function, called space-filling curve, defined on a closed interval of the real line space whose image is a two-dimensional region in the Euclidean plane $\mathbf{R}^2$. There are several versions of space-filling curve theorem. The common version is that there is a surjective continuous map $f : \mathbf{I} \to \mathbf{I}^2$. But for convenience of its proof, we first prove its version given in Theorem 5.23.4 from which the common version follows in Corollary 5.23.5, since a triangle and a square in the Euclidean plane are topologically equivalent.

**Historically**, G. Cantor (1845–1918) proved in 1877 that the interval $\mathbf{I} = [0, 1]$ and the square $\mathbf{I}^2$ have the same number of points by establishing a bijective correspondence between the sets $\mathbf{I}$ and $\mathbf{I}^2$, i.e., $\mathbf{I} \sim \mathbf{I}^2$. He remarked that the **dimension is not a set-theoretic concept**. The difference of their dimensions involves topology. G. Peano (1858–1932) showed again in 1890 that there is a continuous function on $\mathbf{I}$ such that its image is a two-dimensional region, such as a square or a triangle in the Euclidean plane, called a **space-filling curve or Peano curve** named after G. Peano.

**Example 5.23.1** Let $X, Y$ be two topological spaces defined by

$$X = \mathbf{I} = [\mathbf{0}, \mathbf{1}] \text{ and } \mathbf{Y} = \mathbf{I} \times \mathbf{I}.$$

Then a continuous map $f: X \to Y$ geometrically represents a path in the square $Y$. There exist continuous maps $f: X \to Y$ whose image covers the whole square $Y$ by "Space-filling Curve Theorem" (see Corollary 5.23.5). No such map $f$ can be injective, otherwise, $f$ would be homeomorphism by Theorem 5.4.5. But it is not possible, since deletion of a point $p$ from $X$ makes the space $X - \{p\}$ disconnected but it not so for the space $Y - \{f(p)\}$.

**Remark 5.23.2 Analogue of Schroeder–Bernstein theorem for sets is not valid for topological spaces.** This theorem for sets asserts that if $X$ and $Y$ are two sets, such that $X \sim Y_1 \subset Y$ and $Y \sim X_1 \subset X$, then $X \sim Y$. But its analogue result is not valid for topological spaces in the sense that there exist continuous surjective maps $f: X \to Y$ having its inverse map $g: Y \to X$ which is continuous but it not injective showing that such spaces $X$ and $Y$ may not be homeomorphic see Example 5.23.1.

**Remark 5.23.3** Since a triangle and a square in the Euclidean plane are topologically equivalent, Theorem 5.23.4 proves one version of space-filling curve for a triangle, from which the common version for a square follows in Corollary 5.23.5.

**Theorem 5.23.4** *There is a surjective continuous map $f: \mathbf{I} \to \triangle$, where $\triangle$ is a triangle in the Euclidean plane $\mathbf{R}^2$.*

**Proof** Without any loss of generality, we assume that $\triangle$ is an equilateral triangle in the Euclidean plane $\mathbf{R}^2$ having each side is of length of one half unit. Construct a sequence of continuous functions $\{f_n\}$, where $f_n: \mathbf{I} \to \triangle: n = 1, 2, 3, \ldots$ is defined as follows: The first three functions $f_1, f_2, f_3$ are defined as depicted in the Fig. 5.3.

More precisely, the curve $f_1(I)$ joins two vertices of the $\triangle$ by a broken line passing through its center of gravity in the sense that the broken point is the center of gravity of the $\triangle$, as shown in Fig. 5.3(a). To construct $f_2$, the triangle is subdivided into four smaller congruent triangles with the curve $f_2(I)$ as shown in Fig. 5.3 (b)). As $n$ increases, the curve $f_n$ will go on increasing filling up more and more points of the $\triangle$. Given two positive integers $m, n$ with $n \geq m$, and a point $t \in \mathbf{I}$, a triangle is obtained containing both the points $f_n(t)$ and $f_m(t)$ in this triangle, whose sides have length $1/2^m$ units. Hence

$$||f_n(t) - f_m(t)|| \leq 1/2^m, \ \forall \, t \in \mathbf{I},$$

where $||f_n(t) - f_m(t)||$ denotes Euclidean distance between the points $f_n(t)$ and $f_m(t)$ in the Euclidean plane $\mathbf{R}^2$. This asserts that the constructed sequence $\{f_n\}$ is uniformly convergent and the limit function $f: \mathbf{I} \to \triangle$ is continuous, since every function $f_n$ is continuous. To show that $f(\mathbf{I}) = \triangle$, we use the fact that every image point of $f_n$ lies within the distance $1/2^n$ from every point of the $\triangle$ for all $n$. Let $U$ be a nbd of a point $x \in \triangle$ in $\mathbf{R}^2$ and $n_0$ be an integer large enough so that the open ball $B_x(1/2^{n_0-1})$ falls inside the nbd $U$. Select a point $t_0 \in \mathbf{I}$ such that $||x - f_{n_0}(t_0)|| \leq 1/2^{n_0}$. Since $||f_{n_0}(t) - f(t)|| \leq 1/2^{n_0}, \ \forall \, t \in \mathbf{I}$, it follows from triangle inequality of the usual metric on the Euclidean plane $\mathbf{R}^2$ that $||x - f(t_0)|| \leq 1/2^{n_0-1}$. This shows that each point of the triangle $\triangle$ is a limit of the image set $f(\mathbf{I})$. Again, since, $I$ is

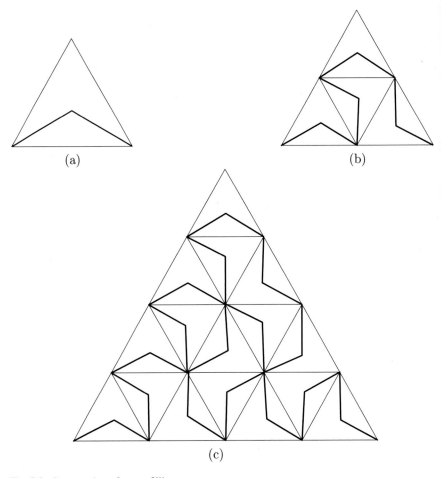

**Fig. 5.3** Construction of space-filling curve

closed in the real line space **R** its continuous image $f(I)$ under the continuous map $f$ is also a closed subset of the Euclidean plane $\mathbf{R}^2$ and hence it contains all its limit points. This asserts that the image of this function $f$ is the whole set $\triangle$.                    □

**Corollary 5.23.5**  (*Space-Filling Curve Theorem*) *There is a surjective continuous map $\psi : \mathbf{I} \to \mathbf{I}^2$.*

***Proof*** Since there exists a homeomorphism $h: \triangle \to \mathbf{I}^2$. it follows from Theorem 5.23.4 that the composite map

$$\psi = h \circ f : \mathbf{I} \to \mathbf{I}^2$$

is a surjective continuous map.                                                              □

**Corollary 5.23.6**  *If there exists a continuous onto map $f: \mathbf{I} \to \mathbf{I}^2$, then $f$ cannot be a homeomorphism.*

***Proof***  This existence of a continuous onto map $f: \mathbf{I} \to \mathbf{I}^2$ is guaranteed by space-filling curve theorem 5.23.5. If $f$ is injective, then $f: \mathbf{I} \to \mathbf{I}^2$ is a continuous bijective map and hence $f$ is a homeomorphism by Theorem 5.4.5. But is not possible, since deletion of any point from $\mathbf{I}^2$. makes the remaining space disconnected (such a point is called a cut point in $\mathbf{I}$ but but no such cut point exists in $\mathbf{I}^2$, because deletion of a point from $\mathbf{R}^2$ keeps the remaining space connected. This asserts that the space $\mathbf{I}$ and $\mathbf{I}^2$ cannot be homeomorphic.                                          $\square$

**Corollary 5.23.7**  *There exists no homeomorphism*

$$f: \mathbf{I} \to \mathbf{I}^2.$$

***Proof***  It follows from Corollary 5.23.6.                                  $\square$

## 5.24   Function Spaces

This section continues the study of topologies generated by functions initiated in Chap. 3 by defining different topologies on the set $\mathcal{F}(X, Y)$ of all set functions $f: X \to Y$ for sets $X$ and $Y$ with specializing the sets $X$ and $Y$. More precisely, various topologies such as compact open topology, uniform convergence topology, point open topology and pointwise convergence topology are defined on $\mathcal{F}(X, Y)$ making $\mathcal{F}(X, Y)$ function spaces. Such spaces play an important role in topology and geometry. For its key role in topology, see **Basic Topology, Volume 3** of the present series of books. The concept of uniform convergence topology on every compact set was born through the study of classical theory of functions. For the study of function spaces $\mathcal{F}(X, Y)$, it becomes necessary to convey the following basic concepts.

**Definition 5.24.1**  Let $X$ and $Y$ be nonempty arbitrary sets. The set $\mathcal{F}(X, Y)$ is endowed with a topology generated by its suitable subcollection as a subbase for a topology $\tau$. Then the topological space $(\mathcal{F}(X, Y), \tau)$ **is called a function space.**

**Definition 5.24.2**  Let $X$ and $Y$ be two nonempty arbitrary sets. Given a nonempty set $Z$, let

$$\psi: Z \times X \to Y$$

be a function of two variables. Keeping the first variable fixed at $z \in Z$, we obtain a function

$$\psi_z: X \to Y, x \mapsto \psi(z, x),$$

which gives a one-parameter family of functions. This defines a function

$$g: Z \to \mathcal{F}(X, Y), z \mapsto \psi_z,$$

called the **associated function of** $\psi$.

**Definition 5.24.3** Let $X$, $Y$ and $Z$ be three nonempty arbitrary sets and

$$g: Z \to \mathcal{F}(X, Y), z \mapsto \psi_z$$

be given in 5.24.2. Then the map

$$h: \mathcal{F}(Z \times X, Y)) \to \mathcal{F}(Z, \mathcal{F}(X, Y)), \ \psi \mapsto g$$

is a bijection, called the **exponential law for sets**.

***Remark 5.24.4*** As the map $h$ given in Definition 5.24.3 is bijective, it follows that $g$ can be recovered from $\psi$ and conversely, where $g$ and $\psi$ are given in Definition 5.24.2. This is called exponential law for sets. Because, using the notation, $Y^X$ for $\mathcal{F}(X, Y)$, **Definition 5.24.3 asserts that** $Y^{Z \times X} = (Y^X)^Z$ (which is equal up to equipotency by a bijection).

**Definition 5.24.5** Given two nonempty sets $X$ and $Y$. The map

$$e: \mathcal{F}(X, Y) \times X \to Y, (f, x) \mapsto f(x)$$

is called the **evaluation map**.

### 5.24.1   Compact Open Topology

This subsection studies compact open topology, which arises in a natural way. If $X$ is a metric space, then the compact open topology on $\mathcal{F}(X, Y)$ for any topological space $Y$ coincides with the topology of uniform convergence (see Definition 5.24.11) on compact subsets (see Exercise 76 of Section 5.28).

**Definition 5.24.6** Let $X$ and $Y$ be topological spaces and $C(X, Y)$ be the set of all continuous functions from $X$ to $Y$. Then a topology $\tau$, called **compact open topology**, is defined on $C(X, Y)$ by taking a subbase for the topology $\tau$ consisting of all subsets of the form

$$V_{K,U} = \{f \in C(X, Y): f(K) \subset U\},$$

where $K$ is a compact subset of $X$ and $U$ is an open subset of $Y$. An open set in the compact open topology $\tau$ on $C(X, Y)$ is an arbitrary union of finite intersections of sets of the form $V_{K,U}$.

Theorem 5.24.7 proves the continuity of the evaluation map $e$ and characterizes the continuity of the function $\psi: Z \times X \to Y$ by its associated function $g$ described in Definition 5.24.2 and also provides the key results to prove **Theorem of Exponential Correspondence** and **Exponential Law** formulated in Theorem 5.24.9.

**Theorem 5.24.7** *Let $Y, Z$ be two arbitrary topological spaces and $X$ be a locally compact Hausdorff space. If $C(X, Y)$ is endowed with compact open topology $\tau$, then (using the notations given in Definition 5.24.2)*

(i) *the evaluation map (function) $e: C(X, Y) \times X \to Y$, $(f, x) \mapsto f(x)$ is continuous;*

(ii) *a map (function) $\psi: Z \times X \to Y$ is continuous iff its associated map (function)*

$$g: Z \to C(X, Y), z \mapsto \psi_z$$

*is continuous, where*

$$\psi_z: X \to Y, \ x \mapsto \psi(z, x).$$

**Proof**    (i) Given an element $(f, x) \in C(X, Y) \times X$, let $U$ be a nbd of $f(x)$ in $Y$. Then $e(f, x) = f(x) \in U$ and hence $x \in f^{-1}(U)$, which is an open set in $X$ by continuity of $f$. As $X$ is a locally compact Hausdorff space by hypothesis, it follows from Definition 5.6.8 that there exists an open set $W$ in $X$ with $\overline{W}$ compact such that

$$x \in W \subset \overline{W} \subset f^{-1}(U).$$

This asserts that $V_{\overline{W}, U} = B$ forms a subbasic open set for the topology $\tau$ on $C(X, Y)$, which contains $f$. Consider the open nbd $B \times W$ of $(f, x)$ in $C(X, Y) \times X$. If $(f', x') \in C(X, Y) \times X$, then

$$e(f', x') = f(x') \in f'(W) \subset f'(\overline{W}) \subset U.$$

This implies that $e$ maps $B \times W$ into $U$. This asserts that $e^{-1}(U)$ is open in $C(X, Y) \times X$ and hence the evaluation map $e$ is continuous.

(ii) First suppose that $\psi$ is continuous. Then given $z \in Z$, the map $f_z: X \to Z \times X, x \to (z, x)$ is continuous and $\psi_z = \psi \circ f_z$. It shows that each $\psi_z$ is continuous. Moreover, if $z \in Z$ and $V_{K,U}$ is any subbasic open nbd of $g(z) = \psi_z$ in $C(X, Y)$, then there exists an open nbd $W$ of $z$ such that $g(W) \subset V_{K,U}$. This asserts that $g$ is continuous. Conversely, assume that $g: Z \to C(X, Y)$ is continuous. Then $\psi$ is the composite map

$$Z \times X \xrightarrow{\ g \times 1_d\ } C(X, Y) \times X \xrightarrow{\ e\ } Y,$$

i.e., $\psi = e \circ (g \times 1_d): Z \times X \to Y$ is a function. Since $e$ and $g$ are continuous, it follows that $\psi$ is continuous.

$\square$

**Corollary 5.24.8** *Let $Y, Z$ be two arbitrary topological spaces and $X$ be a locally compact Hausdorff space. If $C(X, Y)$ is endowed with compact open topology, then a map $k: Z \to C(X, Y)$ is continuous iff the composite map*

$$Z \times X \xrightarrow{\ k \times 1_d\ } C(X, Y) \times X \xrightarrow{\ e\ } Y$$

*is continuous.*

**Proof** It follows from the second part of Theorem 5.24.7 by taking $\psi = e \circ (k \times 1_d): Z \times X \to Y$. Because $k$ is then the associated map of $\psi$.                    □

We summarize the above discussion in the basic and important theorems.

**Theorem 5.24.9** *(Theorem of exponential correspondence) Let $X$ be a locally compact Hausdorff space and $Y, Z$ be arbitrary topological spaces. Then a function $g: Z \to C(X, Y)$, $z \mapsto \psi_z$ is continuous, iff the composite function (map)*

$$\psi: Z \times X \xrightarrow{\ g \times 1_d\ } C(X, Y) \times X \xrightarrow{\ e\ } Y$$

*i.e., $\psi = e \circ (g \times 1_d): Z \times X \to Y$ is continuous, where $\psi_z: X \to Z$, $x \mapsto \psi(z, x)$.*

**Theorem 5.24.10** *(Exponential Law) Let $X$ be a locally compact Hausdorff space and $Z$ be a Hausdorff space. If $Y$ is a topological space, then the function (map)*

$$h: C(Z, C(X, Y)) \to C(Z \times X, Y), \ g \mapsto \psi = e \circ (g \times 1_d)$$

*is a homeomorphism.*

### 5.24.2  Uniform Convergence Topology

This subsection studies compact open topology, which arises in a natural way. For example, if $Y$ is a metric space, then for any topological space $X$, the topology defined by uniform convergence on compact subsets coincides with the compact open topology (see Definition 5.24.6) on $C(X, Y)$.

**Definition 5.24.11** Let $(X, \tau)$ be a compact space and $(Y, d)$ be a metric space. Then the map

$$\psi: C(X, Y) \times C(X, Y) \to \mathbf{R}, \ (f, g) \mapsto \sup_{x \in X}\{d(f(x), g(x))\}$$

defines a metric on $C(X, Y)$ and the corresponding topology $\tau_\psi$ on $C(X, Y)$ induced by the metric $\psi$ is called the **topology of uniform convergence**.

**Remark 5.24.12** The definition of topology of uniform convergence on the collection $\mathcal{B}(X, Y)$ of all bounded functions from an arbitrary set $X$ to a metric space $(Y, d)$ is similar.

**Theorem 5.24.13** *Let $(X, \tau)$ be a compact space, $(Y, d)$ be a metric space and $\mathcal{C}(X, Y)$ be endowed with the topology $\tau_\psi$ of uniform convergence given in Definition 5.24.11. Then the following statements are equivalent in the topological space $(\mathcal{C}(X, Y), \tau_\psi)$.*

(i) *The sequence $\{f_n\}$ of continuous functions in $\mathcal{C}(X, Y)$ converges to $h \in \mathcal{C}(X, Y)$ with respect to the distance function $\psi$ given in Definition 5.24.11 is continuous.*

(ii) *The sequence $\{f_n\}$ of functions converges to $h$ in the usual sense.*

**Proof** (i) $\implies$ (ii): Suppose (i) holds, i.e., the sequence $\{f_n\}$ of functions in $\mathcal{C}(X, Y)$ converges to $h \in \mathcal{C}(X, Y)$. Then given an $\epsilon > 0$, there exists a positive integer $n_0$ such that $\psi(f_n, h) < \epsilon$, whenever $n \geq n_0$. Hence

$$\forall n \geq n_0, \ d(f_n(x), h(x)) \leq \sup_{x \in X}\{d(f_n(x), h(x))\} = \psi(f_n, h) < \epsilon, \ \forall x \in X.$$

This shows that the sequence $\{f_n\}$ uniformly converges to $h$, showing that (i) $\implies$ (ii). Clearly, $h$ is continuous.

(ii) $\implies$ (i) Suppose that (ii) holds. Then given an $\epsilon > 0$, there exists a positive integer $n_0$ such that

$$\forall n \geq n_0, d(f_n(x), h(x)) < \epsilon/2, \ \forall x \in X.$$

This asserts that

$$\forall n \geq n_0, \sup_{x \in X}\{d(f_n(x), h(x))\} \leq \epsilon/2 < \epsilon \implies \psi(f_n, h)) < \epsilon.$$

It shows that the sequence $\{f_n\}$ converges to $h$ and hence (ii) $\implies$ (i). $\quad\square$

**Theorem 5.24.14** *Let $\mathcal{B}(X, Y)$ of all bounded functions from an arbitrary set $X$ to a metric space $(Y, d)$ and $\mathcal{B}(X, Y)$ be endowed with the topology $\tau_\psi$ of uniform convergence. Then the following statements are equivalent in the topological space $(\mathcal{B}(X, Y), \tau_\psi)$.*

(i) *The sequence $\{f_n\}$ of functions in $\mathcal{B}(X, Y)$ converges to $h \in \mathcal{B}(X, Y)$ with respect to the distance function $\psi$ given in Definition 5.24.11.*

(ii) *The sequence $\{f_n\}$ of functions converges to $h$ in the usual sense.*

**Proof** It is similar to the proof of the Theorem 5.24.13. $\quad\square$

### 5.24.3  Point Open Topology

This subsection conveys the concept of point open topology to study the function spaces $F(X, Y)$. by specifying the sets $X$ and $Y$.

**Definition 5.24.15** Let $X$ be an arbitrary set and $Y$ be a topological space. Define a subbase **B** of the product topology on $\mathcal{F} = \Pi\{Y_x : x \in X\}$, where **B** consists of all subsets of $\mathcal{F}(X, Y)$ of the form $\{f \in \mathcal{F}(X, Y) : f(x_0) \in U\}$, i.e., all functions $f \in \mathcal{F}(X, Y)$ such that $f$ maps an arbitrary point $x_0 \in X$ to an arbitrary open set $U$ of $Y$. We call this product topology $\tau_p$ generated by **B**, the **point open topology on** $\mathcal{F}(X, Y)$.

*Example 5.24.16* For $\mathcal{F}(\mathbf{I},\ \mathbf{R})$, the members of the defining subbase **B** for the topology $\tau_p$ are of the form

$$\{f \in \mathcal{F}(\mathbf{I},\ \mathbf{R}) : f(i_0) \in U, i_0 \in \mathbf{I} \text{ and } U \text{ is an open subset of } \mathbf{R}\}.$$

**Geometrically,** the subbase **B** for this topology $\tau_p$ consists of elements, which are functions having graphs passing through the open set $U$ on the vertical line $\mathbf{R}$ standing at the point $i_0$ on the horizontal $x$-axis.

**Theorem 5.24.17** *Let $\tau$ be the point open topology on $\mathcal{C}(X, Y)$. Then the following statements are equivalent in the topological space $(\mathcal{C}(X, Y), \tau)$.*

(i)  *The sequence $\{f_n\}$ of continuous functions in $\mathcal{C}(X, Y)$ converges to $h \in \mathcal{C}(X, Y)$ with respect to the topology $\tau$.*

(ii)  *The sequence $\{f_n\}$ of functions converges pointwise to $h$.*

**Proof** Proceed as in proof of the Theorem 5.24.13.                                              $\square$

**Theorem 5.24.18** *Let $\mathcal{B}(X, Y)$ of all bounded functions from an arbitrary set $X$ to a metric space $(Y, d)$ and $\mathcal{B}(X, Y)$ be endowed with the topology $\tau_\psi$ of uniform convergence. Then the following statements are equivalent in the topological space $(\mathcal{B}(X, Y), \tau_\psi)$.*

(i)  *The sequence $\{f_n\}$ of functions in $\mathcal{B}(X, Y)$ converges to $h \in \mathcal{B}(X, Y)$ with respect to the distance function $\psi$ given in Definition 5.24.11.*

(ii)  *The sequence $\{f_n\}$ of functions converges to $h$ in the usual sense.*

**Proof** It is similar to the proof of the Theorem 5.24.13.                                         $\square$

*Remark 5.24.19* For more study on point open topology, see Exercise 101 of Section 5.28.

### 5.24.4 Pointwise Convergence Topology

This subsection conveys the concept of pointwise convergence topology to study the mapping spaces $M(X, Y)$ by specifying the sets $X$ and $Y$. Let $\{f_n\}$ be a sequence of functions $f_n: X \to Y$ from an arbitrary set $X$ to a topological space $Y$. Then this sequence $\{f_n\}$ is said to converge pointwise to a function $f: X \to Y$, if for every $x_0 \in X$, the sequence $\{f_n(x_0)\}$ converges to $h(x_0)$, i.e., $lim_{n \to \infty} f_n(x_0) = h(x_0)$.

**Definition 5.24.20** Let $X$ be an arbitrary set and $Y$ be a topological space. Given the set $\mathcal{F}(X, Y)$ of all maps from $X$ to $Y$., points $x_i \in X$, $i = 1, 2, \ldots, n$, and open sets $U_i: i = 1, 2, \ldots, n$, taking the sets $\{x_i, U_i\}_{n=1}^n = \{f \in \mathcal{F}(X, Y): f(x_i) \in U_i, i = 1, 2, \ldots, n\}$ as a subbase for a topology $\tau_p$. This topology is called the **pointwise convergence topology on** $\mathcal{F}(X, Y)$.

**Remark 5.24.21** Let $X$ be a topological space. If $(Y, d)$ is a metric space, then the sequence of functions

$$\{f_n: X \to Y\}$$

converges pointwise to $h$ iff for every $x_0 \in X$ and every $\epsilon > 0$, there exists an integer $n_0$ depending on $n_0$ and $\epsilon$ such that

$$d(f_n(x_0), h(x_0)) < \epsilon, \text{ whenever } n \geq n_0.$$

**Example 5.24.22** Let $X = \mathbf{I}$ and $Y = \mathbf{R}$ (real line space). If $f_n : \mathbf{I} \to \mathbf{R}$, $x \mapsto x^n$, then the sequence $\{f_n\}$ converges pointwise to the function

$$h: \mathbf{I} \mapsto \mathbf{R}, \ x \mapsto \begin{cases} 0, & \text{for all } x \in [0, 1) \\ 1, & \text{for } x = 1, \end{cases}$$

which is not continuous, although $\{f_n\}$ is a sequence of continuous functions.

**Remark 5.24.23** For more study on point open topology, see Exercise 102 of Sect. 5.28.

### 5.24.5 Relations on Different Topologies on $\mathcal{F}(X, Y)$

The compact open topology, pointwise convergence topology and uniform convergence topology on the set $\mathcal{F}(X, Y)$ of all mappings from $X$ to $Y$ have some weaker or stronger relations given in Proposition 5.24.24.

**Proposition 5.24.24** *Let* $\tau_c$, $\tau_p$, $\tau_\psi$ *denote the compact open topology, pointwise convergence topology and uniform convergence topology on the set* $\mathcal{F}(X, Y)$ *of all*

*mappings from X to Y respectively (by specifying the sets X and Y ). Then these three
topologies have the following relations for a metric space Y*

  *(i)* $\tau_p \subset \tau_c$;
 *(ii)* $\tau_c \subset \tau_\psi$.

**Proof** It follows from the respective definitions. $\qquad\qquad\qquad\qquad\qquad\square$

### 5.24.6 Ascoli's Theorem on Function Spaces

This subsection proves **Ascoli's Theorem** 5.24.30, which characterizes compact
subsets of a certain class of function spaces. For this purpose we start with the
concepts of uniform boundedness and equicontinuity. Let $\mathcal{C} = \mathcal{C}([0, 1])$ be the set of
real-valued continuous functions on $[0, 1]$. Then it is a real vector space under usual
compositions of addition and scalar multiplication of functions. Using the different
norm functions on it, some important metrics are defined on $\mathcal{C} = \mathcal{C}([0, 1])$. Such
metric spaces are born through the study of the problems in analysis.

For several forms of Ascoli's theorem and Ascoli–Arzela theorem see Exercises
78–80 of Sect. 5.28. Ascoli's theorem is named after Giulio Ascoli (1843–1896). He
introduced the concept of equicontinuity in 1884, which is one of the fundamental
concepts in the theory of real functions.

**Definition 5.24.25** Let $X$ be an arbitrary set and $\mathcal{F} = \{f \colon X \to \mathbf{R}\}$ be the set of all
real-valued functions on $X$. Then $\mathcal{F}$ is said to be **uniformly bounded** if there exists
a real number $K > 0$ if

$$|f(x)| \leq K, \quad \forall f \in \mathcal{F}, \text{ and, } \forall x \in X$$

In particular, a subset $\mathcal{F} \subset \mathcal{C}$ is uniformly bounded if $||f|| \leq K, \quad \forall f \in \mathcal{F}$, i.e., if
$\mathcal{F}$ is bounded in $\mathcal{C}$.

**Example 5.24.26** $\mathcal{F} = \{f_n : [0, 1] \to \mathbf{R} \colon f_n(x) = \cos nx, \quad \forall n = 1, 2, 3, \ldots\} \subset \mathcal{C}$ is
uniformly bounded where $K = 1$ may be taken.

**Definition 5.24.27** Let $(X, d)$ be an arbitrary metric space and $\mathcal{F} = \{f \colon X \to \mathbf{R}\}$
be the set of all real-valued functions on $X$. Then $\mathcal{F}$ is said to be **equicontinuous** if
for every $\epsilon > 0$, there a $\delta > 0$, depending on $\epsilon$ such that

$$d(x, y) < \delta \implies |f(x) - f(y)| < \epsilon, \quad \forall f \in \mathcal{F}.$$

**Remark 5.24.28** Definition 5.24.27 asserts that $\delta$ depends on $\epsilon$ but not on $f$ and
hence every $f \in \mathcal{F}$ is uniformly continuous.

***Example 5.24.29*** The subset $\mathcal{F} = \{f : [0, 1] \to \mathbf{R} : f$ converges uniformly$\}$ of $\mathcal{C}$ is equicontinuous.

Theorem 5.24.30 characterizes compactness of closed subsets of the function space $\mathcal{C}$ with the help of uniform boundedness and equicontinuity properties.

**Theorem 5.24.30** *(Ascoli) Let* $(X, d)$ *be a compact metric space and $\mathcal{F}$ be a closed subspace of the function space* $C(X, \mathbf{R})$ *or* $C(X, \mathbf{C})$. *Then $\mathcal{F}$ is compact iff it is uniformly bounded and equicontinuous.*

***Proof*** First assume that $\mathcal{F}$ is compact. Then it is bounded and hence uniformly bounded in $\mathcal{C}$. To prove that $\mathcal{F}$ is equicontinuous, take an $\epsilon > 0$. Then it follows that there is an $\epsilon/3$-net $\mathcal{N} = \{f_1, f_2, \ldots, f_n\}$ in $\mathcal{F}$. Since each $f_i \in \mathcal{N}$, is uniformly continuous, for $i = 1, 2, 3, \ldots, n$, there exists $\delta_i > 0$ such that

$$d(x, y) < \delta_i \implies |f_i(x) - f_i(y)| < \epsilon/3, \ x, y \in X.$$

Let $\delta = \min\{\delta_1, \delta_2, \ldots, \delta_n\}$. Then for any $f \in \mathcal{F}$, select $f_i \in \mathcal{N}$ such that

$$\|f - f_i\| = sup\{|f(x) - f_i(x)| : x \in X\} < \epsilon/3.$$

Then for any $x, y \in X$

$$d(x, y) < \delta \implies |f(x) - f(y)| \le |f(x) - f_i(x)| + |f_i(x) - f_i(y)|$$
$$+|f_i(y) - f(y)| < \epsilon/3 + \epsilon/3 + \epsilon/3 = \epsilon.$$

This asserts that $\mathcal{F}$ is equicontinuous.

Conversely, let the closed subset $\mathcal{F}$ of the complete metric space $\mathcal{C}$ be uniformly bounded and equicontinuous. To prove that $\mathcal{F}$ is compact, use the uniformly boundedness and equicontinuity properties of $\mathcal{F}$. □

***Remark 5.24.31*** The Gelfand–Kolmogoroff Theorem is a basic theorem in topology. It studies the rings $C(X, \mathbf{R})$ for compact Hausdorff spaces $X$ and says that two compact Hausdorff spaces $X$ and $Y$ are homeomorphic iff the corresponding rings $C(X, \mathbf{R})$ and $C(Y, \mathbf{R})$ are isomorphic. This deep result recovers the topology of $X$ from the ring structure of $C(X, \mathbf{R})$. The study of Gelfand–Kolmogoroff theorem is given in Chap. 6.

# 5.25 Applications

This section conveys some applications of connectedness and compactness expanded in Sects. 5.25.1–5.25.5. with an eye to solve various problems.

## 5.25.1  Geometric Applications

This subsection solves some geometric problems such as lifting problems, compactness of the real projective space $\mathbf{R}P^2$ and connectedness of the $n$-sphere $S^n$ for $n \geq 1$. Let $S^1 = \{z \in \mathbf{C}: |z| = 1\}$ be the unit circle in the complex plane $\mathbf{C}$ and $p \colon \mathbf{R} \to S^1$, $t \mapsto e^{2\pi i t}$ be the exponential map. $S^1$ is algebraically a group under usual multiplication of complex numbers. It is also a topological group (see **Volume 2**), since the multiplication function $S^1 \times S^1 \to S^1$, $(z, w) \mapsto zw$ and the inversion function $S^1 \to S^1$, $z \mapsto z^{-1}$ are both continuous. The additive group $(\mathbf{R}, +)$ is also a topological group under usual addition of real numbers. The map

$$p \colon (\mathbf{R}, +) \to (S^1, \cdot)$$

is a group homomorphism with its kernel, $\ker p = \{t \in \mathbf{R} : p(t) = 1\} = \mathbf{Z}$.


**Definition 5.25.1** Let $(X, x_0)$ be a pointed topological space consisting of a topological space $X$ and a point $x_0 \in X$, called a base point. A base point preserving continuous map

$$f \colon (X, x_0) \to (S^1, 1)$$

is said to have a **lifting** $\tilde{f}$ if

$$\tilde{f} \colon (X, x_0) \to (\mathbf{R}, t_0)\ (t_0 \in \ker p = \mathbf{Z})$$

is a continuous map such that $p \circ \tilde{f} = f$

There is a natural question: Under what condition this lifting exists uniquely? Theorem 5.25.2 and Theorem 5.25.3 provide answers to this question.

**Theorem 5.25.2** *Let $X$ be a connected space and $f \colon (X, x_0) \to (S^1, 1)$ be a given continuous map. If $\tilde{f}, \tilde{g} \colon (X, x_0) \to (\mathbf{R}, t_0)$ are two liftings of $f$, then $\tilde{f} = \tilde{g}$, whenever $\tilde{f}(x_0) = \tilde{g}(x_0)$.*

**Proof** Let $p \colon \mathbf{R} \to S^1$, $x \mapsto e^{2\pi i x}$ be the exponential map and $\tilde{f}, \tilde{g} \colon (X, x_0) \to (\mathbf{R}, t_0)$ be two liftings of $f \colon (X, x_0) \to (S^1, 1)$. Then $p \circ \tilde{f} = f$ and $p \circ \tilde{g} = f$. Define a map

$$\psi \colon (X, x_0) \to (\mathbf{R}, t_0),\ x \mapsto \tilde{f}(x) - \tilde{g}(x).$$

Then $(p \circ \psi)(x) = p(\tilde{f}(x) - \tilde{g}(x)) = e^{2\pi i(\tilde{f}(x) - \tilde{g}(x))} = (p \circ \tilde{f})(x)/(p \circ \tilde{g})(x) = f(x)/f(x) = 1$, $\forall x \in X$. It asserts that $\psi(x) \in \ker p = \mathbf{Z}$. Thus $\psi$ is an integral-valued continuous function. Since $X$ is connected by hypothesis, it follows from discreteness of $\mathbf{Z}$ that $\psi$ is a constant function and hence $\psi(x_0) = \tilde{f}(x_0) - \tilde{g}(x_0) = 0$, since by hypothesis, $\tilde{f}(x_0) = \tilde{g}(x_0)$. Then for the constant function $\psi$ it follows that $\psi(x) = 0$ for all $x \in X$. It proves the uniqueness of the lifting of $f$.                                        $\square$

**Theorem 5.25.3** *Given a convex compact subset of the Euclidean n-space* $\mathbf{R}^n$ *for some n, a continuous map* $f: (X, x_0) \to (S^1, 1)$ *and an integer* $t_0$, *there exists a unique lifting* $\tilde{f} : (X, x_0) \to (\mathbf{R}, t_0)$ *such that* $p \circ f = \tilde{f}$.

**Proof** By hypothesis, $X$ is a convex compact subset of the metric space $\mathbf{R}^n$ with standard metric. Then $f$ is uniformly continuous. Consequently, given an $\epsilon > 0$, whenever, $||x - x'|| < \epsilon$, then $|f(x) - f(x')| < 2$ (diameter of the circle $S^1$). This implies that the points $f(x)$ and $f(x')$ cannot be antipodal points and hence $f(x)/f(x') \neq -1$. Since $X$ is bounded, there exists a positive integer $n$ such that $||x - x'||/n < \epsilon$, $\forall x \in X$. Subdivide the line segment from the point $x_0$ and to the point $x$ for each $x \in X$, (which lies entirely in $X$, since $X$ is convex) into $k$ equal subintervals by inserting the points $x_0, x_1, \ldots, x_k = x$. This asserts that $||x_{i+1} - x_i|| = ||x - x_0||/n < \epsilon$, which shows that $f(x_{i+1})/f(x_i) \neq -1$. Again, for each $i$ satisfying $0 \leq i \leq k - 1$, the function

$$f_i: X \to S^1 - \{-1\}, \ x \mapsto f(x_{i+1})/f(x_i)$$

is continuous, since the usual multiplication $S^1 \times S^1 \to S^1$ and inversion $S^1 \to S^1$ are continuous. Taking $i = 0, 1, 2, \ldots, k - 1$ successively, it follows that

$$f(x) = f(x_0)f_0(x)f_1(x) \cdots f_k(x) \text{ (\textbf{telescopic product} on } S^1).$$

Taking $f(x_0) = z_0$, we get a continuous map

$$\tilde{f}: X \to \mathbf{R}, \ x \mapsto z_0 + \log f_0(x) + \log f_1(x) + \cdots + \log f_{k-1}(x)$$

such that $f_i(x_0) = 1$, $\forall i$ and $p \circ f = \tilde{f}$. If $\tilde{f}$ and $\tilde{g}$ are two liftings of $f$ satisfying the above conditions, then it follows from Theorem 5.25.2 that $\tilde{f}$ and $\tilde{g}$ are equal. $\square$

**Corollary 5.25.4** *Let* $\mathbf{I} = [0, 1]$ *and* $\dot{\mathbf{I}} = \{0, 1\}$. *If* $f: (\mathbf{I}, \dot{\mathbf{I}}) \to (S^1, 1)$ *is continuous, then there exists a unique continuous map* $\tilde{f}: \mathbf{I} \to \mathbf{R}$ *such that* $p \circ f = \tilde{f}$.

**Proof** $\mathbf{I} = [0, 1]$ and $\dot{\mathbf{I}} = \{0, 1\}$ are subspaces of the real line space $\mathbf{R}$. Taking in particular, $X = \mathbf{I}$, corollary follows from Theorem 5.25.2 and Theorem 5.25.3. $\square$

**Definition 5.25.5 (Geometric cone)** Let $X$ be a compact subspace of $\mathbf{R}^n$. The geometric cone on $X$ denoted by **CX** defined by

$$\mathbf{CX} = \{y \in \mathbf{R}^{n+1}: y = tu + (1 - t)x, \ u = (0, 0, \ldots .0, 1) \in \mathbf{R}^{n+1}, \ x \in X, \ 0 \leq t \leq 1\}$$

is built up by all straight-line segments joining the point $u = (0, 0, \ldots .0, 1) \in \mathbf{R}^{n+1}$ to some point $x \in X$.

The concept of geometric cone **CX** on $X$ is now extended to the concept of cone $CX$ on an arbitrary topological space $X$.

**Definition 5.25.6** Given an arbitrary topological space $X$, an equivalence relation $\sim$ on $X \times \mathbf{I}$ is defined by $(x, t) \sim (y, s)$ iff $t = s = 1$. The quotient space $X \times \mathbf{I} / \sim$, denoted by $CX$ is called the **cone** over $X$.

**Remark 5.25.7** Given an arbitrary topological space $X$, the cone $CX$ over $X$ is obtained as the quotient space provided by the partition given by

(i) the subset $X \times \{1\}$;
(ii) sets consisting of a single point $(x, t)$ for $x \in X$ and $t \in \mathbf{I}$.

**Geometrically**, the cone $CX$ over $X$ is obtained from $X \times \mathbf{I}$ by identifying its top $X \times \{1\}$ to a single point and this identified point $[x, 1]$ is its vertex where $[x, t]$ denotes the point of $CX$ corresponding to the point $(x, t) \in X \times \mathbf{I}$ under the identification map

$$p : X \times \mathbf{I} \to CX, \ (x, t) \mapsto [x, t].$$

The space $X$ is embedded as a closed subspace of $CX$ by the map $x \to [x, 1]$.

There is a close relation between the cones $CX$ and $\mathbf{CX}$ on a topological space $X$ and it is proved in Proposition 5.25.8.

**Proposition 5.25.8** *The geometric cone* $\mathbf{CX}$ *is homeomorphic to $CX$ for any compact subspace $X$ of $\mathbf{R}^n$.*

**Proof** Define a map

$$f : X \times \mathbf{I} \to \mathbf{CX}, \ (x, t) \to tu + (1 - t)x : u = (0, 0, \ldots .0, 1) \in \mathbf{R}^{n+1}, \ x \in X, \ t \in \mathbf{I}.$$

Then $f$ is continuous and surjective. Moreover, $f(x, t) = f(y, s)$ iff either $x = y$ and $t = s$ or $t = s = 1$. This shows that the partition of $X \times \mathbf{I}$ provided by $f$ is precisely the same as provided by the identification space $CX$. Again as $X \times I$ is compact, the geometric cone $\mathbf{CX} \subset \mathbf{R}^{n+1}$ is Hausdorff. This asserts by Theorem 5.4.5 that $f$ is an identification map. $\qquad\square$

**Remark 5.25.9** An alternative proof of Theorem 5.17.6 is now given in Theorem 5.25.10 by using the concept of compactness, establishing a link between compactness and connectedness properties.

**Theorem 5.25.10** *The real line space* $\mathbf{R}$ *is connected.*

**Proof** Let $\{U, V\}$ be a nontrivial separation by open sets of the real line space $\mathbf{R}$. Suppose $x, y$ be two real numbers such that $x < y$, $x \in U$, $y \in V$. Let $A = U \cap [x, y]$, $B = V \cap [x, y]$. Since $A, B$ are closed subsets of the compact set $[x, y]$, they are compact. This asserts that the topological product space $A \times B$ is also compact. Consider the continuous map

$$d : A \times B \to \mathbf{R}, \ (a, b) \mapsto |a - b|.$$

Then its continuous image $d(A \times B) = Y \subset \mathbf{R}$ is compact and hence closed in $\mathbf{R}$. This asserts that $Y$ contains its glb $\beta_Y$. Hence there exists some $(a, b) \in A \times B$ such that $d(a, b) = \beta_Y > 0$. If $c = \frac{1}{2}(a + b)$, then $d(a, c) = \frac{1}{2}\beta_Y = d(c, b)$. This gives a contradiction, since the point $c$ belongs to neither $U$ nor $V$. This contradiction forces to prove the proposition. □

**Corollary 5.25.11** *Every open interval* $(a, b)$ *in the real number space* $\mathbf{R}$ *is connected.*

**Proof** Since $(a, b)$ and $\mathbf{R}$ are homeomorphic spaces, this corollary follows from Theorem 5.25.10 and Corollary 5.17.24, since connectedness is a topological property. □

**Proposition 5.25.12** *Let* $X$ *be a topological space and* $A \subset X$ *be connected. If* $B \subset X$ *is a subset such that* $A \subset B \subset \bar{A}$, *then* $B$ *is also connected.*

**Proof** Let $\{U, V\}$ be a separation of the subspace $B$ of the topological $X$. Then it follows that the pair $\{(U \cap A), (V \cap A)\}$ is a separation of the subspace $A$ of $X$. By hypothesis, the subspace $A$ is connected. Hence it follows that either of the sets $U \cap A$ or $V \cap A$ is $\emptyset$. If $V \cap A = \emptyset$, then $A \subset U$ and hence $B \subset \bar{A} \subset \bar{U}$. Since $U$ is closed in $B$, it follows that

$$B \subset B \cap \bar{U} = U.$$

This asserts that $V = \emptyset$ and hence the separation $\{U, V\}$ of $B$ is trivial. This asserts that subspace $B$ is connected. If $U \cap A = \emptyset$, then as above, the subspace $B$ is connected. □

**Corollary 5.25.13** *Every closed interval* $[a, b] \subset \mathbf{R}$ *is connected.*

**Proof** Since the open interval $(a, b)$ is connected and $\overline{(a, b)} = [a, b]$, the corollary follows Proposition 5.25.12. □

**Proposition 5.25.14** *The n-sphere* $S^n$ *is connected for every integer* $n \geq 1$.

**Proof** Let $x \in S^n$ be a point. Then the spaces $S^n - \{x\}$ and $\mathbf{R}^n$ are homeomorphic by stereographic projection. As $\mathbf{R}^n$ is connected, it follows that the space $S^n - \{x\}$ is also connected. Again $\overline{S^n - \{x\}} = S^n$. Hence it follows by Proposition 5.25.12 that $S^n$ is connected. □

**Proposition 5.25.15** *The n-sphere* $S^n$ *is compact for every integer* $n \geq 1$.

**Proof** Since $S^n$ is a closed and bounded subset of $\mathbf{R}^{n+1}$, compactness of $S^n$ follows from Heine–Borel theorem 5.2.4, which asserts that a subset of $\mathbf{R}^{n+1}$ is compact iff it closed and bounded. □

**Proposition 5.25.16** *The real projective space* $\mathbf{R}P^2$ *is the space of all lines through the origin in* $\mathbf{R}^3$. *It is a compact space.*

**Proof** To show that the real projective space $\mathbf{R}P^2$ is compact, consider the projection map

$$p\colon S^2 \to \mathbf{R}P^2, \ x \mapsto \{x, -x\}.$$

Then $p$ is continuous under the quotient topology on $\mathbf{R}P^2$ in the sense that a subset $U \subset \mathbf{R}P^2$ is open iff $p^{-1}(U)$ is open in $S^2$. Since $S^2$ is compact by Proposition 5.25.15 and $p$ is continuous, it follows that $\mathbf{R}P^2 = p(S^2)$ is compact.    □

### 5.25.2 Matrix Algebra from Viewpoint of Connectedness and Compactness

The subsection presents some topological applications in algebra, specially in matrix algebra from the viewpoint of connectedness and compactness. Let $M(n, \mathbf{R})$ be the set of all $n \times n$ matrices over $\mathbf{R}$. It is studied on considering it as a subspace of the Euclidean $\mathbf{n}^2$-space $\mathbf{R}^{n^2}$. Similarly, $M(n, \mathbf{C})$ is studied. Let $\mathbf{R}^* = \mathbf{R} - \{0\}$. Then the subspace $\mathbf{R}^*$ of $\mathbf{R}$ is neither connected nor compact.

**Proposition 5.25.17** *The general linear group $GL(n, \mathbf{R})$ is not connected.*

**Proof** To show that $GL(n, \mathbf{R})$ is not connected, consider the determinant function

$$det\colon GL(n, \mathbf{R}) \to \mathbf{R}^*, \ A \mapsto detA.$$

It is a continuous surjective map whose image $\mathbf{R}^*$ is not connected and hence $GL(n, \mathbf{R})$ fails to be connected.    □

**Proposition 5.25.18** *The general linear group $GL(n, \mathbf{R})$ is not compact.*

**Proof** Consider the determinant function

$$det\colon GL(n, \mathbf{R}) \to \mathbf{R}^*, \ A \mapsto detA.$$

It is a continuous surjective map. As its continuous image is $\mathbf{R}^* = \mathbf{R} - \{0\}$, which is not compact, $GL(n, \mathbf{R})$ fails to be compact.    □

**Proposition 5.25.19** *Let $S = \{M \in M(n, \mathbf{R})\colon M$ is symmetric$\} \subset M(n, \mathbf{R})$. Then the subspace $S$ is connected.*

**Proof** For any $A, B \in S$, and any t satisfying $0 \le t \le 1$,

$$(tA + (1 - t)B) \in S \implies$$

$S$ is path connected and hence $S$ is connected by Corollary 5.22.11.    □

**Proposition 5.25.20** *The subspace* $X = \{M \in M(n, \mathbf{R}): M$ *is symmetric and positive definite* $\}$ *of* $\mathbf{M}(n, \mathbf{R})$ *is connected.*

**Proof** By definition, $X$ is a subspace of $M(n, \mathbf{R})$. If $A \in X$ is positive definite, then for any vector $y \in \mathbf{R}^n$, the real number $y^t A y > 0$. Then for any two matrices $A, B \in X$, and for all $t$ satisfying $0 \le \lambda \le 1$,

$$y^t(\lambda A + (1 - \lambda)B)y = \lambda(y^t A y) + (1 - \lambda)(y^t B y)y > 0$$

asserts that the matrix $\lambda A + (1 - \lambda)B) \in X$, $\forall \lambda$ satisfying $0 \le \lambda \le 1$ and hence $X$ is path connected. This implies that it is connected by Corollary 5.22.11. $\square$

**Proposition 5.25.21** *The subspace*

$$X = \{A \in M(n, \mathbf{R}): x^t A x \ge 0, \ \forall x \in \mathbf{R}^n\},$$

*of* $M(n, \mathbf{R})$ *is both path connected and connected.*

**Proof** For any two matrices $A, B \in X$, and any $\lambda \in [0, 1]$

$$x^t(\lambda A + (1 - \lambda)B)x \ge 0$$

asserts that $X$ is path connected and hence it is also connected by Corollary 5.22.11. $\square$

**Proposition 5.25.22** *The subspace*

$$X = \{M = (m_{ij}) \in M(n, \mathbf{R}): trace\ M = \Sigma_{i=1}^n m_{ii} = 0\}$$

*of* $M(n, \mathbf{R})$ *is connected but not compact.*

**Proof** Consider the trace map $tr$ defined by

$$tr : M(n, \mathbf{R}) \to \mathbf{R}, \ M \mapsto trace\ M.$$

Then $tr$ is a linear transformation. Let $M, N \in X$ be any two arbitrary elements. Then trace $M = 0$ and trace $N = 0$. Hence for all $t$ satisfying $0 \le t \le 1$,

$$tr\ (tM + (1 - t)N)) = t\ trM + (1 - t)\ trN = 0$$

asserts that $X$ is path connected and hence it is also connected Corollary 5.22.11. For the second part consider the matrices of the form

$$M_t = \begin{pmatrix} t & 0 \\ 0 & -t \end{pmatrix},$$

for all $t \in \mathbf{R}$. Then $tr\, M_t = 0$, $\forall t \in \mathbf{R}$. Since the set $\mathbf{R}$ is unbounded, it follows that the given set $X$ is not compact.                                                      □

**Corollary 5.25.23**  *The kernel of the linear homomorphism*

$$tr: M\,(n, \mathbf{R}) \to \mathbf{R}, \ M \mapsto trace\ M.$$

*is connected in* $M\,(n, \mathbf{R})$.

**Proof**  It follows from Proposition 5.25.22.                                        □

## 5.25.3  Topological Study of Algebraic Groups

This subsection gives an application of compactness in group theory by using Tychonoff product theorem.

**Proposition 5.25.24**  *Let $G$ be an algebraically group endowed with a topology such that its group multiplication is continuous in the product topology. If the subsets $A$, $B$ of $G$ are compact, then $AB = \{ab : a \in A, b \in B\}$ is also compact.*

**Proof**  Consider the continuous multiplication

$$m: G \times G \to G, (g, h) \mapsto gh.$$

By hypothesis $A$ and $B$ are compact subsets of $G$. Then their product space $A \times B$ is also compact by Tychonoff Product theorem 5.11.4. Since $AB$ is the image of $A \times B$ under the continuous multiplication $m$, it follows that $AB$ is also compact. □

**Remark 5.25.25**  For more study of topological applications to algebra and geometry, see **Basic Topology, Volume 2 and Volume 3** of the present series of books.

## 5.25.4  Applications to Homeomorphism Problems

This subsection uses two basic topological properties such as connectedness and compactness properties to solve the homeomorphism problem in topology: given two topological spaces are they homeomorphic or not ? It is one of the main problems of topology. Moreover by using the combined properties of compactness and Hausdorff, it is proved that the quotient space $\mathbf{D}^n/S^{n-1}$ is homeomorphic to the $n$-sphere $S^n$.

We now consider some problems given in the following examples.

***Example 5.25.26*** Consider the subspaces of $\mathbf{R}^2$ defined by

(i) $A = \{((x, y) \in \mathbf{R}^2 : xy = 0\}$,
(ii) $B = \{((x, y) \in \mathbf{R}^2 : x + y \geq 0 \text{ and } xy = 0\}$
(iii) $C = \{((x, y) \in \mathbf{R}^2 : xy = 1\}$,
(iv) $D = \{((x, y) \in \mathbf{R}^2 : x + y \geq 0 \text{ and } xy = 1\}$
  Are these spaces homeomorphic? The answer is that all of these spaces are not homeomorphic. Its justification is given below:

The subspace $C$ represents geometrically a hyperbola in $\mathbf{R}^2$ having two connected components described by

$$C = \left\{ \left( x, \frac{1}{x} \right) : x \in \mathbf{R}^+ \right\} \cup \left\{ \left( -x, -\frac{1}{x} \right) : x \in \mathbf{R}^+ \right\}.$$

The other subspaces $A$, $B$ and $D$ of $\mathbf{R}^2$ are also described as

$$A = \{(x, 0) \cup (y, 0) : x, y \in \mathbf{R}\}.$$

$$B = \{(x, 0) \cup (y, 0) : x, y \in \mathbf{R}^+ \cup \{0\}\}.$$

$$D = \left\{ \left( x, \frac{1}{x} \right) : x \in \mathbf{R}^+ \right\}.$$

Suppose there exists a homeomorphism

$$f : A \to C,$$

where $A$ is connected but $C$ is disconnected having two connected components. Since connectedness is a topological property, the image $f(A) = C$ is connected, which is not true. This asserts the topological spaces $A$ and $C$ are not homeomorphic. Proceeding in a similar way it follows that $A$ is nether homeomorphic to $B$ nor $D$.

***Example 5.25.27*** Consider the subspaces of $\mathbf{R}^2$ defined by

(i) $A = \{((x, y) \in \mathbf{R}^2 : x^2 + y^2 = 1\}$,
(ii) $B = \{((x, y) \in \mathbf{R}^2 : ax^2 + by^2 + 2hxy = 1, \text{ and } ab - h^2 < 0\}$
(iii) $C = \{((x, y) \in \mathbf{R}^2 : ax^2 + by^2 + 2hxy = 1, \text{ and } ab - h^2 = 0\}$
  Are these spaces homeomorphic? The answer is not positive. Its justification is given below:

**Geometrically**, the subspace $A$ represents a circle, which is compact. On the other hand, the subspace $B$ represents a hyperbola, which is not compact and the subspace $C$ represents a parabola, which is not compact. This asserts that $A$ is homeomorphic to neither $B$ nor $C$, since compactness is a topological property.

There is a natural problem: which space the quotient space $\mathbf{D}^n/S^{n-1}$ identifies? Theorem 5.25.28 gives its answer.

**Theorem 5.25.28** *There exists a homeomorphism* $\psi : \mathbf{D}^n/S^{n-1} \to S^n$.

***Proof*** Proceed as in Example 5.4.11 to construct a continuous bijective map

$$\psi : \mathbf{D}^n/S^{n-1} \to S^n.$$

Now, use Corollary 5.4.8 to show that $\psi$ is a homeomorphism.                    □

### 5.25.5  Alternative Proof of Heine–Borel Theorem and Compactness of $S^n$

This subsection proves Heine–Borel theorem, whose proof is different from the method prescribed in Theorem 5.2.4. An important consequence of this result proves the compactness of the $n$-sphere $S^n$.

**Theorem 5.25.29  (Generalized Heine–Borel theorem)** *The closed interval* $\mathbf{I} = [0, 1]$ *in the real line space* $\mathbf{R}$ *is compact. In general a subset of the Euclidean $n$-space* $\mathbf{R}^n$ *is compact iff it is closed and bounded.*

***Proof*** **Proof I:** Since the closed interval $[a, b]$ is compact and is homeomorphic to $\mathbf{I}$, by the topological property of compactness, it follows that $\mathbf{I}$ is also compact. It proves the first part of the theorem. Since a compact subset in a metric space is closed and bounded, it follows that it follows that every compact subset in $\mathbf{R}^n$ is closed and bounded. Conversely, let $Y$ be a closed and bounded subset of $\mathbf{R}^n$. Let $K > 0$ be a large real number such that $Y \subset [-K, K]^n$. Since $\mathbf{I}$ is homeomorphic to $\mathbf{J} = [-K, K]$ and $\mathbf{I}$ is compact, it follows that the interval $\mathbf{J}$ is also compact. Hence $[-K, K]^n = \mathbf{J} \times \mathbf{J} \times \cdots \times \mathbf{J}$ (n-product) is also compact by Tychonoff theorem. Since every closed subset of a compact set is compact, it follows that $Y$ is compact.

**Proof II:** A closed and bounded subspace of $\mathbf{R}^n$ is a closed subspace of the product space $\prod_{i=1}^n [a_i, b_i]$ admitting the product topology $\sigma$ coinciding with relative topology as a subspace of the Euclidean topology on $\mathbf{R}^n$. Use Tychonoff theorem on the product space $(X, \tau)$ by using classical Heine–Borel theorem for compactness of the closed intervals $[a_i, b_i]$ in $\mathbf{R}$.

**Proof III:** Let $\mathcal{F}$ be an open covering of $\mathbf{I}$. Define the set

$$Y = \{x \in \mathbf{I} : [0, x] \text{ is covered by finitely many open sets in } \mathcal{F}\}.$$

Then *sup* $Y \in Y$ and *sup* $Y = 1$ asserts that **I** has a finite subcovering of the covering $\mathcal{F}$. □

**Corollary 5.25.30** *Every continuous map $f : S^1 \to \mathbf{R}$ is uniformly continuous.*

**Proof** It follows by using the result that any continuous map on a compact space is uniformly continuous. □

**Corollary 5.25.31** *For every nonconstant continuous map $f : S^1 \to \mathbf{R}$, there are uncountably many pairs of distinct points $x, y \in S^1$ such $f(x) = f(y)$.*

**Proof** Since $S^1$ is compact and $f$ is continuous, its image set $f(S^1) = X \subset \mathbf{R}$ is also compact. Hence it is a closed and bounded subset in $\mathbf{R}$. This shows that $f$ assumes both maximum and minimum values $M$ and $m$ (say) respectively. Let $x_0$ be an arbitrary point between $m$ and $M$. Assume that $f^{-1}(x_0)$ consists of only one point of $S^1$. Suppose that $f^{-1}(x_0) = y_0 \in S^1$. Since $S^1 - \{y_0\}$ is connected and $f$ is continuous, it follows that its continuous image $f(S^1 - \{y_0\}) = X - \{x_0\}$ is connected in $\mathbf{R}$, which is not true. Since $X$ is both compact and connected in $\mathbf{R}$, it follows that $X$ is a closed interval in $\mathbf{R}$. This contradiction implies that the assumption $f^{-1}(x_0)$ consists of only one point of $S^1$ is not true. Since the point $x_0$ is arbitrary, the corollary follows. □

**Proposition 5.25.32** *Let $(X, d)$ be a metric space and $Y \subset X$ be nondegenerate in the sense that card $Y > 1$. If $Y$ is a connected subset of $X$, then the subspace $Y$ is uncountable.*

**Proof** Let $a, b \in Y$ be two distinct points and $f : Y \to \mathbf{R}, x \mapsto d(x, a)$ be a map. Then $f$ is continuous and $f(Y)$ is a connected subset of $\mathbf{R}$ such that $f(a) = 0, f(b) = d(b, a) \neq 0$. Suppose $d(b, a) = k$ is a positive real number. Hence $f(Y)$ must contain the open interval $(0, k)$, which is uncountable. This implies that $Y$ is also uncountable. □

**Proposition 5.25.33** *Let $X$ be nonempty subset of the Euclidean line $\mathbf{R}$. If every continuous function*

$$f : X \to \mathbf{R}$$

*is bounded, then $X$ is compact.*

**Proof** It follows by showing that $X$ is closed and bounded in $\mathbf{R}$. If $X$ is not closed, there exists some point $a \in \overline{X} - X$. Then the function

$$f : X \mapsto \mathbf{R}. \, x \mapsto \frac{1}{x - a}$$

is continuous but it is unbounded, which contradicts our hypothesis. This asserts that $X$ is closed in $\mathbf{R}$. Again, if $X$ is unbounded, then the function

$$f : X \to X, \ x \to x$$

is continuous but it is unbounded, which contradicts our hypothesis. Consequently, $X$ is closed and bounded in $\mathbf{R}$. Hence it follows by Heine-Borel theorem of compactness that $X$ is compact.                                                                                      $\square$

## 5.26   Application in Measure Theory

Measure theory is an important branch of mathematics. This section proves Theorem 5.26.3, which determines the Lebesque measure for a certain class of compact subsets of the Euclidean $n$-space $\mathbf{R}^n$. This theorem is a key result in the study of manifolds (see **Basic Topology, Volume 2** of the present series of books).

**Definition 5.26.1**   A subset $X \subset \mathbf{R}^n$ is said to have **Lebesque measure zero** in $\mathbf{R}^n$, denoted by $\mu(X) = 0$, if for every $\epsilon > 0$, there is a countable family of $n$-dimensional cubes in $\mathbf{R}^n$ such that the sum of their volumes $< \epsilon$. Equivalently, $\mu(X) = 0$ in $\mathbf{R}^n$ iff for any $\epsilon > 0$, there is an open set $U$ such that $X \subset U$ and its volume Vol ( $U$ ) $< \epsilon$.

**Proposition 5.26.2**   *Any countable union of subsets with measure zero in $\mathbf{R}^n$ has also measure zero.*

*Proof*   Let $\{X_i\}$ be a countable family of subsets in $\mathbf{R}^n$ each having measure zero. If $X = X_1 \cup X_2 \cup \cdots$ and $X_i \subset U_i$, where each $U_i$ is open in $\mathbf{R}^n$ and Vol ( $U_i$ ) $< \epsilon/2^i$ for $i = 1, 2, \cdots$, then

$$X \subset U = U_1 \cup U_2 \cup \cdots \text{ and Vol}(U) \leq \Sigma_i \text{Vol } (U_i) < \Sigma_i \epsilon/2^i < \epsilon.$$

Hence the proposition follows from Definition 5.26.1.

$\square$

**Theorem 5.26.3   (Fubini).** *Let $X$ be a compact set in $\mathbf{R}^n$ such that every subset*

$$X_t = X \cap (\{t\} \times \mathbf{R}^{n-1})$$

*has measure zero in the hyperplane $\mathbf{R}^{n-1}$ for each $t \in \mathbf{R}$.. Then the Lebesque measure $\mu(X) = 0$ in $\mathbf{R}^n$.*

*Proof*   Let $\mathbf{I}^n$ be the $n$-cube. Suppose that $X \subset \mathbf{I}^n$. Define a function

$$f : \mathbf{I} \to \mathbf{R}, \ t \mapsto \mu(X \cap ([0, t] \times \mathbf{I}^{n-1})).$$

Given any positive real number $\epsilon$, there exists by hypothesis an open set $U \subset \mathbf{I}^n$ such that

$$X \cap (t \times \mathbf{I}^{n-1}) \subset t \times U, \text{ where Vol}(U) < \epsilon.$$

By hypothesis, $X$ is compact. Hence there exists a real number $t_0 > 0$ such that

$$X \cap ([t - t_0, t + t_0] \times \mathbf{I}^{n-1}) \subset [[t - t_0, t + t_0] \times U.$$

Then for any real number $s$ such that $0 \leq s < t_0$, the set

$$X \cap ([0, t + s] \times \mathbf{I}^{n-1} \subset (X \cap ([0, t] \times \mathbf{I}^{n-1})) \cup ([t, t + s] \times U)$$

can be covered by an open set $V$ with

$$Vol(V) < f(t) + \epsilon s.$$

Hence for all $s$ satisfying $0 \leq s < t_0$, it follows that

$$f(t + s) \leq f(t) + \epsilon s.$$

Proceeding in a similar way, it follows from

$$X \cap ([0, t] \times \mathbf{I}^{n-1}) \subset (X \cap ([0, t - s] \times \mathbf{I}^{n-1})) \cup ([t - s, s] \times U)$$

that

$$f(t) \leq f(t - s) + \epsilon s, \ \forall s \text{ satisfying } 0 \leq s < t_0.$$

It asserts that

$$\left| \frac{f(t + s) - f(t)}{s} \right| \leq \epsilon \ \forall s \text{ satisfying } |s| < t_0.$$

This implies that $f$ is differentiable at $t \in \mathbf{I}$ with its derivative 0. Since $f(0) = 0$, it also follows that $f(1) = 0$. Hence the theorem follows from Definition 5.26.1.

$\square$

**Remark 5.26.4**  An alternative proof of Theorem 5.26.3 is available in Exercise 105 of Sect. 5.28.

## 5.27  Further Applications

This section presents some interesting applications of the Intermediate Value Theorem and Heine- Borel Theorem and compactness property.

## 5.27.1   Some Applications of Real-valued Continuous Functions

We now prove Theorem 5.27.1, which conveys a generalization of the classical Bolzano-Weierstrass theorem saying that every bounded infinite subset of **R** has at least one limit point. Another form of Bolzano-Weierstrass theorem is given in Theorem 5.3.6.

**Theorem 5.27.1 (Generalization of Bolzano-Weierstrass theorem)** *Any infinite subset of a compact space must have a limit point.*

**Proof** Let $(X, \tau)$ be a compact space and $A$ be an infinite subset of $X$ such that $A$ has no limit point. For every $x \in X$, define a nbd $N_x$ of $x$ in $X$ such that

$$N_x \cap A = \begin{cases} \varnothing, & \text{if } x \notin A \\ \{x\}, & \text{if } x \in A \end{cases}$$

otherwise, $x$ would be a limit point of $A$. By hypothesis, $X$ is compact. Hence the open covering $\{N_x : x \in X\}$ of $X$ has a finite subcovering. But each set $N_x$ contains atmost one point of $A$. Hence, it follows that the set $A$ is finite. This contradiction proves the theorem.

**Theorem 5.27.2** *Every real-valued continuous function $f: X \to \mathbf{R}$ defined on a compact space $X$ is bounded and it attains its bounds.*

**Proof** Since $f(X)$ is the continuous image of a compact set $X$, it follows that $f(X)$ is a compact subset of **R**. Hence it is a closed and bounded subset of **R** by Heine–Borel theorem 5.2.4. This implies that $f$ is bounded. Again, since $f(X)$ is closed, both $sup(f(X)) = M$ and $\inf(f(X)) = m$ exist in $f(X)$. Hence there are points $a, b \in X$ such that

$$f(a) = M \text{ and } f(b) = m.$$

$\square$

**Remark 5.27.3** Let $A$ be a nonempty subset of a topological space $X$. Then every accumulation point of $A$ is a limit point of $A$. But in a Hausdorff space and hence in a metric space, the concepts of accumulation point and limit point coincide. For a topological space having a countable open base for its topology, the space is compact iff it satisfies (B–W)-property (see Chap. 7).

**Theorem 5.27.4 (Weierstrass theorem)** *Let $X$ be a compact space and $f: X \to \mathbf{R}$ be a continuous map. Then $f$ is bounded and attains its maximum and minimum values.*

**Proof** Since $X$ is compact and $f$ is continuous, it follows by Corollary 5.8.2 that $f(X)$ is a compact subspace of **R** and hence it is closed and bounded in **R**. This implies that the function $f$ is bounded. As $f(X)$ is closed, it contains all of its limit points. This asserts that $sup_{x \in X} f(x) \in f(X)$ and $inf_{x \in X} f(x) \in f(X)$ exist finitely in $f(X)$.

□

**Corollary 5.27.5** *Every continuous real-valued function on a compact set is bounded.*

**Proof** Let $X$ be a compact space and $f: X \to \mathbf{R}$ be an arbitrary continuous map. Then it is bounded by Theorem 5.27.4. □

The converse of Corollary 5.27.5 is formulated in Proposition 5.27.7.

**Proposition 5.27.6** *Let $X$ be a nonempty closed and bounded subset in $\mathbf{R}^n$ and $f: X \to \mathbf{R}$ be continuous. Then f(X) has a maximum and a minimum.*

**Proof** By hypothesis, $X$ is a nonempty closed and bounded subset in $\mathbf{R}^n$. Hence $X$ is compact. Again, since, $f: X \to \mathbf{R}$ is continuous, its image $f(X)$ is a nonempty bounded set of real numbers and it attains its maximum and also minimum values by Theorem 5.27.4. □

**Proposition 5.27.7** *Let $X$ be a subspace of the real line space $\mathbf{R}$ such that every continuous real-valued function on $X$ is bounded. Then $X$ is compact.*

**Proof** Let every continuous function $f: X \to \mathbf{R}$ be bounded. Suppose $X$ is not bounded. Then the function $f: X \to \mathbf{R}$, $x \mapsto ||x||$ is not bounded on $X$. This implies a contradiction and hence $X$ is bounded. If possible, $X$ is not compact. Then it is not closed by Heine–Borel theorem. Then there exists a point $\alpha$ in $\overline{X} - X$. Under this situation, the function $g: X \to \mathbf{R}$, $x \mapsto ||x - \alpha||^{-1}$ cannot be bounded on $X$. This contradiction shows that $X$ is compact. □

A generalization of Proposition 5.27.7 is given in Proposition 5.27.8.

**Proposition 5.27.8** *Let $X$ be a subspace of the Euclidean space $\mathbf{R}^n$ such that every continuous real-valued function on $X$ is bounded. Then $X$ is compact.*

**Proof** Suppose $X$ is not compact. Then the function $f: X \to \mathbf{R}, x \mapsto ||x||$ is not bounded on $X$. If $X$ is bounded but it is not compact, then it is not closed by Heine–Borel theorem. Then there exists a point $\alpha \in \overline{X} - X$. Under this situation, the function $g: X \to \mathbf{R}, x \mapsto ||x - \alpha||^{-1}$ is not bounded on $X$. □

**Theorem 5.27.9** *Let $X$ be a nonempty compact subset of a complete metric space $(M, d)$. If $f: X \to X$ is a map satisfying the condition (contraction)*

$$d(f(x), f(y)) < d(x, y)$$

*for all $x, y$ ($x \neq y$) $\in X$, then there exists a unique point $x \in X$ such that $f(x) = x$ i.e., $f$ has a unique fixed point.*

**Proof** Let $g$ be a map defined by

$$g: X \to \mathbf{R}, x \mapsto d(x, f(x)).$$

As $f$ is a contraction map, it is continuous and hence $g$ is continuous. Again, since $X$ is compact and $X \neq \emptyset$, the map $g$ attains its minimum $m$ at a point $x_0 \in X$. Then $d(x_0, f(x_0)) = m$. By minimality $m$ of $g$, it implies that

$$d(f(x_0), f(f(x_0))) = g(f(x_0)) \geq m = d(x_0, f(x_0)).$$

Then by contraction theorem (see Chap. 2), it follows that $f(x_0) = x_0$. If $x_1 \in X$ is a fixed point of $f$, then $f(x_1) = x_1$. Since $f$ is a contraction map,

$$d(f(x_0), f(x_1)) = d(x_0, x_1) \implies x_0 = x_1,$$

otherwise, there would be a contradiction by the condition of the given contracting map $f$. $\qquad\square$

**Proposition 5.27.10** *Let $(X, d)$ be a compact metric space. If $f: X \to X$ is an isometry, then the map $f$ is onto.*

**Proof** Let $(X, d)$ be a compact metric space and $f: X \to X$ be an isometry. Then

$$d(x, x') = d((f(x), f(x'))), \forall x, x' \in X.$$

If possible, $f$ is not onto. Then there is a point $x$ which is not in $f(X)$. Since $f(X)$ is closed, there is a real number $\epsilon > 0$ such that $d(x, f(X)) \geq \epsilon$. Again, since by hypothesis, $X$ is compact, by using Bolzano–Weierstrass theorem, the sequence $\{f^n(x)\}$ has a convergent subsequence $\{f^{n_k} f(x)\}$. Again, for $k < m$, it follows that

$$d(f^{n_k}(x), f^{n_m}(x)) = d(x, f^{n_m - n_k}(x)) \geq .\epsilon$$

But it is not possible, since every convergent sequence in $X$ is a Cauchy sequence. This contradiction proves the proposition. $\qquad\square$

**Proposition 5.27.11** *Let $X$ be a connected space and $Y$ be a discrete space. If $f: X \to Y$ is locally constant, then $f$ is constant.*

**Proof** Left as an exercise. $\qquad\square$

**Proposition 5.27.12** *Let $X$ be a compact metric space with metric $d$ and $Y \subset X$ be nonempty. If for a given continuous map $f: X \to \mathbf{R}$, its restriction $f|_Y$ to $Y$ assumes a maximum value on $Y$, then $Y$ is compact.*

*Proof* Let $a \in \overline{Y}$. Define a function

$$h: X \to \mathbf{R}, x \mapsto -d(x, a).$$

Since the restriction function $h|_Y$ attains a maximum value in $Y$ by hypothesis and $a \in \overline{Y}$, it follows that the function $h|_Y$ assumes nonpositive values for attaining values arbitrarily close to 0. Then its maximum value is precisely 0. This implies that $a \in Y$. This asserts that $Y$ is a closed subset of the compact space $X$ and hence $Y$ is also compact. $\square$

**Proposition 5.27.13** $X \subset \mathbf{R}^n$ *be an arbitrary connected, locally path connected subset. Then $X$ is path connected.*

*Proof* Given a point $a \in X$, define a set

$$X_a = \{x \in X \ : \ a \text{ and } x \text{ are connected by a path in } X\} \subset X.$$

As $X$ is locally path connected, then the set $X_a$ is open and its complement in $X$ is also so. These two open sets produce a partition of $X$. As by hypothesis $X$ is connected, this partition is trivial and hence $X_a = X$. $\square$

## 5.27.2 Brouwer Fixed Point Theorem for Dimension 1

Brouwer fixed point theorem named after Luitzen Egbertus Jan Brouwer (1881–1966) asserts that every continuous mapping $f: \mathbf{D}^n \to \mathbf{D}^n$ has a fixed point. This subsection proves Brouwer fixed point theorem for dimension one by using Bolzano theorem 5.18.6. Proof for general case is given in **Basic Topology, Volume 2** by using topology of manifolds and in **Basic Topology, Volume 3** by using algebraic topology.

**Theorem 5.27.14** *(Brower fixed point theorem for dimension 1) Every continuous map $f: \mathbf{I} \to \mathbf{I}$ has a fixed point.*

*Proof* If $f(0) = 0$ or $f(1) = 1$, then the proof is trivial. So without loss of generality, we may assume that $f(0) > 0$ and $f(1) < 1$. Consider the map

$$h: \mathbf{I} \to \mathbf{R}, x \mapsto x - f(x).$$

Then $h$ is a continuous map such that $h(0) \cdot h(1) < 0$. It asserts by Corollary 5.18.6 that there is a point $\alpha \in \mathbf{I}$ such that $h(\alpha) = 0$. This asserts that $f(\alpha) = \alpha$. This shows that $\alpha$ is a fixed point of $f$. $\square$

**Geometrical Interpretation of Theorem** 5.27.14: If $f: \mathbf{I} \to \mathbf{I}$ is a continuous map, then its graph $G_f = \{(x, f(x)): x \in \mathbf{I}\}$ lies entirely in the unit square $\mathbf{I}^2 = \mathbf{I} \times \mathbf{I}$. The point $(\alpha, f(\alpha))$ in Theorem 5.27.14 is the point of intersection of the graph $G_f$ and the line $y = x$.

**Theorem 5.27.15** *Let $X$ and $Y$ be homeomorphic spaces and $f: X \to X$ be any continuous map. Then $f$ has a fixed point iff every continuous map $g: Y \to Y$ has a fixed point.*

**Proof** Let $X$ and $Y$ be homeomorphic spaces. Then there exist homeomorphisms

$$h: X \to Y$$

and

$$k: Y \to X$$

such that

$$h \circ k = 1_Y \text{ and } k \circ h = 1_X.$$

Let $g: Y \to Y$ be an arbitrary continuous map. First suppose that every continuous map $f: X \to X$ has a fixed point. Consider the map

$$f = k \circ g \circ h: X \to X$$

is continuous. Hence by hypothesis, $f(x_0) = x_0$ for some $x_0 \in X$. If $h(x_0) = y_0 \in Y$, then

$$g(y_0) = g(h(x_0)) = (h \circ k)(g(h(x_0))) = h \circ (k \circ g \circ h)(x_0)$$
$$= (h \circ f)(x_0) = h(f(x_0)) = h(x_0) = y_0.$$

asserts that $y_0$ is a fixed point of $g: Y \to Y$. This proves that if $f$ has a fixed point, then $g$ has also a fixed point. Since the hypothesis is symmetric with respect to homeomorphic spaces $X$ and $Y$, proceed as in the first part to show that if every continuous map $g: Y \to Y$ has a fixed point, then each continuous map $f: X \to X$ has also a fixed point. □

**Corollary 5.27.16** *Every continuous map $f: [a, b] \to [a, b]$ has a fixed point.*

**Proof** Since the closed intervals $[a, b]$ and $[0, 1] = \mathbf{I}$ are homeomorphic and every continuous map $g: [0, 1] \to [0, 1]$ has a fixed point by Theorem 5.27.14. Hence it follows that $f: [a, b] \to [a, b]$ has also a fixed point. □

## 5.28  Exercises

1. Let $(X, \tau)$ be a countably compact space. Show that every closed subspace of $(X, \tau)$ is also countably compact.

2. Let $(X, \tau)$ be a topological space. Show that the union of any finite collection of countably compact subspaces of $(X, \tau)$ is also countably compact.

3. Let $X$ be a compact metric space and $f : X \to \mathbf{R}$ be continuous. Show that $f(X)$ is bounded; and $\exists$ (there exist) points $a, b \in X$ such that $f(a) = \inf_{x \in X} f(x)$ and $f(b) = \sup_{x \in X} f(x)$.

4. A metric space $X$ has the finite intersection property for closed sets in $X$ if every decreasing sequence of closed, nonempty sets has nonempty intersection. Show that a metric space is sequentially compact iff it has the finite intersection property for closed sets.

5. Show that the open interval $(2,4)$ is not homeomorphic to the closed interval $[2,4]$ in the real line space $\mathbf{R}$.
   [Hint: Use the fact that compactness is a topological property. Here, $[2,4]$ is compact but $(2,4)$ is not so.]

6. Let $M$ be a subset of a metric space $(X, d)$. Show that the following statements are equivalent:

   (i) $M$ is compact.
   (ii) $M$ is countably compact.
   (iii) $M$ is sequentially compact.

7. Let $f : \mathbf{I} \to \mathbf{R}^n$ be a continuous map which is injective. Show that the spaces $\mathbf{I}$ and $f(\mathbf{I}) \subset \mathbf{R}^n$ are homeomorphic.
   [Hint: Use the result that every continuous surjective map from a compact space to a Hausdorff space is a homeomorphism. Here, $\mathbf{I}$ is compact and $\mathbf{R}^n$ is a metric space and hence it is Hausdorff.]

8. Show that the circle $S^1$ is not homeomorphic to the real line space $\mathbf{R}$.
   [Hint: $S^1$ is compact but $\mathbf{R}$ is not so.]

9. Show that the Sierpinski space $(S, \tau_S)$

   (i) is compact;
   (ii) is connected;
   (iii) is path connected;
   (iv) is locally path connected.
   [Hint: $(S, \tau_S)$ is connected, since the only sets which are both open and closed in this space are $\emptyset$ and $S$.]

10. Show that

    (i) if $X$ is a path connected space and $f : X \to Y$ is a continuous map, then $f(X) \subset Y$ is also path connected;
    (ii) if $f : X \to Y$ is a homeomorphism, then $X$ is path connected iff $Y$ is path connected;
    (iii) if $\{X_i\}$ is a family of path connected subspaces of $X$ such that their intersection $\bigcap X_i \neq \emptyset$, then their union $\bigcup X_i$ also path connected.

11. In the Euclidean plane $\mathbf{R}^2$, consider two subspaces

(i) $X_1 = \{(x,\ sin\ \frac{\pi}{x}) \in \mathbf{R}^2 : 0 < x \leq 1\}$;

(ii) $X_2 = \{(0, y) \in \mathbf{R}^2 : -1 \leq y \leq 1\}$.

Show that their union $X = X_1 \cup X_2$ is connected but it is not path connected. [Hint: The subspace $X_1 \subset \mathbf{R}^2$ is connected, since it is the image of $(0, 1]$ of a continuous map, It is not possible to join points of $X_1$ to a point of $X_2$.]

12. Let $f : X \to \mathbf{I}$ be continuous map. Show that $X$ is connected iff $f$ is a constant function given by $f(x) = 0\ or\ 1,\ \forall x \in X$.

[Hint: Use the result that $f(X) \subset \mathbf{I}$ is connected.]

13. Show the general linear group $GL(n, \mathbf{R})$, considered as a subspace of $\mathbf{R}^{n^2}$ is not connected.

14. Show that in the Euclidean plane $\mathbf{R}^2$, the graph of the parabola $y = x^2$ is not homeomorphic to the graph of hyperbola $x^2 - y^2 = 1$, though both of them are noncompact.

[Hint: Use the fact that one is path connected but the other is not so.]

15. Let $(X, d)$ be a complete metric space and $A$ be a subspace of $X$. Show that $\overline{A}$ is compact iff $A$ is totally bounded in $X$, i.e., iff every subset of the metric space $X$ is contained in the union of finitely many open balls of radius $r$, for any $r > 0$.

16. For a metric space $X$, show that the following statements are equivalent:

   (i) **(Heine–Borel Property)** $X$ is compact;

   (ii) **(Bolzano–Weierstrass Property)** $X$ is (B–W) (limit point) compact;

   (iii) **(Sequentially Compactness Property)** Every sequence in $X$ has a convergent subsequence.

17. Show that

   (i) every compact space is countably compact;

   (ii) every sequentially compact space is countably compact;

   (iii) every sequentially compact subset of a metric space is totally bounded.

18. Let $C(X, \mathbf{R})$ be the ring of all real-valued continuous functions from a topological space $X$ to the real line space $\mathbf{R}$. In particular, if $X$ is a finite subspace of $\mathbf{R}$ and $A$ is a proper ideal of the ring $C(X, \mathbf{R})$, show that there is a nonempty subset $S$ of $X$ such that $A$ consists of precisely the functions in $C(X, \mathbf{R})$ which vanish on $S$.

19. Let $X$ and $Y$ be two topological spaces. Then a continuous map $f : X \to Y$ is said to be **proper** if $f^{-1}(A)$ is a compact subset of $X$ for every compact subset $A$ of $Y$. Show that

   (i) if $f : X \to Y$ is a closed map such that for every point $y \in Y$, the set $f^{-1}(y)$ is compact in $X$, then the map $f$ is proper;

   (ii) if $X$ is compact, then the projection map

$$p : X \times Y \to Y,\ (x, y) \mapsto y$$

   is proper.

20. Let $X$ be a compact space. Show that every infinite subset of $X$ has a limit point. [Hint: It is sufficient to show that if a subset $A$ of $X$ has no limit point, then the set $A$ is finite (see proof of Theorem 5.3.3).]

21. Let $X$ be a compact space and $Y$ be a Hausdorff space. Show that every continuous map $f: X \to Y$ is a closed map.

22. Let $X$ be a compact space and $f: X \to \mathbf{R}$ be continuous. Show that $f$ assumes its maximum and minimum (values).

23. Show that if $f: \mathbf{I} \to \mathbf{R}$ is a continuous map, then $f(\mathbf{I})$ is a segment.

24. Show that the projective plane is homeomorphic to the mapping cone of the map

$$f: S^1 \to S^1, \; z \mapsto z^2$$

in the complex plane.

25. Let $X$ be a compact Hausdorff space. Show that $X$ is also a regular space.

26. Let $(X, d)$ be metric space and $A$ be a compact subset of $X$. Show that

(i) $A$ is closed and bounded;
(ii) for any closed set $F$ contained in $A$, the set $F$ is compact;
(iii) any continuous image of $A$ to a topological space is a compact subset;
(iv) **(Extreme value property)** if $X$ is itself compact, then given a continuous function $f: X \to \mathbf{R}$, there are points $x_1, x_2 \in X$ such that

$$f(x_1) \leq f(x) \leq f(x_2), \; \forall x \in X.$$

27. Let $(X, d)$ be a metric space and $A$ be a closed subspace of $X$. Show that the following statements are equivalent:

(i) $A$ is compact;
(ii) for any collection $\mathcal{F}$ of closed subsets of $A$ having finite intersection property, the intersection $\bigcap F_{F \in \mathcal{F}} \neq \emptyset$;
(iii) every sequence in $A$ has a convergent subsequence;
(iv) every infinite subset of $A$ has a limit point;
(v) $(A, d)$ is a complete metric space that is totally bounded, i.e., given an $\epsilon > 0$, there exist points $x_1, x_2, \ldots, x_n \in A$ such that

$$A \subset \bigcup_{i=1}^{n} B_{x_i}(\epsilon).$$

[Hint: Equivalence of statements (i) and (ii) follows from Theorem 5.9.6. To prove $(iii) \implies (iv)$, take an infinite subset $B$ of $A$. Then $\mathbf{B}$ has a sequence of distinct points $\{x_n\}$. Hence by (iii), this sequence $\{x_n\}$ has a convergent subsequence, which converges to a point $x \in \mathbf{B}$. This shows that $x$ is a limit point of $B$. To prove that $(iv) \implies (iii)$, take an infinite sequence $\{x_n\}$ of distinct points in $A$. Then by assumption $(iv)$, the sequence $\{x_n\}$ has a limit point $x$ say. Since $A$ is closed by hypothesis, $x \in A$. Hence, it follows that the sequence $\{x_n\}$ has a convergent subsequence. This gives equivalence of statements (iii) and (iv).]

28. Deduce from (v) of Exercise 27 that the closed interval $[a, b] \subset \mathbf{R}$ (with usual topology) is compact.
    [Hint: $[a, b]$ is a closed subset of the complete metric space $\mathbf{R}$ and hence it is complete. It is totally bounded, since for any $\epsilon > 0$, there are $x_i \in [a, b]$ such that
    $$a = x_1 < x_2 < \cdots < x_n = b$$
    with $x_j - x_{j-1} < \epsilon$. This asserts that
    $$[a, b] \subset \bigcup_{j=1}^{j=n} B_{x_j}(\epsilon).]$$

29. Let $X$ be a topological space. If $X_0$ is a connected subset and $\{X_j : j \in \mathbf{J}\}$ is a family of connected subsets of $X$ such that $X_0 \cap X_j \neq \emptyset$, $\forall j \in \mathbf{J}$. Show that $X_0 \bigcup (\bigcup_{j \in \mathbf{J}} X_j)$ is connected. Hence show that $\mathbf{R}^n$ is connected.
    [Hint: To show that $\mathbf{R}^n$ is connected. let $X_0 = \mathbf{0} = (0, 0, \dots 0)$ and $\{X_j : j \in \mathbf{J}\}$ denote the set of all lines in $\mathbf{R}^n$ through the origin $\mathbf{0}$ by taking the indexing set the unit sphere in $\mathbf{R}^n$. Hence $\mathbf{R}^n = X_0 \bigcup (\bigcup_{j \in \mathbf{J}} X_j)$ is connected by the first part.]

30. Let $X$ be a topological space. Show that

    (i) if $\{X_k : k \in \mathbf{K}\}$ is a family of path connected subsets of $X$ such that $X_k \cap X_l \neq \emptyset \, \forall k, l \in \mathbf{K}$, then $A = \bigcup_{k \in \mathbf{K}} X_k$ is also path connected;
    (ii) if $\{X_k : k \in \mathbf{N}\}$ is a sequence of path connected subsets of $X$ such that $X_k \cap X_{k+1} \neq \emptyset$, $\forall k \in \mathbf{N}$, then $A = \bigcup_{k \in \mathbf{N}} X_k$ is also path connected.

31. Let $(X, \tau)$ be a topological space. Show that

    (i) $X$ is locally connected iff for each open set $U \subset X$, every connected component of $U$ is open in $X$;
    (ii) $X$ is locally path connected iff for each open set $U \subset X$, every path component of $U$ is open in $X$;
    (iii) (a) if every connected path component of $X$ is contained is a connected component of $X$ and
         (b) $X$ is locally path connected,
         then the connected components and connected path components of $X$ are identical.

32. Let $X$ be a compact space and $Y$ be a Hausdorff space. If $f : X \to Y$ is a one–one continuous map, show that the spaces $X$ and $f(X)$ are homeomorphic.

33. Let $X, Y$ be topological spaces and $f : X \to Y$ be a continuous map. Given a sequentially compact subset $S$ of $X$, show that $f(S)$ is also a sequentially compact subset of $Y$.

34. Let $(X, \tau)$ be a topological space. Show that it is countably compact iff it satisfies any one of the following conditions:

   (i) Each countable family of closed sets in $X$ having the finite intersection property (FIP) has a nonempty intersection in $X$;
   (ii) Each descending chain of nonempty closed sets in $X$ :

$$A_1 \supset A_2 \supset \cdots \supset$$

   has a nonempty intersection in $X$ **(Cantor's intersection theorem)**;
   (iii) Each infinite sequence $\{x_n\}$ in $X$ has a cluster point $p \in X$ in the sense that given any open set $U$ containing $p$ and any positive integer $m$, there exists a positive integer $n_0$ such that $x_m \in U$, $\forall m > n_0$.
   (iv) Each infinite set $A \subset X$ has an $\omega$-accumulation point $p \in X$ in the sense that every nbd of $p$ intersects $A$ in infinitely many points.

35. Let $\mathbf{RP}^2$ be the real projective plane (space of lines through the origin in $\mathbf{R}^3$) and $SO(3, \mathbf{R})$ be the group of orthogonal transformations of $\mathbf{R}^3$ of determinant 1. Let $X \subset SO(3, \mathbf{R})$ be the subset of nonidentity symmetric transformations with the usual subspace topology inherited from $\mathbf{R}^{3^2}$. Show that

   (i) $SO(3, \mathbf{R})$ and $\mathbf{RP}^2$ are compact spaces under their usual topologies;
   (ii) $\mathbf{RP}^2$ and $X$ are homeomorphic spaces.

36. Let $(X, \tau)$ be a topological space and $\mathcal{F}$ be a family of subsets of $X$ with subspace topology such that

$$X = \bigcup \{S : S \in \mathcal{F}\}.$$

   If each subset $S \in \mathcal{F}$ is connected and no pair of elements of $\mathcal{F}$ are separated from each other in $X$, show that the topological space $(X, \tau)$ is connected.

37. Show that cardinality of connected components of a topological space $(X, \tau)$ is a topological characteristic of $X$ in the sense that this cardinality is the same as the cardinality of connected components of any topological space $Y$ that is homeomorphic to $X$.

38. Let $(X, \tau)$ be a topological space. Show that the following statements are equivalent:

   (i) $X$ is a connected space.
   (ii) If a subset $Y$ of $X$ is open and closed in $X$, then $Y = \emptyset$ or $Y = X$.
   (iii) If $X = Y \cup Z$, where $\bar{Y} \cap Z = \emptyset$ and $Y \cap \bar{Z} = \emptyset$, then $Y = \emptyset$ or $Z = \emptyset$.
   (iv) Every continuous map $f : X \to \{1, 2\}$ is a constant map.

39. Show that a topological space $X$ is connected iff given any two points in $X$ they lie in some connected subset of $X$, i.e., iff $X$ has only one connected component.

40. Show that each subspace of the Sorgenfrey line, which is compact is also countable.
    [Hint: Consider the identity map from the Sorgenfrey line onto the Euclidean line **R**.]

41. Show that the Hilbert space is connected

42. Show that in the Hilbert metric space, the unit closed ball

$$\mathbf{D}^\infty = \{(x_n)_{n=1}^\infty \in \mathbf{R}^\infty : [\Sigma_{n=1}^\infty x_n^2]^{\frac{1}{2}} \leq 1\}$$

   (i) is closed and bounded;

  (ii) is not compact.
    [Hint: Consider $\{(x_n)_{n=1}^\infty \in \mathbf{R}^\infty : x_k = 1, x_n = 0, \forall n \neq k\}$. It has no accumulation point.]

43. Let $X_1, X_2, X_3$ be three subsets of a topological space $(X, \tau)$. Show that

   (i) $X_1 \cup X_2$ is connected if $X_1$ and $X_2$ are both connected in $X$ and $X_1 \cap X_2 \neq \emptyset$;

  (ii) $X_1 \cup X_2 \cup X_3$ is connected if all of $X_1, X_2, X_3$ are connected and $X_1 \cap X_3 \neq \emptyset$ and $X_2 \cap X_3 \neq \emptyset$.

44. If $\{(X_a, \tau_a) : a \in \mathbf{A}\}$ is a family of connected spaces, then that their product space $\Pi_{a \in \mathbf{A}}(X_a, \tau_a)$ is also connected.

45. Show the subspace $X$ of $\mathbf{R}^2$ defined by

$$X = \{(t, 1/t) : t \in (0, 1]\} \subset \mathbf{R}$$

is connected.
    [Hint: Define $f : (0, 1] \mapsto \mathbf{R}^2 : t \mapsto (t, 1/t)$. Then $X$ is the continuous image of $f$ in $\mathbf{R}^2$.]

46. Let $C_x$ be a connected component of a topological space $(X, \tau)$ at an arbitrary point $x \in X$. Show that $C_x$ is a closed set of $X$.
    [Hint: Use the results that $C_x$ is the largest connected set containing the point $x$ of $X$ and $\overline{C_x}$ is also connected. This asserts that $\overline{C_x} = C_x$.]

47. Given a metric space $(X, d)$, the metric $d$ is said to an $M$-**metric**, if every open ball $B_x(\epsilon)$ in $(X, d)$ is connected for every $x \in X$ and every $\epsilon > 0$. Show that a metrizable space $(X, \tau)$ is connected and locally connected iff there exists an $M$-metric $d$ on $X$ such that $\tau = \tau_d$ (the topology induced by $d$).

48. Let $\{X_a, \tau_a : a \in \mathbf{A}\}$ be a family of topological spaces and $(X, \tau)$ be their product space. If $(X, \tau)$ is locally connected, show that

   (i) every $X_a$ is locally connected;

  (ii) every $X_a$ excepting only for a finite number of $a$'s is connected.

49. Let $X$ be a complete metric space and $Y$ be a closed subset of $X$. Show that the following statements are equivalent.

(i) $Y$ is compact.

(ii) Every infinite subset of $Y$ has a limit point in $Y$.

(iii) $Y$ is totally bounded.

50. Show that the circle $S^1$ is connected.

[Hint: Consider the mapping $f : \mathbf{I} \to S^1$, $t \mapsto (\cos 2\pi t, \sin 2\pi t)$].

51. Let $(X, \tau)$ be a topological space. Show that it is locally compact Hausdorff iff there exists a unique space $X^+$ (up to homeomorphism) such that

(i) $X$ is a subspace of $X^+$;

(ii) the cardinality of the set $X^+ - X$ is one and

(iii) $X^+$ is a compact Hausdorff space.

52. Show that

(i) the infinite dimensional Euclidean space $\mathbf{R}^\infty$ is not locally compact;

(ii) the space the $\mathcal{C}(\mathbf{I}, \mathbf{R})$ (with compact open topology) is not locally compact.

[Hint: Consider the sequence of functions $\{f_n\}$ in $\mathcal{C}(\mathbf{I}, \mathbf{R})$ defined by

$$f_n : \mathbf{I} \mapsto \mathbf{R}.\, x \mapsto \begin{cases} nx, & \text{for all } x \in [0, 1/n] \\ 1, & \text{for all } x \in [1/n, 1] \quad .] \end{cases}$$

53. Let $(X, \tau)$ be a complete regular space and $C \subset X$ be compact. If $U \supset C$ be an open set in $X$, show that there exists a continuous map $f : X \to \mathbf{I}$ such that

$$f(x) = \begin{cases} 0, & \text{for all } x \in C \\ 1, & \text{for all } x \in X - U \end{cases} .$$

54. Let $(X, \tau)$ be topological space. Prove the following statements:

(i) given a point $x \in X$ and a subset $A \subset X$, the point $x \in \bar{A}$ iff there exists a net $\psi$ in $A$ which converges to the point $x$;

(ii) given a point $x \in X$ and a subset $A \subset X$, the point $x$ is an accumulation point of $A$ iff there exists a net $\psi$ in $A - \{x\}$ which converges to the point $x$.

(iii) The space $(X, \tau)$ is Hausdorff iff any two limits of any convergent net are equal, i.e., every net in $(X, \tau)$ converge at most one point.

(iv) The space $X$ is compact iff every net in $(X, \tau)$ has a subset which converges to point of $X$.

55. Let $(X, \tau)$ be topological space and $\psi : \mathcal{D} \to X$ be a net in $(X, \tau)$. Show that a point $a \in X$ is a cluster point of $\psi$ iff there exists a subnet $\beta : \mathcal{F} \to X$ converging to the point $a$.

56. **(Characterization of compactness by net)** Let $(X, \tau)$ be a topological space. Show that it is compact iff every net $\psi : \mathcal{D} \to X$ in $X$ has a subnet which converges to a point in $X$.

57. **(Characterization of continuity by net)** Let $f: X \to Y$ be a function of topological spaces. Show that it is continuous at a point $a \in X$ iff for every net $\psi: \mathcal{D} \to X$ in $X$ converging to the point $a$, the composite function $\beta = f \circ \psi: \mathcal{D} \to Y$ converges to the point $f(a)$.

58. Let $\mathcal{F} = \{X_a: a \in A\}$ be a family of connected spaces. If $(X, \tau)$ is the topological product space of this family of connected spaces, show that the space $X$ is also connected. Use this result to prove that

   (i) the Euclidean $n$-space $\mathbf{R}^n$ is connected for every integer $n \geq 1$;
   (ii) the $n$-cube $\mathbf{I}^n$ is connected for every integer $n \geq 1$;
   (iii) $S^n$ is connected for every integer $n \geq 1$.
   [Hint: For any point $x \in S^n$, the subspace $S^n - \{x\}$ is homeomorphic to $\mathbf{R}^n$ by stereographic projection from $x$. Since the closure $\overline{S^n - \{x\}} = S^n$, it follows that $S^n$ is connected.]

59. Let $X$ be a compact subset of the metric space $(\mathbf{R}^n, d)$ and $\{B_k\}$ be an open covering of $X$. Show that there is an $\epsilon > 0$ such that the open $\epsilon$- ball $B_a(\epsilon)$ centered at a point $a \in X$ with radius $\epsilon$, which is contained in one of the balls $B_k's$.

60. **(Harr–Konig theorem)** Show that a linearly ordered set $X$ with its interval topology is compact iff it is order-complete in the sense that every nonempty subset of $X$ has a least upper bound (lub) and a greatest lower bound ( glb).
   [Hint: Let $X$ be linearly ordered set having a lub and a glb. If $\tau$ is the interval topology on $X$, then $X$ is compact. Conversely, if $\tau$ is the interval topology on $X$ and $X$ has no lub and no glb, then $X$ cannot be compact.]

61. Let $(X, \tau)$ be a metrizable space and $\mathcal{C}$ be an open covering of $X$. Show that there is an open covering $\mathcal{L}$ which refines $\mathcal{C}$ and is countably locally finite.

62. Let $(X, \tau)$ be a regular space. Show that the following statements are equivalent:

   (i) An open covering $\mathcal{C}$ of $X$ which is countably locally finite.
   (ii) An covering $\mathcal{C}$ of $X$ which is locally finite.
   (iii) A closed covering $\mathcal{D}$ of $X$ which is locally finite.
   (iv) An open covering $\mathcal{C}$ of $X$ which is locally finite.

63. Let $(X, \tau)$ be a paracompact space and $\mathcal{L}$ be a locally finite open covering of $X$. Show that for every open set $U \in \mathcal{L}$, there is an open set $V_U$ such that

$$\overline{V_U} \subset U \text{ and } \mathcal{V} = \{V_U: U \in \mathcal{L}\}\text{is also a locally finite open covering of}X.$$

64. Let $(X, \tau)$ be a normal space and $\mathcal{L}$ be a locally finite open covering of $X$. If $\{V_U: U \in \mathcal{L}\}$ is another open covering of X such that $\mathcal{V} = \overline{V_U} \subset U$, show that there exists a partition of unity on $X$ subordinate to the covering $\mathcal{L}$.

65. Let $(X, \tau)$ be a paracompact Hausdorff space and $\mathcal{S}$ be a family of subsets of $X$ such that for every subset $S \in \mathcal{S}$, there is a real number $\epsilon(S) > 0$. If $\mathcal{S}$ is locally finite, show that there is a real-valued continuous function

$$f: X \to \mathbf{R}: f(x) > 0, \quad \forall x \in X \text{ and} f(x) \leq \epsilon(S), \forall S \in \mathcal{S}.$$

66. Let $X \cup \{\infty\}$ be the one-point compactification of $X$. Show that $X \cup \{\infty\}$ is Hausdorff iff $X$ is locally compact and Hausdorff. Hence prove that the one-point compactification $\mathbf{Q} \cup \{\infty\}$ of $\mathbf{Q}$ is not Hausdorff.
    [Hint: $\mathbf{Q}$ is not locally compact.]

67. Show the one-point compactification of an open disk $\mathbf{D}^2$ in the Euclidean space $\mathbf{R}^2$ is homeomorphic to the 2-sphere $S^2$.

68. Prove that a homeomorphism $h: \mathbf{R}^3 \to \mathbf{R}^3$ has a unique extension to a homeomorphism $\tilde{h}: S^3 \to S^3$.
    [Hint: Use the result that the 3-sphere $S^3$ is the one-point compactification $\mathbf{R}^3 \cup \{\infty\}$ of $\mathbf{R}^3$.]

69. Let $X$ and $Y$ be locally compact Hausdorff spaces and $f: X \to Y$ be a continuous surjective map. Show that $f$ has a continuous extension

$$\tilde{f}: X \cup \{\infty\} \to Y \cup \{\infty\}$$

    iff $f^{-1}(A)$ is compact for every compact subset $A$ of $Y$. Hence show that homeomorphic spaces have homeomorphic one-point compactifications. Is its converse necessarily true? Justify your answer.
    [Hint: Converse is not true.]

70. Show that every compact Hausdorff space is normal.

71. Let $(X, \tau)$ be a compact Hausdorff space and $A$ be a subset of $X$. Show that $A$ is closed in $X$ iff $A$ is compact.

72. Show that a compact metric space is sequentially compact and conversely, a sequentially compact metric space is compact.

73. A topological space $(X, \tau)$ is said be said to be **locally $n$-Euclidean**, if every point $x \in X$ has an open nbd homeomorphic to an open subset of $\mathbf{R}^n$. It is said to be an **$n$-manifold**, if it is locally $n$-Euclidean and is Hausdorff having a countable basis. Show that $S^2$ is a 2-manifold.

74. Let $X$ and $Y$ be topological spaces and $C(X, Y)$ be the set of all continuous maps from $X$ to $Y$ endowed with the compact open topology. Show that

    (i) the space $C(X, Y)$ is Hausdorff if the space $Y$ is so;
    (ii) the space $C(X, Y)$ is regular if the space $Y$ is so.

75. Let $(X.\tau)$ be a topological space and $(Y, d)$ be metric space. If the set $C(X, Y)$ of all continuous maps from $X$ to $Y$ is endowed with both the compact open topology $\sigma_\tau$ and uniform convergence topology $\sigma_d$, show that $\sigma_\tau = \sigma_d$.

76. Let $(X, d)$ be a compact metric space and $A$ be a closed subset of $C(X, \mathbf{R})$. Show that

    (i) $A$ is compact if $A$ is equicontinuous and
    (ii) the set $A_x = \{f(x): f \in A\}$ of real numbers is bounded in $\mathbf{R}$ for every $x \in X$ (called **pointwise bounded**).

77. Let $X, Y$ and $Z$ be three topological spaces. Define a map

$$\psi \colon C(X, Y) \times C(Y, Z) \to C(X, Z) \colon (f, g) \mapsto g \circ f.$$

If $X, Z$ are Hausdorff spaces, and $Y$ is locally compact, show that the map $\psi$ is continuous.

78. **(Ascoli's theorem)** Let $(X, d)$ be a compact metric space. Show that a closed subspace $A$ of

   (i) $C(X, \mathbf{R})$ is compact iff it is bounded and equicontinuous;
   (ii) $C(X, \mathbf{C})$ is compact iff it is bounded and equicontinuous

79. **(Classical Version of Ascoli's theorem)** Let $(X, \tau)$ be a compact metric space and $C(X, \mathbf{R}^n)$ be the function space with uniform topology induced by the Euclidean metric $d$ on $\mathbf{R}^n$. Show that a subspace $\mathcal{F}$ of $C(X, \mathbf{R}^n)$ has compact closure iff it is equicontinuous and pointwise bounded under the metric $d$.

80. **(Ascoli–Arzela)** Let $(X, \tau)$ be a compact space and

   $$\mathcal{F} = \{f_n \in C(X, \mathbf{R}^m) \colon f_n \text{ is pointwise bounded and equicontinuous}\}.$$

   Show that $\mathcal{F}$ has a uniformly convergent subsequence.

81. Given a locally connected space $(X, \tau)$, show that every connected component of an open subspace $Y$ of $X$ is open in $(X, \tau)$. Hence show that each connected component of $X$ is open.

82. Let $(Y, d)$ be a complete metric space. Show that under the metric given in Definition 5.24.11, $C(X, Y)$ is a complete metric space.

83. Given a Hausdorff space $Y$, show that an embedding $k \colon Y \to C(X, Y)$ is a closed map in the sense that $k(Y) \subset C(X, Y)$ is a closed set in $C(X, Y)$.

84. Let $(X, \tau)$ be a completely regular space $T_1$ space and $C(X, \mathbf{R})$ be the set of all real valued continuous functions on $X$. Show that there exists a function $f \in C(X, \mathbf{R})$ such that for points $x \neq y \in X$, the points $f(x) \neq f(y) \in \mathbf{R}$, i.e., distinct points in $X$ have distinct images in $\mathbf{R}$ under $f$.

85. Let $(X, \tau)$ be a completely regular space. Show that the space $X$ is homeomorphic to a subspace of a product space of closed intervals.
   [Hint: Let $C(X, \mathbf{I})$ be the set of all continuous functions from $X$ into the closed interval $\mathbf{I} = [0, 1]$. Considering $C(X, I)$ as an indexing set, for each $f \in C(X, I)$, take $\mathbf{I}_f = [0, 1]$ and $Y = \Pi_{f \in C(X, \mathbf{I})} \mathbf{I}_f$. Define a map

   $$\psi \colon X \to Y \colon x \mapsto \psi_x \text{ where } \psi_x \colon C(X, \mathbf{I}) \to \mathbf{I}, \ f \mapsto f(x).$$

   Then $\psi$ is a homeomorphism onto $\psi(X)$.]

86. Show that

   (i) every closed subspace of a paracompact space is paracompact;
   (ii) a subspace of a paracompact space is not necessarily paracompact.

87. Let $X$ and $Y$ be two homeomorphic spaces. Show that a continuous map $f \colon X \to X$ has a fixed point iff every continuous map $g \colon Y \to Y$ has a fixed point.

88. Let $X$ be a connected space and $a \in X$ be a point such that $X - \{a\}$ is totally connected. Show that $X - \{x\}$ is connected for every $x \in X - \{a\}$.

89. (i) Let $X$ be a topological space and $A \subset X$ be connected. If $B \subset X$ is such that $A \subset B \subset \overline{A}$, show that $B$ and $\overline{A}$ are both connected.

    (ii) Hence show that every connected component of a topological space is a closed set.

90. Show that every locally compact topological space is regular.

91. Let $X$ be a locally compact space and $X = \bigcup_{n=1}^{\infty} X_n$, where each $X_n$ is a compact set. Show that $X$ is paracompact.

92. Let $X$ be a locally compact space having a countable base. Show that $X$ is paracompact.

93. Let $X$ be a compact metric space and $Y$ be a metric space. Show that every continuous map $f : X \to Y$ is uniformly continuous.

94. Let $X$ be a compact metric space and $Y$ be a metric space and $f : X \to Y$ be a continuous onto mapping. Show that the inverse mapping $f^{-1} : Y \to X$ is a continuous surjective map.

95. Let $X \subset \mathbf{R}$ be a noncompact set. Show that

    (i) there exists a real-valued function $f : X \to \mathbf{R}$ which is not bounded;
    (ii) there exists a real-valued function $f : X \to \mathbf{R}$ which has no maximum;
    (iii) if $X$ is also bounded, then there exists a real-valued function $f : X \to \mathbf{R}$ which not uniformly continuous.

96. Show that the real projective space $\mathbf{R}P^n$ is compact.
    [Hint: $\mathbf{R}P^n$ is the quotient space of the compact space $S^n$ obtained by identifying the diametrically the opposite points of $S^n$.]

97. Let $(X, \tau)$ be any compact Hausdorff space and $C(X, \mathbf{R})$ be the space of all continuous functions $f : X \to \mathbf{R}$. Define for $f \in C(X, \mathbf{R})$, its norm

$$||f|| = max_{x \in \mathbf{X}} \{|f(x)|\}$$

and define a metric

$$d : C(X, \mathbf{R}) \times C(X, \mathbf{R}) \to \mathbf{R}, \ (f, g) \mapsto ||f - g||.$$

For $X = [0, 1]$, show that

    (i) $C(\mathbf{X}, \mathbf{R})$ is a Banach space;
    (ii) $(C((X, \mathbf{R}), d)$ is a complete metric space.
    (iii) $(C((X, \mathbf{R}), d)$ is a paracompact space.
    [Hint: For (ii), use the result that a uniformly convergent sequence of continuous functions converges to a continuous map and for (iii) apply Theorem 5.10.9.]

98. Show that a topological space $(X, \tau)$ is a Baire space iff given any countable family $\{X_n\}$ of open sets in $(X, \tau)$, every one of which is dense in $(X, \tau)$ is such

that their intersection

$$\bigcap_{n=1}^{\infty} X_n \neq \emptyset \text{ and it is dense in} X.$$

99. Show that

(i) if $(X, \tau)$ is a compact Hausdorff space, then it is a Baire space;

(ii) if $(X, \tau)$ is a locally compact Hausdorff space, then it is a Baire space;

(iii) if $(X, d)$ is complete metric space, then it is a Baire space of second category.

100. **(An alternative form of Baire Category theorem)** Let $X$ be a complete metric space. Show that in $X$, the following statements are equivalent.

(i) Every countable union of nowhere dense subsets has empty interior.

(ii) Every countable union of nowhere dense closed subsets of $X$ has empty interior.

(iii) If a countable union has nonempty interior, then the closure of some subset in the union has nonempty interior.

(iv) Every countable intersection of open and dense subsets of $X$ is also dense.

101. Let $X$ be an arbitrary set and $(Y, \sigma)$ be a topological space. Let $\mathcal{C}$ be a subfamily of $F(X, Y)$. Show that $\mathcal{C}$ is compact with respect to the point open topology $\tau$, if

(i) $\mathcal{C}$ is a closed subspace of $(F(X, Y), \tau)$;

(ii) for every point $x \in X$, the closure $\overline{\{f(x): f \in \mathcal{C}\}}$ is compact in $(Y, \sigma)$.

(iii) Moreover, if $(Y, \sigma)$ is Hausdorff, then $\mathcal{C}$ is compact under the point open topology $\tau$ iff $\mathcal{C}$ is closed and for every point

$$x \in X, \text{ the closure } \overline{\{f(x): f \in \mathcal{C}\}} \text{ is compact.}$$

102. Let $X$ be an arbitrary set and $Y$ be a topological space. Show that a sequence of functions $\{f_n\}$ in $\mathcal{F}(X, Y)$ converges to a function $h \in \mathcal{F}(X, Y)$ with respect to the point open topology on $\mathcal{F}(X, Y)$, iff $\{f_n\}$ converges pointwise to h.

103. Show that the following statements are equivalent in any metrizable space $(X, \tau)$:

(i) $X$ is a compact space;

(ii) $X$ is a limit point (B–W) compact space in the sense that every infinite subset of $X$ has a limit point;

(iii) $X$ is a sequentially compact space;

104. Let $(X, \tau)$ be a topological space. It is said to be $\sigma$-**compact,** if it is representable as a union of countably many compact subspaces. Prove that a locally compact Hausdorff space is paracompact iff it is a disjoint union of open $\sigma$-compact subsets of $X$.

105. Let $\mathbf{R}$ be the Euclidean line and $\mathbf{R}^n$ be the $n$-dimensional Euclidean space. Prove the following statements to have an alternative proof of Theorem 5.26.3:

  (i) Let $\mathbf{I}_1, \mathbf{I}_2, \cdots, \mathbf{I}_n$ be a covering of the closed interval $\mathbf{I} = [a, b] \subset \mathbf{R}$. Then there exists another covering $\mathbf{I}_1{}', \mathbf{I}_2{}', \ldots, \mathbf{I}_n{}'$ of $[a, b]$ such that each $\mathbf{I}_j{}' \subset \mathbf{I}_i$ for some $i$ and $\Sigma_{j=1}^n length(\mathbf{I}_j{}') < 2(b - a)$.

  (ii) For any subset $A \subset \mathbf{R}$, let $V_A$ represent the product space $V_A = A \times \mathbf{R}^{n-1} \subset \mathbf{R}^n$. Suppose $X \subset \mathbf{R}^n$ is compact and for any $p \in \mathbf{R}$ the set $V_p$ represents the vertical slice $\{p\} \times \mathbf{R}^{n-1} \subset \mathbf{R}^n$. If $X \cap V_p \subset U$ for some open set $U$ of $V_p$, then there exists $X \cap V_A \subset A \times U$ for some suitably small interval $A \subset \mathbf{R}$ containing the point $p$.

  (iii) (Fubini) If $X \subset \mathbf{R}^n$ is closed and $X \cap V_p$ has measure zero in $V_p$ for every $p \in \mathbf{R}^n$, then $X$ has measure zero in $\mathbf{R}^n$.
  [Hint: Use (i) and (ii) to prove (iii).]

## *Multiple Choice Exercises*

Identify the correct alternative (s) (there may be none or more than one) from the following list of exercises:

1. Let $(X, \tau)$ be a Hausdorff space. Then

  (i) any point $x \in X$ and any compact subset $A \subset X$, not containing the point $x$ can be separated by disjoint open sets in $(X, \tau)$;
  (ii) every compact subspace of $(X, \tau)$ is closed;
  (iii) every one-to-one continuous map from a compact space to $(X, \tau)$ is a homeomorphism.

2. Given two subsets $A$ and $B$ of the Euclidean line $\mathbf{R}$, let $X$ be the set defined by

$$X = \{x \in \mathbf{R} : x = a + b, \ a \in A, \ b \in B\}.$$

  (i) If $A$ is closed in $\mathbf{R}$ and $B$ is compact, then $X$ is closed in $\mathbf{R}$.
  (ii) If $A$ is closed in $\mathbf{R}$ and $B$ is compact, then $X$ is compact.
  (iii) If $A$ and $B$ are both closed in $\mathbf{R}$, then $X$ is also closed

3. Consider the following subspaces $X_1, X_2, X_3$ of the Euclidean plane $\mathbf{R}^2$.

  (i) If $X_1 = \{(x, y) \in \mathbf{R}^2 : x^2 + y^2 = 1\}$, then $X_1$ is compact.
  (ii) If $X_2 = \{(x, y) \in \mathbf{R}^2 : x^2 + y^2 < 1\}$, then $X_2$ is compact.
  (iii) If $X_3 = \{(x, y) \in \mathbf{R}^2 : xy = 1\}$, then $X_3$ is compact.

4. Consider the following subspaces $X_1, X_2, X_3$ of the Euclidean plane $\mathbf{R}^2$.

  (i) If $X_1 = \{(x, y) \in \mathbf{R}^2 : 0 \le 1, \ 0 < y \le 1\}$, then $X_1$ is locally compact.
  (ii) If $X_2 = \{(x, y) \in \mathbf{R}^2 : x^2 + y^2 + 103xy > 5\}$, then $X_2$ is locally compact.

(iii) If $X_3 = \{(x, y) \in \mathbf{R}^2 : x,\ y \text{ are irrational}\}$, then $X_3$ is locally compact.

5. Let $f \colon \mathbf{R} \to \mathbf{R}$ be a continuous map and $G_f = \{(x, f(x)) \colon x \in \mathbf{R}\}$ be its graph in the Euclidean plane $\mathbf{R}^2$. Then

   (i) $G_f$ is connected in $\mathbf{R}^2$;
   (ii) $G_f$ is open in $\mathbf{R}^2$;.
   (iii) $G_f$ is closed in $\mathbf{R}^2$.

6. Consider the following subspaces $X_1, X_2, X_3$ of the Euclidean plane $\mathbf{R}^2$. Then

   (i) $X_1 = \{(x, y) \in \mathbf{R}^2 : 1 < x^2 + y^2 < 2\}$ is connected.
   (ii) $X_2 = \{(x, y) \in \mathbf{R}^2 : x^2 y^2 = 1\}$ is connected.
   (iii) $X_3 = \{(x, y) \in \mathbf{R}^2 : x^2 + y^2 = 2\}$ is not connected.

7. Consider the following subspaces $X_1, X_2, X_3$ of the Euclidean plane $\mathbf{R}^2$. Then

   (i) $X_1 = \{(x, y) \in \mathbf{R}^2 : x,\ y \text{ are integers}\}$ is locally compact.
   (ii) $X_2 = \{(x, y) \in \mathbf{R}^2 : x^2 + y^2 = 1 \text{ and } xy \neq 0\}$ is compact.
   (iii) $X_3 = \{(x, y) \in \mathbf{R}^2 : x,\ y \text{ are irrationals}\}$ is locally compact.

# References

Adhikari, A., Adhikari, M.R.: Basic Topology, Volume 2: Topological Groups, Topology of Manifolds and Lie Groups. Springer, India (2022)

Adhikari, M.R.: Basic Topology, Volume 3: Algebraic Topology and Topology of Fiber Bundles. Springer, India (2022)

Adhikari, M.R.: Basic Algebraic Topolgy and its Applications. Springer, India (2016)

Adhikari, M.R., Adhikari, A.: Basic Modern Algebra with Applications. Springer, New Delhi, New York, Heidelberg (2014)

Alexandrov, P.S.: Introduction to Set Theory and General Topology. Moscow (1979)

Armstrong, M.A.: Basic Topology. Springer, New York (1983)

Bredon, G.E.: Topology and Geometry. Springer, New York (1993)

Borisovich, Y.C.U., Blznyakov, N., Formenko, T.: Introduction to Topology. Translated from the Russia by Oleg Efimov. Mir Publishers, Moscow (1985)

Brown, R.: Topology: A Geometric Account of General Topology, Homotopy Types, and the Fundamental Groupoid. Wiley, New York (1988)

Chatterjee, B.C., Ganguly, S., Adhikari, M.R.: Introduction to Topology. Asian Books, New Delhi (2002)

Dugundji, J.: Topology. Allyn & Bacon, Newton, MA (1966)

Fuks, D.B., Rokhlin, V.A.: Beginner's Course in Topology. Springer, New York (1984)

Hu, S.T.: Introduction to General Topology. Holden-Day, San Francisco (1966)

Heritt, E.: The role of compactness in analysis. American Math Monthly **67**(6), 499–516 (1960)

Johnstone, P.T.: Stone Spaces. Cambridge University Press, Cambridge (1996)

Kalajdzievski, S.: An Illustrated Introduction to Topology and Homotopy. CRC Press Taylor & Francis Group (2015)

Kelly, J.L.: General Topology, Van Nostrand, New York, 1955. Springer, New York (1975)

Mendelson, B.: Introduction to Topology College Mathematical Series. Allyn and Bacon, Boston (1962)

Munkres, J.R.: Topology. Prentice-Hall, New Jersey (2000)

Patterson, E.M.: Topology. Oliver and Boyd (1959)

Rotman, J.J.: An Introduction to Algebraic Topology. Springer, New York (1988)

Rudin, M.E.: A new proof that metric spaces are paracompact. Proc. Am. Math. Soc. **20**, 603 (1969)

Singer, I.M., Thorpe, J.A.: Lecture Notes on Elementary Topology and Geometry. Springer, New York (1967)

Steenrod, N.E.: A convenient category of topological spaces. Michigan Math. J. **14**(2), 133–152 (1967)

Stone, A.H.: Paracompactness and product spaces. Bull. Am. Math. Soc. **54**, 977–982 (1948)

Stone, M.H.: Compactification of Topological Spaces. Ann. Soc. Pol. Math. **21**, 153–160 (1948)

# Chapter 6
# Real-Valued Continuous Functions

This chapter continues the study of continuous functions from a topological space to the real line space **R**, called the **real-valued continuous functions,** or, simply, **real functions**; such functions play a central role in topology and analysis. This chapter also studies uniform convergence of real-valued functions and characterizes normal spaces through separation by real-valued continuous functions. It is not true that a nonconstant real-valued continuous function can always be defined on a given space. But on normal spaces, in particular, on metric spaces, there always exist nonconstant real-valued continuous functions by **Urysohn lemma** 6.2.8, named after P. S. Urysohn (1998–1924), which is an outstanding result characterizing normal spaces by the real-valued continuous functions. His approach by using dyadic rational numbers studied in this chapter, is more general and different from Urysohn lemma for metric spaces proved in Chap. 2. This lemma is used to prove **Tietze extension theorem** 6.5.1.

A characterization of completely regular spaces by real-valued continuous functions is given in Theorem 6.3.11. A deep result of this chapter is the **Gelfand–Kolmogoroff Theorem** 6.7.7, which proves that two compact Hausdorff spaces $X$ and $Y$ are homeomorphic iff the corresponding rings $\mathcal{C}(X, \mathbf{R})$ and $\mathcal{C}(Y, \mathbf{R})$ are isomorphic. This result recovers the topology of $X$ from the ring structure of $\mathcal{C}(X, \mathbf{R})$. Moreover, Corollary 6.7.8 characterizes compact Hausdorff spaces in terms of algebras. Another problem in topology is to find conditions under which a topological space is metrizable. **Urysohn metrization theorem** 6.7.9 provides an affirmative answer of this problem. Two embedding problems are solved in Sect. 6.8.1. Moreover, Theorem 6.8.14 determines completely the compact subsets of the Euclidean $n$-space $\mathbf{R}^n$ by real-valued continuous functions. Further applications are available in Sect. 6.8.

For this chapter the paper (Hewitt 1946) and the books (Bredon 1983; Chatterjee et al. 2002; Patterson 1959; Singer and Thorpe 1967; Adhikari 2016, 2022; Adhikari and Adhikari 2014, 2022; Alexandrov 1979; Borisovich et al. 1985; Brown 1988; Dugundji 1966; Fuks and Rokhlin 1984; Hewitt 1966; Hu 1966; Kelly 1975; Mendelson 1962; Munkres 2000; Stephen 1970), and some others are referred in the Bibliography.

© The Author(s), under exclusive license to Springer Nature Singapore Pte Ltd. 2022    391
A. Adhikari and M. R. Adhikari, *Basic Topology 1*,
https://doi.org/10.1007/978-981-16-6509-7_6

# 6.1   Real-Valued Continuous Functions: Introductory Concepts

This section is devoted to the study of real-valued continuous functions, which are special type of continuous functions from a topological space to the real line space. Urysohn Lemma 6.2.8 provides a vast supply of real-valued continuous function from a normal space and hence from metric spaces, metrizable spaces, and also from compact Hausdorff spaces. Throughout this chapter, the topological spaces **R** and **I** = [0, 1] represent the real line space (**R**, $\sigma$) with natural topology $\sigma$ and (**I**, $\sigma_I$) the subspace of (**R**, $\sigma$), respectively.

## 6.1.1   Continuity of Real-Valued Functions

This subsection addresses the continuity of real-valued functions on topological spaces.

**Definition 6.1.1**   Let $(X, \tau)$ be a topological space. Then a function

$$f : (X, \ \tau) \to (\mathbf{R}, \ \sigma) \text{ (or into } (\mathbf{I}, \ \sigma_I))$$

is said to be a **real-valued function or a real function on** $X$.

**Proposition 6.1.2**   *Let $(X, \tau)$ be a topological space and $(\mathbf{R}, \sigma)$ be the real number space with usual topology $\sigma$. Then a function $f : X \to \mathbf{R}$ is said to be continuous at a point $a \in X$; if corresponding to a given real number $\epsilon > 0$, there exists an open nbd $U$ of $a$ such that*

$$|f(x) - f(a)| < \epsilon, \ \forall x \in U.$$

***Proof***   The open intervals of **R** together with the empty set $\emptyset$ form an open base for the natural topology $\sigma$ on **R**. Since any open nbd $V$ of the point $f(a) \in \mathbf{R}$ contains an open interval $(f(a) - \epsilon, \ f(a) + \epsilon)$ containing the point $f(a)$, it follows that there exists an open nbd $U$ of $a$ such that

$$|f(x) - f(a)| < \epsilon, \forall x \in U.$$

$\square$

Proposition 6.1.3 generalizes Lebesgue sets of a continuous function (see Chap. 3) and is used in subsequent discussion.

**Proposition 6.1.3**   *Let $(X, \tau)$ be a topological space and $(\mathbf{R}, \sigma)$ be the real number space with usual topology $\sigma$. If $a, b \ (a < b) \in \mathbf{R}$ and $f : (X, \tau) \to (\mathbf{R}, \sigma)$ is a continuous function, then*

*(i) all the three subsets $X_1$, $X_2$ and $X_3$ of $X$ defined by*

$$X_1 = \{x \in X: f(x) \le a\},$$
$$X_2 = \{x \in X: f(x) \ge a\} \text{ and}$$
$$X_3 = \{x \in X: a \le f(x) \le b\}$$

*are closed subsets of $(X, \tau)$ ; and*
*(ii) all the three subsets $U_1$, $U_2$, and $U_3$ of $X$ defined by*

$$U_1 = \{x \in X: f(x) < a\},$$
$$U_2 = \{x \in X: f(x) > a\} \text{ and}$$
$$U_3 = \{x \in X: a < f(x) < b\}$$

*are open subsets of $(X, \tau)$.*

***Proof*** (i) The improper intervals $(-\infty, a]$, $[a, +\infty)$ , which are half infinite intervals and the closed interval $[a, b]$, are closed sets in $(\mathbf{R}, \ \sigma)$, and $f$ is continuous by hypothesis. Hence it follows that the sets

$$X_1 = f^{-1}((-\infty, a]), \ X_2 = f^{-1}([a, +\infty)), \ X_3 = f^{-1}(([a, b])$$

are closed sets in $(X, \tau)$.
(ii) Similarly, (ii) is proved.

$\square$

**Theorem 6.1.4** *Let $(X, \ \tau)$ be an arbitrary topological space. If*

$$f: (X, \ \tau) \to (\mathbf{R}, \ \sigma)$$

*and*

$$g: (X, \ \tau) \to (\mathbf{R}, \ \sigma)$$

*are two continuous functions, then each of the functions*

*(i) $f + g: (X, \ \tau) \to (\mathbf{R}, \ \sigma)$, $x \mapsto f(x) + g(x)$;*
*(ii) $f - g: (X, \ \tau) \to (\mathbf{R}, \ \sigma)$, $x \mapsto f(x) - g(x)$;*
*(iii) $f/g: (X, \ \tau) \to (\mathbf{R}, \ \sigma)$, $x \mapsto (f(x)/g(x), provided, g(x) \ne 0, \ \forall x \in X$;*
*(iv) $fg: (X, \ \tau) \to (\mathbf{R}, \ \sigma)$, $x \mapsto f(x)g(x)$;*
*(v) $cf: (X, \ \tau) \to (\mathbf{R}, \ \sigma)$, $x \mapsto cf(x)$, $\forall c \in \mathbf{R}$*

*is continuous*

***Proof*** Left as an exercise. $\square$

**Theorem 6.1.5** *Let* $(X, \tau)$ *be an arbitrary topological space and* $f : (X, \tau) \to$ $(\mathbf{R}, \sigma)$ *be a continuous function and* $g : (\mathbf{R}, \sigma) \to (\mathbf{R}, \sigma)$ *be a continuous function. Then their composite function*

$$g \circ f : (X, \tau) \to (\mathbf{R}, \sigma), x \mapsto g(f(x))$$

*is also continuous.*

**Proof** It follows from continuity of composite map of two continuous maps (see Chap. 3). □

Theorem 6.1.6 is an immediate application of Theorems 6.1.5 and 6.1.4.

**Theorem 6.1.6** *Let* $(X, \tau)$ *be an arbitrary topological space,* $f : (X, \tau) \to (\mathbf{R}, \sigma)$ *and* $g : (X, \tau) \to (\mathbf{R}, \sigma)$ *be continuous functions. Then the functions*

*(i)*

$$|f| : (X, \tau) \to (\mathbf{R}, \sigma), \ x \mapsto |f(x)|$$

*(ii)*

$$\max \ (f, g) : (X, \tau) \to (\mathbf{R}, \sigma), \ x \mapsto \max\{f(x), g(x)\}$$

*and*

*(iii)*

$$\min \ (f, g) : (X, \tau) \to (\mathbf{R}, \sigma), \ x \mapsto \min\{f(x), g(x)\}$$

*are continuous.*

**Proof** (i) The map $|f| : (X, \tau) \to (\mathbf{R}, \sigma)$, $x \mapsto |f(x)|$ is continuous, since $|f|$ is the composite of two continuous maps: $f_1 : (X, \tau) \to (\mathbf{R}, \sigma)$, $x \mapsto f(x)$ and $f_2 : (\mathbf{R}, \sigma) \to (\mathbf{R}, \sigma)$, $f(x) \mapsto |f(x)|$, and hence $|f| = f_2 \circ f_1$ implies that $f$ is continuous by Theorem 6.1.5.

(ii) Since $\max\{f(x), g(x)\} = \frac{1}{2}[(f(x) + g(x)) + |f(x) - g(x)|]$ is the sum of two continuous functions, it is also continuous by Theorem 6.1.4.

(iii) Since $\min\{f(x), g(x)\} = \frac{1}{2}[(f(x) + g(x)) - |f(x) - g(x)|]$ is the difference of two continuous functions, it is also continuous by Theorem 6.1.4. □

**Proposition 6.1.7** *Let* $f : (X, \tau) \to (\mathbf{I}, \sigma_{\mathbf{I}})$ *be a function such that for all real numbers with* $0 < a, b < 1$, *the inverse images* $f^{-1}([0, b))$ *and* $f^{-1}((a, 1])$ *are open sets in* $(X, \tau)$. *Then* $f$ *is continuous.*

**Proof** Since the intervals $[0, b)$ and $(a, 1]$ constitute a subbase for the subspace topology $\sigma_{\mathbf{I}}$ and the inverse images $f^{-1}([0, b))$ and $f^{-1}((a, 1])$ are open sets in $(X, \tau)$ by hypothesis, it follows that $f$ is continuous. □

### 6.1.2 Uniform Convergence of Real-Valued Functions

This subsection studies uniform convergence of infinite sequences of real-valued functions on a topological space with a view to examine the continuity of the limit function of a sequence of uniformly convergent continuous real-valued functions. Cauchy's criterion for uniform convergence of a sequence of real functions in topological setting is proved in Theorem 6.1.9.

**Definition 6.1.8** Given a topological space $(X, \tau)$, an infinite sequence of real functions $\{f_n\}$ on $X$ is said to **converge uniformly** to a real function $f$ on $X$ if for every positive real number $\epsilon$, there is a positive integer $n_0$ such that

$$|f_n(x) - f(x)| < \epsilon, \ \forall x \in X \text{ and } \forall n \geq n_0.$$

On the other hand, an infinite sequence of real functions $\{f_n\}$ on $X$ is said to be a **Cauchy sequence**, if for every positive real number $\epsilon$, there is a positive integer $n_0$ such that

$$|f_n(x) - f_m(x)| < \epsilon, \ \forall x \in X \text{ and } \forall m, n \geq n_0.$$

**Theorem 6.1.9** (Cauchy's criterion for uniform convergence) *Given a topological space $(X, \tau)$, an infinite sequence of real functions $\{f_n\}$ on $X$, converges uniformly on $X$, iff for every positive real number $\epsilon$, there is a positive integer $n_0$ such that*

$$|f_n(x) - f_m(x)| < \epsilon. \forall x \in X, \text{ and } \forall m, n > n_0 \qquad (6.1)$$

**Proof** Let $(X, \tau)$ be a topological space and $(\mathbf{R}, \sigma)$ be the real line space such that the condition (6.1) holds for the sequence $\{f_n\}$ of real functions on $X$. Then for every $x \in X$, the sequence $\{f_n(x)\}$ is a Cauchy sequence in the space $(\mathbf{R}, \sigma)$, which is a complete metric space. This asserts that the sequence $\{f_n(x)\}$ is convergent. If $f(x)$ is its limit, then the sequence $\{f_n\}$ on $X$ converges to $f$. We claim that this convergence is uniform. To show it, given an $\epsilon > 0$, suppose there exists a positive integer $n_0$ such that

$$|f_n(x) - f_m(x)| < \epsilon, \ \forall x \in X \text{ and } \forall m, n \geq n_0.$$

Keeping $n$ fixed and allowing $m \to \infty$, it follows that $f_m(x) \to f(x)$, *as $m \to \infty$* and hence

$$|f_n(x) - f(x)| < \epsilon, \ \forall x \in X \text{ and } \forall n \geq n_0.$$

Consequently, the condition (6.1) asserts that the sequence $\{f_n(x)\}$ converges uniformly to $f$.

Conversely, suppose the sequence $\{f_n\}$ on $X$ converges uniformly to $f$. Then given a positive real number $\epsilon$, there is a positive integer $n_0$ such that

$$|f_n(x) - f(x)| < \epsilon/2, \ \forall\, n \geq n_0 \text{ and } \forall\, x \in X.$$

Hence it follows that for all $m, n \geq n_0$,

$$|f_n(x) - f_m(x)| \leq |f_n(x) - f(x)| + |f(x) - f_m(x)| < \epsilon$$

for all $x \in X$, which gives the condition (6.1).                                      $\square$

**Theorem 6.1.10** *Let* $(X, \tau)$ *be a topological space and* $\{f_n\}$ *be a uniformly convergent sequence of continuous real functions on* $X$. *If* $f$ *is the limit of the sequence* $\{f_n\}$, *then* $f$ *is also a continuous real function on* $X$.

**Proof** Let $\epsilon > 0$ and $y \in X$ be an arbitrary point. By hypothesis, $\{f_n\}$ is uniformly convergent with $f$ as its limit. Then it follows that there is a positive integer $n_0$ such that

$$|f_{n_0}(x) - f(x)| < \epsilon/3, \ \forall\, x \in X \tag{6.2}$$

and hence

$$|f_{n_0}(y) - f(y)| < \epsilon/3.$$

Again, as the function $f_{n_0}$ is continuous at the point $y$, there exists an open set $U_y$ containing $y$ such that

$$|f_{n_0}(x) - f_{n_0}(y)| < \epsilon/3, \ \forall\, x \in U_y.$$

Hence it follows that

$$|f(x) - f(y)| \leq |f(x) - f_{n_0}(x)| + |f_{n_0}(x) - f_{n_0}(y)|$$
$$+ |f_{n_0}(y) - f(y)| < \epsilon/3 + \epsilon/3 + \epsilon/3 = \epsilon \ \forall\, x \in U_y.$$

This asserts that the real function $f$ is continuous on $X$, since $y \in X$ is an arbitrary element of $X$.

$\square$

**Definition 6.1.11** Let $(X, \tau)$ be a topological space and $\{f_n : X \to \mathbf{R}\}$ be an infinite sequence of real functions. Then the infinite series $\Sigma_{n=1}^{\infty} f_n$ or simply, $\Sigma f_n$ of the sequence of functions $\{f_n : X \to \mathbf{R} : n = 1, 2, \ldots\}$ is said to converge to the real function $f$ on $X$; if the sequence of partial sums

$$\{s_n = f_1 + f_2 + \cdots + f_n : n = 1, 2, \ldots\}$$

converges uniformly to $f$ on $X$, i.e., for every $\epsilon > 0$, there exists a positive integer $n_0$ such that

$$|S_n(x) - f(x)| < \epsilon, \ \forall\, n \geq n_0, \text{ and } \forall\, x \in X.$$

**Theorem 6.1.12** (Weierstrass M- test) *Let $(X, \tau)$ be a topological space and $\{f_n : X \to \mathbf{R}; n = 1, 2, \ldots\}$ be a given infinite sequence of real functions such that*

$$|f_n(x)| < M_n, \ \forall \, n \in \mathbf{N} \ and \ \forall \, x \in X,$$

*where $M_n$ are positive constants and the series $\Sigma M_n$ is convergent. Then the infinite series $\Sigma f_n$ converges uniformly on $X$.*

**Proof** Since the infinite series $\Sigma M_n$ of positive constants is convergent, it follows that given an $\epsilon > 0$, there is a positive integer $n_0$ such that

$$\Sigma_{k=t}^{n} M_k < \epsilon, \ \forall \, n > t > n_0.$$

Hence for the sequence $\{s_n\}$ of partial sums,

$$|s_n(x) - s_{t-1}(x)| = |f_t(x) + f_{t+1}(x) + \cdots + f_n(x)|$$
$$\leq \Sigma_{k=t}^{n} |f_k(x)| \leq \Sigma_{k=t}^{n} M_k < \epsilon, \ \forall \, x \in X \, \forall \, n > t > n_0$$

asserts by Theorem 6.1.9 that this sequence is uniformly convergent on $X$, and hence, the given infinite series $\Sigma f_n$ converges uniformly on $X$. $\qquad \Box$

**Theorem 6.1.13** *Let $(X, \tau)$ be a topological space and $\{f_n : X \to \mathbf{R} : n = 1, 2, \ldots\}$ be a given infinite sequence of continuous functions on $X$. If the infinite series $\Sigma f_n$ converges uniformly in $X$, then the sum of the series is also a continuous real function on $X$.*

**Proof** By hypothesis, every $f_n$ is continuous on $X$. Hence every partial sum $s_n = f_1 + f_2 + \cdots + f_n$ is continuous for $n = 1, 2, \ldots, n$. Again, the infinite series $\Sigma f_n$ is uniformly convergent on $X$ iff the infinite sequence of partial sums $\{s_n\}$ is uniformly convergent on $X$. Hence the theorem follows by using Theorem 6.1.10. $\qquad \Box$

### 6.1.3  $\mathcal{C}(X, \mathbf{R})$ and $\mathcal{B}(X, \mathbf{R})$

This subsection studies two special type of metric spaces $\mathcal{C}(X, \mathbf{R})$ and $\mathcal{B}(X, \mathbf{R})$ (defined in Chap. 2), which play a key role in the study of real functions, where $\mathcal{C}(X, \mathbf{R})$ denotes the set of all continuous real functions on a compact topological space $X$ and $\mathcal{B}(X, \mathbf{R})$ denotes the set of all bounded continuous real functions on $X$. If $X$ is compact, then $\mathcal{C}(X, \mathbf{R}) = \mathcal{B}(X, \mathbf{R})$. Consider the metrics

$$d : \mathcal{C}(X, \mathbf{R}) \times \mathcal{C}(X, \mathbf{R}) \to \mathbf{R}, \ (f, g) \mapsto \sup\{|f(x) - g(x)| : x \in X\},$$

$$\rho: (\mathcal{B}(X, \mathbf{R}) \times \mathcal{B}(X, \mathbf{R}) \to \mathbf{R}, \ (f, g) \mapsto \sup\{|f(x) - g(x)|: x \in X\}.$$

Then $d$ and $\rho$ induce topologies $\tau_d$ and $\tau_\rho$ on $\mathcal{C}(X, \mathbf{R})$ and $\mathcal{B}(X, \mathbf{R})$, respectively. The spaces $(\mathcal{C}(X, \mathbf{R}), \tau_d)$ and $(\mathcal{B}(X, \mathbf{R}), \tau_\rho)$ form important classes of topological spaces.

**Remark 6.1.14** The constant functions in $\mathcal{C}(X, \mathbf{R})$ are continuous. It is a natural question: does there exist nonconstant continuous functions in $\mathcal{C}(X, \mathbf{R})$? The answer is affirmative for specifying spaces $X$. For example, if $X$ is a metric space, then Urysohn lemma asserts that for every pair of disjoint closed sets of $X$, there exists a nonconstant continuous function $f: X \to [0, 1]$ (see Chap. 2).

**Theorem 6.1.15** *Let* $(X, \tau)$ *be a topological space and* $\{f_n\}$ *be a sequence of functions in* $\mathcal{B}(X, \mathbf{R})$. *Then*

(i) $\{f_n\}$ *converges to a function* $f$ *in the metric space* $(\mathcal{B}(X, \mathbf{R}), \rho)$ *iff the sequence converges uniformly to* $f$ *on* $X$;

(ii) $\{f_n\}$ *is a Cauchy sequence in the metric space* $(\mathcal{B}(X, \mathbf{R}), \rho)$ *iff it is a uniformly Cauchy sequence in the sense that given an* $\epsilon > 0$, *there is an integer* $n_0$ *such that*

$$|f_n(x) - f_m(x)| < \epsilon, \ \forall x \in X, \ and \ \forall m, n \geq n_0.$$

**Proof**   (i) Let $\{f_n\}$ converge to a function $f \in \mathcal{B}(X, \mathbf{R})$. Then $\rho(f_n, f) \to 0$, and hence given an $\epsilon > 0$, there exists a positive integer $n_0$ such that

$$\rho(f_n, f) < \epsilon, \ \forall n \geq n_0.$$

This implies by definition of $\rho$ that

$$|f_n(x) - f(x)| < \epsilon, \ \forall n > n_0, \ \text{and} \ \forall x \in X.$$

This proves that $f_n \to f$ uniformly on $X$. Conversely, assume that $\{f_n\}$ converges uniformly to a function $f \in \mathcal{B}(X, \mathbf{R})$. Then given an $\epsilon > 0$, there is a positive integer $n_0$ such that

$$\rho(f_n, f) < \epsilon, \ \forall n \geq n_0$$

and hence $f_n \to f$ in $\mathcal{B}(X, \mathbf{R})$.

(ii) It is similar to the proof of (i).

$\square$

**Theorem 6.1.16** *For any topological space $(X, \tau)$, the metric space $(\mathcal{B}(X, \mathbf{R}), \rho)$ is complete.*

**Proof** Let $\{f_n\}$ be a Cauchy sequence in the metric space $(\mathcal{B}(X, \mathbf{R}), \rho)$. Then it follows from the definition of the metric $\rho$ that

$$|f_n(x) - f_m(x)| \leq \rho(f_n, f_m), \ \forall x \in X.$$

This implies that for every $x \in X$, the sequence $\{f_n(x)\}$ is a Cauchy sequence in $\mathbf{R}$. Let $f(x) = lim_{n \to \infty} f_n(x)$. Then $\rho(f_n, f) \to 0$ and $f$ is continuous. This implies that $f_n \to f$ and $f \in \mathcal{B}(X, \mathbf{R})$, and hence, the metric space $(\mathcal{B}(X, \mathbf{R}), \rho)$ is complete. $\qquad \square$

Theorem 6.1.17 essentially due to Italian mathematician U. Dini (1845–1918), gives a result involving uniform convergence of increasing sequences in $\mathcal{C}(X, \mathbf{R})$ for every compact space $X$.

**Theorem 6.1.17 (Dini)** *Let $(X, \tau)$ be a compact space, $\{f_n\}$ be an increasing sequence in $\mathcal{C}(X, \mathbf{R})$ and $f \in \mathcal{C}(X, \mathbf{R})$. If $f_n(x) \to f(x), \ \forall x \in X$, then $f_n \to f$ uniformly on $X$.*

**Proof** Given an $\epsilon > 0$, construct the set

$$X_n = \{x \in X : f(x) < f_n(x) + \epsilon\} \subset X, \ \forall n \geq 1.$$

Since by hypothesis, $\{f_n\}$ is an increasing sequence in $\mathcal{C}(X, \mathbf{R})$ with $n$ increasing, it follows that $X_n \subset X_{n+1}, \ \forall n \geq 1$. Again, since $f_n(x) \to f(x), \ \forall x \in X$, it follows that $X = \bigcup_{n=1}^{\infty} X_n$. As both $f$ and $f_n$ are continuous, it follows that $X_n$ is open in $(X, \tau), \ \forall n \geq 1$. Since $X$ is compact by hypothesis, the open covering $\{X_n\}$ has a finite subcovering. Then by ascending properties of these sets, it follows that there is a single positive integer $n_0$ such that $X_{n_0} = X$. This implies that

$$X_n \subset X_{n_0}, \ \forall n \geq n_0.$$

Hence the inequality

$$f_n(x) \leq f(x) < f_n(x) + \epsilon, \ \forall n > n_0$$

asserts that $f_n \to f$ uniformly on $X$. $\qquad \square$

## 6.2   Urysohn Lemma: Separation of Disjoint Subsets of Topological Spaces by Real-Valued Continuous Functions

This section conveys the concept of separation of disjoint subsets of a topological space by real-valued continuous functions and presents more results on continuous functions from a topological space to the real line space and applies them to study normal spaces. For example, Urysohn Lemma 6.2.8 is an important result which characterizes normal spaces by real-valued continuous functions by using dyadic rational numbers.

**Definition 6.2.1** Let $(X, \tau)$ be a topological space and A,B be two disjoint subsets of $X$. Then these two subsets are said to be **separated** by a real-valued continuous function if there exists a continuous function

$$f: X \to \mathbf{R} \text{ such that } f(x) = 0, \ \forall x \in A, \ f(x) = 1,$$
$$\forall x \in B, \text{ and } 0 \leq f(x) \leq 1, \ \forall x \in X.$$

i.e., if there exists a continuous function $f$ such that

$$f: X \to \mathbf{R}. \ x \mapsto \begin{cases} 0, \text{ for all } x \in A \\ 1, \text{ for all } x \in B \end{cases}$$

and

$$0 \leq f(x) \leq 1 \text{ for all } x \in X.$$

Such a function $f$ (if it exists) is called **Urysohn function** or **characteristic function** corresponding to the pair of disjoint subsets $A$ and $B$ in $(X, \tau)$, which is sometimes denoted by $\kappa_{(A,B)}$.

### 6.2.1   Existence of Real-Valued Continuous Functions

This subsection solves the natural question: what are the topological assumptions that ensure the existence of sufficiently many continuous real functions on a topological space ? In Chap. 4, different classes of topological spaces such as $T_i$-spaces, regular, normal, and completely normal spaces are studied by using the separation axioms. On the other hand, the existence of continuous real functions on a topological space is closely related to the separation axioms satisfied by the space. Let $X$ be the trivial topological space having only two open sets such as $\emptyset$ and $X$ itself. Then the only real continuous functions on $X$ are the constant functions. Again, given a topological $X$ with the property that for every pair of distinct elements $x_1, x_2 \in X$, there exists a continuous function $f: X \to \mathbf{R}$ such that $f(x_1) \neq f(x_2)$, then the space $X$ is Hausdorff (see Exercise 6 of Sect. 6.9).

**Theorem 6.2.2** *(Existence of continuous real functions) Let $(X, \tau)$ be a topological space and $U_r$ be an open set in $(X, \tau)$ for each dyadic rational number $r = m/2^n$ ( $0 \leq m \leq 2^n$) such that if $r < t$, then $\overline{U_r} \subset U_t$. Then the function*

$$f: X \to \mathbf{R}, \; x \mapsto \begin{cases} glb \; \{r: x \in U_r\}, & if \, x \in \bigcup_r U_r \\ 1, & if \, x \notin \bigcup_r U_r \end{cases},$$

*is continuous.*

**Proof** Let $r$ be a dyadic rational number. For any such $r$,

   (i) if $f(x) < r$, then $x \in U_r$; otherwise, if $x \notin U_r$, then $f(x) \geq r$;
   (ii) if $f(x) \leq r$, then $x \in U_r$, otherwise, $f(x) > r$. Then $x \notin U_r$ implies that $x \in X - U_r$. This shows that for any real number $s$, the set

$$f^{-1}(-\infty, s) = \{x \in X: f(x) < s\} = \bigcup \{U_r: r < s\},$$

is open in $(X, \tau)$ and the set

$$f^{-1}(w, +\infty) = \{x \in X: f(x) > w\} = \bigcup \{X - U_r: r > w\} = \bigcup \{X - \overline{U_t}: t > w\}.$$

is also open in $(X, \tau)$. Again, since the half infinite intervals $(-\infty, s)$ and $(w, +\infty)$, which are improper intervals, give a subbasis for the usual topology of $\mathbf{R}$, it follows that the function

$$f: X \to \mathbf{R}$$

is continuous.

$\square$

Recall that a topological space $(X, \tau)$ is normal iff for every closed set $A$ and open set $U$ containing $A$, there exists an open set $V$ in $(X, \tau)$ such that

$$A \subset V \subset \overline{V} \subset U.$$

**Lemma 6.2.3** (One form of Urysohn Lemma) *Let $(X, \tau)$ be a normal space, $A$ be a closed set and $U$ be an open set containing $A$ such that $A \subset U$. Then there exists a continuous map $f: X \to \mathbf{I}$ such that*

$$f(x) = \begin{cases} 0, \textit{for all } x \in U \\ 1, \textit{for all } x \in X - U. \end{cases}$$

**Proof** To prove this lemma, we use the notations defined in Theorem 6.2.2. By hypothesis, $(X, \tau)$ is normal, $A$ is a closed set, and $U$ is an open set such that $A \subset U$. We here take $U_1 = U$ and $\overline{U_0} = A$. Then by normality criterion of $(X, \tau)$, for the closed set $A$ and open set $U$ containing $A$, there exists an open set $V = U_{1/2}$ in $(X, \tau)$ such that

$$A = \overline{U_0} \subset U_{1/2} \text{ and } \overline{U_{1/2}} \subset U_1 = U.$$

Again, for the closed set $\overline{U_{1/2}}$ and the open set $U_1 = U$ containing the closed set $\overline{U_{1/2}}$, we have

$$\overline{U_1/2} \subset U_{3/4} \text{ and } \overline{U_{3/4}} \subset U_1 = U,$$

Proceeding in this way, we have

$$\overline{U_0} \subset U_{1/4} \text{ and } \overline{U_{1/4}} \subset U_{1/2},$$

$$\overline{U_1/2} \subset U_{3/4} \text{ and } \overline{U_{3/4}} \subset U_1 = U,$$

and so on. Consequently, the existence of the requisite continuous map $X \to \mathbf{I}$ follows from Theorem 6.2.2.                                                                                          □

**Remark 6.2.4** Another form of Urysohn lemma is available in Urysohn lemma 6.2.8.

**Theorem 6.2.5** *Given a topological space $(X, \tau)$, if two disjoint subsets $A$ and $B$ of $X$ are separated by a continuous real function, then $A$ and $B$ are also strongly separated by a pair of disjoint open sets in $(X, \tau)$.*

**Proof** Given a topological space $(X, \tau)$, let the two disjoint subsets $A$ and $B$ of $X$ be separated by a continuous real function. Then there is a Urysohn function $f$ corresponding to the pair of disjoint subsets $A$ and $B$ in $X$. Let $\sigma_{\mathbf{I}}$ be the topology on $\mathbf{I} = [0, 1]$ relative to the usual topology $\sigma$ of the real number space $\mathbf{R}$. Then the half-open intervals $[0, 1/4)$ and $(3/4, 1]$ are open sets in the space $(\mathbf{I}, \sigma_{\mathbf{I}})$. By using the continuity of $f$, it follows that $f^{-1}([0, 1/4) = U$, and $f^{-1}((3/4, 1] = V$ are two open sets in $(X, \tau)$. Since $[0, 1/4) \cap ((3/4, 1] = \emptyset$, it also follows that $f^{-1}([0, 1/4) \cap f^{-1}((3/4, 1] = \emptyset$. Consequently, under the given hypothesis, there exist two disjoint open sets $U$ and $V$ in $(X, \tau)$ such that

$$A \subset U \text{ and } B \subset V.$$

This asserts that the disjoint subsets $A$ and $B$ of $X$ are strongly separated by disjoint open sets $U$ and $V$ in $(X, \tau)$,                                                                          □

***Example 6.2.6*** The converse of Theorem 6.2.5 is not necessarily true. In support consider the **Niemytzki's tangent disk topology** $\tau$ on $X = \{(x, y) \in \mathbf{R}^2 : y \geq 0\}$ (see Chap. 3).

## 6.2.2   Functionally Separable Sets

This subsection conveys in Definition 6.2.7 the concept of functionally separable sets, which is used in the study of real-valued continuous functions. This concept coincides with the concept given in Definition 6.2.1, but it is expressed in a different language for Urysohn lemma.

**Definition 6.2.7** Let $(X, \tau)$ be topological space. Two subsets A and B of $X$ are said to be **functionally separable** if there exists a continuous map $f : X \to \mathbf{I}$ such that

$$f(x) = \begin{cases} 0, \text{ for all } x \in A \\ 1, \text{ for all } x \in B \end{cases}$$

and

$$0 \leq f(x) \leq 1 \text{ for all } x \in X.$$

## 6.2.3   Urysohn Lemma and Characterization of Normal Spaces

This subsection proves Urysohn Lemma 6.2.8, which is a key result in topology and characterizes normal spaces by real-valued continuous functions. As every metric space is normal and every compact Hausdorff space is also normal, Urysohn lemma can be conveniently used to study such spaces by the concept of functionally separability. Tietze Extension Theorem 6.5.1 is a natural generalization of Urysohn Lemma 6.2.8. On the other hand, $T_i$-spaces (for $i = 1, 2, 3$) cannot be characterized by using the concept of separation by continuous real functions (Hewitt 1946). So we now proceed to study normal spaces by Urysohn lemma. The existence of continuous functions on a topological space is closely related to the axioms of separation satisfied by a particular type of topological spaces prescribed in Urysohn Lemma 6.2.8.

**Lemma 6.2.8** (*Urysohn Lemma*) *A topological space* $(X, \tau)$ *is normal if and only if corresponding to every pair of disjoint closed sets* $P, Q$ *in* $(X, \tau)$, *there exists a continuous map*

$$f : (X, \tau) \to (\mathbf{I}, \sigma_{\mathbf{I}})$$

*such that*

$$f(x) = \begin{cases} 0, \text{for all } x \in P \\ 1, \text{for all } x \in Q \end{cases}$$

*and*

$$0 \le f(x) \le 1 \text{ for all } x \in X.$$

**Proof** Existence of such an $f$ is established in Theorem 6.2.2. First assume that a continuous function $f: (X, \tau) \to (\mathbf{I}, \sigma_{\mathbf{I}})$ satisfies the given condition for every pair disjoint closed sets in $(X, \tau)$. This implies that they are separated by the continuous function $f$. Hence they are also strongly separated by the continuous function $f$ by Theorem 6.2.5. This proves that $(X, \tau)$ is normal.

To prove it independently, consider the subspaces $\mathbf{I}_1$, $\mathbf{I}_2$ of $\mathbf{I}$ defined by

$$\mathbf{I}_1 = \{t \in \mathbf{I}: 0 \le t < 1/4\}, \ \mathbf{I}_2 = \{t \in \mathbf{I}: 3/4 < t \le 1\}.$$

Since $\mathbf{I}_1$ and $\mathbf{I}_2$ are half-open sets in $(\mathbf{I}, \sigma_{\mathbf{I}})$ and $f$ is continuous, it follows that $f^{-1}(\mathbf{I}_1)$ and $f^{-1}(\mathbf{I}_2)$ are open sets in $(X, \tau)$. As $\mathbf{I}_1 \cap \mathbf{I}_2 = \emptyset$, it follows that $f^{-1}(\mathbf{I}_1) \cap f^{-1}(\mathbf{I}_2) = \emptyset$. Again since $f(P) = 0$ and $f(Q) = 1$ by hypothesis, it follows that $P \subset f^{-1}(\mathbf{I}_1)$ and $Q \subset f^{-1}(\mathbf{I}_2)$. This asserts that the space $X$ is normal.

Conversely, suppose that $(X, \tau)$ is a normal space. We prove that for every pair of disjoint closed sets $P$, $Q$ in $(X, \tau)$, there is a continuous map $f: (X, \tau) \to (\mathbf{I}, \sigma_{\mathbf{I}})$ such that

$$f(x) = \begin{cases} 0, \text{ for all } x \in P \\ 1, \text{ for all } x \in Q \end{cases}$$

and

$$0 \le f(x) \le 1 \text{ for all } x \in X.$$

This part follows immediately from Lemma 6.2.3.

**An alternative** proof of this part (last part). It is proved at two stages. At the first stage, we construct $f$, and at the second stage, we prove its continuity.

(i) **Construction of** $f$: Corresponding to each dyadic proper fraction of the form $t = r/2^n$, for a fixed $n$ and $r = 1, 2, \ldots, 2^n - 1$, an open set $V(r/2^n)$ is defined such that $\bar{V}(t_1) \subset V(t_2)$, whenever $t_1 < t_2$, and each set $V(r/2^n)$ contains $P$ but does not meet $Q$. Take $V(1) = X - Q$. By hypothesis, it follows that the closed set $P$ is contained in the open set $X - Q$. As $X$ is a normal space by hypothesis, then there exists an open set, abbreviated, $V(1/2)$ by normality criterion such that

$$P \subset V(1/2), \text{ and } \overline{V(1/2)} \subset X - Q.$$

Again as the closed set $\overline{V(1/2)}$ is a subset of the open set $X - Q$, by the same argument, there exist open sets abbreviated $V(1/4)$ and $V(3/4)$ such that

$$P \subset V(1/4), \ \overline{V(1/4)} \subset V(1/2) \text{ and } \overline{V(1/2)} \subset V(3/4), \ \overline{V(3/4)} \subset X - Q.$$

Proceeding in a similar way, we can define open sets $V(1/8)$, $V(3/8)$, $V(5/8)$, $V(7/8)$ for the pairs of sets

$$(P, \ V(1/4)), \ (\overline{V(1/4)}, (V(1/2)), \ (\overline{V(1/2)}, (V(3/4)), \ (\overline{V(3/4)}, X - P),$$

respectively, such that

$$P \subset V(1/8), \ \overline{V(1/8)} \subset V(1/4), \ \overline{V(1/4)} \subset V(3/8), \dots, \overline{V(7/8)} \subset X - Q.$$

Continuing this process, we obtain at the $n$-th stage the open sets

$$V(1/2^n), \ V(3/2^n), V(5/2^n)$$

such that
$$P \subset V(1/2^{n-1})), \ (\overline{V(1/2^{n-1})}) \subset V(2/2^{n-1}),$$

corresponding to the pairs of sets

$$(P, V(1/2^{n-1})), \ (\overline{V(1/2^{n-1})}, V(2/2^{n-1})).$$

This shows that for every dyadic proper fraction $r/2^n$ for a fixed $n$ and $r = 1, 2, \dots, 2^n - 1$
there exists an open set $V(r/2^n)$ such that

(i) $P \subset V(r/2^n)$ for every such open set $V(r/2^n)$;
(ii) $\overline{V(r/2^n)} \subset X - Q$;
(iii) for $r_1/2^n < r_2/2^m$, $\overline{V(r_1/2^n)} \subset V(r_2/2^m)$.

Define

$$f: X \to \mathbf{I}: x \mapsto \begin{cases} \text{glb} \{r/2^n : x \in V(r/2^n)\}, \text{if} x \in \bigcup_{r/2^n} V(r/2^n) \\ 1, \text{if } x \notin \bigcup_{r/2^n} V(r/2^n). \end{cases}$$

Since $0 < r/2^n < 1$, it follows that $0 \le f(x) < 1, \forall x \in X$. Again, for every $x \in P$, $x$ belongs to every $V(r/2^n)$ and hence $f(x) = 0$. Moreover, for any $x \in Q$, $x \notin V(r/2^n)$ implies $f(x) = 1$. Hence it follows that

$$f: X \to \mathbf{I}$$

is a function such that

$$f(x) = \begin{cases} 0, \text{ for all } x \in P \\ 1, \text{ for all } x \in Q. \end{cases}$$

and

$$0 \leq f(x) \leq 1 \text{ for all } x \in X.$$

(ii) **Continuity of f** : Define

$$\mathbf{B} = \{[0, x), (z, t), (y, 1]: x, y, z, t \in \mathbf{I} \text{ are irrational}\}.$$

Then **B** forms a basis of open sets for the subspace topology on $\mathbf{I} = [0, 1]$
As $(z, t) = [0, t) \cap (z, 1]$, it follows that

$$f^{-1}(z, t) = f^{-1}[0, t) \cap f^{-1}(z, 1].$$

To show the continuity of $f$, it is sufficient to prove that $f^{-1}[0, t)$ and $f^{-1}[z, 1]$ are both open sets in $(X, \tau)$ for all irrational numbers $z$ and $t$. Since,

$$f^{-1}[0, t) = \bigcup_{r/2^n < t} V(r/2^n) \text{ and } f^{-1}(z, 1] = \bigcup_{r/2^n > t} V(r/2^n),$$

it follows that $f^{-1}[0, t)$ and $f^{-1}(z, 1]$ are both open sets for all irrational numbers $z$ and $t$. This proves the continuity of $f$.

□

The Urysohn Lemma 6.2.8 can be restated in Lemma 6.2.9.

**Lemma 6.2.9** *(Alternative Form of Urysohn Lemma ) A topological space $(X, \tau)$ is normal if and only if every pair of disjoint closed sets $P$, $Q$ in $(X, \tau)$ are separated by a continuous real-valued function $f: (X, \tau) \to (\mathbf{R}, \sigma)$ such that*

$$f(x) = \begin{cases} 0, \text{for all } x \in P \\ 1, \text{for all } x \in Q \end{cases}$$

*and*

$$0 \leq f(x) \leq 1 \text{ for all } x \in X.$$

**Corollary 6.2.10** *Let $(X, \tau)$ be a normal space and $P$, $Q \subset X$ be two disjoint closed subsets of $(X, \tau)$. Then there exists continuous function $h: X \to [a, b] \subset \mathbf{R}$ such that $h(P) = a$, and $h(Q) = b$.*

***Proof*** By hypothesis, $P$ and $Q$ are disjoint closed sets in the normal space $(X, \tau)$. Hence it follows by Urysohn Lemma 6.2.8 that there exists a continuous function $f : X \to \mathbf{I}$ such that

$$f(x) = \begin{cases} 0, \text{ for all } x \in P \\ 1, \text{ for all } x \in Q \end{cases}$$

and

$$0 \leq f(x) \leq 1 \text{ for all } x \in X.$$

Define a continuous map

$$g : \mathbf{I} \to [a, b], \ t \mapsto a + (b - a)t.$$

Hence the composite function $h = g \circ f : X \to [a, b]$ is a continuous function satisfying the required property. Because, $h(p) = (g \circ f)(p) = g(f(p)) = g(0) = a, \ \forall p \in P \implies h(P) = a$. Similarly, it follows that $h(Q) = b$. $\qquad \square$

Urysohn Lemma 6.2.9 has also an alternative form in terms of a continuous extension given in Corollary 6.2.11.

**Corollary 6.2.11** *Let $(X, \tau)$ be a normal space, $P, Q$ be two disjoint closed subsets in $(X, \tau)$. If*

*(i) $Y = P \cup Q$; and*
*(ii) $f : Y \to \mathbf{I}$ is a function satisfying the conditions $f(P) = 0$ and $f(Q) = 1$,*

*then $f$ has a continuous extension $\tilde{f} : X \to \mathbf{I}$ over $X$.*

***Proof*** By hypothesis, $(X, \tau)$ is normal, $P, Q \subset X$ are two disjoint closed subsets of $X$. If $Y = P \cup Q$ and $f : Y \to [0, 1]$ is defined by

$$f(x) = \begin{cases} 0, \text{ for all } x \in P \\ 1, \text{ for all } x \in Q \end{cases},$$

then $f$ is continuous. Because, $P$ and $Q$ are both open and closed in $Y$, they are open by its normality hypothesis of $(X, \tau)$. Since by hypothesis, $P$ and $Q$ are two disjoint closed set of the normal space $(X, \tau)$, by Urysohn lemma, there exists a continuous function $\tilde{f} : X \to [0, 1]$ such that

$$\tilde{f}(x) = \begin{cases} 0, \text{ for all } x \in P \\ 1, \text{ for all } x \in Q \end{cases}.$$

This shows that $\tilde{f} : X \to \mathbf{I} = [0, 1]$ is a continuous extension of $f$ over $X$. $\qquad \square$

***Example 6.2.12***  The hypothesis in Urysohn lemma 6.2.9 asserts that the sets $P$ and $Q$ are to be closed in $(X, \tau)$ is necessary. For example, the sets $P = (0, 1)$ and $Q = (1, 2)$ are disjoint open subsets of the normal space $X = \mathbf{R}$, but there exists no continuous function $f: X \to [0, 1]$ such that

$$f(x) = \begin{cases} 0, \text{ for all } x \in P \\ 1, \text{ for all } x \in Q \end{cases}$$

***Example 6.2.13***  Let $(X, \tau)$ be a topological space. Suppose for any two disjoint closed sets $A$ and $B$ in $(X, \tau)$, there exists a continuous function $f: X \to \mathbf{I}$ such that

$$f(x) = \begin{cases} 0, \text{ for all } x \in A \\ 1, \text{ for all } x \in B \end{cases}$$

and

$$0 \leq f(x) \leq 1 \text{ for all } x \in X,$$

then the disjoint open sets

$$U = f^{-1}\left([0, \frac{1}{2})\right) \text{ and } V = f^{-1}\left((\frac{1}{2}, 1]\right)$$

in $(X, \tau)$ contain $A$ and $B$, respectively, and separate them strongly. The openness of $U$ and $V$ follows from the fact that the lower-limit topology and the upper-limit topology on $\mathbf{R}$ are both strictly stronger than the usual topology on $\mathbf{R}$. It proves one part of Urysohn lemma.

## 6.3   More on Completely Regular and Tychonoff Spaces

This section continues the study of completely regular spaces initiated in Chap. 4 and gives an emphasis on Tychonoff spaces with the help of real-valued continuous functions. Tychonoff spaces form a special class of completely regular spaces.

Example 6.2.13 raises the problems:

(i) Can the proof of the Urysohn lemma be generalized for a completely regular space?
(ii) Does there exist any link between a completely regular space and a normal space?

***Remark 6.3.1***  A positive answer is established by Urysohn lemma by considering the defining properties of a completely regular space, which is a Hausdorff space $(X, \tau)$ such that given any closed set $A$ in $(X, \tau)$ and any point $a \in X$ not lying in $A$, (i.e., $a \notin A$), there is a real-valued continuous function $f: X \to \mathbf{R}$ such that

$$f(x) = \begin{cases} 0, \text{ for } x = a \\ 1, \text{ for all } x \in A \end{cases}$$

and

$$0 \leq f(x) \leq 1 \text{ for all } x \in X,$$

i.e., $A$ and $\{a\}$ are separated by a continuous real-valued function on $X$.

### 6.3.1  More on Completely Regular Spaces

This subsection continues the study of completely regular spaces which is versatile, because such spaces include metric spaces, normal spaces, and also locally compact Hausdorff spaces. This establishes a close link by Proposition 6.3.5 between completely regular spaces and real-valued continuous functions on such spaces.

**Theorem 6.3.2** *Given any collection $\{(X_i, \tau_i) : i \in \mathbf{A}\}$ of completely regular spaces, their product space $(X, \tau) = \Pi_{i \in \mathbf{A}} (X_i, \tau_i)$ is also completely regular under product topology.*

**Proof** Let $(X, \tau)$ be the product space of the collection $\{(X_i, \tau_i) : i \in \mathbf{A}\}$ of completely regular spaces. For any point $a = (a_i) \in X$, let $A$ be a closed set in the product space $(X, \tau)$ not containing the point $a$. If $\Pi \, U_i$ is a basis element containing the point $a$, which does not meet the set $A$, then $U_i = X_i$ except for finitely many values of $i$, say $i = i_1, i_2, \ldots, i_n$. Let $\pi_{i_k} : X \to X_{i_k}$ be the projection maps. Since by hypothesis, each $(X_i, \tau_i)$ is completely regular, given $k = 1, 2, \ldots, n$, select continuous maps

$$f_k : X_{i_k} \to [0, 1] \text{ with } f_k(a_{i_k}) = 1, \text{ and } f_k(X - U_{i_k}) = 0.$$

Define functions

$$\psi_k : X \mapsto \mathbf{R}, x \mapsto (f_k \circ \pi_{i_k})(x) = f_k(\pi_{i_k}(x)), \, \forall \, k = 1, 2, \ldots, n.$$

Then each $\psi_k$, being the composition of two continuous maps, is a real-valued continuous map and is such that it vanishes outside $\pi_{i_k}^{-1}(U_{i_k})$. Hence it follows that the function

$$f : X \to [0, 1], x \mapsto \psi_1(x)\psi_2(x) \cdots \psi_n(x)$$

is the required continuous function such that

$$f(a) = 1, \text{ and } f(x) = 0, \, \forall \, x \in X - \Pi \, U_i \text{ (i.e., outside } \Pi \, U_i).$$

$\square$

Proposition 6.3.3 proves that the property of completely regularity of topological spaces is hereditary.

**Proposition 6.3.3** *(Hereditary property of completely regular spaces) Let $(X, \tau)$ be a completely regular space and $(Y, \tau_Y)$ be a subspace of $(X, \tau)$. Then $(Y, \tau_Y)$ is also completely regular.*

***Proof*** By hypothesis, $(X, \tau)$ is completely regular space and $(Y, \tau_Y)$ is a subspace of $(X, \tau)$. Then by definition, $(X, \tau)$ is a $T_1$-space, and hence, its subspace $(Y, \tau_Y)$ is also a $T_1$-space. Let $A$ be closed in $Y$, $y \in Y$ and $y \notin A$. Then $A = Y \cap B$ for some closed set $B$ in $X$. Moreover, $y \notin B$, otherwise, $y \in Y \cap B = A$ would imply a contradiction. By hypothesis, $X$ is completely regular, and hence, the closed set $B$ and the point $\{y\}$ are separated by a continuous real function

$$f : X \to \mathbf{R}, \ x \mapsto \begin{cases} 0, \text{for } x = y \\ 1, \text{for all } x \in B \end{cases}$$

and

$$0 \le f(x) \le 1 \text{ for all } x \in X.$$

Then its restriction $g = f|_Y$ to $Y$ is also continuous and separates $y$ and $A$. It implies that $(Y, \tau_Y)$ is also completely regular. $\qquad\square$

**Corollary 6.3.4** *Let $(X, \tau)$ be a locally compact Hausdorff space. Then it is completely regular.*

***Proof*** By hypothesis, $(X, \tau)$ is locally compact Hausdorff. Then its 1-point compactification $\tilde{X}$ is compact and Hausdorff. This asserts that it is normal. Then $\tilde{X}$ is completely regular. Since $X$ is a subspace of the completely regular space $\tilde{X}$, it follows by Proposition 6.3.3 that $(X, \tau)$ is completely regular. $\qquad\square$

**Proposition 6.3.5** *Let $(X, \tau)$ be a topological space and $A$ be any closed set in $X$. If for every point $x \in X - A$, there is a continuous function $f : X \to \mathbf{R}$ such that $f(y) = 0$, $\forall y \in A$ but $f(x) \ne 0$. Then the space $X$ is completely regular.*

***Proof*** By hypothesis, $A$ is a closed set of $(X, \tau)$. Suppose that for every point $x \in X - A$, there is a function

$$g : X \to \mathbf{R}$$

such that $g(y) = 0$, $\forall y \in A$ but $g(x) \ne 0 \in \mathbf{R}$. Define a continuous map

$$f : X \to \mathbf{R}, y \mapsto g(x)^{-1} g(y).$$

This map shows that there is a continuous map $f$ such that

$$f: X \to \mathbf{R}, \ y \mapsto \begin{cases} g(y), & \text{for all } y \in A \\ g(x)^{-1}g(y), & \text{for all } y \in X - A, \end{cases}$$

Then

$$0 \le f(y) \le 1 \text{for all} y \in X.$$

This asserts that $f$ is a real-valued continuous function such that

$$f: X \to \mathbf{R}, y \mapsto \begin{cases} 0, \text{for all } y \in A \\ 1, \text{for all } x \in X - A, \end{cases}$$

and

$$0 \le f(x) \le 1 \text{ for all } x \in X.$$

This implies that $(X, \tau)$ is completely regular. $\qquad\square$

**Definition 6.3.6** Let $X, Y$ be two nonempty sets and $\mathcal{F} = \{f_k : X \to Y : k \in \mathbf{K}\}$ be a family of functions from $X$ to $Y$. Then the family $\mathcal{F}$ is said to **separate points** of $X$ if for every pair of distinct points $x, y \in X$, there is a function $f \in \mathcal{F}$ such that $f(x) \ne f(y)$.

*Example 6.3.7* The family of real-valued continuous functions

$$\mathcal{F} = \{f_n : \mathbf{R} \to \mathbf{R}, x \mapsto \sin \ nx, \ \forall n \in \mathbf{N}\}$$

does not separate points. Because, for every function $f_n \in \mathcal{F}$ and for the pair of distinct points $0, \pi \in \mathbf{R}$, its image points $f_n(0) = 0 = f(\pi)$. This shows that the family $\mathcal{F}$ can not separate points.

Proposition 6.3.8 determines the **Hausdorff structure** of $X$ from $\mathcal{C}(X, \mathbf{R})$ provided $\mathcal{C}(X, \mathbf{R})$ separates points in $X$.

**Proposition 6.3.8** *Let $\mathcal{C}(X, \mathbf{R})$ be the space of all real-valued continuous functions on a topological space $(X, \tau)$. If the family $\mathcal{C}(X, \mathbf{R})$ separates points, then the topological space $(X, \tau)$ is Hausdorff.*

*Proof* Since by hypothesis, the family $\mathcal{C}(X, \mathbf{R})$ separates points, given two distinct points $x, y \in X$, there is continuous function $f: X \to \mathbf{R}$ such that $f(x) \ne f(y)$. Again since, $\mathbf{R}$ is Hausdorff, there are disjoint open subsets $U$ and $V$ of $\mathbf{R}$ such that $f(x) \in U$ and $f(y) \in V$. Hence it follows that the disjoint open sets $f^{-1}(U)$ and $f^{-1}(V)$ in $X$ are such that $x \in f^{-1}(U)$ and $y \in f^{-1}(V)$. This implies that $X$ is Hausdorff. $\qquad\square$

**Proposition 6.3.9** *Let $(X, \tau)$ be a completely regular $T_1$ space and $\mathcal{C}(X, \mathbf{R})$ be the set of all real-valued continuous functions on $X$. Then $\mathcal{C}(X, \mathbf{R})$ separates points.*

**Proof** Let $(X, \tau)$ be a completely regular $T_1$ space and $x, y \in X$ be two points such that $x \neq y$. By hypothesis $X$ is $T_1$. Hence $\{y\}$ is closed and $x$ is not an element of $\{y\}$. Again, since $X$ is completely regular, there exists a continuous function

$$f: X \to \mathbf{R}$$

such that $f(x) = 0$ and $f(\{y\}) = f(y) = 1$. This asserts that $f(x) \neq f(y)$.   □

**Corollary 6.3.10** *Let $(X, \tau)$ be a completely regular $T_1$ space and $C(X, \mathbf{R})$ be the set of all real-valued continuous functions on $X$. Then there exists an embedding $f: X \to \mathbf{R}$.*

**Proof** It follows from Proposition 6.3.9 that there exists an injective continuous map $f: X \to \mathbf{R}$, and hence, $f$ is an embedding (see Theorem 6.8.4 for an independent proof).   □

### 6.3.2  Characterization of Completely Regular Spaces by Real-Valued Continuous Functions

This subsection gives a characterization of completely regular spaces $X$ by real-valued continuous functions on $X$ in Theorem 6.3.11 and another characterization of completely regularity property of a normal space by its regularity property in Theorem 6.3.14.

**Theorem 6.3.11** (Characterization of completely regular space) *A topological space $(X, \tau)$ is completely regular iff for each point $a \in X$ and each member $V$ in a subbase $\mathcal{B}$ of $\tau$ with the property that if $a \in V$, there is a real-valued continuous function $f: X \to \mathbf{R}$ such that*

$$f(x) = \begin{cases} 0, \text{for } x = a \\ 1, \text{for all } x \in X - V \end{cases}$$

*and*

$$0 \leq f(x) \leq 1 \text{ for all } x \in X.$$

**Proof** Suppose $(X, \tau)$ is a topological space such that for every point $a \in X$ and every member $V$ in a subbase $\mathcal{B}$ for the $\tau$ with the property that if $a \in V$, there is a real-valued continuous function $f: X \to \mathbf{R}$ such that

$$f(x) = \begin{cases} 0, \text{for} x = a \\ 1, \text{for all } x \in X - V \end{cases}$$

and

$$0 \leq f(x) \leq 1 \text{ for all } x \in X.$$

We claim that the topological space $(X, \tau)$ is completely regular. To prove this, let $A$ be any closed set in $(X, \tau)$ and $a \in X$ be a point such that $a \notin A$. Then $a \in X - A$ and there exists an open set $U \in \mathcal{S}$ such that $a \in U \subset X - A$, where $\mathcal{S}$ is a base generated by $\mathcal{B}$. Since, $U$ is an intersection of finitely many members of $\mathcal{B}$, the open set $U$ can be expressed as

$$U = V_1 \cap V_2 \cap \cdots \cap V_n, \text{ where } V_i \in \mathcal{B}, \text{ for } i = 1, 2, \ldots, n.$$

Then $a \in V_i$ for each $i = 1, 2, \ldots, n$. Hence, by hypothesis, there is a real-valued continuous function $f_i \colon X \to \mathbf{R}$, for each $i = 1, 2, \ldots, n$, such that

$$f_i(x) = \begin{cases} 0, \text{ for } x = a \\ 1, \text{ for all } x \in X - V_i \end{cases}$$

and

$$0 \leq f_i(x) \leq 1 \text{ for all } x \in X.$$

Define a function

$$h \colon X \to \mathbf{I}, x \mapsto \sup\{f_i(x) \colon i = 1, 2, \ldots, n\}.$$

Then $h$ is a real-valued continuous function on $X$ such that

$$h(x) = \begin{cases} 0, \text{ for } x = a \\ 1, \text{ for all } x \in A \end{cases}$$

and

$$0 \leq h(x) \leq 1 \text{ for all } x \in X.$$

Moreover, $h(a) = 0$ and for any $x \in A$, $x \in A \subset X - U$ (since $U \subset X - A$), and hence the point $x \in X - V_j$ for some index $j$, since

$$X - U = (X - V_1) \cup (X - V_2) \cup \cdots \cup (X - V_n).$$

This asserts that the topological space $(X, \tau)$ is completely regular.

Conversely, suppose that the topological space $(X, \tau)$ is completely regular and $a \in V$, where $V \in \mathcal{B}$ (a subbase for $\tau$). Then $X - V$ is a closed set in $(X, \tau)$ such that $a \notin X - V$. Hence there exists a real-valued continuous function $f \colon X \to \mathbf{R}$ such that

$$f(x) = \begin{cases} 0, \text{ for } x = a \\ 1, \text{ for all } x \in X - V \end{cases}$$

and
$$0 \le f(x) \le 1 \text{ for all } x \in X.$$

$\square$

***Example 6.3.12*** A normal space may not be completely regular. For example, Niemytzki tangent disk topology (see Chap. 3) is completely regular, but it is not normal .

***Remark 6.3.13*** Example 6.3.12 raises the problem: does there exist any situation under which the concepts of normality and completely regularity coincide? Theorem 6.3.14 solves this problem by characterizing completely regularity property of normal spaces in terms of its regularity property.

**Theorem 6.3.14** *A normal space* $(X, \tau)$ *is completely regular iff it is regular.*

***Proof*** Let $(X, \tau)$ be a normal space. If it is completely regular, then it is regular by their definitions. Conversely, suppose that $(X, \tau)$ is both normal and regular. We claim that it is completely regular. Let $A$ be a closed subset in $(X, \tau)$ and $y \in X$ be a point such that $y \notin A$. Then $X - A$ is an open set in $(X, \tau)$ such that $y \in X - A$. Since by hypothesis $(X, \tau)$ is regular, it follows that there exists an open set $U$ such that $y \in U$ and $\overline{U} \subset X$ -A. Again, since by hypothesis $(X, \tau)$ is normal, it follows that there is a Urysohn function corresponding to the pair of disjoint closed sets $\overline{U}$ and $A$ in $(X, \tau)$ such that

$$f(x) = \begin{cases} 0, \text{ for all } x \in \overline{U} \\ 1, \text{ for all } x \in A \end{cases}$$

and
$$0 \le f(x) \le 1 \text{ for all } x \in X.$$

This proves that $(X, \tau)$ is completely regular.

$\square$

### 6.3.3  *Tychonoff Spaces*

This subsection continues a study of Tychonoff spaces which started at Chap. 5. Completely regular $T_1$-spaces called Tychonoff spaces form a special class of completely regular spaces. For example, every $T_1$ normal space is a Tychonoff space by Proposition 6.3.17. For more study of Tychonoff spaces see Exercises 7 and 8 of Sect. 6.9.

**Definition 6.3.15** A topological space $(X, \tau)$ is said to be a **Tychonoff space** if it is a completely regular $T_1$-space.

***Example 6.3.16***  (i)  The subspace $\mathbf{I} = [0, 1]$ of the real line space $\mathbf{R}$ is a Tychonoff space;

(ii)  Niemytzki tangent disk topology is Tychonoff, but it is not normal.

Proposition 6.3.17 provides a vast supply of Tychonoff spaces.

**Proposition 6.3.17**  *Every $T_1$ normal space $(X, \tau)$ (i.e., $T_4$ -space) is a Tychonoff space.*

***Proof*** Let $(X, \tau)$ be a $T_1$ normal space, $A$ be any closed set in $(X, \tau)$, and $a \in X$ be a point such that $a \notin A$. As by hypothesis, $X$ is a $T_1$- space, the one-pointic set $\{a\}$ is closed in $X$. Hence corresponding to the pair of closed sets $A$ and $\{a\}$, there exists a real-valued continuous function $f : X \to \mathbf{R}$ by Urysohn lemma such that

$$f(x) = \begin{cases} 0, \text{ for } x = a \\ 1, \text{ for all } x \in A \end{cases}$$

and

$$0 \leq f(x) \leq 1 \text{ for all } x \in X.$$

This shows that $(X, \tau)$ is a completely regular space, which is also a $T_1$-space by hypothesis. Hence it is a Tychonoff space. $\qquad \square$

# 6.4  $G_\delta$-sets, Perfectly Normal Spaces, and Urysohn Functions

This section characterizes perfectly normal spaces by $G_\delta$-sets and studies the existence of Urysohn functions by $G_\delta$-sets.

## 6.4.1  $G_\delta$-sets and $F_\delta$-sets

This subsection conveys the concepts of $G_\delta$-sets and $F_\delta$-sets. The motivation of these concepts comes from the observation: the intersection of infinitely many open sets may not be an open set. For example, in the real line space $(\mathbf{R}, \sigma)$, every open interval being an open set, each $U_n = (-\frac{1}{n}, \frac{1}{n}) : n = 1, 2, 3, \ldots$ is an open set, but their intersection $\bigcap_1^\infty \{U_n\} = \{0\}$ is not open, since its complement $\mathbf{R} - \{0\}$ is not closed, because 0 is a point of accumulation of $\mathbf{R} - \{0\}$, which lies outside the set $\mathbf{R} - \{0\}$. Dually, the union of infinitely many closed sets may not be a closed set. This observation leads to define the concepts of $G_\delta$-sets and $F_\delta$-sets in a topological spaces $(X, \tau)$. For characterization of Urysohn functions by $G_\delta$-sets. see Exercise 10 of Sect. 6.9.

**Definition 6.4.1** Let $(X, \tau)$ be a topological space.

(i) The intersection of a countable family of open sets is called a $G_\delta$-**set** in $(X, \tau)$.
(ii) The union of a countable family of closed sets is called an $F_\delta$-**set** in $(X, \tau)$.

**Proposition 6.4.2** *Let $(X, \tau)$ be a topological space. Then*

*(i) The complement of a $G_\delta$-set in $(X, \tau)$ is an $F_\delta$-set in $(X, \tau)$.*
*(ii) The complement of an $F_\delta$-set in $(X, \tau)$ is a $G_\delta$-set in $(X, \tau)$.*
*(iii) The intersection of two $F_\delta$-sets is also an $F_\delta$-set in $(X, \tau)$.*
*(iv) The union of two $G_\delta$-sets is also an $G_\delta$-set in $(X, \tau)$.*
*(v) The union of a countable family of $F_\delta$-sets is an $F_\delta$-set in $(X, \tau)$.*
*(vi) The intersection of a countable family of $G_\delta$-sets is a $G_\delta$-set in $(X, \tau)$.*

**Proof** It follows from Definition 6.4.1 by using the duality principle.          □

**Example 6.4.3** In the real line space $(\mathbf{R}, \sigma)$,

(i) $\mathbf{Q}$ is an $F_\delta$-set but it is not a $G_\delta$-set;
(ii) On the other hand, dually, the subset $P = \mathbf{R} - \mathbf{Q}$ of irrational numbers is a $G_\delta$-set but it is not an $F_\delta$-set;
(iii) Every closed set is a $G_\delta$-set;
(iv) Every open set is an $F_\delta$-set.

## 6.4.2   Perfectly Normal Spaces

This subsection introduces the concept of perfectly normal spaces by imposing certain conditions in addition to the conditions prescribed in Urysohn lemma. This lemma characterizes normal spaces $(X, \tau)$ by showing the existence of a real-valued continuous function $f$, called **Urysohn function**, corresponding to every pair of disjoint closed sets $A$ and $B$ in $(X, \tau)$ such that

$$f(x) = \begin{cases} 0, \text{ for all } x \in A \\ 1, \text{ for all } x \in B \end{cases}$$

and

$$0 \leq f(x) \leq 1 \text{ for all } x \in X.$$

**Remark 6.4.4** The motivation of the concept of perfectly normal spaces comes from the observation: given two disjoint closed sets $A$ and $B$ in a topological space $(X, \tau)$, there may exist points $x \in X - A$ and $y \in X - B$ such that $f(x) = 0$ and $f(y) = 1$. It shows that although the relations $f^{-1}(0) \supset A$ and $f^{-1}(1) \supset B$ hold, the relations

$f^{-1}(0) = A$ and $f^{-1}(1) = B$ may not hold. This leads to the concept of perfectly normal spaces by imposing stronger conditions than those for normal spaces.

**Definition 6.4.5** Let $(X, \tau)$ be a topological space. It is said to be **perfectly normal** if corresponding to every pair of disjoint closed sets $P$ and $Q$ in $(X, \tau)$, there exists a real-valued continuous function $f: X \to \mathbf{R}$ such that

$$0 \leq f(x) \leq 1, \ \forall x \in X, \ f^{-1}(0) = A \text{ and } f^{-1}(1) = B.$$

**Proposition 6.4.6**  *Every perfectly normal space is a normal space.*

**Proof**  Urysohn Lemma 6.2.8 asserts that every perfectly normal space is a normal space.                                                                    ☐

**Example 6.4.7**  Every metrizable space is perfectly normal by Proposition 6.4.13.

**Remark 6.4.8**  For characterization of a perfectly normal space see Sect. 6.4.3 and Exercise 12 of Sect. 6.9.

## 6.4.3  Existence of Urysohn Function and Characterization of Perfectly Normal Spaces

This subsection gives a necessary and sufficient condition for existence of Urysohn function corresponding to a pair of disjoint closed sets of a normal space by using the concept of $G_\delta$-sets and characterizes perfectly normal spaces.

**Theorem 6.4.9**  *(Existence of Urysohn function)*  *Let $(X, \tau)$ be a normal space and $A, B$ be a pair of disjoint closed sets in $(X, \tau)$. Then corresponding to this pair $A, B$, there exists a Urysohn function $f: X \to \mathbf{R}$ satisfying the additional conditions*

$$f^{-1}(0) = A \text{ and } f^{-1}(1) = B$$

*iff $A$ and $B$ are both $G_\delta$-sets.*

**Proof**  First suppose that corresponding to the pair $A, B$ of disjoint closed sets in $(X, \tau)$, there exists a Urysohn function $f: X \to \mathbf{R}$ satisfying the additional conditions $f^{-1}(0) = A$ and $f^{-1}(1) = B$. Define

$$U_n = \{x \in X : f(x) < \frac{1}{n}\} \subset X, \ n = 1, 2, \ldots$$

and

$$V_n = \{x \in X : f(x) > \frac{n}{n+1}\} \subset X, \ n = 1, 2, \ldots.$$

This shows that each of $U_n$ and $V_n$ are open sets in $(X, \tau)$. Then each of the sets

$$A = \bigcap_1^\infty \{U_n\} \text{ and } B = \bigcap_1^\infty \{V_n\}$$

is the intersection of a countable collection of open sets, and they are both $G_\delta$-sets in $(X, \tau)$.

Conversely, let $A$ and $B$ be both $G_\delta$-sets. Then corresponding to the pair $A$, $B$ of disjoint closed sets in $(X, \tau)$, there exists a Urysohn function $h: X \to \mathbf{R}$ satisfying $h^{-1}(0) = A$ and another Urysohn function $g: X \to \mathbf{R}$ satisfying $g^{-1}(0) = B$. Define a function

$$f: X \to \mathbf{R}, \quad x \mapsto \frac{h(x)}{h(x) + g(x)}.$$

Clearly, $f$ is well-defined, since $h(x) + g(x) \neq 0$, $\forall x \in X$, since the sets $A$ and $B$ are disjoint.
Then $f$ is a continuous function such that

(i) $0 \leq f(x) \leq 1$, $\forall x \in X$;
(ii) $f^{-1}(0) = A$, since $f(x) = 0$ iff $h(x) = 0$;
(iii) $f^{-1}(1) = B$, since $f(x) = 1$ iff $g(x) = 0$.

□

Corollary 6.4.10 characterizes perfectly normal spaces. For its another characterization see Exercise 12 of Sect. 6.9.

**Corollary 6.4.10** *Let $(X, \tau)$ be a topological space. Then it is perfectly normal iff*

*(i) $(X, \tau)$ is normal and*
*(ii) every closed set in $(X, \tau)$ is a $G_\delta$-set.*

**Proof** It follows from Theorem 6.4.9.                                      □

**Remark 6.4.11** Corollary 6.4.10 proves the equivalence of the two definitions of a perfectly normal space given in Definitions 6.4.12 and 6.4.5.

**Definition 6.4.12 (Alternative definition)** A topological space is said to be **perfectly normal** if

(i) it is normal and
(ii) its every closed subset is a $G_\delta$-set.

Proposition 6.4.13 provides a vast supply of perfectly normal spaces.

**Proposition 6.4.13** *Every metrizable space is perfectly normal.*

**Proof** Let $(X, \tau)$ be a metrizable space and $d: X \times X \to \mathbf{R}$ be a metric such that its induced topology $\tau_d = \tau$. Since every metrizable space is normal, $(X, \tau)$ is a normal space. Let $P$ be any closed set in $(X, \tau)$. Then for every $n \in \mathbf{N}$,

$$U_n = \{x \in X : d(x, P) < \frac{1}{2^n}\} \text{ is an open set.}$$

Since by hypothesis, $P$ is closed in $(X, \tau)$, it follows that

$$\bigcap\{U_n : n \in \mathbf{N}\} = P.$$

This implies that $P$ is a $G_\delta$-set. It concludes by Corollary 6.4.10 that $(X, \tau)$ is perfectly normal. □

**Example 6.4.14** Every perfectly normal space is completely regular, and it is also completely normal. But its converse is not true.

(i) For example, consider the set $\mathbf{Z}$ of integers and the topology $\tau$ on $\mathbf{Z}$ consisting of $\mathbf{Z}, \emptyset$ and all even integers. Then $(\mathbf{Z}, \tau)$ is completely normal but is not perfectly normal. Because, the set of odd integers is closed in $(\mathbf{Z}, \tau)$ but it is not a $G_\delta$-set in $(\mathbf{Z}, \tau)$.

(ii) Moreover, by using Corollary 6.4.10, it follows that the topological space $(\mathbf{Z}, \tau)$ is not regular.

# 6.5 Tietze Extension Theorem: Characterization of Normal Spaces

This section proves Tietze Extension Theorem 6.5.1, which characterizes normal spaces, by using Urysohn lemma. There are several extension problems in topology. For example, if $1_{S^1} : S^1 \to S^1$ is the identity map on the circle $S^1 = \{(x, y) \in \mathbf{R}^2 : x^2 + y^2 = 1\}$, then it has no continuous extension over the entire closed disk $\mathbf{D}^2 = \{(x, y) \in \mathbf{R}^2 : x^2 + y^2 \leq 1\}$ (see **Basic Topology: Volume 3** of the present series of books). So, it is a natural problem in topology: given a topological space $(X, \tau)$, when is a continuous function defined on a subspace $A \subset X$ continuously extendable over the whole space $X$ ? Tietze Extension Theorem 6.5.1 partially solves this problem for real-valued continuous functions on normal spaces. This theorem has many applications in extension problems.

Theorem 6.5.1 now called Tietze extension theorem is named after Heinerich Franz Friedrich Tietze (1880–1964). Historically, he proved a theorem in 1915, for a metric space given in Corollary 6.5.3. Urysohn published its general version in an article in 1925. Tietze also made significant contribution to topology and introduced Tietze transformation between presentations of groups.

Tietze Extension Theorem 6.5.1 is a natural generalization of Urysohn Lemma 6.2.8, where Urysohn Lemma finds its significant application. It characterizes normal spaces $X$ in terms of continuous extension over $X$ of every continuous function $f: Y \to \mathbf{R}$, on every closed subset $Y \subset X$. An alternative form of Tietze Extension Theorem 6.5.1 is given in Exercise 14 of Sect. 6.9.

**Theorem 6.5.1** *(Tietze Extension Theorem)* *A topological space $(X, \tau)$ is normal iff for every closed set $Y$ in $(X, \tau)$, every continuous map $f: Y \to \mathbf{I}$ has a continuous extension over $X$.*

**Proof** Let $(X, \tau)$ be a topological space such that the given conditions hold. Claim that $(X, \tau)$ is normal. By hypothesis, it is assumed that for any closed set $Y$ in $(X, \tau)$, every continuous map $f: Y \to \mathbf{I}$ has a continuous extension over $X$. Let $A, B$ be a pair of disjoint closed sets in $(X, \tau)$. If $Y = A \cup B$, then $Y$ is a closed subset in $(X, \tau)$. Consider a function $f$ on the topological space $Y$ with relative topology $\tau_Y$ induced from the topology $\tau$ defined by

$$f: Y \to \mathbf{I}, \ y \mapsto \begin{cases} 0, \text{for all } y \in A \\ 1, \text{for all } y \in B \end{cases}$$

Since $A$ and $B$ are disjoint closed sets in $(X, \tau)$, it follows that $f$ is a real-valued continuous function. Hence by hypothesis $f$ has a continuous extension $F$ over $X$. This shows that

$$F: X \to \mathbf{I}$$

is a continuous map such that $F(x) = f(x)$, $\forall x \in Y$. Define two subsets $U$ and $V$ in $X$

$$U = \{x \in X: F(x) < 1/2\}$$

and

$$V = \{x \in X: F(x) > 1/2\}.$$

Then $U$ and $V$ are open sets in $(X, \tau)$ by Proposition 6.1.3, since $F$ is continuous. Hence they are two disjoint open sets in $(X, \tau)$ such that $A \subset U$ and $B \subset V$. This asserts that $(X, \tau)$ is a normal space.

Conversely, let $Y$ be a closed subspace of a normal space $(X, \tau)$. Then it is to be proved that every continuous map

$$f: Y \to \mathbf{I}$$

has a continuous extension over $X$. Let $A$ and $B$ be a pair of disjoint closed sets in the normal space $(X, \tau)$. Then for this pair of closed sets, there exists an Urysohn function, also called a characteristic function $\kappa_{(A,B)}$ by Urysohn lemma 6.2.8 such that

$$\kappa_{(A,B)}: X \to \mathbf{I}, \; x \mapsto \begin{cases} 0, \text{ for all } x \in A \\ 1, \text{ for all } x \in B \end{cases}$$

Let $f: Y \to \mathbf{I}$ be an arbitrary continuous map. Define maps

$$f_n: Y \to \mathbf{I} \text{ with } f_0 = f \text{ and } g_n: X \to \mathbf{I}, \; \forall n \geq 1$$

inductively as follows (assume that $f_n$ is already defined). Let $A_n$ and $B_n$ be defined as follows:

$$A_n = \{x \in Y: f_n(x) \leq 1/3 \cdot (2/3)^n\}$$

and

$$B_n = \{x \in Y: f_n(x) \geq 2/3 \cdot (2/3)^n\}.$$

Then $A_n$ and $B_n$ are two disjoint closed sets in $(X, \tau)$ by Proposition 6.1.3. Let

$$\kappa_{(A_n,B_n)}: X \to \mathbf{I}$$

be the corresponding characteristic function. Suppose

$$g_n = (1/3)\,(2/3)^n \cdot \kappa_{(A_n,B_n)}: X \to \mathbf{I}, \; x \mapsto (1/3)\,(2/3)^n \kappa_{(A_n,B_n)}(x).$$

Define

$$f_{n+1}: Y \to \mathbf{I}, \; x \mapsto f_n(x) - g_n(x), \; \forall x \in Y.$$

Then $g_n(x) \leq f_n(x), \; \forall x \in Y$, because, if $x \in A_n$, then $\kappa_{A,B}(x) = 0$ $\implies g_n(x) = 0$ and hence $g_n(x) \leq f_n(x), \; \forall x \in A_n$ and if $x \notin A_n$, then

$$f_n(x) > (1/3) \cdot (2/3)^n \text{ and } g_n(x) \leq (1/3) \cdot (2/3)^n.$$

Hence it follows that $g_n(x) \leq f_n(x), \; \forall x \in Y$. This asserts that the map

$$f_{n+1}: Y \to \mathbf{I}, \; x \mapsto f_n(x) - g_n(x)$$

is well-defined, since $g_n(x) \leq f_n(x), \; \forall x \in Y$. This gives the inductive definition of the maps $f_n$ and $g_n$. Again,

$$0 \leq g_n(x) \leq (1/3)(2/3)^n, \; \forall x \in X, \text{ since } g_n(x) = (1/3)(2/3)^n \cdot \kappa_{(A_n,B_n)}(x), \; \forall \in X.$$

It asserts that the series $\Sigma g_n(x)$ is uniformly convergent on $X$, and hence, it follows that

$$\lim_{n\to\infty} [g_o(x) + g_1(x) + \cdots + g_n(x)]$$

exists for all $x \in X$. Hence it defines a continuous function $F: X \to \mathbf{I}$. To prove that $F$ is a continuous extension of $f: Y \to I$ over $X$, it is sufficient to show that $F(x) = f(x)$, $\forall x \in Y$. To show it we use the inequalities

$$0 \leq f_n(x) \leq (2/3)^n$$

obtained by induction on $n$. Again,

$$f_{k+1}: Y \to \mathbf{I}, \ x \mapsto f_k(x) - g_k(x) \text{implies} \ f_{k+1}(x) = f_k(x) - g_k(x), \ \forall x \in Y, \ k = 0, 1, \ldots, n.$$

Consequently, $g_k(x) = f_k(x) - f_{k+1}(x) \ \forall x \in Y, \ k = 0, 1, \ldots, n.$

It asserts by taking summation that

$$\Sigma_{k=0}^n g_k(x) = f(x) - f_{n+1}(x), \ \text{since} \ f_0(x) = f(x), \ \forall x \in Y.$$

Hence passing to the limit as $n \to \infty$, it follows that

$$F(x) = f(x), \ \forall x \in Y,$$

because, as $n \to \infty$, LHS gives $F(x)$ and as $n \to \infty$, RHS gives f(x), since

$$0 \leq f_{n+1}(x) \leq (2/3)^{n+1}$$

and hence $f_{n+1}(x) \to 0$, $\forall x \in Y$ as $n \to \infty$,
This concludes that the continuous function

$$f: Y \to \mathbf{I}$$

has a continuous extension

$$F: X \to \mathbf{I}$$

over $X$.                                                                                           $\square$

**Remark 6.5.2** Tietze proved his extension theorem in 1915 where X is a metric space, now given in Corollary 6.5.3. The general version as given in Corollary 6.5.3 was first time found in an article published by Urysohn in 1925.

**Corollary 6.5.3** (Tietze Extension Theorem) *If A is a closed subspace of a metric space X, then every continuous map $f: A \to \mathbf{I}$ has a continuous extension over X.*

**Proof** It follows from Theorem 6.5.1 as particular case, when $X$ is a metric space.                                                                            □

**Corollary 6.5.4** (Tietze Extension Theorem) *If $A$ is a closed subspace of a normal space $X$, then every continuous map $f : A \to \mathbf{I}$ has a continuous extension over $X$.*

**Proof** It follows from one part of Theorem 6.5.1.                            □

**Example 6.5.5** The condition in Theorem 6.5.1 saying that $Y$ is to be closed in the topological space $(X, \tau)$ is necessary. Because on default of this condition, the theorem fails. For example, if $Y = (0, \infty)$, then $Y$ is not closed in the real space $(\mathbf{R}, \sigma)$. The function $f$ defined by

$$f(x) = \sin(1/x)$$

is continuous on the subspace $(Y, \sigma_Y)$ of $(\mathbf{R}, \sigma)$ but this function has no continuous extension over $(\mathbf{R}, \sigma)$, because $\lim_{x \to +0} f(x)$ is indeterminate. For another example in this respect, see Exercise 17 of Sect. 6.9.

**Example 6.5.6** Sorgenfrey line $(\mathbf{R}_l, \sigma_l)$ is a normal space. The **Sorgenfrey plane $\mathbf{R}_l^2$** is the product space $(\mathbf{R}, \sigma_l) \times (\mathbf{R}, \sigma_l)$ with product topology having a base consisting of all sets of the form $\{[x, y) \times [z, w)\}$ in the plane. The set of all points in $\mathbf{R}_l^2$ with rational coordinates is dense in $\mathbf{R}_l^2$. But its subspace $A = \{t \times (-t) : t \in \mathbf{R}_l\}$ has the discrete topology. Geometrically, $A$ represents the line $x + y = 0$ in the Sorgenfrey plane $\mathbf{R}_l^2$. Proposition 6.5.7 shows that the product of two normal spaces may not be normal.

**Proposition 6.5.7** *Sorgenfrey plane $\mathbf{R}_l^2$ with Sorgenfrey product topology is not normal.*

**Proof** Suppose this space $\mathbf{R}_l^2$ is normal. Consider $A = \{(x, y) \in \mathbf{R}^2 : y = -x\} \subset \mathbf{R}^2$ endowed with discrete topology induced by Sorgenfrey product topology. Then every map $f : A \to \mathbf{R}$ is continuous. Define

$$P = \{z \in A : z \text{ is a rational point}\}$$

and

$$Q = \{z \in A : z \text{ is an irrational point}\}$$

Then $P$ and $Q$ are disjoint closed sets in $\mathbf{R}_l^2$. Consider in particular, the continuous map

$$f : A \to \mathbf{R}, \ z \mapsto \begin{cases} 2, \text{ for all } y \in P \\ -2, \text{ for all } y \in Q \end{cases}$$

Use Tietze extension theorem on $f$ to have a continuous extension $\tilde{f} : \mathbf{R}_l^2 \to \mathbf{R}$. But this leads to assert that $P$ and $Q$ cannot be separated by disjoint open sets, producing a contradiction of our assumption that $\mathbf{R}_l^2$ is normal. This contradiction asserts that $\mathbf{R}_l^2$ is not normal.                                                                □

## 6.6   Rings $\mathcal{C}(X, \mathbf{R})$ for Compact Hausdorff Spaces $X$

This section studies rings $\mathcal{C}(X, \mathbf{R})$ for compact Hausdorff spaces $X$, and it establishes a 1-1 correspondence between the points of $X$ and maximal ideals of the ring $\mathcal{C}(X, \mathbf{R})$ in Theorem 6.6.8. For **Gelfand–Kolmogoff Theorem** see Sect. 6.7. This theorem says that two compact Hausdorff spaces $X$ and $Y$ are homeomorphic iff the corresponding rings $\mathcal{C}(X, \mathbf{R})$ and $\mathcal{C}(Y, \mathbf{R})$ are isomorphic. This deep result recovers the topology of $X$ from the ring structure of $\mathcal{C}(X, \mathbf{R})$.

### *Ideals of Rings $\mathcal{C}(X, \mathbf{R})$ for Compact Hausdorff Spaces $X$*

This subsection addresses the maximal ideals of the ring $\mathcal{C}(X, \mathbf{R})$ of all real-valued continuous functions on $X$, when $X$ is a compact Hausdorff space in general and $X = [0, 1]$ in particular. Theorem 6.6.5 proves that there is a bijective correspondence between the points of a compact Hausdorff space $X$ and the maximal ideals of the ring $\mathcal{C}(X, \mathbf{R})$. Basic topological tools used in this theorem is the Urysohn lemma.

**Definition 6.6.1**  Given a topological space $X$, let $\mathcal{C}(X, \mathbf{R})$ be the set of all real-valued continuous functions on $X$. Then $\mathcal{C}(X, \mathbf{R})$ is a commutative ring (not an integral domain) under

(i)  pointwise addition

$$f + g: X \to \mathbf{R}, \ x \mapsto f(x) + g(x) \in \mathbf{R}$$

and
(ii)  pointwise multiplication

$$fg: X \to \mathbf{R}, \ x \mapsto f(x)g(x) \in \mathbf{R}.$$

It is called the **ring of real-valued continuous functions** on $X$. The zero element $O$ and identity element $\alpha$ of this ring are the functions defined by

$$0(x) = 0 \text{ and } \alpha(x) = 1 \ \forall x \in X.$$

Moreover, $\mathcal{C}(X, \mathbf{R})$ forms a vector space over $\mathbf{R}$, which is also a **real algebra**.

Proposition 6.6.2 expresses $C$ in the language of category theory.

**Proposition 6.6.2** *Let* $\mathcal{T}op$ *be the category of topological spaces and their continuous maps and* $\mathcal{R}ing$ *be the category of rings and their homomorphisms. Then*

$$C: \mathcal{T}op \to \mathcal{R}ing$$

*is a contravariant functor.*

**Proof** The object function is defined by assigning to every object $X \in \mathcal{T}op$ the corresponding ring $C(X, \mathbf{R}) \in \mathcal{R}ing$ and the morphism function is defined by assigning to every morphism $\psi: X \to Y \in \mathcal{T}op$ the corresponding ring homomorphism

$$C(\psi): C(Y. \mathbf{R}) \to C(X. \mathbf{R}), \quad f \mapsto f \circ \psi.$$

Hence it follows that $C$ is a contravariant functor. □

**Remark 6.6.3** Proposition 6.6.2 shows that for any topological space $X$, under pointwise addition and pointwise multiplication, $C(X, \mathbf{R})$ is a ring. On the other hand, Gelfand–Kolmogoroff Theorem 6.7.7 asserts that two compact Hausdorff spaces $X$ and $Y$ are homeomorphic iff the corresponding rings $C(X, \mathbf{R})$ and $C(Y, \mathbf{R})$ are isomorphic. Hence, given two compact Hausdorff spaces $X$ and $Y$, an isomorphism of the corresponding rings $C(X, \mathbf{R})$ and $C(Y, \mathbf{R})$ implies that the spaces $X$ and $Y$ are homeomorphic and conversely a homeomorphism of two given compact Hausdorff spaces $X$ and $Y$ implies that the corresponding rings $C(X, \mathbf{R})$ and $C(Y, \mathbf{R})$ are isomorphic.

**Definition 6.6.4** A proper ideal $M$ of a ring $R$ is said to be a **maximal ideal** of $R$ if there is no proper ideal of $R$ strictly containing $M$ in the sense that there is no ideal $A$ of $R$ such that

$$M \subset_{\neq} A \subset_{\neq} R.$$

The ring $C(X, \mathbf{R})$ is now studied when $X = \mathbf{I} = [0, 1]$ with subspace topology induced by the usual topology from $\mathbf{R}$.

**Theorem 6.6.5** *Let $R$ be the ring $C([0, 1], \mathbf{R})$ and $\mathcal{M}$ be the set of all maximal ideals in $R$. Then the maximal ideals in the ring $R$ correspond to the points in $[0, 1]$, in the sense that there exists a bijection $\phi: [0, 1] \to \mathcal{M}$.*

**Proof** From ring theory, it follows that for any point $t \in \mathbf{I} = [0, 1]$, $M_t = \{f \in R: f(t) = 0\}$ is a maximal ideal of $R$ and hence $M_t \in \mathcal{M}$. To prove the theorem, it is sufficient to prove that the map

$$\psi: \mathbf{I} \to \mathcal{M}, \quad t \mapsto M_t = \{f \in R: f(t) = 0\}$$

is a bijection. Let $p, q \in [0, 1]$ be two distinct points. Since every compact Hausdorff space is normal and $(\mathbf{I}, \sigma_\mathbf{I})$ is a compact Hausdorff space as a subspace of the real line space $(\mathbf{R}, \sigma)$, the space $(\mathbf{I}, \sigma_{\mathbf{I}})$ is normal and the one-pointic sets $\{p\}$ and $\{q\}$ are two disjoint closed subsets in $(\mathbf{I}, \sigma_{\mathbf{I}})$. Hence it follows by Urysohn lemma that there exists a continuous real function

$$f : \mathbf{I} \to \mathbf{R}, \ t \mapsto \begin{cases} 0, \text{if } t = p \\ 1, \text{if } t = q \end{cases}$$

and

$$0 \leq f(t) \leq 1 \text{ for all } t \in \mathbf{I}.$$

This asserts that the map $\psi$ is injective. To prove that $\psi$ is surjective, take any $M \in \mathcal{M}$. Then for this $M$, there is a point $t \in \mathbf{I}$ at which every function $f \in M$ vanishes. Otherwise, by compactness of $\mathbf{I}$, we would reach at a contradiction. This implies that $M = M_t$. This concludes that the map $\psi$ is surjective, and hence, $\psi$ is a bijection.                                                                                 $\square$

**Corollary 6.6.6** *Let $R$ be the ring $\mathcal{C}([0, 1], \mathbf{R})$ of real-valued continuous functions on $\mathbf{I} = [0, 1]$. Then the maximal ideals in $R$ correspond to the points in $\mathbf{I}$.*

**Proof** Let $\mathcal{F}$ be the set of all maximal ideals of the ring $R$. Since the map

$$\psi : \mathbf{I} \to \mathcal{F}, t \mapsto M_t = \{f \in R : f(t) = 0\}$$

is a bijection by Theorem 6.6.5, it follows that the maximal ideals in the ring $R$ correspond to the points in $\mathbf{I}$.                                                                 $\square$

**Remark 6.6.7** There is a natural problem: what is the generalization of Theorem 6.6.5 ? Theorem 6.6.8 solves this problem communicating a positive answer for an arbitrary compact Hausdorff space $X$ in place of the space $\mathbf{I} = [0, 1]$.

**Theorem 6.6.8** *Let $X$ be a given compact Hausdorff space and $R = \mathcal{C}(X, \mathbf{R})$ be the ring of all real- valued continuous functions on the compact space $X$. Then corresponding to each point $x \in X$, there is a unique maximal ideal $M_x$ of $R$ and conversely.*

**Proof** Let $\mathcal{M}$ be the set of all maximal ideals of the ring $R$. To prove the theorem, it is sufficient to prove that the map

$$\psi : X \to \mathcal{M}, \ x \mapsto M_x = \{f \in R : f(x) = 0\}$$

is a bijection. Its proof is similar to the proof of the Theorem 6.6.5.                     $\square$

**Corollary 6.6.9** *Let $R$ be the ring $\mathcal{C}(X, \mathbf{R})$ of real -valued continuous functions on a compact Hausdorff space $X$. Then the maximal ideals in $R$ correspond to the points in $x \in X$.*

**Proof** Proceed as in Corollary 6.6.6.                                                         $\square$

## 6.7  The Gelfand–Kolmogoroff Theorem

This section continues the study of the rings $C(X, \mathbf{R})$ for compact Hausdorff spaces $X$ and proves the Gelfand–Kolmogoroff Theorem 6.7.3. It has another form given in Theorem 6.7.7, which asserts that two compact Hausdorff spaces $X$ and $Y$ are homeomorphic iff the corresponding rings $C(X, \mathbf{R})$ and $C(Y, \mathbf{R})$ are isomorphic. This deep result recovers the topology of $X$ from the ring structure of $C(X, \mathbf{R})$.

**Definition 6.7.1** Let $X$ and $Y$ be topological spaces. Then every continuous map

$$\psi: X \to Y$$

induces a homomorphism of rings

$$\psi^*: C(Y, \mathbf{R}) \to C(X, \mathbf{R}), \ g \mapsto g \circ \psi.$$

**Remark 6.7.2** Let $\alpha: X \to \mathbf{R}, \ x \mapsto 1 \in \mathbf{R}$. Then the ring $C(X, \mathbf{R})$ with identity element $\alpha$ is completely determined. Is its converse true? Its positive answer is available in this section asserting that whenever $X$ is a compact Hausdorff space , the ring $C(X, \mathbf{R})$ determines the space $X$ uniquely up to homeomorphism. Here, $C(X, \mathbf{R})$, is endowed with the discrete topology.

**Theorem 6.7.3** *Let $X$ be a compact Hausdorff space, $C(X, \mathbf{R})$ be the ring with identity element $\alpha$ and $\mathcal{H}(X)$ be the set of all nonzero ring homomorphisms $h: C(X, \mathbf{R}) \to \mathbf{R}$. If $\mathcal{H}(X)$ is endowed with the function-space topology of pointwise convergence, then*

*(i)  the only nonzero homomorphisms $h: C(X, \mathbf{R}) \to \mathbf{R}$ are the evaluation maps*

$$h_x: C(X, \mathbf{R}) \to \mathbf{R}, \ f \mapsto f(x).$$

*(ii)  (**Gelfand–Kolmogoroff**) the map*

$$\mu: X \to \mathcal{H}(X), \ x \mapsto h_x$$

*is a homeomorphism.*

**Proof**  (i)  Since $\mathbf{R}$ is a field, the kernel, ker $h$ is a maximal ideal in $C(X, \mathbf{R})$. Hence by Theorem 6.6.8, there exists a unique point $x_0 \in X$ such that

$$ker(h) = M_{x_0} = \{f \in C(X, \mathbf{R}): f(x_0) = 0\}.$$

If $f \in C(X, \mathbf{R})$ and $f(x_0) = c_0 \in \mathbf{R}$, then $f - c_0\alpha \in \ker h \implies h(f - c_0\alpha)$ $= 0$ shows that

$$h(f) = h(c_0\alpha) = c_0 = f(x_0) \text{ and } h = h_{x_0},$$

since $h_{x_0}(f) = f(x_0) = h(f)$, $\forall f \in \mathcal{C}(X, \mathbf{R})$.

(ii) By using the first part, it follows that the map

$$\mu: X \to \mathcal{H}(X), \ x \mapsto h_x \text{ is surjective.}$$

Since $X$ is a compact Hausdorff space by hypothesis, it is normal. For any two points $x_0 \neq x_1 \in X$, the sets $\{x_0\}$ and $\{x_1\}$ are disjoint closed sets in $X$. Then by Urysohn lemma there exists an $f \in \mathcal{C}(X, \mathbf{R})$ is such that

$$f(x) = \begin{cases} 0, \text{ for } x = x_0 \\ 1, \text{ for } x = x_1 \end{cases}$$

and

$$0 \leq f(x) \leq 1 \text{ for all } x \in X.$$

This shows that $\mu$ is injective. Consequently, $\mu$ is bijective. To show that it is continuous, let $U = (f, V)$ be a subbasis nbd of the point $\mu(x_0)$. Then

$$\mu^{-1}(U) = \{x \in X : h_x(f) \in V\} = \{x \in X : f(x) \in V\} = f^{-1}(V)$$

is an open set. Since $X$ is compact and Hausdorff by hypotheses and the topological space $\mathcal{H}(X)$ is Hausdorff, it follows that $\mu$ is a homeomorphism.

$\square$

## Characterization of Compact Hausdorff Spaces by Rings of Continuous Functions

This subsection characterizes compact Hausdorff spaces $X$ by rings $C(X, \mathbf{R})$. Its motivation comes from the given problem. Proposition 6.6.2 transfers a problem of topology to algebra. It raises the following problem: is its converse true ? More precisely, given a ring $\mathcal{C}(X, \mathbf{R})$, is it possible to recover the topological structure of $X$ from the ring structure of $\mathcal{C}(X, \mathbf{R})$? Its positive answer is available in Gelfand–Kolmogoroff Theorem 6.7.7, which asserts that two compact Hausdorff spaces $X$ and $Y$ are homeomorphic iff the corresponding rings $C(X, \mathbf{R})$ and $C(Y, \mathbf{R})$ are isomorphic, and hence, it gives an interplay between topology and algebra.

**Fig. 6.1** Commutative
diagram connecting $\psi$ and
$\psi^{**}$

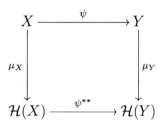

**Definition 6.7.4** Let $\mathcal{H}(X)$ be the space of all nonzero ring homomorphisms $h: \mathcal{C}(X, \mathbf{R}) \to \mathbf{R}$ endowed with the function-space topology of pointwise convergence for every compact Hausdorff space $X$. Then the ring homomorphism

$$\psi^*: \mathcal{C}(Y, \mathbf{R}) \to \mathcal{C}(X, \mathbf{R}), \ g \mapsto g \circ \psi,$$

induced by a continuous map $\psi: X \to Y$, induces a second map

$$\psi^{**}: \mathcal{H}(X) \to \mathcal{H}(Y), \ h \mapsto h \circ \psi^*.$$

Each $\psi^{**}(h): \mathcal{C}(Y, \mathbf{R}) \to \mathbf{R}$ is a nonzero homomorphism, because $(h \circ \psi^*)(\alpha) = h(\alpha) = 1$. Since the rings $\mathcal{C}(Y, \mathbf{R})$ are endowed with the discrete topology, it follows that $\psi^*$ and hence $\psi^{**}$ are both continuous. They are such that

$$\psi^{**} \circ \mu_X = \mu_Y \circ \psi: X \to \mathcal{H}(Y),$$

where $\mu_X: X \to \mathcal{H}(X)$ and $\mu_Y: Y \to \mathcal{H}(Y)$ are homeomorphisms provided by Theorem 6.7.3. It asserts that the the diagram in Fig. 6.1 connecting $\psi$ and $\psi^{**}$ is commutative.

**Proposition 6.7.5** *Let $X$ and $Y$ be two compact Hausdorff spaces and $g: \mathcal{C}(Y, \mathbf{R}) \to \mathcal{C}(X, \mathbf{R})$ be any ring homomorphism. If $g(\alpha) = \alpha$ for identity element $\alpha \in \mathcal{C}(Y, \mathbf{R})$ then there exists a unique continuous map*

$$\psi: X \to Y \ such \ that \ \psi^* = g.$$

***Proof*** Let $\psi: X \to Y$ be defined by

$$\psi = \mu_Y^{-1} \circ \psi^{**} \circ \mu_X.$$

Since $\psi^{**}: \mathcal{H}(X) \to \mathcal{H}(Y)$ is continuous and $\mu_X$, $\mu_Y$ are both homeomorphisms, it follows that $\psi$ is a continuous map. Moreover, $\psi(x) = y$ iff $g(k)(x) = k(y)$, $\forall k \in \mathcal{C}(Y, \mathbf{R})$. If

$$\psi^*: \mathcal{C}(Y, \mathbf{R}) \to \mathcal{C}(X, \mathbf{R})$$

is the induced map, then

$$\psi^*(k)(x) = k \circ \psi(x) = k(y) = g(k)(x), \, \forall x \in X \implies \psi^*(k)$$
$$= g(k), \, \forall k \in C(Y, \mathbf{R}) \implies \psi^* = g.$$

To prove the uniqueness of $\psi$, let

$$\psi, \lambda: X \to Y$$

be two continuous maps such that $\psi(x) \neq \lambda(x)$ for some point $x \in X$. Then it follows by using the given conditions on $Y$ that there is an $f \in C(Y, \mathbf{R})$ such that $f(\lambda(x)) \neq f(\psi(x))$ and hence $\lambda^* \neq \psi^*$.      $\square$

***Remark 6.7.6*** Gelfand–Kolmogoroff Theorem 6.7.7 asserts that given two compact Hausdorff spaces, $X$ and $Y$, an isomorphism between the corresponding rings $C(X, \mathbf{R})$ and $C(X, \mathbf{R})$ implies that the spaces $X$ and $Y$ are homeomorphic and conversely, a homeomorphism between two given compact Hausdorff spaces $X$ and $Y$ implies an isomorphism between the corresponding rings $C(X, \mathbf{R})$ and $C(Y, \mathbf{R})$. As a vector space, $C(X, \mathbf{R})$ forms a real algebra with multiplication $(f.g)(x) = f(x)g(x) \in \mathbf{R}$. Hence it characterizes compact Hausdorff spaces in terms of algebras (see Corollary 6.7.8).

**Theorem 6.7.7** (Another form of Gelfand–Kolmogoroff) *Two compact Hausdorff spaces $X$ and $Y$ are homeomorphic iff the corresponding rings $C(X, \mathbf{R})$ and $C(Y, \mathbf{R})$ are isomorphic.*

***Proof*** Let $X$ and $Y$ be two compact Hausdorff spaces and $\psi: X \to Y$ be a homeomorphism with its inverse homeomorphism $\phi: Y \to X$ such that $\psi \circ \phi = 1_Y$ and $\phi \circ \psi = 1_X$. Then $\psi$ induces a ring homomorphism

$$\psi^*: C(Y, \mathbf{R}) \to C(X, \mathbf{R}), \, f \mapsto f \circ \psi.$$

Then by the functorial property of the contravariant functor $C$ (see Proposition 6.6.2), it follows that the induced homomorphisms satisfy the relation:

$$(\psi \circ \phi)^* = \phi^* \circ \psi^* = \text{identity}$$

and

$$(\phi \circ \psi)^* = \psi^* \circ \phi^* = \text{identity}.$$

It implies that $\psi^*$ is an isomorphism of rings.

Conversely, let $h: C(Y, \mathbf{R}) \to C(X, \mathbf{R})$ be a ring isomorphism. Since any homomorphism

$$h: C(Y, \mathbf{R}) \to C(X, \mathbf{R})$$

satisfies the property $h(\alpha) = \alpha$, where $\alpha \in \mathcal{C}(Y, \mathbf{R})$ is the identity element and every isomorphism $h$ satisfies the conditions prescribed in Proposition 6.7.5, it follows that there exists a continuous map

$$\psi: X \to Y : \psi = \mu_Y^{-1} \circ \psi^{**} \circ \mu_X \text{ with } \psi^* = h.$$

As $\psi$ is a continuous bijective map from a compact space to a Hausdorff space, it follows that $\psi$ is a homeomorphism. $\qquad \square$

Corollary 6.7.8 characterizes compact Hausdorff spaces in terms of algebras and recovers the topology of $X$ from the algebra $\mathcal{C}(X, \mathbf{R})$.

**Corollary 6.7.8** (An alternative form of Gelfand–Kolmogoroff) *Two compact Hausdorff spaces $X$ and $Y$ are homeomorphic iff the corresponding algebras $\mathcal{C}(X, \mathbf{R})$ and $\mathcal{C}(Y, \mathbf{R})$ are isomorphic.*

**Proof** Consider $\mathcal{C}(X, \mathbf{R})$ and $\mathcal{C}(Y, \mathbf{R})$ as algebras over $\mathbf{R}$. Since an isomorphism between algebras is a ring isomorphism which preserves the vector space structures, to prove the corollary, proceed as in proof of Theorem 6.7.7. $\qquad \square$

**Theorem 6.7.9** *In the language of category theory,*

$$\mathcal{C}: \mathcal{T}op \to \mathcal{R}ings$$

*is a contravariant functor from the category of topological spaces and their continuous maps to the category of rings and their homomorphisms such that*

(i) *if $X$ and $Y$ be two homeomorphic spaces, then the corresponding rings $\mathcal{C}(X, \mathbf{R})$ and $\mathcal{C}(Y, \mathbf{R})$ are isomorphic and*

(ii) *in particular, two compact Hausdorff spaces $X$ and $Y$ are homeomorphic iff the corresponding algebras $\mathcal{C}(X, \mathbf{R})$ and $\mathcal{C}(Y, \mathbf{R})$ are isomorphic.*

**Proof** By hypothesis, $\mathcal{T}op$ is the category of topological spaces and their continuous maps and $\mathcal{R}ings$ is the category of rings and their homomorphisms. Define object function $X \mapsto \mathcal{C}(X, \mathbf{R})$ for each object $X \in \mathcal{T}op$. If $\alpha: X \to Y \in \mathcal{T}op$, then $\mathcal{C}(f)$ defines a ring homomorphism

$$\mathcal{C}(\alpha) = \alpha^*: \mathcal{C}(Y, \mathbf{R}) \to \mathcal{C}(X, \mathbf{R}), \ f \mapsto f \circ \alpha$$

which defines the morphism function $\alpha \mapsto \alpha^*$. Hence it follows that $\mathcal{C}$ is a contravariant functor.

(i) It follows from the functorial properties of $\mathcal{C}$.

(ii) It follows from Corollary 6.7.8. $\qquad \square$

***Remark 6.7.10*** The functor $\mathcal{C}$ defined in Theorem 6.7.9 is similar to cohomology functor (studied in **Basic Topology: Volume 3** of the present book series) in the sense that

(i)  both are contravariant functors from the category of topological spaces and their continuous maps to the category of rings and their homomorphisms and
(ii) both of them establish a key link between topology and algebra.

## 6.8  Applications

This section presents some interesting applications arising out of some problems not discussed in previous chapters. The applications include a study of embedding problems in Sect. 6.8.1, a proof of Weierstrass Theorem 6.7.9 in topological setting and application of Baire category theorem in Sect. 6.8.3 and some other applications in Sects. 6.8.4 and 6.8.4.

### *6.8.1  Embedding Problems: Urysohn Metrization Theorem*

The concept of embedding is very important in mathematics. A topological space $(X, \tau)$ is said to be embedded in a topological space $(Y, \sigma)$, if there exists a homeomorphism $f$ from $X$ onto a subspace of $Y$, called an embedding $f : X \to Y$. Then the space $X$ may be considered as a subspace of $Y$. This subsection solves a natural problem: can every completely regular $T_1$-space be embedded in the real line space ? It also solves another problem: can every completely regular space be embedded in $\mathbf{I}^K$ for some indexing set $K$? Finally, Urysohn metrization theorem providing sufficient conditions for metrizabilty of a topological space is proved by using embedding theorems in this subsection.

**Definition 6.8.1** Let $(X, \tau)$ and $(Y, \sigma)$ be two topological spaces. A continuous map

$$f : (X, \tau) \to (Y, \sigma)$$

is said to be an **embedding** if $f$ is injective and it defines a homeomorphism from $X$ onto $f(X)$.

***Example 6.8.2*** Let $(Y, \tau_Y)$ be any subspace of a topological space $(X, \tau)$. Then the inclusion map

$$i : Y \hookrightarrow X$$

is an embedding of Y in X, which says that every inclusion map in topological spaces is an embedding.

**Proposition 6.8.3** *Let $(X, \tau)$ be a regular space with a countable basis and $\mathbf{R}^\infty$ be the infinite dimensional Euclidean space. If $\mathbf{R}^\infty$ is endowed with the product topology, then the map*

$$f: X \to \mathbf{R}^\infty, \quad x \mapsto (f_1(x), f_2(x), \ldots)$$

*is an embedding of $X$ in $\mathbf{R}^\infty$.*

**Proof** Let $(X, \tau)$ be a regular space with a countable basis. Then there exists a countable family of continuous functions $\{f_n: X \to [0, 1]\}$ with the property that given a point $a \in X$ and a nbd U of $a$ in $X$, there is an integer $n$ such that

$$f_n(a) > 0 \text{ and } f_n(x) = 0, \ \forall x \in X - U$$

by using Exercise 20 of Sect. 6.9. Hence it follows that the map

$$f: X \to \mathbf{R}^\infty, \quad x \mapsto (f_1(x), f_2(x), \ldots)$$

is continuous, since $\mathbf{R}^\infty$ has the product topology and each $f_n$ is continuous. Moreover, $f$ is injective, since given two points $x \neq y$ in $X$, there exists a positive integer $n$ such that

$$f_n(x) > 0 \text{ and } f_n(y) = 0 \implies f(x) \neq f(y).$$

It asserts that $f$ is a continuous injective map. Let the subset $Y = f(X) \subset \mathbf{R}^\infty$ have the subspace topology inherited from the product topology of $\mathbf{R}^\infty$. This shows that $f$ is a continuous bijective map from $X$ onto $Y$. Since given any open set $U$ in $(X, \tau)$, the set $f(U)$ is open in the subspace $Y$, it follows that $f$ is homeomorphism from $X$ onto the subspace $Y$ of $\mathbf{R}^\infty$. This asserts that $f$ is an embedding of $X$ in $\mathbf{R}^\infty$. $\square$

Theorem 6.8.4 solves an embedding problem (unpublished work of M. R. Adhikari).

**Theorem 6.8.4** *Let $(X, \tau)$ be a completely regular $T_1$-space and $\mathcal{C}(X, \mathbf{R})$ be the set of all real -valued continuous functions on $(X, \tau)$. Then the topological space $(X, \tau)$ can be embedded in the real line space $\mathbf{R}$.*

**Proof** Let $(X, \tau)$ be a completely regular $T_1$- space and $\mathcal{C}(X, \mathbf{R})$ be the set of all real -valued continuous functions on $(X, \tau)$. Since the space $(X, \tau)$ is $T_1$, for distinct points $x$ and $y$ in $X$, the one-pointic set $\{y\}$ is closed and $x$ is not an element of $\{y\}$. Again, since $(X, \tau)$ is completely regular by hypothesis , there exists a continuous function

$$f \in \mathcal{C}(X, \mathbf{R}) \text{ such that } f(x) = 0 \text{ and } f(\{y\}) = f(y) = 1.$$

This asserts that $f(x) \neq f(y)$, and hence, it is proved that $f: X \to \mathbf{R}$ is an embedding. $\square$

**Remark 6.8.5** The basic tool used to prove Urysohn metrization theorem is the separation of disjoint closed sets by real-valued continuous functions. This tool is now also utilized to prove the Embedding Theorem 6.8.6.

**Theorem 6.8.6** *(Embedding theorem) Let $(X, \tau)$ be a completely regular $T_1$-space and $\mathcal{C} = \{f_a : X \to \mathbf{R} : a \in \mathbf{A}\}$ be an indexed family of continuous maps from $X$ to $\mathbf{R}$ such that for each point $p \in X$ and each nbd $U$ of $p$ in $X$, there is an index $a \in \mathbf{A}$ such that*

$$f_a(p) > 0, \text{ and } f_a(x) = 0, \forall x \in X - U.$$

*Then the map*

$$f : X \to \mathbf{R}^{\mathbf{A}}, x \mapsto (f_a(x))_{a \in \mathbf{A}}$$

*is an embedding of $X$ in $\mathbf{R}^{\mathbf{A}}$. In particular, if $f_a$ maps $X$ into $\mathbf{I} = [0, 1]$, for all $a \in \mathbf{A}$, then the above defined map*

$$f : X \to \mathbf{I}^{\mathbf{A}}$$

*is an embedding of $X$ in $\mathbf{I}^{\mathbf{A}}$.*

**Proof** It follows by proceeding as in Theorem 6.8.3 by replacing $\mathbf{R}^{\infty}$ by $\mathbf{R}^{\mathbf{A}}$ and adjusting the proof accordingly. □

**Corollary 6.8.7** *Every completely regular $T_1$-space can be embedded in $\mathbf{I}^K$ for some indexing set $K$.*

**Proof** Let $(X, \tau)$ be a completely regular $T_1$ space and $\mathcal{C} = \{f_i : X \to \mathbf{I} : i \in \mathbf{K}\}$ be the family of all continuous functions from $X$ to $\mathbf{I}$. Then the corollary follows immediately from Theorem 6.8.6.. □

**Remark 6.8.8** The above discussion is summarized in the Urysohn metrization theorem providing a sufficient condition for metrizability of a certain class of regular spaces, which is a **basic theorem in topology**, because, by Proposition 6.8.3, every regular space $(X, \tau)$ with a countable basis is embedded in $\mathbf{R}^{\infty}$ with product topology, which is a metrizable space.

**Theorem 6.8.9** *(Urysohn metrization theorem) Let $(X, \tau)$ be a regular space having a countable basis. Then $(X, \tau)$ is metrizable.*

**Proof** It follows from Remark 6.8.8. □

**Remark 6.8.10** Urysohn metrization theorem 6.7.9 provides a sufficient condition for metrizability of a topological space having a countable basis. Its another form asserting that every second countable and completely regular space is metrizable is available in Chap. 7.

## 6.8.2 Application to Analysis

This subsection gives some applications related to classical analysis. For example, Weierstrass Theorem 6.7.9, which is a basic theorem in analysis, is proved in a topological setting. Moreover, a characterization of compact subsets of the Euclidean $n$-space $\mathbf{R}^n$ by real-valued continuous functions is given in Theorem 6.8.14.

**Theorem 6.8.11** (Weierstrass theorem) *Let* $(X, \tau)$ *be a compact space and* $f: X \to \mathbf{R}$ *be a real-valued continuous map. Then* $f$ *is bounded and attains its maximum and minimum values.*

**Proof** Let $(X, \tau)$ be a compact space and $f: X \to \mathbf{R}$ be a real-valued continuous map. Then $f(X)$ is a compact subspace of $\mathbf{R}$ and hence it is closed and bounded. This implies that the function $f$ is bounded. As $f(X)$ is closed, it contains all of its limit points. This asserts that both $sup_{x \in X} f(x) \in f(X)$ and $\inf_{x \in X} f(x) \in f(X)$ exist. $\qquad\square$

**Remark 6.8.12** Weierstrass theorem 6.7.9 asserts that a continuous real-valued function on a compact set is bounded. Its converse is also true by Proposition 6.8.13.

**Proposition 6.8.13** *Let* $X \subset \mathbf{R}^n$ *be a subspace such that every real-valued continuous function on* $X$ *is bounded. Then* $X$ *is compact.*

**Proof** Suppose $X$ is bounded but it is not compact. Then it is not closed by Heine-Borel theorem. Hence, there exists a point $\alpha$ in $\overline{X} - X$ such that the function $g: X \to \mathbf{R}, x \mapsto ||x - \alpha||^{-1}$ is not bounded on $X$, which gives a contradiction. This contradiction proves that $X$ is compact. $\qquad\square$

Theorem 6.8.14 determines completely the compact subsets of the Euclidean $n$-space $\mathbf{R}^n$ by real-valued continuous functions.

**Theorem 6.8.14** *Let* $X$ *be a nonempty subset of* $\mathbf{R}^n$. *Then* $X$ *is compact iff every real-valued continuous function on* $X$ *is bounded.*

**Proof** It follows by using Weierstrass theorem 6.7.9 and Proposition 6.8.13. $\qquad\square$

**Definition 6.8.15** Let $\mathcal{F} = \{f_i: X \to \mathbf{R}\}$ be a collection of real-valued functions from a metric space $X$ with metric $d$. Then this collection is said to be **equicontinuous** if for every $\epsilon > 0$, there exists a $\delta > 0$ (depending on $\epsilon$) such that whenever

$$d(x_1, x_2) < \delta, \text{ then } |f(x_1) - f(x_2)| < \epsilon, \forall f \in \mathcal{F}.$$

**Proposition 6.8.16** *Let* $F: \mathbf{I} \times \mathbf{I} \to \mathbf{R}$ *be a continuous function and* $\mathcal{F}$ *be the family of functions*

$$f: \mathbf{I} \to \mathbf{R}, x \mapsto \int_0^1 g(z) F(x, z) dz,$$

*where* $g: \mathbf{I} \to \mathbf{R}$ *is a continuous function such that* $|g(x)| \leq 1$, $\forall x \in \mathbf{I}$ *and* $f \in \mathcal{F}$. *Then the family* $\mathcal{F}$ *is equicontinuous.*

**Proof** As by hypothesis, $F$ is continuous on the compact set $\mathbf{I} \times \mathbf{I}$, it is uniformly continuous. So, given an $\epsilon > 0$, there exists a $\delta > 0$ such that whenever

$$[(x_1 - x_2)^2 + (y_1 - y_2)^2]^{\frac{1}{2}} < \delta, \text{ then } |F(x_1, y_1) - F(x_2, y_2)| < \epsilon.$$

Let $f \in \mathcal{F}$ and $x_1, x_2 \in \mathbf{I}$ be two points such that $|x_1 - x_2| < \delta$. Then

$$|f(x_1) - f(x_2)| = |\int_0^1 g(z)[F(x_1, z) - F(x_2, z)]dz| < \int_0^1 \epsilon dz = \epsilon$$

asserts that the family $\mathcal{F}$ is equicontinuous, since the function $f \in \mathcal{F}$ is arbitrary. $\square$

**Proposition 6.8.17** *Let* $f: \mathbf{R} \to \mathbf{R}$ *be a continuous map such that it maps open sets to open sets. Then* $f$ *is monotonic.*

**Proof** If possible, under the given hypothesis, $f$ is not monotonic. So, assume that there are three points $x, y, z \in \mathbf{R}$ such that $x < y < z$ and $f(x) < f(y) > f(z)$. By using Weierstrass Theorem 6.7.9, the map $f$ has a maximum value $K$ say, in $[x, z]$, which will not be attained at $x$ or $z$. This implies that $f(x, z)$ is not open, since it contains $K$ but does not contain $K + \epsilon$ for any $\epsilon > 0$. This contradiction implies that $f$ is monotonic. $\square$

**Proposition 6.8.18** *Let* $f, g: \mathbf{I} \to [0, \infty)$ *be two continuous functions such that*

$$\sup_{x \in \mathbf{I}} f(x) = \sup_{x \in \mathbf{I}} g(x).$$

*Then there is a point* $x_0 \in \mathbf{I}$ *with the property that*

$$f(x_0)^2 + 100f(x_0) = g(x_0)^2 + 100 g(x_0).$$

**Proof** Suppose

$$\sup_{x \in \mathbf{I}} f(x) = \sup_{x \in \mathbf{I}} g(x) = K.$$

As $f, g$ are continuous on the compact set $I$, there exist points $x_1, x_2 \in \mathbf{I}$ such that $f(x_1) = g(x_2) = K$. Define a continuous function

$$h: \mathbf{I} \to [0, \infty), x \mapsto f(x) - g(x).$$

Then

$$h(x_1) = f(x_1) - g(x_1) = K - g(x_1) \geq 0$$

and

$$h(x_2) = f(x_2) - g(x_2) = f(x_2) - K \leq 0.$$

By continuity of $h$, it follows that there exists a point $x_0 \in [x_1, x_2]$ such that $h(x_0) = 0$ and hence $f(x_0) = g(x_0)$ proves the proposition. □

## 6.8.3  Application of Baire Category Theorem

This subsection applies Baire category theorem to prove uniform boundedness principle for real-valued continuous functions on a metric space. Baire category theorem asserts that a complete metric space is of the second category in the sense that it cannot be expressed as a union of countable number of nowhere dense sets. On the other hand a subset $S$ of a topological space $X$ is said to of the first category if it can be expressed as a countable union of nowhere dense sets. Otherwise $S$ is said to be of second category.

**Theorem 6.8.19** (Uniform boundedness principle) *Let* $(X, d)$ *be a complete metric space and* $\mathcal{C}$ *be a family of real-valued continuous functions on* $X$ *such that for each point* $x \in X$, *there exists a constant* $K_x$ *with the property that*

$$|f(x)| \leq K_x, \ \forall \ f \in \mathcal{C}.$$

*Then there exist a nonempty open set* $U \subset X$ *and a constant* $K$ *such that*

$$|f(x)| \leq K, \ \forall \ f \in \mathcal{C}, \ and \ \forall \ x \in U.$$

**Proof** Baire category theorem is applied to prove this theorem. For each $n \in \mathbf{N}$ and for each $f \in \mathcal{C}$, construct the set

$$X_{f,n} = \{x \in X : |f(x)| \leq n\} \subset X.$$

Since $X_{f,n} = f^{-1}([-n, n])$, the set $X_{f,n}$ is closed in $X$. Again construct the set

$$X_n = \bigcap_{f \in \mathcal{C}} X_{f,n} = \{x \in X : |f(x)| \leq n, \ \forall \ f \in \mathcal{C}\}.$$

Since, each $X_{f,n}$ is a closed set in $X$, the set $X_n$ is the intersection of a family of closed sets and hence $X_n$ is a closed set in $X$. Again, since if $x \in X$, then $x \in X_n$, $\forall \, n \geq K_x$, it follows that $\bigcup_{n=1}^{\infty} X_n = X$. Hence by Baire category theorem, every $X_n$ is not nowhere dense. This asserts that there exists some $X_n$ such that $\bar{X}_n = X_n$, which will contain a nonempty open set $U \subset X$ such that

$$|f(x)| \leq n, \ \forall \ f \in \mathcal{C} \ and \ \forall \ x \in U.$$

Hence, the theorem follows by taking $K = n$. □

### 6.8.4   Extreme Value Theorem

Real-valued continuous functions establish a close link with certain class of topological spaces.

**Proposition 6.8.20**  *Let $X$ be a compact space and $f: X \to \mathbf{R}$ be a continuous map. Then there exist points $\alpha, \beta \in X$ such that*

$$f(X) \subset [f(\alpha), f(\beta)] \subset \mathbf{R}$$

**Proof**  Since by hypothesis, $f$ is continuous and $X$ is compact, $f(X) \subset \mathbf{R}$ is compact. Hence $f(X)$ is a closed and bounded subset of $\mathbf{R}$. This asserts that $f(X)$ has the lub and glb. Again, since $f(X)$ is closed, there exist points $\alpha, \beta \in X$ such that such that

$$f(X) \subset [f(\alpha), f(\beta)] \subset \mathbf{R}$$

$\square$

**Corollary 6.8.21**  (Extreme value theorem) *Let $[a, b] \subset \mathbf{R}$ have the subspace topology and $f: [a, b] \to \mathbf{R}$ be continuous. Then $f$ attains its absolute maximum value $M$ and absolute minimum value $m$ in the sense that*

$$M = lub_{x \in [a,b]} \, f(x) \text{ and } m = glb_{x \in [a,b]} \, f(x).$$

**Proof**  Since the interval $[a, b] \subset \mathbf{R}$ is compact in the real line space $\mathbf{R}$, there exist points $\alpha$ and $\beta$ in $[a, b]$ such that

$$f(\beta) < f(x) < f(\alpha) \, \forall \, x \in [a, b].$$

This asserts that
$$f(\alpha) = M \text{ and } f(\beta) = m.$$

$\square$

### 6.8.5   Other Applications

This subsection gives some other application on the metric space $X = \mathcal{C}([0, 1])$ endowed with sup metric in Theorem 6.8.22 and also prove topological properties of metric spaces in an alternative way.

**Theorem 6.8.22** *Let $X = \mathcal{C}([0, 1])$ be the metric space of all continuous real-valued functions on $\mathbf{I} = [0, 1]$ endowed with sup metric. Then there exists a function $f \in X$ such that the derivative of $f$ does not exist at any point $t \in \mathbf{I}$.*

**Proof** Given any positive integer $n$, define

$$X_n = \{f \in X : |\frac{f(t+s) - f(t)}{s}| \leq n \text{ for some } t \text{ and } \forall s \text{ such that } t + s \in \mathbf{I}\}.$$

As $X$ is a complete metric space, it is a space of the second category by Baire category theorem and hence $\bigcup_{n=1}^{\infty} X_n \neq X$. This shows that there exists an $f \in X$ such that $f \notin X_n$ for any $n$. Hence for this function $f$,

$$|\frac{f(t+s) - f(t)}{s}| > n, \forall t \in \mathbf{I} \text{ and some } s \in \mathbf{I}, \text{ which depends on } t \text{ and } n.$$

Hence it follows that

(i) for every fixed $t \in \mathbf{I}$, $s \to 0$, as $n \to \infty$ and
(ii) $\lim_{s \to 0} \sup |\frac{f(t+s) - f(t)}{s}|$ is not finite, which implies that the derivative of $f$ does not exist at any point $t \in \mathbf{I}$.

$\square$

**Remark 6.8.23** It is already been proved that every metric space is normal. The same result is proved in Proposition 6.8.24 by using Urysohn lemma. This lemma is also applicable for metrizable spaces, in view of Proposition 6.8.26.

**Proposition 6.8.24** *Every metric space is normal.*

**Proof** (**Proof by Uryshon lemma**): Let $(X, d)$ be a metric space and $P, Q$ be two disjoint closed subsets of $X$. Consider the function

$$f : X \to \mathbf{R}, \ x \mapsto \frac{d(x, P)}{d(x, P) + d(x, Q)}.$$

Clearly, $f$ is a well-defined continuous function such that

$$f(x) = \begin{cases} 0, & \forall x \in P \\ 1, & \forall x \in Q \end{cases}$$

and

$$0 \leq f(x) \leq 1, \ \forall x \in X.$$

Hence by Urysohn lemma 6.2.8, it follows that $(X, d)$ is normal. $\square$

***Remark 6.8.25*** Urysohn lemma is also applicable for metrizable spaces, in view of Proposition 6.8.26.

**Proposition 6.8.26** *Every metrizable space is a normal space.*

**Proof** $(X, \tau)$ be a metrizable space. Then there exists a metric $d$ on $X$ such that the topology $\tau_d$ induced by $d$ on $X$ coincides with $\tau$, i.e., $\tau = \tau_d$. We claim that it is a normal space.

**Proof I**: By hypothesis, $(X, \tau)$ is a metrizable space. Hence the proposition follows from Proposition 6.8.24.

**Proof II**: Let $P$ and $Q$ be a pair of disjoint closed sets in $(X, \tau_d)$. Using the metric $d$ on $X$, define two subsets of $X$:

$$U = \{x \in X : d(x, P) < d(x, Q)\}$$

and

$$V = \{x \in X : d(x, P) > d(x, Q)\}.$$

Then $U$ and $V$ are two open sets in $(X, \tau_d)$ by Proposition 6.1.3, and hence, they are disjoint open sets such that $P \subset U$ and $Q \subset V$. This asserts that $(X, \tau_d)$ is a normal space. $\qquad\square$

**Corollary 6.8.27** *Every metrizable space is normal and Hausdorff.*

**Corollary 6.8.28** *The subspace* $\mathbf{I} = [0, 1]$ *of the real line space* $\mathbf{R}$ *is a normal Hausdorff space.*

**Proof** Since $\mathbf{R}$ is both a normal and Hausdorff space, it follows that its subspace $\mathbf{I}$ is also so. $\qquad\square$

## 6.9   Exercises

1. **(Weierstrass $M$-test)** Let $\{f_n\}$ be an infinite sequence of real functions on a topological space $(X, \tau)$ such that for all $x \in X$,

$$|f_n(x)| \leq M_n, \ n = 1, 2, \ldots,$$

where the series $\Sigma M_n$ of positive constants $M_n$ is convergent. Show that the infinite series $\Sigma f_n$ converges uniformly on $(X, \tau)$.
   [Hint: Use Theorem 6.1.9 .]
2. If $\{f_n\}$ is an infinite sequence of continuous real functions on a topological space $(X, \tau)$ and if $\Sigma f_n$ converges uniformly on $(X, \tau)$, show that the sum of the series is also a continuous real function on $(X, \tau)$.
   [Hint: Use Theorem 6.1.10 and Exercise 1.]

3. **(Tietze-Urysohn theorem)** Let $(X, \tau)$ be a normal space and $A$ be a closed subset of $(X, \tau)$. Show that given any bounded continuous function $f: A \to \mathbf{R}$, there exists a continuous function $\psi: X \to \mathbf{R}$ such that

   (i) $\psi|_A = f$ and
   (ii) $\sup_{x \in X} |\psi(x)| = \sup_{x \in A} |f(x)|$.

4. Prove Urysohn lemma 6.2.8 from Tietze Extension Theorem 6.5.1.

5. **(Characterization of connectedness** in terms of real-valued continuous functions) Let $(X, \tau)$ be a topological space and $f: X \to \mathbf{R}$ be a function. Then the support of $f$, is defined to be the set

$$S_f = \{x \in X: f(x) \neq 0\} \subset X.$$

   Show that a topological space $(X, \tau)$ is connected iff the support $S_f$ of every real-valued continuous function $f: X \to \mathbf{R}$ is disconnected.

6. Let $(X, \tau)$ be a topological space such that for every pair of distinct points $x_1, x_2 \in X$, there is a real-valued continuous function $f$ on $X$ with the property $f(x_1) \neq f(x_2)$. Show that the space $(X, \tau)$ is Hausdorff.
   [Hint: Let $x_1, x_2 \in X$ be a pair of distinct points, $U$ & $V$ be two disjoint open sets in $\mathbf{R}$ and $f: X \to \mathbf{R}$ be a continuous map such that $f(x_1) \in U$ and $f(x_2) \in V$. Then $f^{-1}(U)$ and $f^{-1}(V)$ are two disjoint open sets in $X$ such that $x_1 \in f^{-1}(U)$ and $x_2 \in f^{-1}(V)$.]

7. Let $(X, \tau)$ be a topological space. Show that

   (i) if $(X, \tau)$ is completely regular space and $(Y, \tau_Y)$ is a subspace of $(X, \tau)$, then $(Y, \tau_Y)$ is also completely regular.
   (ii) if $(X, \tau)$ is a Tychonoff space, then $(Y, \tau_Y)$ is also Tychonoff.

8. (i) Show that the topological product of any family $\{(X_\alpha, \tau_\alpha): \alpha \in \mathbf{A}\}$ of topological spaces is Tychonoff iff every $(X_\alpha, \tau_\alpha)$ is a Tychonoff space.
   (ii) Show by using (i) that the unit cube $\mathbf{I}^\mathbf{A}$ is a Tychonoff space.
   (iii) Show by using (ii) that a topological space $(X, \tau)$ is Tychonoff iff $(X, \tau)$ is homeomorphic to a subspace of the unit cube $\mathbf{I}^\mathbf{A}$.

9. Given a topological space $(X, \tau)$, let $\mathcal{C}(X, \mathbf{R})$ be the space of all real-valued continuous functions on $(X, \tau)$. For $f, g \in \mathcal{C}(X, \mathbf{R})$, show that each of the the functions

   (i) $f \wedge g: X \to \mathbf{R}$, $x \mapsto \min\{f(x), g(x)\}$
   (ii) $f \vee g: X \to \mathbf{R}$, $x \mapsto max\{f(x), g(x)\}$
   (iii) $|f|: X \to \mathbf{R}$, $x \mapsto |f(x)|$

   is continuous.
   [Hint: Consider the product $\phi \circ \psi$ of two continuous maps $\psi: X \to \mathbf{R}^2$, $x \mapsto (f(x), g(x))$ and $\phi: \mathbf{R}^2 \to \mathbf{R}$, $(t, s) \mapsto \min\{t, s\}$ to show that $f \wedge g$ is continuous.]

10. Let $(X, \tau)$ be a normal space and $A$, $B$ be two disjoint closed sets in $(X, \tau)$. Show that corresponding to this pair of closed sets, there is a Urysohn function

$$f: X \to \mathbf{I}$$

satisfying the properties $f^{-1}(0) = A$ and $f^{-1}(1) = B$, iff $A$ and $B$ are both $G_\delta$-sets.

11. Show that

   (i) Every perfectly normal space is completely normal.
   (ii) Every locally compact Hausdorff space is completely regular.
   (iii) Sorgenberg plane $\mathbf{R}^2$ with Sorgenberg topology is not normal.

[Hint: Suppose this space $\mathbf{R}^2$ is normal. Consider $A = \{(x, y) \in \mathbf{R}^2 : y = -x\} \subset \mathbf{R}^2$ endowed with discrete topology induced by Sorgenberg topology. Then every map $f: A \to \mathbf{R}$ is continuous. Use Tietze extension theorem on $f$ to have a continuous extension $\tilde{f} : \mathbf{R}^2 \to \mathbf{R}$. But this leads to a contradiction.]

12. Let $(X, \tau)$ be a topological space and $(\mathbf{I}, \sigma)$, be the subspace of the real line space, where $\mathbf{I} = [0, 1]$. Show that $(X, \tau)$ is perfectly normal iff for any nonempty closed set $P$ in $(X, \tau)$ and any point $a \in X - P$, there exists a real continuous function

$$f: (X, \tau) \to (\mathbf{I}, \sigma)$$

such that

$$f(a) = 1 \text{ and } f^{-1}(0) = P.$$

13. Show that the normed linear space of bounded uniformly continuous real-valued functions on $\mathbf{R}$, equipped with sup-norm is complete.

14. **(Another form of Tietze extension theorem)** Let $(X, \tau)$ be a normal space and $A$ be a closed subset of $(X, \tau)$. If the closed interval $[-1, 1]$ is endowed with the subspace topology of the real line space $\mathbf{R}$ and $f: A \to [-1, 1]$ is continuous, show that $f$ has a continuous extension

$$\tilde{f}: X \to [-1, 1] \text{over} X.$$

[Hint: Let $C = f^{-1}([\frac{1}{3}, 1])$ and $D = f^{-1}([-1, -\frac{1}{3}])$. Then corresponding to the disjoint closed subsets $C$ and $D$ in $(X, \tau)$, there exists by Corollary 6.2.10, a continuous function

$$f_1: X \to \left[ -\frac{1}{3}, \frac{1}{3} \right]$$

such that $f(C) = \frac{1}{3}$, $f(D) = -\frac{1}{3}$ and $|f(x) - f_1(x)| \le \frac{2}{3}$, $\forall x \in A$. By induction, construct the functions

$$f_n: X \to \left[ -\frac{2^{n-1}}{3^n}, \frac{2^{n-1}}{3^n} \right], \quad \forall n \in \mathbf{N}$$

such that

$$|f(x) - \Sigma_{i=1}^{n} f_i(x)| \le \left(\frac{2}{3}\right)^n, \forall x \in A.$$

Finally, apply Weierstrass $M$-test to show that the series $\Sigma_{i=1}^{\infty} f_i(x)$ converges uniformly to a continuous function

$$\tilde{f} \colon X \to [-1, 1],$$

which is our required continuous extension of $f$.]

15. Let $(X, \tau)$ be a normal space and $(\mathbf{R}, \sigma)$ be the real line space. Given a closed subset $A$ of $(X, \tau)$, if $f \colon A \to \mathbf{R}$ is a continuous function such that $|f(x)| \le M, \forall x \in A$, show that there exists a continuous function $g \colon (X, \tau) \to (\mathbf{R}, \sigma)$ such that

    (i) $|g(x)| \le M/3, \forall x \in X$, and
    (ii) $|f(x) - g(x)| \le 2M/3, \forall x \in A$.

16. **(Existence of Partition of unity)** Let $(X, \tau)$ be a normal space and $\mathcal{U} = \{U_1, U_2, \dots . U_n\}$ be an open covering of a closed subset $A$ of $(X, \tau)$. Show that there are continuous functions

$$\psi_i \colon (X, \tau) \to (\mathbf{R}, \sigma)$$

such that

    (i)   $0 \le \psi_j(x) \le 1$, $\forall x \in X$ and $\forall j$ satisfying $1 \le j \le n$;
    (ii)  $\psi_j(x) = 0$, $\forall x \notin U_j$ and $\forall j$ satisfying $1 \le j \le n$;
    (iii) $\Sigma_{j=1}^{n} \psi_j(x) = 1 \ \forall x \in A$; and
    (iv)  $\Sigma_{j=1}^{n} \psi_j(x) \le 1 \ \forall x \in X$.

    Such indexed family of continuous function $\psi_i$ is called a **partition of unity** on $X$, dominated by (or subordinate to) an open covering $\mathcal{U}$ of $A$.

17. Consider the subspace $F = (0, 1]$ of $(\mathbf{I}, \sigma_I)$ and the function $f \colon F \to \mathbf{R}, x \mapsto sin(1/x)$. Show that $f$ is continuous but $f$ has no continuous extension over $\mathbf{I} = [0, 1]$.

18. Let $(X, d)$ be a metric space and $P, Q$ be two disjoint closed sets in $X$. Show that there exists a continuous function a continuous function $f$ such that

$$f \colon X \to \mathbf{R}. x \mapsto \begin{cases} 1, \text{ for all } x \in P \\ -1, \text{ for all } x \in Q \end{cases}$$

and

$$-1 \le f(x) \le 1 \text{ for all } x \in X - P \cup Q.$$

[Hint: Define

$$f: X \to x \mapsto \frac{d(x, Q) - d(X, P)}{d(x, Q) + d(X, P)}.$$

Then $f$ is well-defined and continuous.]

19. Let $\{f_n\}$ be a sequence of real-valued continuous functions on the closed interval $[a, b]$ of the Euclidean line. Show that

   (i) (**Arzela–Ascoli theorem**)If this sequence is uniformly bounded and equicontinuous, then there exists a subsequence of the sequence $\{f_n\}$ such that this subsequence converges informally.

   (ii) The converse of **Arzela–Ascoli theorem** is also true in the sense if every subsequence of $\{f_n\}$ has a uniformly convergent subsequence, then $\{f_n\}$ is uniformly bounded and equicontinuous.

20. Let $(X, \tau)$ be a regular space with a countable basis. Show that there exists a countable family of continuous functions $\{f_n : X \to [0, 1]\}$ with the property that given a point $a \in X$ and a nbd U of a, there an integer $n$ such that

$$f_n(a) > 0 \text{ and } f_n(x) = 0, \forall x \in X - U.$$

## *Multiple Choice Exercises*

Identify the correct alternative (s) (there may be none or more than one) from the following list of exercises:

1. (i) Every metrizable space is not normal.
   (ii) If $(X, \tau)$ is a normal space and $(Y, \tau_Y)$ is a closed subspace of $(X, \tau)$ under the induced topology $\tau_Y$, then $(X, \tau)$ is also normal.
   (iii) If If $(X, \tau)$ is a normal space having at least two distinct elements, then there exists a nonconstant real-valued continuous function on $X$.

2. Let $(X, \tau)$ be a nonempty compact Hausdorff space and $C(X, \mathbf{R})$ be the set of all real-valued continuous functions on $(X, \tau)$.

   (i) If $X$ has at least $n$ distinct elements, then the dimension of the linear space $C(X, \mathbf{R})$ over $\mathbf{R}$ is $n$.

   (ii) If $X_1$ and $X_2$ are two nonempty disjoint closed sets in $(X, \tau)$, then there exists a function $f \in C(X, \mathbf{R})$ such that

$$f: X \to \mathbf{R}, x \mapsto \begin{cases} -3, \text{if } x \in X_1 \\ 4, \text{if } x \in\in X_2. \end{cases}$$

   (iii) If $A$ is a nonempty closed subset of $(X, \tau)$ and $h: A \to \mathbf{R}$ is a continuous function, then there exists a function $f \in C(X, \mathbf{R})$ such that

$$f(x) = h(x), \forall \in A.$$

3. Let $S^1$ be the unit circle in the Euclidean plane $\mathbf{R}^2$.

   (i) There exists a continuous one-one function $f: S^1 \to S^1$.

   (ii) For every continuous one-one map $f: S^1 \to \mathbf{R}$, there exist an uncountable number of pairs of distinct points $x, y \in S^1$ such that $f(x) = f(y)$.

   (iii) There exists a continuous and one–one and onto map $f: S^1 \to \mathbf{R}^2$.

4. Consider the four subspaces of the Euclidean plane of $\mathbf{R}^2$.

$$A = \{(x, y) \in \mathbf{R}^2 : x^2 + y^2 \leq 1\};$$

$$B = \{(x, y) \in \mathbf{R}^2 : x^2 + y^2 < 1\};$$

$$C = \{(x, y) \in \mathbf{R}^2 : x^2 + y^2 = 3/2\}$$

and

$$D = \{(x, y) \in \mathbf{R}^2 : x^2 + y^2 \geq 2\}.$$

Examine the the validity of following statements.

   (i) For any continuous function $h : A \to \mathbf{R}$, there exists a continuous function $f: R^2 \to \mathbf{R}$ such that

$$f(x) = h(x), \quad \forall x \in A.$$

   (ii) For any continuous function $h : B \to \mathbf{R}$, there exists a continuous function $f: \mathbf{R}^2 \to \mathbf{R}$ such that

$$f(x) = h(x), \quad \forall x \in B.$$

   (iii) There exists a continuous function $f$ such that

$$f: \mathbf{R}^2 \to \mathbf{R}, \ x \mapsto \begin{cases} 1, \text{ for all } x \in C \\ 0, \text{ for all } x \in A \cup D. \end{cases}$$

5. Let $[a, b]$ be a closed interval in the Euclidean line and $\mathcal{C}[a, b]$ be the ring of real-valued continuous functions on $[a, b]$.

   (i) If $X = \{f \in \mathcal{C}[-5, 5]: f(5) = f(-5) = 0\}$, then there exists a continuous function $f \in X$ such that $f(x) = 2, \forall x \in [-1, 0]$ and $f(x) = 3, \forall x \in [1, 2] \cup [3, 4]$.

   (ii) If $X = \{f \in \mathcal{C}[-5, 5]: f(5) = f(-5) = 0\}$, then for every $f \in X$, there exists a point $p \in (-5, 5)$ such that $f(p) = p$.

   (iii) For every point $x \in [-5, 5]$, there exists a unique maximal ideal $M_x$ in the ring $\mathcal{C}[-5, 5]$.

(iv) Corresponding to a maximal ideal $M$ in the ring $C[-5, 5]$, there is a unique point $x \in [-5, 5]$ such that the maximal ideal $M_x$ (if (iii) is true) coincides with $M$.

# References

Adhikari, A., Adhikari, M.R.: Basic Topology, Volume 2: Topological Groups, Topology of Manifolds and Lie Groups. Springer, India (2022)

Adhikari, M.R.: Basic Topology, Volume 3: Algebraic Topology and Topology of Fiber Bundles. Springer, India (2022)

Adhikari, M.R.: Basic Algebraic Topology and its Applications. Springer, India (2016)

Adhikari, M.R., Adhikari, A.: Basic Modern Algebra with Applications. Springer, New Delhi, New York, Heidelberg (2014)

Alexandrov, P.S.: Introduction to Set Theory and General Topology. Moscow (1979)

Bredon, G.E.: Topology and Geometry. Springer-Verlag, New York (1983)

Borisovich, Y.C.U., Blznyakov, N., Formenko, T.: Introduction to Topology. Mir Publishers, Moscow, Translated from the Russia by Oleg Efimov (1985)

Brown, R.: Topology: A Geometric Account of General Topology, Homotropy Types, and the Fundamental Groupoid. Wiley, New York (1988)

Chatterjee, B.C., Ganguly, S., Adhikari, M.R.: Introduction to Topology. Asian Books, New Delhi (2002)

Dugundji, J.: Topology. Allyn & Bacon, Newton, MA (1966)

Fuks, D.B., Rokhlin, V.A.: Beginner's Course in Topology. Springer-Verlag, New York (1984)

Hewitt, F.: On two problems of Urysohn. Ann. Math. **47**(2), 503–509 (1946)

Hewitt, F.: Introduction to General Topology. Holden-Day, San Francisco (1966)

Hu, S.T.: Introduction to General Topology. Holden-Day, San Francisco (1966)

Kelly, J.L.: General Topology. Van Nostrand, New York, 1955; Springer-Verlag, New York (1975)

Mendelson, B.: Introduction to Topology College Mathematical Series. Allyn and Bacon, Boston (1962)

Munkres, J.R.: Topology. Prentice-Hall, New Jersey (2000)

Patterson, E.M.: Topology. Oliver and Boyd (1959)

Singer, I.M., Thorpe, J.A.: Lecture Notes on Elementary Topology and Geometry. Springer-Verlag, New York (1967)

Stephen, W.: General Topology. Addison-Wesley (1970)

# Chapter 7
# Countability, Separability and Embedding

This chapter continues the study of special classes of topological spaces such as spaces satisfying either of the two axioms of countability formulated by Hausdorff in 1914 or satisfying the axiom of separability introduced by Frechét in 1906, both initiated in Chap. 3, which do not arise from the study of calculus and analysis in a natural way. They arise through a deep study of topology. The axiom of first countability arose through the study of convergent sequences. For example, Theorem 7.11.4 characterizes Hausdorff property of a topological space satisfying the first axiom of countability by a convergent sequence. For a metric space the concept of compactness and the Bolzano–Weierstrass property (B–W) property coincides (see Chap. 5).

The present chapter characterizes compactness property in Theorem 7.1.8 by the (B–W) property for more general topological spaces such as a space having a countable open base. It is proved in Theorem 7.5.5 that every topological space can be embedded in a separable space. The classical Urysohn Metrization Theorem 7.10.5, which asserts that every topological space satisfying the second axioms of countability and regularity can be embedded in a metric space, is also proved in this chapter, which implies that such a topological space is metrizable. This theorem gives a sufficient condition of metrizability of a topological space. This chapter also studies Lindelöf spaces from the viewpoint of countability and separability. Various applications of the concepts of this chapter are also available in Sect. 7.11.1.

Motivation of the study of the concepts of countability and separability of topological spaces comes from the following natural problems:

(i) A metric space $M$ is compact iff every infinite subset $X \subset M$ has at least one accumulation point. Is this characterization of compactness valid for an arbitrary topological space?

Its partial solution is available in Theorem 7.1.8 for a topological space having a countable open base.

© The Author(s), under exclusive license to Springer Nature Singapore Pte Ltd. 2022
A. Adhikari and M. R. Adhikari, *Basic Topology 1*,
https://doi.org/10.1007/978-981-16-6509-7_7

(ii) Does there exist a certain connection between compactness and the Bolzano–Weierstrass property (B–W property) of an arbitrary topological space?
Its partial solution is available in Theorem 7.1.8 for a topological space having a countable open base.

(iii) Does there exist a sufficient condition for metrizability of an arbitrary topological space?
Its partial solution is available in Theorem 7.10.5 for a second countable and completely regular space.

(iv) Can the Hausdorff property of a topological space be characterized by its convergent sequences?
Its partial solution is available in Theorem 7.11.4 for a first countable space.

(v) Can an accumulation point of a subset $A$ of a topological space be characterized by an infinite sequence of points in $A$?
Its partial solution is available in Theorem 7.11.6 for a first countable space.

(vi) Is every topological space embeddable in the Hilbert cube $I^N$?
Its partial solution is available in Theorem 7.10.7 for a completely regular second countable space.

(vii) Is every topological space embeddable in a separable space?
Its partial solution is available in Theorem 7.5.5.

So, for a deep study in topology, it is necessary to further study the countability, separability and Lindelöf properties of topological spaces. Moreover, two embedding problems are solved in Theorem 7.5.5.

For this chapter the books (Bredon 1983; Chatterjee et al. 2002; Munkres 2000; Adhikari 2016, 2022; AdhikariAdhikari 2014; Adhikari and Adhikari 2022; Borisovich et al. 1985; Dugundji 1966; Kalajdzievski 2015; Mendelson 1962; Morris 2007; Stephen 1970) and some others are referred in the Bibliography.

## 7.1 Characterization of Compactness by Bolzano–Weierstrass Property

For a metric space the compactness property and the property that its every infinite subset has at least one accumulation point coincide. It raises the natural question: whether these two properties are the same for an arbitrary topological space? To have a partial answer, this section characterizes compactness property of topological spaces having countable open base for its topology by Bolzano–Weierstrass (B–W) property. This establishes a close relation between compactness property and the Bolzano–Weierstrass (B–W) property in such spaces. On the other hand, given a metric space $(X, d)$, the compactness of its associated topological space $(X, \tau_d)$ has been characterized in Chap. 5 by using (B–W) property. There are topological spaces which are not compact, but its every infinite subset has an accumulation point.

**Definition 7.1.1** A topological space $(X, \tau)$ is said to have a **countable open base** or simply countable base at a point $x \in X$, if there is a countable family $\mathcal{B}_x$ of open nbds of $x$ such that every nbd of $x$ in $(X, \tau)$ contains at least one member of the family $\mathcal{B}_x$.

**Remark 7.1.2**  A subset $X$ of a metric space $M$ is said to have the Bolzano–Weierstrass property if every sequence in $X$ has a convergent subsequence in the sense that it has a subsequence which converges to a point in $X$. For the general case, it is defined in Definition 7.1.3.

**Definition 7.1.3**  A topological space $(X, \tau)$ is said to have **Bolzano–Weierstrass (B–W) property** if every countable open covering of $X$ has a finite subcovering.

**Theorem 7.1.4**  *Let $(X, \tau)$ be a topological space and $Y$ be a subspace of $X$ such that every infinite subset of $Y$ has a point of accumulation in $Y$. Then every countable open covering of $Y$ has a finite subcovering.*

**Proof**  Let $(X, \tau)$ be a topological space and $Y$ be a subspace of $X$ such that every infinite subset of $Y$ has a point of accumulation in $Y$. Let $\{U_n : n \in \mathbf{N}\}$ be a countable open covering of $Y$, where each $U_n$ is an open subset of $(X, \tau)$. Suppose there is no finite subcovering of $Y$. Then for each $m \in \mathbf{N}$, the open set $U_m^* = \bigcup_{n=1}^{m} U_n$ can not cover $Y$. This shows that for each integer $m$, there is a point $x_m \in Y$ such that $x_m$ is not an element of $U_m^*$. Consider the subset $S = \{x_1, x_2, \ldots, x_m, \ldots\} \subset Y$. Then the set $S$ is infinite, and it has a limit point $x \in Y$ by hypothesis. Consequently, $x \in U_t$ for some index $t$. Then there are infinitely many points of $S$ which will lie in $U_t$, since $U_t$ is a nbd of the point $x$. This shows that for some choice of the integer $m > t$, the point

$$x_m \in U_t \subset U_t^* \subset U_m^*.$$

This contradicts the choice of $x_m$. This contradiction asserts that there is a finite subcollection of open sets $\{U_n : n \in \mathbf{N}\}$ covering $Y$.   □

**Remark 7.1.5**  Theorem 7.1.4 asserts that (B–W) property of a topological space implies its compactness property. So, it has become necessary to search for a sufficient condition such that its every open covering has a countable subcovering. Lindelöf Theorem 7.1.6 provides such a sufficient condition.

**Theorem 7.1.6  (Lindelöf)** *If a topological space $(X, \tau)$ has a countable open base for its topology $\tau$, then every open covering of the space $X$ has a countable subcovering.*

**Proof**  Let $(X, \tau)$ be a topological space and $\mathcal{B} = \{B_k : k \in \mathbf{K}\}$ be a countable open base for the topology $\tau$. Then for every point $x \in X$ and every open set $U \in \tau$ containing the point $x$, there exists an element $B_k \in \mathcal{B}$ such that

$$x \in B_k \subset U.$$

This asserts that $U = \bigcup \{B_k : k \in \mathbf{K}'\}$ for some subset $\mathbf{K}' \subset \mathbf{K}$. Suppose, $\{U_i : i \in \mathbf{K}\}$ is an open covering of $X$. Then there exists a countable subset $\mathbf{K}^* \subset \mathbf{K}''$ such that

$\{U_i : i \in \mathbf{K}^*\}$ also forms a countable covering of $X$, because, for every point $x \in X$ and every open set $U_j$ containing $x$, there is a member $B_k \in \mathcal{B}$ such that

$$x \in B_k \subset U_j.$$

$\square$

**Remark 7.1.7** The above discussion is summarized in the following basic and important result characterizing compactness of topological spaces with the help of Bolzano–Weierstrass (B–W) property.

**Theorem 7.1.8** *Let* $(X, \tau)$ *be a topological space having a countable open base for the topology* $\tau$. *Then the space* $(X, \tau)$ *is compact iff it has the Bolzano–Weierstrass (B–W) property.*

**Proof** Let $(X, \tau)$ satisfy (B–W) property. If it has a countable open base for the topology $\tau$, then every open covering of the space $X$ has a countable subcovering by Theorem 7.1.6. This shows that every countable open covering of $X$ has also a finite subcovering of $X$ by (B–W) property of $X$. This proves that $(X, \tau)$ is compact.

Conversely, let $(X, \tau)$ be compact. Then every open covering of $X$ has a finite subcovering. This implies that every countable open covering of $X$ has also a finite subcovering. This asserts that $(X, \tau)$ satisfies (B–W) property.                    $\square$

## 7.2   Countability and Separability

This section continues the study of countability and separability initiated in Chap. 3 for topological spaces which are closely related to metrizable and Lindelöf spaces. There is a close link between the concepts of countability and separability. The concept of a countable open base at every point in a topological space is applied to study important class of spaces such as metrizable, first countable, second countable, separable and Lindelöf spaces.

Recall the following definitions for smooth study of this chapter:

**Definition 7.2.1** A topological space $(X, \tau)$ is said to satisfy the **first axiom of countability**, if there exists a countable open base at every point in $X$. A space $(X, \tau)$ satisfying the first axiom of countability is said to be a first countable (or locally separable) space.

**Definition 7.2.2** A topological space $(X, \tau)$ is said to satisfy the **second axiom of countability,** if there exists a countable open base for the topology $\tau$. A space $(X, \tau)$ satisfying the second axiom of countability is said to be a second countable (or completely separable) space.

**Example 7.2.3**    (i) The real line space $(\mathbf{R}, \sigma)$ is a second countable space, since the collection of all open intervals $(a, b)$ with rational end points forms a countable open base for the topology $\sigma$.

(ii) The Euclidean space $\mathbf{R}^n$ with the usual product topology is a second countable space, since the collection of all products $\Pi_{n \in \mathbf{Z}} I_n$, where open intervals $(a, b)$ with rational end points for finitely many values of $n$ and $I_n = \mathbf{R}$ for all other values of $n$ form a countable open base for the product topology.

**Definition 7.2.4** A topological space $(X, \tau)$ is said to be **separable** if there is a countable dense subset $Y$ in the topological space $(X, \tau)$ in the sense that $\overline{Y} = X$ for some subset $Y \subset X$.

For the metric space, it is the convention to use the following definition.

**Definition 7.2.5** A metric space $X$ is said to be separable if there exists a sequence $\{x_n\}$ in $X$, which is dense in $X$, equivalently, if there is a countable subset $Y$ in $X$ such that $\overline{Y} = X$.

**Example 7.2.6** (i) The real line space $\mathbf{R}$ is a separable, since $\mathbf{Q}$ is countable and $\overline{\mathbf{Q}} = \mathbf{R}$.

(ii) $\mathbf{R}^n$ is separable, since sequence of points having rational coordinates is countable.

**Theorem 7.2.7** *Every metrizable space is first countable.*

**Proof** Let $(X, \tau)$ be a metrizable space. Then there is a metric $d$ on $X$ such that its induced topology $\tau(d) = \tau$. Consider the collection of open balls

$$\mathcal{S} = \{B_x(1/n) : x \in X, n \in \mathbf{N}\}$$

and

$$\mathcal{B} = \{B_x(\epsilon) : x \in X, \epsilon > 0\}.$$

Then $\mathcal{B}$ along with $\emptyset$ gives the topology $\tau(d)$ on $X$. Claim that $\mathcal{S}$ forms a base for the topology $\tau(d)$ on $X$. Given an open set $U \in \tau(d)$ and a point $x \in U$, there is an $\epsilon > 0$ such that $x \in B_x(\epsilon) \subset U$. Take $n \in \mathbf{N}$ such that $1/n < \epsilon$. Then

$$x \in B_x(1/n) \subset B_x(\epsilon) \subset U.$$

Hence it follows that $\mathcal{S}$ forms a base for the topology $\tau(d)$ on $X$. This shows that for every point $x \in X$, the collection

$$\mathcal{S} = \{B_x(1/n) : n \in \mathbf{N}\}$$

forms a countable base at the point $x$ for the topology $\tau(d)$ on $X$. This proves that $(X, \tau)$ is a first countable space. $\square$

**Remark 7.2.8** There is a natural problem: does there exist a necessary and sufficient condition for a topological space to be metrizable? Urysohn's metrization theorem which asserts that every second countable normal $T_1$ space is metrizable (see Chap. 6). It gives a partial solution of the above problem in 1924. For its complete solution the book (Kelly 1975) is referred.

## 7.2.1  Lindelöf Space and Lindelöf Theorem

This section studies Lindelöf spaces through the concepts of countability and separability. A topological space $(X, \tau)$ is said to be Lindelöf if every open covering of $X$ has a countable subcovering and it is countably compact if every countable open covering of $X$ has a finite subcovering. Clearly, $(X, \tau)$ is compact iff it is both Lindelöf and countably compact. **Historically,** the term "Lindelöf space" was given by Kuratowski and Sierpinski in 1921. Theorem 7.2.10 called the Lindelöf theorem in honor of E.L. Lindelöf (1870–1946) provides a rich supply of interesting Lindelöf spaces. His original version was on open subsets of $\mathbf{R}^n$. The other form of Lindelöf theorem is given in Theorem 7.2.12.

**Definition 7.2.9** A topological space $(X, \tau)$ is said to be a **Lindelöf space** if every open covering of $X$ has a countable subcovering.

**Theorem 7.2.10** *(Lindelöf) Every second countable space is Lindelöf.*

**Proof** Let $(X, \tau)$ be a second countable space, $\mathbf{B} = \{U_i : i = 1, 2, \ldots\}$ be a countable open base for the topology $\tau$ and $\mathcal{C} = \{V_\alpha : \alpha \in \mathbf{A}\}$, where $\mathbf{A}$ is an indexing set, be an open covering of $X$. Let $x \in X$, and $V \in \mathcal{C}$ be such that $x \in V$. Then there exists a set $U \in \mathbf{B}$, such that $x \in U \subset V$. Since $x \in V \in \mathcal{C}$, if $x$ runs over $X$ and $V$ runs over $\mathcal{C}$, then the corresponding family of sets $\{U\}$ form a countable subfamily $\mathbf{U}$ of $\mathbf{B}$. If $\mathbf{U} = \{U_1, U_2, \ldots\}$, then $U_i$ is contained in some $V \in \mathcal{C}$. Without loss of generality, we may assume that $U_i \subset V_i$ for all $i = 1, 2, \ldots$ Then the family of open sets $\{V_i : i = 1, 2, \ldots\}$, thus determined, forms a countable subcovering of $\mathcal{C}$, because, every point $x \in X$ is in some $U_i$, and hence the point $x$ is in the corresponding open set $V_i$. This asserts that the space $(X, \tau)$ is a Lindelöf space.                      □

**Corollary 7.2.11** *The real number space $(\mathbf{R}, \sigma)$, with the usual topology $\sigma$, is second countable. It is also first countable, separable and Lindelöf.*

**Proof** The empty set $\emptyset$ and the collection of open intervals $\{(a, b) : a < b, \text{ and } a, b \text{ are rational numbers}\}$ form a countable open base for the topology $\sigma$ on $\mathbf{R}$. This shows that $(\mathbf{R}, \sigma)$ is a second countable space. Hence it follows by Theorem 7.2.10 that $(\mathbf{R}, \sigma)$ is Lindelöf. It is also first countable, separable.                      □

**Theorem 7.2.12** *(Another Form of Lindelöf Theorem)  Let $(X, \tau)$ be a second countable space and $U$ be a nonempty open set in $(X, \tau)$. If $U$ is expressed as a union of a family $\{U_i\}$ of open sets, then $U$ can also be expressed as a countable union of $U_i$'s.*

**Proof** By hypothesis, $(X, \tau)$ is a second countable space and $U$ be a nonempty open set in $X$. Then there exists a countable open base $\mathbf{B} = \{U_n\}$ for the topology $\tau$ on $X$. Let $x \in U$ be an arbitrary point. Then this point $x \in U_i$ for some $i$ and there exists a basic open set $U_m \in \mathbf{B}$ such that

$$x \in U_m \subset U_i.$$

As the point $x$ is an arbitrary point of $U$, it follows that there exists a countable subfamily $\mathbf{B}'$ of the open base $\mathbf{B} = \{U_n\}$, whose union is $U$. Again, for each open set $V_i$ belonging in this subfamily $\mathbf{B}'$, we can choose an open set $U_i \in \mathbf{B}$ such that $V_i \subset U_i$. Then the family $\{U_i\}$ obtained in this way is countable, whose union is $U$. $\qquad\square$

**Remark 7.2.13** Theorem 7.2.10 asserts that every second countable space is Lindelöf. It is also first countable and separable by Theorem 7.2.14.

**Theorem 7.2.14** *Every second countable space is*

(i) *first countable;*
(ii) *separable.*

**Proof** Let $(X, \tau)$ be a second countable space, and let $\mathbf{B} = \{U_i : i = 1, 2, \ldots\}$ be a countable open base for the topology $\tau$.

(i) Let $x \in X$ be an arbitrary point. Then the collection $\mathbf{B_x}$ of all those $U_i \in \mathbf{B}$, which contain the point $x$, forms a countable open base at the point $x$ for the topology $\tau$, because for any open set $U$ containing $x$, there exists an open set $U_j \in \mathbf{B}$, such that $x \in U_j \subset U$ and this $U_j \in \mathbf{B_x}$. This shows that there exists a countable open base at every point $x \in X$. This asserts that the space $(X, \tau)$ is also a first countable space.

(ii) Select a point $x_i$ from each $U_i$, for $i = 1, 2, \ldots$ such that $Y = \{x_i : i = 1, 2, \ldots\}$ is an infinite sequence. Then the set $Y$ is a countable subset of $X$. We claim that $\bar{Y} = X$. Let $p \in X - Y$, and $V$ be any open set containing the point $p$. Then there exists an open set $U_j \in \mathbf{B}$, such that $p \in U_j \subset V$. The points $p \neq x_j$, since $x_j \in Y$ and $p \notin Y$. This implies that $V$ intersects $Y$ at the point $x_j \neq p$ and hence $p \in Y'$ (derived set of $Y$). This shows that any point $x \in X$ is either in $Y$, or is a limiting point of $Y$. This asserts that $x \in \bar{Y}$, $\forall x \in X$ and hence $X = \bar{Y}$. Consequently, the space $(X, \tau)$ is separable.

$\qquad\square$

**Remark 7.2.15** Theorem 7.2.16 shows that the converses of the statements (i) and (ii) of Theorem 7.2.14 are not true.

**Theorem 7.2.16** *The topological space* $(\mathbf{R}, \sigma_c)$*, with the cofinite topology* $\sigma_c$

(i) *is separable;*
(ii) *is Lindelöf; but*
(iii) *is not first countable.*

**Proof** (i) The space $(\mathbf{R}, \sigma_c)$ is separable, because the rational numbers form a countable dense subset in this space.

(ii) To prove that the space $(\mathbf{R}, \sigma_c)$ is Lindelöf, consider an arbitrary open covering $\mathcal{C}$ of this space and any $U \in \mathcal{C}$. Then the complement of $U$ in $\mathbf{R}$ is finite, and hence there exists a finite subfamily of $\mathcal{C}$, which forms an open covering of

$X - U$. This asserts that the open set $U$ together with this finite collection of members of $\mathcal{C}$ form a countable open subcovering of $(\mathbf{R}, \sigma_c)$. This proves that the space $(\mathbf{R}, \sigma_c)$ is Lindelöf.

(iii) If possible, assume that the space $(\mathbf{R}, \sigma_c)$ is first countable. Then for each $x \in \mathbf{R}$, there exists a countable open base $\mathcal{B}_x = \{U_i : i = 1, 2, \ldots\}$ about the point $x$. Then each set $(\mathbf{R} - U_i)$ is finite set, for $i = 1, 2, \ldots$ and hence

$$\bigcup_{i=1}^{\infty}(\mathbf{R} - U_i) = \bigcup\{(\mathbf{R} - U_i) : i = 1, 2, \ldots\}$$

is a countable set. Let $X$ be a finite subset of $\mathbf{R}$, not containing $x$. Then $x \in \mathbf{R} - X \in \sigma_c$ and there exists an $U_j \in \mathcal{B}_x$, such that

$$x \in U_j \subset \mathbf{R} - X.$$

This shows that every finite subset $X$ of $\mathbf{R}$ which does not contain the point $x$, is contained in a set $\mathbf{R} - U_j$, for some $j$. The union of all such finite subsets $X$ is the set $\mathbf{R} - \{x\}$, which is uncountable, but it is contained in the union $\bigcup_{i=1}^{\infty}(\mathbf{R} - U_i)$, which is countable. This contradiction proves that the space $(\mathbf{R}, \sigma_c)$ is not first countable.

$\square$

Proposition 7.2.17 provides an example of first countable Lindelöf space, which is not separable.

**Proposition 7.2.17** *Let $\mathbf{R}$ be the set of real numbers. If $\sigma$ consists of*

(i) *the empty set $\emptyset$;*
(ii) *all those subsets of $\mathbf{R}$, which do not contain 0;*
(iii) *the 3 subsets $\mathbf{R} - \{1, 2\}$, $\mathbf{R} - \{1\}$, $\mathbf{R} - \{2\}$;*
(iv) *the whole set $\mathbf{R}$,*

*then $(\mathbf{R}, \sigma)$ is a first countable, Lindelöf space, but it is not separable.*

**Proof** Clearly, $(\mathbf{R}, \sigma)$ is a topological space. For any point $x \in \mathbf{R}$, Define

$$U_x = \begin{cases} \{x\} & \text{if } x \neq 0 \\ \mathbf{R} - \{1, 2\} & \text{if } x = 0 \end{cases},$$

i.e.,

$$U_x = \{x\}, \text{ if } x \neq 0, \text{ and } U_x = \mathbf{R} - \{1, 2\}, \text{ if } x = 0.$$

Then $\{U_x\}$ forms a countable open base at the point $x \in \mathbf{R}$ for the topology $\sigma$. This asserts that the space $(\mathbf{R}, \sigma)$ is first countable. Again, any open covering $\mathcal{C}$ of $\mathbf{R}$ includes at least one of the sets given in (iii) and (iv) (such that the point 0 is covered). Let $U \in \mathcal{C}$ be an open set of this type. Then the set $X - U$ has at most two points 1 and 2. If $U_1$, $U_2 \in \mathcal{C}$ are two open sets such that $1 \in U_1$ and $2 \in U_2$, then the collection $\{U, U_1, U_2\}$ forms a finite and hence obviously a countable subcovering of $\mathcal{C}$ for $\mathbf{R}$.Thus proves that $(\mathbf{R}, \sigma)$ is a Lindelöf space. Finally, since every one-pointic subset of $\mathbf{R} - \{0\}$ is open, any subset of $\mathbf{R}$, which is dense in $(\mathbf{R}, \sigma)$, contains the uncountable set $\mathbf{R} - \{0\}$, it follows that the space $(\mathbf{R}, \sigma)$ is not separable. This completes the proof of the proposition. $\square$

**Remark 7.2.18** $G_\delta$-sets and $F_\delta$-sets defined in Chap. 6, are closely related to regular and Lindelöf spaces (see Exercises 24 and 25 of Sect. 7.12).

### 7.2.2 Countable Topological Spaces

This section studies countable topological spaces from the viewpoint of the first countability and second countability properties.

**Definition 7.2.19** A topological space $(X, \tau)$ is said to be a **countable space** if $X$ is a countable set.

**Proposition 7.2.20** $(X, \tau)$ *be an arbitrary countable space. Then* $(X, \tau)$

(i) *is separable;*
(ii) *is Lindelöf;*
(iii) *is not necessarily first countable.*

**Proof** Left as an exercise. $\square$

**Proposition 7.2.21** *Any countable space which is also first countable is second countable.*

**Proof** Let $(X, \tau)$ be an arbitrary countable space, which also first countable and $x \in X$. If $\mathbf{B}_x$ is a countable open base in $(X, \tau)$ about the point $x$, then

$$\{U : U \in \mathbf{B}_x \text{ and } x \in X\}$$

forms a countable open base for the topology $\tau$. This asserts the space $(X, \tau)$ is also second countable. $\square$

## 7.3 Subspaces of First and Second Countable Space

This section studies subspaces of first and second countable spaces and proves that the properties of first and second countability of topological spaces are hereditary.

**Theorem 7.3.1**    *(i)  Every subspace of a first countable space is first countable.*
*(ii)  Every subspace of a second countable space is second countable.*

**Proof**    (i)  Let $(X, \tau)$ be a first countable space and $(Y, \tau_Y)$ be an arbitrary subspace of $(X, \tau)$. Then for every point $y \in Y$, there exists a countable open base $\{U_1(y), U_2(y), \ldots\}$ about the point $y$ for the topology $\tau$ in $(X, \tau)$, and subsequently there is a countable open base $\{V_1(y), V_2(y), \ldots\}$ about $y$ for the topology $\tau_Y$ in $(Y, \tau_Y)$, where $V_i(y) = Y \cap U_i(y)$, for $i = 1, 2, \ldots$ This asserts that $(Y, \tau_Y)$ is also first countable.

(ii)  Let $(X, \tau)$ be a second countable space with $U_1, U_2, \ldots$ a countable open base for the topology $\tau$ and $(Y, \tau_Y)$ be an arbitrary subspace of $(X, \tau)$. Then it follows that the collection of sets $\{Y \cap U_1, Y \cap U_2, \ldots\}$ forms a countable open base for the subspace topology $\tau_Y$. This proves that $(Y, \tau_Y)$ is also second countable.
$\square$

**Remark 7.3.2**  It follows from Theorem 7.3.1 that the properties of first and second countability of topological spaces are hereditary. On the other hand, the properties of separability and Lindelöf of topological spaces are not hereditary, because a subspace of a separable space may not be separable (see Example 7.6.5) and a subspace of a Lindelöf space may not be Lindelöf (see Example 7.6.6).

## 7.4   Properties of Appert's Space and Sorgenfrey Line

This section studies Appert's space (see Definition 7.4.1) and Sorgenfrey line from the viewpoint of countability and separability and other properties. For example, the Appert's space is a countable space, which is not first countable; on the other hand, Sorgenfrey line is Lindelöf and is also separable.

### 7.4.1   Properties of Appert's Space

This section studies Appert's space. This space provides an example of a countable space which is both separable and Lindelöf. This space is both hereditarily separable and hereditarily Lindelöf.

**Definition 7.4.1**  For a subset $X$ of $\mathbf{N}$ of natural numbers, let $N(n, X)$ be the number of integers in $X$ which are less than a given integer n and $\tau$ be the family of those subsets $U$ of $\mathbf{N}$, for which either

(i)  $1 \notin U$ or
(ii)  $1 \in U$ and $\lim_{n \to \infty} \frac{N(n, U)}{n} = 1$, where $N(n, U)$ is the number of integers in $U$ which are less than the given integer $n$.

Then $\tau$ forms a topology on $X$ and the corresponding topological space $(X, \tau)$ is known as the **Appert's space**.

Theorem 7.4.2 proves some properties of Appert's space.

**Theorem 7.4.2** *Let $(\mathbf{N}, \tau)$ be the Appert's space. This space*

*(i) is countable;*
*(ii) is separable and Lindelöf;*
*(iii) is neither first countable nor second countable.*

**Proof** Let $(\mathbf{N}, \tau)$ be the Appert's space.

(i) The Appert's space $(\mathbf{N}, \tau)$ is clearly a countable space.
(ii) It follows immediately, since every countable space is separable and Lindelöff.
(iii) Suppose $(\mathbf{N}, \tau)$ is first countable. Let $\Omega = \{U_n : n = 1, 2, \ldots\}$ be a countable open base about the point 1. Since every $U_n$ is an infinite subset, select a point $x_n \in U_n$ such that $x_n > 10^n$. Define

$$U = \mathbf{N} - \{x_n : n = 1, 2, \ldots\}.$$

As $N(n, U) \geq n - \log_{10} n$, then it follows that

$$\lim_{n \to \infty} \frac{N(n, U)}{n} \geq \lim_{n \to \infty} \frac{(n - \log_{10} n)}{n} = 1.$$

This shows that $U$ is an open set containing 1. Again, since $\Omega$ forms an open base about the point 1, there is an open set $U_r \in \Omega$ such that

$$1 \in U_r \subset U.$$

But it is not tenable, because, $x_r \in U_r$ but $x_r \notin U$. This asserts that the Appert's space $(X, \tau)$ is not first countable, and hence it is not second countable.

$\square$

**Corollary 7.4.3** *Appert's space is both hereditarily separable and hereditarily Lindelöf.*

**Proof** Since every countable space is separable and Lindelöf by Proposition 7.2.20 and since every subspace of a countable space is also a countable space, it follows that the Appert's space is both hereditarily separable and hereditarily Lindelöf. $\square$

## 7.4.2 More Properties of Sorgenfrey Line

This section proves more properties of Sorgenfrey line in Theorem 7.4.4 and in Corollary 7.4.5, in addition to its properties discussed earlier.

**Theorem 7.4.4** *Let* $(Y, \sigma_Y)$ *be a subspace of the Sorgenfrey line* $(\mathbf{R}, \sigma_l)$. *Then* $(Y, \sigma_Y)$

(i) *is Lindelöf, and*
(ii) *is also separable.*

***Proof***  (i) Let $(Y, \sigma_Y)$ be a subspace of $(\mathbf{R}, \sigma_l)$ and $\mathcal{V} = \{V_k : k \in \mathbf{K}\}$ be an open covering of $Y$ in $(Y, \sigma_Y)$. If $V_k = Y \cap W_k$, $W_k \in \sigma_l : k \in \mathbf{K}$, then $\mathcal{W} = \{W_k : k \in \mathbf{K}\}$ forms an open covering of the subset $Y$ in $(\mathbf{R}, \sigma_l)$ and hence $Y \subset \bigcup\{W_k : k \in \mathbf{K}\}$. Now, for every point $x \in Y$, there exists a half-open interval $U_x = [x, x + \delta_x)$ such that $x \in U_x \subset W_v$, for some $v \in \mathbf{K}$. Then the family $\{U_x : x \in Y\}$ of open sets forms an open covering of $Y$. Claim that there exists a countable subfamily $\{U_{x_i}\}$, such that $\bigcup\{U_{x_i}\} = \bigcup\{U_x\}$. Given a fixed positive integer $n$, and a subfamily $\{U_x^n\}$ consisting of all those half-open intervals $U_x^n \in \{U_x\}$, for which $\delta_x > 1/n$. If $y \in U^n = \bigcup\{U_x^n\}$ is a point such that it is not an interior point of any $U_x^n$, then there exists an interval of length $\geq 1/n$, located at the left of $y$ and containing no point with $\delta_x > 1/n$. It asserts that the set $X$ consisting of all such points $y$, is countable and it can be covered by the countable family $\{U_y^n : y \in X\}$. If the left end points of all the remaining intervals $U_x^n$ are deleted, then a family of open intervals $\mathbf{I}_x^n$ is obtained. Since the real number space $(\mathbf{R}, \sigma)$ (with usual topology $\sigma$) is first countable, it follows that it is second countable by Corollary 7.2.11, and hence it is Lindelöf by Theorem 7.2.14. The set $\mathbf{I}^n = \bigcup\{\mathbf{I}_x^n\}$ is open and it can be covered by a countable subfamily of $\{\mathbf{I}_x^n\}$. Consequently, $U^n$ has also a a countable subcovering $\{U_x^n\}$. Hence $\bigcup\{U_x\} = \bigcup\{U^n\}$ asserts that $Y \subset \bigcup\{U_x\}$ is also covered by a countable subcovering $\{U_{x_i}\}$ of $\{U_x\}$, and hence it has also by a countable subcovering $\{W_i\}$ of $\mathcal{W}$. This implies that $\{V_i\}$ forms a countable subcovering of the covering $\mathcal{V}$ of $(Y, \sigma_Y)$. Hence $(Y, \sigma_Y)$ is a Lindelöf space. This shows that $(\mathbf{R}, \sigma_l)$ is a hereditarily Lindelöf space.

(ii) Let $(Y, \sigma_Y)$ be a subspace of $(\mathbf{R}, \sigma_l)$ and $Y_1$ be the set consisting of all those points in $Y$, which are not accumulation points of $Y$ from the right. This implies that the set $Y_1$ is at most countable. Take $Y_2 = Y - Y_1$. Hence every point of $Y_2$ is an accumulation point of $Y$ from the right. Suppose $Y_0$ is a countable dense subset of $Y$ in the usual Euclidean subspace topology of $Y$. Then $Y_1 \cup Y_0 = Y^*$ is a countable dense subset of $Y$. For any $p \in Y - Y_1 = Y_2$, consider the half-open interval $[p, p + \epsilon)$, for any $\epsilon > 0$. Since $p$ is an accumulation point of $Y$ from the right, and the subspace $(Y, \sigma_Y)$ is a $T_1$-space, there are infinitely many points in $[p, p + \epsilon) \cap Y$. Since $(p, p + \epsilon) \cap Y$ contains points of $Y^*$, the set $[p, p + \epsilon) \cap Y$ contains points of $Y^*$ for an arbitrary $\epsilon > 0$. This implies that $p \in \bar{Y}^*$, (closure $\bar{Y}^*$ is with respect to the topology $\sigma_Y$ on $Y$). Consequently, the subspace $(Y, \sigma_Y)$ is separable. This asserts that $(\mathbf{R}, \sigma_l)$ is a hereditarily separable space.

$\square$

**Corollary 7.4.5** *Let X be a subspace of the Sorgenfrey line* $(\mathbf{R}, \sigma_l)$ *and Y consist of all those points of X, which are not accumulation points of X from the right in* $(\mathbf{R}, \sigma_l)$. *Then Y is at most countable.*

**Proof** Assume that the set $Y$ is uncountable. Consider the $(Y, \sigma_Y)$ as a subspace of $(X, \sigma_X)$ with the topology induced from Sorgenfrey line space $(\mathbf{R}, \sigma_l)$. Then for any $y \in Y$, there is an half-open interval $[y, z_y)$ with the property that

$$[y, z_y) \cap X = \{y\}.$$

Again, for any $y \in Y$, $\{y\} \in \sigma_Y \implies$ the family $\{\{y\} : y \in Y\}$ forms an open covering of $Y$ which has no finite subcovering. This contradicts the result that the space $(Y, \sigma_Y)$ is Lindelöf in Theorem 7.4.4. This contradiction proves that $Y$ is at most a countable set.

$\square$

## 7.5 Topological Embedding Problems

This section solves some embedding problems. It is proved in Theorem 7.5.5 that every topological space is embedded in a separable space and every $T_1$-space is embedded in a Lindelöf space. Moreover, an immediate application of Urysohn Embedding Theorem 7.10.7 asserts that every completely regular second countable space can be embedded in the Hilbert cube $\mathbf{I}^{\mathbf{N}}$ (see Theorem 7.10.7).

**Example 7.5.1** Let $X$ and $Y$ be topological spaces and $y_0 \in Y$ be a fixed point. Then the map

$$f : X \to X \times Y, x \mapsto (x, y_0)$$

is an embedding.

**Example 7.5.2** Let $(\mathbf{R}, \sigma)$ be the real line space with usual topology $\sigma$. It is homeomorphic to every open interval $(a, b) \subset \mathbf{R}$ with relative topology induced by $\sigma$ on $(a, b)$. Hence it follows that the subspace $(a, b)$ is embedded in $\mathbf{R}$.

**Example 7.5.3** Familiar examples of embeddings:

(i) Jordan curve is an embedding $f : S^1 \to \mathbf{R}^2$ and
(ii) topological knots is an embedding $f : S^1 \to \mathbf{R}^3$.
   Both of them are studied in **Basic Topology: Volume 3** of the present series of books.

**Remark 7.5.4** Every embedding theorem in topology is important, for example, Embedding Theorem 7.5.5 is utilized to show that Lindelöf and separability properties of topological spaces are not hereditary (see Examples 7.6.5 and 7.6.6).

Theorem 7.5.5 proves that every topological space is embeddable in a separable space and every $T_i$-space is embeddable in a Lindelöf space.

**Theorem 7.5.5** *(Embedding theorem)*

(i) *Every topological space can be embedded in a separable space.*
(ii) *Every $T_1$-space can be embedded in a Lindelöf space.*

**Proof**   (i) Let $(X, \tau)$ be a given topological space, and the point $y \notin X$ be arbitrary. Endow a topology $\sigma$ on the set $Y = X \cup \{y\}$ by declaring open sets consisting of empty set $\emptyset$ and the subsets $U \cup \{y\}$, for all $U \in \tau$. Hence it follows that the given topological space $(X, \tau)$ is a subspace of $(Y, \sigma)$, because, $\tau = \sigma_X$, the induced topology on $X$ by the $\sigma$. Moreover, $\{y\}$ is a countable dense subset of $(Y, \sigma)$. This asserts that the space $(Y, \sigma)$ is separable. This proves that the topological space $(X, \tau)$ is embedded in the separable space $(Y, \sigma)$.

(ii) Let $(X, \tau)$ be a given $T_1$-space and the point $y \notin X$. Endow a topology $\sigma$ on the set $Y = X \cup \{y\}$ by nbd filters, i.e., nbds of a point $x \in Y$ are to be the same as its nbds in $(X, \tau)$ if $x \in X$, and the nbds of y are the sets $\{y\} \cup A$, where $X - A$ are closed Lindelöff subsets of $X$. Hence it follows that $(Y, \sigma)$ is a Lindelöf space, having $(X, \tau)$ as a subspace. Then it follows that the given topological space $(X, \tau)$ is embedded in $(Y, \sigma)$.

$\square$

## 7.6   More on Separability and Lindelöf Properties

This section proves in Corollary 7.6.3 that separability of a topological space is a topological property like compactness and connectedness proved in Chap. 5. But this property is not hereditary (see Example 7.6.5). Again, Corollary 7.6.4 shows that Lindelöf property of a topological space is also topological. This property is also not hereditary (see Example 7.6.6).

The results that separability and Lindelöf are both topological properties follow as a consequence of Theorem 7.6.1.

**Theorem 7.6.1** *Let $f : (X, \tau) \to (Y, \sigma)$ be a continuous onto map.*

(i) *If the space $(X, \tau)$ is separable, then $(Y, \sigma)$ is also so.*
(ii) *If the space $(X, \tau)$ is Lindelöf, then $(Y, \sigma)$ is also so.*

**Proof** By hypothesis, $f : (X, \tau) \to (Y, \sigma)$ be a continuous onto map. Then $f(X) = Y$.

(i) Let $(X, \tau)$ be a separable space. If $C$ is a countable dense subset of $(X, \tau)$, then $f(C)$ is a countable subset of $Y$ and $f(C)$ is dense in $(Y, \sigma)$. Hence it follows that the space $(Y, \sigma)$ is also separable.

(ii)  Let $(X, \tau)$ be a Lindelöf space and **U** be an open covering of Y in $(Y, \sigma)$. Since f is continuous and onto, it follows that

$$\mathbf{V} = \{f^{-1}(U) : U \in \mathbf{U}\}$$

forms an open covering of $X$ in $(X, \tau)$. Again, since the space $(X, \tau)$ is Lindelöf, it has a countable subcovering **H** of **V**. Then the family

$$\{f(U) : U \in \mathbf{H}\}$$

of open sets is a countable subcovering of **U**, which forms a countable open covering of $Y$ in $(Y, \sigma)$. This asserts that $(Y, \sigma)$ is also a Lindelöf space.

$\square$

**Corollary 7.6.2**    *(i)  Every continuous image of a separable space is separable.*
*(ii)  Every continuous image of a Lindelöf space is Lindelöf.*

**Proof**  It follows from Theorem 7.6.1.                                             $\square$

**Corollary 7.6.3**  *Separability of topological spaces is a topological property.*

**Proof**  It follows from Theorem 7.6.1(i).                                          $\square$

**Corollary 7.6.4**  ***Lindelöf property*** *of topological spaces is a topological property.*

**Proof**  It follows from Theorem 7.6.1(ii).                                         $\square$

***Example 7.6.5***  Separability is a topological property but it is not a hereditary property. The first part is proved by Corollary 7.6.3. For the second part, let $X$ be an uncountable set and $\tau$ be the discrete topology on $X$. Then $(X, \tau)$ is a nonseparable space. This topological space can be embedded in a separable space by Embedding Theorem 7.5.5(i). This asserts that a subspace of a separable space of being separable is not hereditary.

***Example 7.6.6***  Lindelöf property is topological but it is not hereditary: The first part is proved by Corollary 7.6.4. For the second part, take any $T_1$-space $(X, \tau)$ which is not Lindelöf. Then this topological space can be embedded in a Lindelöff space by Embedding Theorem 7.5.5(ii). For a specific example, consider the topological space $(\mathbf{R}, \sigma)$ described in Proposition 7.2.17. Let $\mathbf{R}^* = \mathbf{R} - \{0\}$ Then every one-pointic set $\{x\}$ in the subspace $(\mathbf{R}^*, \sigma_{\mathbf{R}^*})$ is open. Again the collection of all one-pointic sets $\{x\}$ constitutes an uncountable open covering of $\mathbf{R}^*$ in the space $(\mathbf{R}^*, \sigma_{\mathbf{R}^*})$. But this open covering has no countable open subcovering. This asserts that the space $(\mathbf{R}, \sigma)$ is Lindelöf but its subspace $(\mathbf{R}^*, \sigma_{\mathbf{R}^*})$ is not Lindelöf.

**Theorem 7.6.7**  *Let $f : (X, \tau) \to (Y, \sigma)$ be a continuous onto open map.*

(i) *If the space* $(X, \tau)$ *is first countable, then* $(Y, \sigma)$ *is also first countable.*

(ii) *If the space* $(X, \tau)$ *is second countable, then* $(Y, \sigma)$ *is also second countable.*

**Proof** Let f be a continuous open map of a topological space $(X, \tau)$ onto a topological space $(Y, \sigma)$. Then $f(X) = Y$.

(i) Let $(X, \tau)$ be first countable, $y \in Y$, and $x \in f^{-1}(y)$. Then given any countable family $\{U_i : i = 1, 2, \ldots\}$ of open sets, containing $x$ in $(X, \tau)$, there exists a countable family of open sets $\{f(U_i) : i = 1, 2, \ldots\}$, containing the point $f(x) = y$ in $(Y, \sigma)$, since $f$ is an open onto map by hypothesis. To show that this family $\{f(U_i) : i = 1, 2, \ldots\}$ forms a countable open base about the point $y$ in $(Y, \sigma)$, let $V$ be any open set containing $y$ in $(Y, \sigma)$. Since $f$ is a continuous map, $f^{-1}(V)$ is an open set containing $x$ in $(X, \tau)$, and hence there exists an open set $U_j \in \tau$ such that

$$x \in U_j \subset f^{-1}(V) \implies y = f(x) \subset f(U_j) \subset V.$$

This proves that there exists a countable open base about the point $y$. Since the point $y \in Y$ is arbitrary, it follows that the space $(Y, \sigma)$ is a first countable space.

(ii) Let $(X, \tau)$ be a second countable space and $\{U_i : i = 1, 2, \ldots\}$ be a countable open base for $\tau$. Since by hypothesis $f$ is an open mapping, it follows that $\{f(U_i) : i = 1, 2, \ldots\}$ forms a countable family $\mathcal{F}$ of open sets in $(Y, \sigma)$. We claim that $\mathcal{F}$ forms an open base for $\sigma$. To show it, take any $V \in \sigma$. Then $f^{-1}(V) \in \tau$, since f is continuous by hypothesis, and hence it can be represented as a union: $f^{-1}(V) = \bigcup\{U_k : k \in \mathbf{N}_1 \subset \mathbf{N}\}$ for some subset $\mathbf{N}_1$ of the set of all natural numbers $\mathbf{N}$. As $f$ is surjective by hypothesis, it follows that

$$V = f(f^{-1}(V)) = \bigcup\{f(U_k) : k \in \mathbf{N}_1 \subset \mathbf{N}\}.$$

This shows that

$$\{f(U_i) : i = 1, 2, \ldots\}$$

forms a countable open base for $\sigma$ and hence $(Y, \sigma)$ is a second countable space.

$\square$

**Corollary 7.6.8** *Every image under a continuous open map of a first countable (second countable) space is a first countable (second countable) space.*

**Proof** It follows from Theorem 7.6.7.  $\square$

**Proposition 7.6.9** *Every separable metric space is second countable.*

**Proof** Let $(X, d)$ be a metric space and $\tau_d$ be the topology on $X$ induced by $d$. Suppose the space $(X, \tau_d)$ is separable and $Y = \{y_1, y_2, \ldots\}$ is a countable dense subset in $X$. Consider the family of open spheres

$$\mathcal{F} = \{X_{n,t} = \{x \in X : d(x, y_n) < 1/t\} : n, t = 1, 2, \ldots, \}.$$

Then $\mathcal{F}$ forms an open base for the topology $\tau_d$ induced by $d$ on $X$. This concludes that every separable metric space is second countable.

□

**Theorem 7.6.10** *Let $(X, \tau)$ be the topological product of a family of topological spaces $\{(X_\alpha, \tau_\alpha) : \alpha \in \mathbf{A}\}$. If $(X, \tau)$ is a Lindelöf space, then every coordinate space $(X_\alpha, \tau_\alpha)$ lso a Lindelöf space.*

**Proof** Let the product space $(X, \tau)$ be Lindelöf. Since,

(i) Every coordinate space $(X_\alpha, \tau_\alpha)$ is homeomorphic to a closed subspace of the product space $(X, \tau)$;
(ii) every closed subspace of a Lindelöf space is also a Lindelöf space and
(iii) the property that a space is a Lindelöf space is a topological property,

It follows that the component space $(X_a, \tau_a)$ is Lindelöf, for every $a \in \mathbf{A}$.        □

**Example 7.6.11** The converse of Theorem 7.6.10 is not true. For example, consider the product space $(\mathbf{R}, \tau_1) \times (\mathbf{R}, \tau_1)$, which is not a Lindelöf space; on the other hand Theorem 7.4.4 asserts that the Sorgenfrey line $(\mathbf{R}, \tau_1)$ is a Lindelöff space. The product space $\mathbf{R}_l^2$ is called the Sorgenfrey plane with the topology $\tau$ having a base consisting of all sets of the form $\{[x, y) \times [z, w)\}$ in the plane. Its subspace $A = \{(t, -t) : t \in \mathbf{R}\}$ is a closed subset of the product space $(\mathbf{R}, \tau_1) \times (\mathbf{R}, \tau_1)$. **Geometrically,** $A$ represents the line $x + y = 0$ in the Sorgenfrey plane $\mathbf{R}_l^2$. The space $\mathbf{R}_l^2$ can be covered by the open set $\mathbf{R}_l^2$ and the basis elements of the form $\{[x, y) \times [-x, w)\}$ in the plane. Then each of the basis elements intersects the line $A$ in at most one point. Since $A$ is uncountable, no countable subfamily of this basis element can cover the Sorgenfrey plane $\mathbf{R}_l^2$. This asserts that the product space $(\mathbf{R}, \tau_1) \times (\mathbf{R}, \tau_1)$ is not Lindelöf.

**Example 7.6.12** The Sorgenfrey line $(\mathbf{R}, \tau_1)$ is first countable and separable by Theorem 7.4.4 of Sect. 7.12. It is also Lindelöf. Hence, by Exercise 18 of Sect. 7.12, the product space $(\mathbf{R}, \tau_1) \times (\mathbf{R}, \tau_1)$ is first countable and separable but it is not Lindelöf by Theorem 7.6.10.

**Remark 7.6.13** More properties of the topological product space from the viewpoint of countability and separability are available in Exercises 17 and 18 of Sects. 7.12 and Theorem 7.4.4.

## 7.7 Convergence of Sequences in First Countable Spaces

This section studies convergence of sequences in first countable spaces with a view to characterize Hausdorff property of a first countable space with the help of convergent sequence having a unique limit in the space (see Theorem 7.11.4).

**Proposition 7.7.1** *Let $(X, \tau)$ be a first countable space and $x \in X$. If $\mathcal{B} = \{U_i : i = 1, 2, \ldots\}$ is a countable open base about the point $x$ in $(X, \tau)$, then there is an infinite subsequence $\{W_i : i = 1, 2, \ldots\}$ of the sequence $\{U_i : i = 1, 2, \ldots\}$ such that*

(i)  *For every open set $V$ in $(X, \tau)$ containing the point $x$, there is a suffix $k$ such that*

$$W_i \subset V, \ \forall i \geq k.$$

(ii)  *Moreover, if $(X, \tau)$ is $T_1$, then $\bigcap_{n=1}^{\infty} \{W_i\} = \{x\}$.*

**Proof**  By the given condition, $U_1 \cap U_2 \cap \cdots \cap U_k$ is an open set containing the point $x$ in $(X, \tau)$. By hypothesis, $\mathcal{B} = \{U_i\}$ is a countable open base about the point $x$ in $(X, \tau)$. Hence there is some $W_k \in \{U_i\}$ such that

$$x \in W_k \subset U_1 \cap U_2 \cap \cdots \cap U_k, \ \forall k = 1, 2, \ldots.$$

Then the sequence $\{W_n\}$ has the desired properties. Because,

(i)  for any open set $V$ in $(X, \tau)$ containing the point $x$, there exists an open set $U_k \in \mathcal{B}$ such that

$$x \in U_k \subset V.$$

Moreover,

$$W_i \subset U_k, \ \forall i \geq k \implies W_i \subset V, \ \forall i \geq k.$$

(ii)  Suppose $Y = \bigcap \{W_i\}$. Then $x \in W_i$, $\forall i$ implies $x \in Y$. Since by hypothesis, $(X, \tau)$ is $T_1$, then for any pair of distinct elements $x, y \in X$, there exists an open sets $V$ in $(X, \tau)$ such that

$$x \in V \text{ but } y \notin V.$$

This shows that there exists some index $k$ such that $W_i \subset V$ for all $i \geq k$. Hence $y \notin W_i$, $\forall i \geq k$. This implies that $y \notin Y$. This proves that

$$Y = \bigcap_{n=1}^{\infty} \{W_i\} = \bigcap \{W_i : i = 1, 2, \ldots\} = \{x\}.$$

$\square$

**Remark 7.7.2**  Proposition 7.7.1 is applied to prove Theorem 7.11.4 to obtain a characterization of first countable Hausdorff spaces.

# 7.8 Cardinality of Open Sets in a Second Countable Space

This section studies cardinality of open sets in second countable spaces and also second countable Hausdorff spaces. More precisely, Theorem 7.8.1 proves that the cardinality of the set of all open sets in a second countable space is at most equal to the power of continuum $c$; on the other hand, Theorem 7.8.3 determines exactly the cardinality of the set of all open sets in a a second countable Hausdorff space.

**Theorem 7.8.1** *Let $(X, \tau)$ be a second countable space. The cardinality of the set of all open sets in $(X, \tau)$ is at most equal to the power of continuum $c$.*

*Proof* Let $(X, \tau)$ be a second countable space $U \in \tau$ be an arbitrary open set. Then there exists a countable open base

$$\mathbf{B} = \{U_n : n = 1, 2, \ldots\}$$

for the topology $\tau$ such that $U$ is the union of a subfamily $\mathbf{B}'$ of $\mathbf{B}$. This asserts that the cardinality $\text{card}(\tau)$ of the set of all open sets in $\tau$ is not greater than that of the set of all subfamilies $\mathbf{B}'$ of $\mathbf{B}$, which is $c$. This shows that $\text{card}(\tau) \leq c$. □

**Corollary 7.8.2** *Let $(X, \tau)$ be a second countable space. Then the cardinality of the set of all closed sets in $(X, \tau)$ is at most equal to the power of continuum $c$.*

*Proof* As every closed set in $(X, \tau)$ is the complement of an open set in $X$, the corollary follows from Theorem 7.8.1. □

Theorem 7.8.3 determines exactly the cardinality of the set of all open sets in a a second countable Hausdorff space.

**Theorem 7.8.3** *Let $(X, \tau)$ be a second countable Hausdorff space. Then the cardinality of the set of all open sets in $(X, \tau)$ is $c$.*

*Proof* Suppose $(X, \tau)$ is a second countable Hausdorff space. Then by using Exercise 2 of Sect. 7.12, it follows that there is an infinite sequence $\mathbf{S} = \{U_1, U_2, \ldots\}$ of mutually disjoint nonempty open sets $U_1, U_2, \ldots$ in $(X, \tau)$, where different subsequences of $\mathbf{S}$ determine different open sets as their unions. Hence, it follows that

$$\text{card } \tau \geq c,$$

since the set of all open sets subsets of a countable set has cardinality $c$ and all the open sets in $(X, \tau)$ may not be obtained as unions of the subsequences $\mathbf{S}$. Again, by using Theorem 7.8.1, it follows that

$$\text{card } \tau \leq c.$$

These two inequalities assert that

card $\tau = \mathbf{c}$.

$\square$

**Corollary 7.8.4** *Let* $(X, \tau)$ *be a second countable Hausdorff space. Then the cardinality of the set of all closed sets in* $(X, \tau)$ *is* $\mathbf{c}$.

*Proof* As every closed set in $X$ is the complement of an open set in $X$, the corollary follows from Theorem 7.8.3.                                              $\square$

*Remark 7.8.5*  For cardinality of the set of all points in a second countable $T_1$-space, see Exercise 26 of Sect. 7.12.

## 7.9  Points of Condensation of Uncountable Subsets of Second Countable Spaces

This section conveys the concept of point of condensation in a topological space and shows its existence by proving that every uncountable subset of a second countable space has a point of condensation.

**Definition 7.9.1** Let $(X, \tau)$ be a topological space and $A \subset X$ be an uncountable set. Then a point $p \in X$ is said to be a **point of condensation** of $A$ in $(X, \tau)$ if every nbd of $p$ intersects $A$ in an uncountably infinite set of points.

Theorem 7.9.2 provides examples of points of condensation in terms of Definition 7.9.1.

**Theorem 7.9.2** *Let* $(X, \tau)$ *be a second countable space. Then its every uncountable subset* $A$ *has a point of condensation.*

*Proof* Let $X$ be a second countable space with topology $\tau$ and $\mathbf{B} = \{U_n : n = 1, 2, \ldots\}$ be a countable open base for the topology $\tau$. Let $A \subset X$ be a subset such that $A$ has no point of condensation. Then for every $a \in A$, the point $a$ is not a point of condensation of $A$. This asserts that there exists an open set $V$ in $(X, \tau)$ such that $a \in V$ and $V \cap A$ is at most countable. Then there exists some suffix $N_a$ such that $a \in U_{N_a} \subset V$. This shows that $A \cap U_{N_a}$ is at most countable. Represent $A$ as

$$A = \bigcup \{a \in A\} \subset \bigcup \{A \cap U_{N_a} : a \in A\}.$$

Since there can exist at most a countable number of different suffices, it follows that the set $A$ is at most a countable union of countable subsets of $X$. This implies that $A$ is at most a countable subset of $X$. Hence it follows that if $A$ is uncountable, then it has a point of condensation.

$\square$

**Corollary 7.9.3** *Let $(X, \tau)$ be a second countable space and $D$ be an uncountable subset of $X$. If $A \subset X$ consists of all points of condensation of $D$ lying in $D$, then the set $A$ is an uncountable set such that $\overline{A} = A$.*

**Proof** Using Exercise 27 of Sect. 7.12, the corollary follows from Theorem 7.9.2.                                                                                    □

## 7.10 More on Urysohn Metrization Theorem

This section proves Urysohn metrization theorem which gives a sufficient condition for metrizability of topological spaces and its alternative forms which are different from forms given in Chap. 6. The basic tool used to prove Urysohn metrization theorem in Chap. 6 is the separation of disjoint closed sets by real-valued continuous functions. On the other hand, Urysohn metrization Theorem 7.10.4 is proved in this chapter by using the result that every regular and second countable space is normal (see Theorem 7.10.1).

Theorem 7.10.1 proves that the regularity and second countability properties of a topological space taken together imply its normality property.

**Theorem 7.10.1** *Every regular and second countable space is normal.*

**Proof** $(X, \tau)$ be a regular and second countable space. We prove that the space $(X, \tau)$ is normal. Let $P$ and $Q$ be any pair of disjoint closed sets in $(X, \tau)$ and $\mathcal{B}$ form a countable open base for the topology $\tau$. Since every point $p \in P$ is also in the open set $X - Q$, by regularity criterion, there exists an open set $U_p \in \mathcal{B}$ such that

$$p \in U_p \subset \overline{U}_p \subset X - Q.$$

Again, since $\mathcal{B}$ is countable, we can enumerate all the members in $\{U_p : p \in P\}$ to obtain a countable family $\{U_n : n = 1, 2, \ldots\}$ of open sets from the members of $\mathcal{B}$ such that

$$P \subset \bigcup_{n=1}^{\infty} \overline{U}_n, \text{ and } \overline{U}_n \cap Q = \emptyset, \forall n \in \mathbf{N}.$$

Similarly, we obtain a countable family $\{V_n = 1, 2, \ldots\}$ of open sets from the members of $\mathcal{B}$ such that

$$Q \subset \bigcup_{n=1}^{\infty} \overline{V}_n, \text{ and } \overline{V}_n \cap P = \emptyset, \forall n \in \mathbf{N}.$$

If $U_1^* = V_1 - \overline{U}_1$ and $V_1^* = U_1 - \overline{V}_1$, then

$$U_1^* \cap V_1^* = \emptyset, \ U_1^* \in \tau, \ V_1^* \in \tau, \ U_1^* \cap Q = V_1 \cap Q \text{ and } V_1^* \cap P = U_1 \cap P.$$

Define inductively,

$$U_n^* = V_n - \bigcup_{i=1}^{n} \overline{U_i} \text{ and } V_n^* = U_n - \bigcup_{i=1}^{n} \overline{V_i}.$$

Then

$$U_n^*, \ V_n^* \in \tau, \ U_n^* \cap P = U_n \cap P, \text{ and } V_n^* \cap Q = V_n \cap Q.$$

If $U = \bigcup_{n=1}^{\infty} U_n,^*$ and $V = \bigcup_{n=1}^{\infty} V_n,^*$, then

$$U, \ V \in \tau, \ P \subset V, \ Q \subset U, \text{ and } U \cap V = \emptyset.$$

This asserts that the closed sets $P$ and $Q$ are strongly separated by open sets $U$ and $V$ in $(X, \tau)$, and hence it is proved that the space $(X, \tau)$ is normal.

$\square$

**Proposition 7.10.2** *Given a regular space $(X, \tau)$ with a countable basis $\mathcal{B}$, there exists a countable family of continuous functions $f_k : X \to [0, 1]$ with the property that for any point $a \in X$ and any nbd $U$ of $a$, there is an index $k$ such that $f_k(a) > 0$ and $f_k \equiv 0$ outside the nbd $U$.*

**Proof** Let $(X, \tau)$ be a regular space with a countable basis $\mathcal{B} = \{U_n : n = 1, 2, \ldots\}$. Given every pair $(n, m)$ of positive integers two open sets $U_n$ and $U_m$ are chosen from the countable basis $\mathcal{B}$ such that $\overline{U_n} \subset U_m$. Since by Theorem 7.10.1, the space $(X, \tau)$ is normal, hence by Urysohn lemma, corresponding to the closed sets $\overline{U_n}$ and $X - U_m$, there exists a continuous function

$$h_{n,m} : X \to [0, 1]$$

such that

$$h_{n,m}(x) = \begin{cases} 0, \text{ for all } x \in X - U_m \\ 1, \text{ for all } x \in \overline{U_n} \end{cases}$$

and

$$0 \leq h_{n,m}(x) \leq 1 \text{ for all } x \in X.$$

Then the family $\{h_{n,m}\}$ of continuous functions gives the required function. To show it, select an open set $U_m \in \mathcal{B}$ such that $x \in U_m \subset U$. Hence by regularity property of $(X, \tau)$, given a point $x \in X$ and a nbd $U$ of $x$, there exists an open set $U_n \in \mathcal{B}$ such that

$$x \in U_n \subset \overline{U_n} \subset U_m.$$

This shows that the function $h_{n,m}$ defined as above on $X$ is such that

$$h_{n,m}(x) > 0$$

and

$$h_{n,m} \equiv 0 \text{ outside } U.$$

The family $\{h_{n,m}\}$ is indexed by a subset of $\mathbf{N} \times \mathbf{N}$. This produces the required family of continuous functions $f_k : X \to [0, 1]$ such that for any point $a \in X$ and any nbd $U_a$ of $a$, there is an index $k$ with the property that $f_k(a) > 0$ and $f_k \equiv 0$ outside the nbd $U$. □

**Remark 7.10.3** Urysohn Metrization Theorem 7.10.4 provides a sufficient condition for metrizability of regular spaces having countable bases. On the hand, its alternative form formulated in Theorem 7.10.5 provides a sufficient condition for metrizability of second countable and completely regular spaces.

**Corollary 7.10.4** *(Urysohn metrization theorem) Let $(X, \tau)$ be a regular space having a countable basis. Then $(X, \tau)$ is metrizable.*

**Proof** Urysohn metrization theorem follows from Theorem 7.10.1. □

**Theorem 7.10.5** *(An alternative form of Urysohn metrization Theorem) Let $(X, \tau)$ be a second countable and completely regular space. Then $(X, \tau)$ is metrizable.*

**Proof** Let $(X, \tau)$ be a second countable *and completely* regular space with a countable basis $\mathcal{B}$. Construct the countable family of functions $\{f_n : X \to [0, 1]\}$ by Proposition 7.10.2 such that for every point $x \in X$, and any nbd $U$ of $x$, there is an index $n \in \mathbf{N}$ such that $f_n(x) > 0$ and $f_n \equiv 0$ outside $U$. Let $Y = \mathbf{I}^\infty$ with the product topology be the Hilbert cube (see Chap. 3). Define a map

$$H : X \to Y, x \mapsto (f_1(x), f_2(x), f_3(x), \ldots).$$

Then $H$ is continuous, since $Y = \mathbf{I}^\infty$ has the product topology and each of its components $f_n$ is continuous. $H$ is injective, since for distinct points $x, y \in X$, there is an index $k \in \mathbf{N}$ such that $f_k(x) > 0$ and $f_k(y) = 0$ and hence $H(x) \neq H(y)$. This asserts that $H$ is a homeomorphism of $X$ onto its image $H(X) \subset Y$. This says that $H$ is an embedding of $X$ in $\mathbf{I}^{\mathbf{N}}$ and thus the topological space $(X, \tau)$ is embedded into the Hilbert cube $\mathbf{I}^\infty$, which is metrizable. □

**Definition 7.10.6** A topological space $(X, \tau)$ is called a **Frechét space** if every one-pointic set is closed in $(X, \tau)$. It is a $T_1$-space.

**Theorem 7.10.7** *(Urysohn Embedding Theorem) Every completely regular second countable space can be embedded in the Hilbert cube $\mathbf{I}^{\mathbf{N}}$.*

*Proof* **Proof I:** It follows from the proof of the Urysohn Metrization Theorem 7.10.4.
**Proof II:** Let $(X, \tau)$ be a completely regular second countable Frechét space.
It follows from the proof of the Urysohn Metrization Theorem 7.10.4 that $(X, \tau)$
is metrizable. Let $\{f_k : X \to \mathbf{I}\}_{k \in \mathbf{K}}$ be a family of continuous functions $f_k : X \to$
$\mathbf{R} :\in \mathbf{K}$ with indexing set $\mathbf{K}$ such that for any point $a \in X$ and any nbd $U_a$ of $a$,
there is an index $k$ with the property that $f_k(a) > 0$ and $f_k \equiv 0$ outside the nbd $U$
(existence of such functions follows from Proposition 7.10.2). Define the function

$$H : X \to \mathbf{I}^{\mathbf{N}}, \ x \mapsto (f_1(x), f_2(x), f_3(x), \ldots).$$

Then $H$ is continuous, since $Y = \mathbf{I}^{\infty}$ has the product topology and each of its components $f_n$ is continuous. Moreover $H$ is injective, since for distinct points $x, y \in X$,
there is an index $k \in \mathbf{K}$ such that $f_k(x) \neq f_k(y)$. This asserts that $H$ is a continuous
injective map from $X$ into $\mathbf{I}^{\mathbf{N}}$. This proves that $H$ embeds $X$ in the Hilbert cube $\mathbf{I}^{\mathbf{N}}$.
$\square$

Theorem 7.10.8 giving an embedding of a regular second countable Frechét space and
Theorem 7.10.9 characterizing completely regular spaces, follow as a consequence
of the above discussion.

**Theorem 7.10.8** *(One form of Urysohn Embedding Theorem) Every regular second countable Frechét space can be embedded in the Hilbert cube* $\mathbf{I}^{\mathbf{N}}$.

**Theorem 7.10.9** *A topological space* $(X, \tau)$ *is completely regular iff space* $(X, \tau)$
*is homeomorphic to a subspace of* $\mathbf{I}^{\mathbf{K}}$ *for some indexing set* $\mathbf{K}$.

*Remark 7.10.10* A family of continuous functions defined in Urysohn Embedding
Theorem 7.10.7 is said to separate points from the closed sets in $(X, \tau)$.

# 7.11  Applications

This section conveys various applications of the concepts and results discussed in
this chapter.

## 7.11.1  Applications of Lindelöf Theorem

This section presents an important application of Lindelöf Theorem and gives a
characterization of Hausdorff spaces in terms of convergent sequences.
Theorem 7.11.1 gives an important application of Lindelöf Theorem 7.2.12.

**Theorem 7.11.1** *Let* $(X, \tau)$ *be a second countable space and* $\{U_i : i \in A\}$ *be an
arbitrary open base for the topology* $\tau$. *Then* $\{U_i : i \in A\}$ *has a countable subfamily
forming also an open base for the same topology* $\tau$.

*Proof* Let $\mathbf{B} = \{U_n\}$ is an arbitrary countable open base for the topology $\tau$ on $X$. By hypothesis, $\{U_i : i \in \mathbf{A}\}$ is an arbitrary open base for the topology $\tau$. Then every nonempty open set $U_n \in \mathbf{B}$ is a union of $U_i$'s. Hence it follows by Theorem 7.2.12 that this $U_n$ can be represented as a countable subfamily of the family $\{U_i : i \in \mathbf{A}\}$. By this process, a countable family of countable collection of $U_i's$ is obtained. This asserts that the union of this family of collections thus obtained, forms a countable subfamily $\mathbf{B}'$ of the given open base $\{U_i : i \in \mathbf{A}\}$ such that $\mathbf{B}'$ also forms an open base for the topology $\tau$ on $X$. □

**Corollary 7.11.2** *If $(X, \tau)$ has a countable base for $\tau$, then every open covering of $X$ contains a countable subcovering.*

*Proof* It follows from Theorem 7.11.1. □

**Remark 7.11.3** Some authors call Corollary 7.11.2 Lindelöf Theorem, which asserts every open covering of a topological space satisfying the second axiom of countability has a countable subcovering.

## 7.11.2  A Charaterization of First Countable Hausdorff Spaces

Theorem 7.11.4 gives a characterization of Hausdorff property of a first countable spaces by convergent sequences.

**Theorem 7.11.4** *Let $(X, \tau)$ be a first countable space. Then it is Hausdorff iff every convergent sequence in $X$ has a unique limit in $X$.*

*Proof* First suppose that $(X, \tau)$ is a first countable Hausdorff space. Since every convergent sequence in a Hausdorff space has a unique limit, then every convergent sequence in $(X, \tau)$ has a unique limit in $X$.

Conversely, suppose the topological space $(X, \tau)$ is a first countable space such that every convergent sequence in $X$ has a unique limit in $X$. If it is not Hausdorff, then there exists a pair of distinct points $a, b \in X$ which are not strongly separated in $(X, \tau)$. By hypothesis $(X, \tau)$ is first countable. Then apply Proposition 7.7.1 to find a convergent sequence $\{x_n\}$ in $X$ such that

$$\lim_{n \to \infty} x_n \to a, \text{ and } \lim_{n \to \infty} x_n \to b.$$

This implies that there exists a convergent sequence $\{x_n\}$ in $X$ whose limit is not unique. This contradicts the assumption that every convergent sequence in $(X, \tau)$ has a unique limit in $X$. This contradiction proves that the space $(X, \tau)$ is Hausdorff. □

**Proposition 7.11.5** *(i) The Euclidean n-space $\mathbf{R}^n$ is second countable, and also first countable, separable and Lindelöf.*

*(ii)  The unit cube $\mathbf{I}^d$ is second countable, and also first countable, separable and Lindelöf.*

**Proof**  It follows from Theorems 7.2.14 and 18, since the real number space $(\mathbf{R}, \sigma)$ and the closed unit interval $(\mathbf{I} = [0, 1], \sigma_\mathbf{I})$ are second countable.                    $\square$

**Theorem 7.11.6**  *Let $(X, \tau)$ be a first countable space. Then a point $x \in X$, is an accumulation point of a subset $A \subset X$ iff there is an infinite sequence of points $x_1, x_2, \ldots, (x_i \neq x)$ in $A$ such that*

$$\lim_{n \to \infty} x_n = x.$$

**Proof**  Let $(X, \tau)$ be a first countable space and $x \in X$ be an accumulation point of a subset $A$ of $X$. Then by Proposition 7.7.1, there is a sequence of open sets $\{U_i : i = 1.2.\ldots\}$, which are members of an open base for the topology $\tau$ about the point $x$ such that for every open set $V$ containing the point $x$, there an integer $k$ with the property

$$U_i \subset V, \quad \forall i \geq k.$$

Since, $x \in U_n$ and $x$ is an accumulation point of $A$, there exists a point $x_n (\neq x) \in A$ such that $x_n \in U_n$. For $n = 1, 2, \ldots$, a corresponding sequence of points $x_1, x_2, \ldots$ is thus obtained in $A$. Hence it follows that

$$\lim_{n \to \infty} x_n = x.$$

Conversely, suppose that $(X, \tau)$ is a first countable space such that there is an infinite sequence of points $x_1, x_2, \ldots, (x_i \neq x)$ in $A \subset X$ for some $x \in X$ such that

$$\lim_{n \to \infty} x_n = x.$$

To show that $x$ is an accumulation point of $A$, let $V$ be an open set containing the limit point $x$ of the sequence $\{x_n\}$. Then there exists an integer $k$ such that $x_n \in V, \forall n \geq k$. Hence by the given condition, it follows that $x$ is an accumulation point of the subset $A$.

$\square$

## 7.12  Exercises

1. If a topological space $(X, \tau)$ has a countable base for its topology $\tau$, show that there exists a countable dense subset of $X$.
2. Let $(X, \tau)$ be a Hausdorff space. Show that there exists an infinite sequence of mutually disjoint nonempty open sets in $(X, \tau)$.

3. Let $(X, \tau)$ be a regular Lindelöf space. Show that it is a normal space.
4. Let $(X, \tau)$ be a discrete space. Show that it is separable iff $X$ is countable.
5. Show that any subspace of a separable space may not be separable but every open subspace of a separable space is separable.
   [Hint: Use Example 7.6.5 for the first part.]
6. Show that every subspace of a Lindelöf space may not be Lindelöf but every closed subspace of a Lindelöf space is Lindelöf.
7. Let $(X, \tau)$ be a second countable and completely regular space and $B$ be a countable open base for its topology $\tau$. Suppose for each pair of open sets $U, V \in B$ with $\bar{U} \subset V$, there exist a function

$$f : X \to \mathbf{I} : x \mapsto \begin{cases} 0, \text{ for all } x \in U \\ 1, \text{ for all } x \in X - V \end{cases}$$

If $\mathcal{F}$ is the set of all such maps (may be empty), then for each point $x \in X$ and for each closed set $A \subset X$ with $x$ not in $A$, show that there is a function $g \in \mathcal{F}$ such that

   (i) $g$ vanishes on a nbd $U_x$ of $x$ i.e., $g \equiv 0$ on $U_x$ and
   (ii)

$$g \equiv 1_d \text{ on } A.$$

8. **(Heine's continuity criterion)** Let $(X, \tau)$ be a first countable space, $C$ be a subset of $X$ and $(Y, \sigma)$ be a given topological space. Show that a function $f : (C, \tau_C) \to (Y, \sigma)$ is continuous at a point $x \in C$ iff for any infinite sequence of points $\{x_n\}$ in $C$,

$$\lim_{n \to \infty} x_n = x \implies \lim_{n \to \infty} f(x_n) = f(x) \text{ in } (Y, \sigma)$$

9. Show by an example that a subspace of a topological space with a countable dense subset may not have a countable dense subset.
   [Hint: The set of points having rational coordinates is dense in the Sorgenfrey plane $\mathbf{R}_l^2$ but the subspace $A = \{(x, -x)\} \subset \mathbf{R}_l^2$ is uncountable and it has the discrete topology. Hence it has no countable dense subset.]
10. Show the product space of a Lindelöf space and a compact space is Lindeöff.
11. Show that every metric space is first countable. Hence prove that the Euclidean $n$-space $\mathbf{R}^n$ is first countable.
12. Show that there are metric spaces which are not second countable.
13. Show that uncountable sets with cofinite topology are not necessarily first countable.
14. Let $(X, \tau)$ be a first countable space and $x \in X$ be an arbitrary point. Show that at $x$, there exists a countable local base $B_x = \{U_n : n \in \mathbf{N}\}$ such that

$$U_1 \supset U_2 \supset U_3 \supset \cdots \supset .$$

15. Let $\mathcal{C}$ be the family of all complements of finite or infinite countable sets in $X$. Then, $\mathcal{C}$ together with $\emptyset$ forms a topology, called **the countable complement topology** on $X$. Show that the countable complement topology on $\mathbf{R}$ is not first countable.

16. $\mathbf{R}/\sim$ be the quotient space obtained by the equivalence relation $x \sim y$ iff $x - y \in \mathbf{Z}$. (**Geometrically**, $\mathbf{R}/\sim$ represents the wedge of a countable many circles having only point in common).
    Show that

    (i) The quotient space $\mathbf{R}/\sim$ can not be embedded in the Euclidean plane $\mathbf{R}^2$;
    (ii) The quotient space $\mathbf{R}/\sim$ is not first countable.

17. Let $\{(X_i, \tau_i) : i = 1, 2, \ldots, n\}$ be a finite family of topological spaces and $(X, \tau) = \Pi_{n=1}^{n}(X_i, \tau_i)$ be their product space. Show that

    (i) If every coordinate space $(X_i, \tau_i)$ is separable, then their product space $(X, \tau)$ is also separable;
    (ii) If every coordinate space $(X_i, \tau_i)$ is second countable, then their product space $(X, \tau)$ is also second countable.

18. Let $\{(X_a, \tau_a) : a \in \mathbf{A}\}$ be any family of topological spaces and $(X, \tau)$ be their topological product space with natural projection

$$p_a : \Pi X_a = X \to X_a, \forall a \in \mathbf{A}.$$

Show that the product space $(X, \tau)$ is

    (i) first countable iff, every coordinate space $(X_\alpha, \tau_\alpha)$ is first countable and all but a countable number of the topological spaces $(X_\alpha, \tau_\alpha)$ are indiscrete;
    (ii) second countable iff every coordinate space $(X_\alpha, \tau_\alpha)$ is second countable, and all but a countable number of the topological spaces $(X_\alpha, \tau_\alpha)$ are indiscrete;
    (iii) separable iff every coordinate space $(X_\alpha, \tau_\alpha)$ is a separable Hausdorff space having at least two elements and the cardinality card $(\mathbf{A}) \leq c$.

19. (**Box topology**) Let $X_n$ be a copy of the discrete space over the set $\{0, 2\}$ for each $n \in \mathbf{N}$. Then $\Pi_{n \in \mathbf{N}} X_i$ and $\mathbf{R}$ have the same cardinality. Define nonempty open sets in the product space $\Pi_{n \in \mathbf{N}} X_n$ as the unions of the sets of the form $\Pi_{i \in \mathbf{A}} U_i$, where $U_n = X_i$ for finitely many $n$ and $U_i$ is singleton for others. If the indexing set is $\mathbf{A}$, then the family of sets of the form $\Pi_{n \in \mathbf{A}} U_i$ where $U_i$ is open in $X_i$ forms an open base for the topology $\tau$ on $X = \Pi_{i \in \mathbf{A}} X_n$. If $\mathbf{A}$ is an infinite set, then corresponding topology is called the **box topology,** which is different from the product topology. Show that the box topology on the set $\Pi_i \mathbf{R}_i$ is not first countable.

20. (**Bolzano–Weierstrass space**) A topological $x$ is said to be a Bolzano–Weierstrass space (**B–W space**), if every infinite subset of $X$ has a point of accumulation point in $X$. If a topological space $X$ is first countable and a B–W-space, show that $X$

is sequentially compact in the sense that its every sequence has a convergent subsequence.

21. Examine whether the $l_2$-space with $l_2$-metric (see Chap. 2) has a countable basis.

22. Let $(X, \tau)$ be a compact metrizable space. Show that the space $(X, \tau)$ has countable open base.

23. If a metrizable space has a countable dense set, show that it has a countable open base.

24. Let $(X, \tau)$ be a regular space having an open base which is countably locally finite. Show that

   (i) $(X, \tau)$ is normal,
   (ii) every closed set in $(X, \tau)$ is a $G_\delta$-set in $(X, \tau)$.

25. Let $(X, \tau)$ be a regular second countable space. Show that every closed set in $(X, \tau)$ is a $G_\delta$-subset of $(X, \tau)$.

26. Let $S$ be the set of all points in a second countable $T_1$ space. Show that card $S$ is at most equal to $\mathbf{c}$ (i.e., at most power of continuum).

27. Let $(X, \tau)$ be a second countable space and $D$ be an uncountable subset of $X$. If $A \subset X$ consists of all those points of $D$, which are not points of condensation of $D$, then $A$ is at most countable.

28. **(Tychonoff)** Show that every regular second countable space is perfectly normal.

29. Show that every regular Lindelöf space is paracompact.

30. Show that every regular Lindelöf space is normal.

31. Let $(X, \tau)$ be a Lindelöf space and $(Y, \sigma)$ be a topological space. If $f : (X, \tau) \to (Y, \sigma)$ is a continuous onto mapping, show that $(Y, \sigma)$ is also a Lindelöf space.

32. Let $C[a, b] = X$ be the set of all real-valued continuous functions $f$ on $[a, b]$ with norm $||f|| = \int_a^b f(t))dt$. Show that under the induced topology

   (i) $X$ is a separable space;
   (ii) $X$ is a second countable space.

33. Show that the set $\mathbf{R}$ equipped with Euclidean topology $\sigma$ i.e., the real line space $(\mathbf{R}, \sigma)$ is Lindelöf but the same set $\mathbf{R}$ endowed with the Sorgenfrey topology $\tau_1$ i.e., Sorgenfrey line space $(\mathbf{R}, \tau_1)$ is not so.
   [Hint: Sorgenfrey topology $\tau_1$ is not second countable.]

34. Show that the Sorgenfrey line space is

   (i) first countable and separable.
   (ii) Hausdorff but not metrizable.

35. Show that every closed subspace of a Lindelöf space is Lindelöf.

36. Let $(X, \tau)$ be a topological space such that it is both regular and Lindelöf. Show that the space $(X, \tau)$ is normal.

37. Let $(X, \tau)$ be a topological space such that it is both regular and second countable. Show that the space $(X, \tau)$ is normal.

38. Show by an example that the topological product of two Lindelöf spaces may not be Lindelöf.

39. Show that **Niemytzki's tangent dick topology (or Niemytzki's topology)** (see Chap. 3) is separable but it is not second countable.
40. **(Cantor–Bendixon theorem)** Let $(X, \tau)$ be a second countable space and $A$ be a closed set in $(X, \tau)$. Show that $A$ can be represented uniquely as the union of a perfect set and a set that is at most countable.

## *Multiple Choice Exercises*

Identify the correct alternative (s) (there may be none or more than one) from the following list of exercises:

1. Let **R** be the set of all real numbers.

   (i) **R** endowed with topology generated by the closed-open interval $\{[a, b) : b \in \mathbf{Q}\}$ is first countable.
   (ii) **R** endowed with usual topology is not first countable.
   (iii) **R** endowed with discrete topology is separable.

2. The Euclidean space $\mathbf{R}^3$ is

   (i) second countable;
   (ii) separable;
   (iii) first countable;
   (iv) not Lindeöf.

3. (i) Every subspace of a first countable space is first countable.
   (ii) Every subspace of a second countable space is second countable.
   (iii) Separability of topological spaces is a topological property.
4. The real number space $(\mathbf{R}, \sigma)$, with the usual topology $\sigma$, is

   (i) second countable;
   (ii) separable;
   (iii) not Lindelöf.

5. (i) **R** endowed with usual topology is first countable,
   (ii) $\mathbf{R}^2$ endowed with the topology generated by half-open rectangles,

   $$[a, b) \times [c, d) = \{(x, y) \in \mathbf{R}^2 : a \le x < b; c \le y < d\}$$

   is separable.
   (iii) Every topological space can be embedded in a separable space.
6. Let $(\mathbf{R}, \sigma_c)$ denote the topological space $(\mathbf{R}, \sigma_c)$, with the cofinite topology $\sigma_c$. Then $(\mathbf{R}, \sigma_c)$

   (i) is separable;
   (ii) is Lindelöf; but
   (iii) is not first countable.

7. (i) Lindelöf property of topological spaces is not hereditary.
   (ii) Lindelöf property of topological spaces is a topological property.
   (iii) Separability is a topological property but it is not a hereditary property.

8. Let $(\mathbf{R}, \tau_l)$ denote the Sorgenfrey line space. This space

   (i) first countable;
   (ii) separable;
   (iii) not Lindelöf.

# References

Adhikari, M.R.: Basic Algebraic Topology and its Applications. Springer, India (2016)

Adhikari, M.R.: Basic Topology, Volume 3: Algebraic Topology and Topology of Fiber Bundles. Springer, India (2022)

Adhikari, A., Adhikari, M.R.: Basic Topology, Volume 2: Topological Groups, Topology of Manifolds and Lie Groups. Springer, India (2022)

Adhikari, M.R., Adhikari, A.: Basic Modern Algebra with Applications. Springer, New Delhi, NewYork, Heidelberg (2014)

Borisovich, Y.C.U., Blznyakov, N., Formenko, T.: Introduction to Topology. Mir Publishers, Moscow (1985). Translated from the Russia by Oleg Efimov

Bredon, G.E.: Topology and Geometry. Springer-Verlag, New York (1983)

Chatterjee, B.C., Ganguly, S., Adhikari, M.R.: Introduction to Topology. Asian Books, New Delhi (2002)

Dugundji, J.: Topology. Allyn & Bacon, Newton, MA (1966)

Kalajdzievski, K.: An Illustrated Introduction to Topology and Homotopy. CRC Press (2015)

Kelly, J.L.: General Topology, Van Nostrand, New York, 1955. Springer-Verlag, New York (1975)

Mendelson, B.: Introduction to Topology. College Mathematical Series. Allyn and Bacon, Boston (1962)

Morris, S.A.: Topology without tears. http://www.econphd.net/notes.htm (2007)

Munkres, J.R.: Topology. Prentice-Hall, New Jersey (2000)

Stephen, W.: General Topology. Addison-Wesley (1970)

# Chapter 8
# Brief History of Topology I: Motivation of the Subject with Historical Development

The subject **Topology** has become one of the most exciting and influential fields of study in modern mathematics, because of its beauty and scope. Topology starts where sets have some cohesive properties leading to define continuity of functions. This chapter conveys **the history of emergence of the concepts leading to the development of topology** as a subject with their motivations with an emphasis on general topology. Modern mathematics studies development of classical mathematical ideas, emergent areas leading to new areas of mathematics and their interrelationship, fundamental results and their applications in mathematics and beyond it.

Just after the concept of homeomorphism is clearly defined, the subject of topology begins to study those properties of geometric figures which are preserved by homeomorphisms with an eye to classify topological spaces up to homeomorphism, which stands the ultimate problem in topology, where a geometric figure is considered to be a point set in the Euclidean space $\mathbf{R}^n$. But this undertaking becomes hopeless, when there exists no homeomorphism between two given topological spaces.

The subjects **Algebraic Topology (studied in Basic Topology, Volume 3) and Differential Topology (studied in Basic Topology, Volume 2)** were born to solve the problems of impossibility in many cases with a shift of the problem by associating **invariant objects** in the sense that homeomorphic spaces have the same object (up to equivalence). Initially these objects were integers, and subsequent research reveals that more fruitful and interesting results can be obtained from the algebraic invariant structures such as groups and rings. For example, homology and homotopy groups are very important algebraic invariants which provide strong tools to study the structure of topological spaces. The development of homotopy theory started in the middle 1950s. The concepts born in the development of homology and homotopy theories to solve topological problems applications have found outstanding applications to other areas of mathematics leading to the starting points of many theories such as category theory, homological algebra and K-theory, to mention a few (Adhikari, 2016). This is a remarkable feature in the history of topology.

© The Author(s), under exclusive license to Springer Nature Singapore Pte Ltd. 2022
A. Adhikari and M. R. Adhikari, *Basic Topology 1*,
https://doi.org/10.1007/978-981-16-6509-7_8

Classical analysis begins with calculus. The real and complex number systems are the main texts of analysis. While developing classical analysis, prominent mathematicians such as B. Riemann (1826–1866), K. Weierstrass (1815–1897), G. Cantor (1845–1918), H. Lebesgue (1875–1941), D. Hilbert (1862–1943), J. Fourier (1768–1830) and many others identified the basic principles on which the analysis is founded. These principles play a key role in the development of modern analysis, algebra, topology, integration and measure theories.

Today, topology is a key subject interlinking modern analysis, geometry and algebra.The origin of a systematic study of topology may be traced back to the monumental work of Henri Poincaré (1854–1912) in his **Analysis situs** Paris, 1895 together with his first note on topology published in 1892 organized first time the subject topology, now, called algebraic or combinatorial topology. But the beginning of general topology or "point set topology" dates when M. Fréchet (1878–1973) published his paper in 1906 on the treatment of the subject in an abstract setting. He described in his thesis of 1906 a set of axioms on limit of sequences. Set topology developed at first as a branch of calculus, which was invented independently by Issac Newton (1643–1727) and G. Leibniz (1646–1716) based on the concept of limit. Topology lies today on a mathematical foundation. Its main objective is to extend the concept of continuity and dimension theory born in Euclidean spaces. The twentieth century witnessed its greatest development. Before Poincaré some scattered work was done by L. Euler (1707–1783,), B. Riemann (1826–1866), J. B. Listing (1808–1882), C. F. Klein (1849–1925), David Hilbert (1862–1943) and some others.

**Historically, the word "topology" comes from the Greek words** "$\lambda o' \gamma o \zeta$" which means a study and "$\tau o' \pi o \zeta$" which means a place, with an alternative name "analysis situs" aiming at the study of situations. This subject arising as a branch of geometry plays a key role in modern mathematical analysis, because of its study of continuous deformations such as stretching, twisting, crumpling and bending which are allowed, where tearing or gluing are not allowed . The concept of congruence in Euclidean geometry is a special type of equivalence between geometrical objects which is considered identical except for position in Euclidean space. The concept of similarity in Euclidean geometry indicates figures of the same shape, but not necessarily of the same size. Topology develops as a field of study out of geometry and set theory, through the concepts such as space, dimension and continuous transformation. These ideas go back to G. Leibniz who indicated a new geometry of topological type, called "geometry of place" as early as 1679 . The term topology was first coined by J. B. Listing (1808–1882) in 1847, although it was not popular until Felix Hausdorff (1868–1942) developed this subject in his book **Grundzüge der Mengenlehre** of 1914, which stemmed from analysis. His land-marking work sets out the journey of general topology. By the middle of the twentieth century, topology had become a major branch of mathematics.

## 8.1 Early Development of Combinatorial Topology by the Nineteenth-Century Analysts

At the beginning there was no branches of topology like today, and it was considered one unit named combinatorial topology or simply topology. Some concepts in combinatorial topology were found in the study of nineteenth-century analysts. For example,

(i) the investigation of Fourier series by Augustin Cauchy () and Karl Weierstrass () in which the sequence of functions converge to another function is analogous to the study of convergence of sequences of points in a topological space;

(ii) the study of convergence by Georg Cantor () providing a close relation between Fourier series and set theory has established an important mathematical setting for many problems of analysis which provides a general setting for many mathematical concepts closely related to convergence of sequences.

(iii) The German mathematician David Hilbert () proposed an axiomatic setting for general geometry in 1899 , which is other than the geometry studied by the ancient Greeks. Moreover, he gave some axioms in 1910 for neighborhoods of points in an abstract set, which generalizes properties of small disks centered at points in the Euclidean plane.

(iv) Georg Cantor and Richard Dedekind had just made significant advances in placing analysis on a more firmly-based set-theoretic foundation.

(v) The French mathematician Maurice Frenchét () gave a consistent set of axioms for convergence in an abstract set and also axioms for a metric space endowed with a distance metric, called a distance function.

## 8.2 Basic Concepts of Topology Found in the Eighteenth and the Nineteenth Centuries

Many basic concepts studied now in topology had been used in the 18th and nineteenth century by mathematicians like L. Euler ( 1707–1783,), B. Riemann (1826–1866), K. Weierstrass (1815–1897), J. B. Listing (1802–1882), C. F. Klein (1849–1925), H Poincare (1854–1912), G. Cantor (1845–1918), David Hilbert (1862–1943) and some others in a scattered way. Before them the concept of geometry of topological spaces was used as early as 1679 by G. Leibniz ( 1646–1716). But a systematic study of algebraic topology began as an important part of mathematics through the work of Poincare in his "Analysis situs", Paris, 1895. This topology was born through his work on the theory of integral calculus in higher dimensions was earlier called combinatorial topology. Before him, K. Weierstrass studied the concept of limit (as used today in calculus) during the 1860s and reconstructed real number system and proved its certain properties, now called topological.

The development of bold theory of point sets made by Georg Cantor (1845–1918) during the period 1874–1895 has built a new house of topology. F. Hausdorff and others provided a foundation of set-theoretic topology during 1900–1910. L. E. J. Brouwer is the first mathematician who instigated the concept of dimension through fruitful combination of combinatorial and set-theoretic approach of topology. This unified approach of topology was laid on a strong foundation during 1915–1930 by J. W. Alexander, P. L. Alexandroff, S. Lefschetz and some other mathematicians. Topology was called analysis till 1930. S. Lefschetz first published a book in 1930 with this title, and since then the name became popularized (though the term topology was coined by Listing) and this subject witnessed its rapid development. Though the term topology was coined by Listing, S. Lefschetz was the first to publish a book in 1930 with this title. Since then the name became popular. It reformulates the study of differential geometry through the work of H. Whitney on fiber bundles. The work of H. Hopf in 1930s on Lie groups made a revolution in modern algebra with development of new branch of algebra called homological algebra. Further development in 1940s was found through the work of S. Eilenberg and S. MacLane. The term topology was still combinatorial in 1942, and it became algebraic by 1944.

## 8.3   Main Objective of the Study of Topology

This section begins with the motivation of the study of topology. Two natural questions arise:

1. what is the subject topology ?
2. why we study this subject ?
3. what is the main problem of study in topology ?

1. There are many different answers of (1). One may call the subject topology as a qualitative study of geometry without reference to distance in the sense that if one geometric object is obtained from another geometric object by a continuous deformation, then these two geometric objects are considered to be topologically same, called **homeomorphic**. So it is also called a rubber sheet geometry. Accordingly, the geometric objects such as a circle, ellipse and a square are topologically the same, though they are geometrically different.
   H Poincaré remarked in 1912 "— In this discipline, two figures are equivalent whenever one can pass from one to the other by a continuous deformation; whatever else the law of this deformation may be , it must be continuous. Thus a circle is equivalent to an ellipse or even to an arbitrary closed curve, but it is not equivalent to a straight line segment since this segment is not closed. A sphere is equivalent to a convex surface; it is not equivalent to a torus since there is a hole in a torus and in a sphere there is not" .
2. There are also many different answers of (2). The simplest answer is topology is both highly elegant and useful which come from beauty, scope and power of the subject. Its beauty comes from both its various interesting geometric constructions. Its usefulness comes from the basic properties of continuous functions and

geometric objects with their applications in mathematics and also beyond mathematics. For example, it facilitates a study of practically all branches of mathematics, including algebra, real analysis, complex analysis, functional analysis, graph theory, number theory, dynamical systems and differential equations and many more.

3. The main problem in topology is the classification problem of topological spaces up to homeomorphism. To solve this classification problem, given two topological spaces, either we have to find an explicit expression of a homeomorphism between these two spaces or we have to show that it is not possible to construct such a homeomorphism. Algebraic and differential topology were born to prove this impossibility. The usual technique is to assign "invariant" objects which are shared by homeomorphic spaces (i.e., same for homeomorphic spaces). The earliest invariant objects were Euler characteristics which are integers. Subsequently, integral invariants are generalized by inventing algebraic invariants such as groups, rings and modules which offer more information about the structure of the concerned topological spaces. For example, **fundamental group, homotopy and homology groups** (studied in Basic Topology, Volume 3) provide deep insight into the structure of the topological spaces. Homology underwent developed first since its invention by H. Poincaré in 1895; on the other hand, homotopy did not develop until 1930. Since then, there has been an explosive development of homotopy theory, and its connection with homology theory has become a central theme of topology. Many concepts initially introduced in homotopy and homology theories such as $K$-theory, Brouwer fixed-point theorem and so on have found surprising applications to other areas of mathematics and also beyond mathematics.

## 8.4 Inauguration of General Topology

Prior to the work of Frechét in 1906 and that of Hausdorff, the general concept of topological space was not formally defined but the work of Weierstrass and Cantor led to some topological concepts, now called open sets, closed sets, nbds, continuous maps in $\mathbf{R}^n$ and in their subspaces. Riemann extended these concepts in $n$-dimensional manifolds which are locally homeomorphic to $\mathbf{R}^n$ intuitively without any formal definition. Such manifolds are now called $C^r$-manifold for $r \geq 1$.

More precisely, the beginning of general topology dates in 1860s in the work of K. Weierstrass, where he studied the concept of the limit of a function and proved certain properties of number system, which are now called **topological**. The set-theoretic aspect begins where sets admit some cohesive properties leading to define continuity of functions. The sets having such properties are then called **topological spaces**, whose study got momentum through the bold development of set theory created by Georg Cantor (1845–1918) around 1880. This aspect of topology is now known as **general topology or set topology** formed a firm foundation through the work of Felix Hausdorff (1869–1942), Maurice Frechét and others during 1900–

1910. Early development of general topology during 1900–1930 witnessed several concepts such as metric spaces, nbd systems, limit points, sequential limits and closures. Historically, metric spaces developed by F. Hausdorff (1869–1942) in his book **Grundzüge der Mengenlehre** published in 1914 are considered as a milestone of point set topology, where this branch of topology sets out its journey. This book is stemmed from analysis. His land-marking work sets out the systematic study of general topology.

Although P.S. Alexandroff (1896–1982) introduced in 1995 the present axioms for a topology on an abstract set and used the term **topological** in his research papers, based on which the field of general topology was born through the work of Hausdorff. This makes the official inauguration of general topology or point set topology. The book **Grundzüge der Mengenlehre** of Hausdorff published in 1914 and the work of K. Kuratowski (1896–1980) **Sur l' opération $\overline{A}$ de l' analysis situs** published in 1922 are considered as the origin and foundation of general topology. The term **Topology** was coined by J. B. Listing (1808–1882) in 1830s, but Felix Hausdorff (1869–1942) popularized the term topology in 1914 and developed this subject in his book **Grundzüge der Mengenlehre** of 1914, which stemmed from analysis. His land-marking work sets out the systematic journey of general topology.

## 8.5   Beginning of Combinatorial Topology

An older name for the algebraic topology was combinatorial topology in the sense that the investigating topological spaces were constructed from simpler topological spaces by some technique. Algebraic topology was born as a subject through the work of H. Poincaréé based on the idea of dividing a topological space into geometric elements corresponding to the vertices, edges and faces of polyhedra and their higher-dimensional analogues. Such investigation presents many topological invariants including the Euler characteristic. Historically, fundamental group and homology groups are the first important topological invariants of homotopy and homology theories which came from such a search embedded in the work of H. Poincaré (1854–1912) in his land-marking "Analysis situs", Paris, 1895. He invented homology theory, now called,**simplicial homology in 1895** with an aim to study geometric properties of a topological space by converting topological problems to algebraic ones for the first time in the history of topology. The term **"Homotopy"** was first used by M. Dehn and P. Heegaard in 1907. L. E. J. Brouwer (1881–1967) gave the precise definition of continuous deformation by using the concept of homotopy of continuous maps. **The Jordan curve theorem** stated by Jordan in 1892 is a classical theorem. Its first rigorous proof given by Oswald Veblen (1880–1960) in 1905 is one of the remarkable developments of algebraic topology. W. Hurewicz made significant contributions to algebraic topology. The invention of the higher homotopy groups $\pi_n$ by W. Hurewicz in 1935–1936 is a natural generalization of the fundamental group to higher-dimensional analogue of the fundamental group. More precisely, $\pi_n$ is a sequence of covariant functors defined by Hurewicz from topology to algebra by extending the concept of fundamental group formulated by $\pi_n(X) = [S^n, X]$. Lens spaces defined by H. Tietze (1880–1964) in 1908 form an important class of three manifolds in the study of their homotopy classification. Historically, the terminology algebraic topology was first given somewhat later in 1936 by the S. Lefschetz (1884–1972), though his research in this major area of topology began in 1895 through the work of Poincaréé. Algebraic topology and its applications are closely related to other fields of mathematics such as modern algebra, geometry and analysis. On the other hand, general topology travels mainly in the premises of analysis and moves away from combinatorial topology.

### 8.5.1 Combined Aspect of Combinatorial and Set-Theoretic Topologies

L. E. J. Brouwer (1881–1967) gave the precise definition of continuous deformation by using the concept of homotopy of continuous maps (see Basic Topology, Volume III). A union of combinatorial and set-theoretic aspects of topology was achieved first by L. E. J. Brouwer through his investigation of the concept of dimension during 1908–1912 . The unified theory was laid on a solid foundation in the period 1915–1930 by J. W. Alexander ( 1888–1971), P. S. Alexandrov (1896–1982), S. Lefschetz ( 1884–1972) and others. Until 1930 topology was called "analysis situs (position analysis)". Analysis situs conveyed the qualitative properties of geometric figures both in the ordinary space as well as in the space of more than *three* dimensions. It was Lefschetz who first used and popularized the name topology by publishing a book with this title in 1930.

### 8.5.2 Combinatorial Topology Versus General Topology

Topologists were successful in investigating during 1920s and 1939s to convert topological problems to algebraic problems , which led to rename algebraic topology of combinatorial topology. The systematic study of algebraic topology as a subject began with precise formulations and correct proofs at the turn of the *nineteenth* to *twentieth* century (1895–1904) through the work of Henri Poincaré (1854–1912) in his land-marking "Analysis situs", 1895, where he used algebraic ideas in combinatorial topology. The development of combinatorial topology continued through the joint work of M. Dehn (1878–1952) and P. Heegard (1871–1948) classification theorems for two-dimensional surfaces in 1907. The importance of assigning algebraic objects to topological objects was clearly established after by Poincaré, was established by the Dutch mathematician L. E. J. Brouwer ((1881–1967) in his fixed-point theorem. But the terminology algebraic topology was first used somewhat later in 1936 by the Russian-born American mathematician Solomon Lefschetz (1884–1972), although research in this major area of topology was started earlier in the twentieth century. B. Riemann (1826–1866) studied surfaces (now known as Riemann surfaces) related to complex analysis and used combinatorial topology as a tool for investigating functions. Möbius ( 1790–1868) published his work on one-sided surface in 1858, known as Möbius band or Möbius strip. This surface may be constructed as a quotient space $M$ obtained by gluing together the ends of a long rectangular strip of paper with a half twist. The identification topology on $M$ coincides with the subspace topology induced on $M$ by the usual topology on $\mathbf{R}^3$. The Möbiuls strip $M$ is embedded in $\mathbf{R}^3$. The surfaces containing subspaces homeomorphic to the Möbius strip are called nonorientable surfaces and play an important role in the classification of two-dimensional surfaces. On the other hand, C. F. Klein (1849–1925) also published his separate work on one-sided surface in 1882, known as Klein bottle. For their construction see Chap. 1. The Klein bottle is a one-sided surface that is closed in the sense that it is , without any one-dimensional boundaries, and it cannot live in $\mathbf{R}^3$ without intersecting itself. This has created interest of mathematicians as other previously well-known surfaces lived only in $\mathbf{R}^3$. The term topology was still combinatorial in 1942, and it became algebraic by 1944.

## 8.6  Historical Development of the Basic Topics in General Topology

General topology or point set topology is a branch of topology which studies the basic set-theoretic concepts such as continuity, compactness and connectedness and constructions such as quotient spaces used in topology. After the publication of the paper **"Généralisation d'un theoréme de Weierstrass**, C.R.Acad. Sei. 139, 1904, 848–849 of Frechét in 1904 and publication of Hausdorff's classic book **Grundzüge der Mengenlehre, Leipzig, 1914** of Hausdorff , there have been various developments in different directions. It provides foundational aspects of most other fields of topology such as differential topology and topological algebra both studied in Volume 2 and algebraic topology and geometric topology both studied in Volume 3 of the present book **Basic Topology**.

The following subsections convey the motivation of the basic topics and historical development discussed in this book titled **Basic Topology: Volume 1**, starting from metric spaces

### 8.6.1   History of Metric Spaces

The concept of a metric introduced by M. Fréchet (1878–1973) in 1906 is an abstraction of distance in the Euclidean space born through the well-known properties of the Euclidean distance in an abstract setting, and it provides a rich supply of continuous functions. While developing classical analysis, prominent mathematicians such as B. Riemann (1826–1866), K. Weierstrass ( 1815–1897), G. Cantor (1845–1918), H. Lebesgue (1875–1941), D. Hilbert (1862–1943), J. Fourier (1768–1830) and many others identified the basic principles on which the analysis is founded. These principles play a key role in the development of modern analysis, algebra, topology, integration and measure theories. Urysohn lemma for metric spaces facilitates to provide a vast number of continuous functions, and metric spaces provide a rich supply of topological spaces and most of the applications of topology to analysis arise through metric spaces. Normed linear spaces form a special class of metric spaces which provide Banach and Hilbert spaces, and new concepts born through this development form the foundation of modern analysis and generalize many concepts of classical analysis in a more general setting. In many areas of mathematics such as in geometry and analysis, the concept of distance is generalized in an abstract setting by introducing the concept of metric spaces, which facilitates a study of continuous functions defined on abstract sets as well as convergent sequences on these sets. The Euclidean metric defined by the distance between two given points represents the usual length of the straight line segment joining these two pints. On the other hand, a metric in elliptic geometry is defined by the distance on a sphere measured by an angle, and in hyperbolic geometry, a metric space of velocities is defined by special relativity. The special structure of a metric space induces a topology enjoying special properties (see Sect. 8.7) having many applications of topology in modern analysis.

## 8.6.2 Early Development of General Topology

General topology addresses the basic set-theoretic definitions and constructions used in topology. It formulates the basic concepts used in all other branches of topology which are the concepts of continuity, compactness, connectedness, etc., thereby it establishes the foundational aspects of topology and investigates properties of topological spaces and concepts inherent to topological spaces. This aspect of topology formed a firm foundation through the work of Felix Hausdorff (1869–1942) Maurice Fréchet, and others during 1900–1910, though the beginning of general topology dates in 1860s in the work of K. Weierstrass, where he studied the concept of the limit of a function and proved certain properties of number system, which are now called **topological**. Early development of general topology during 1900–1930 witnessed several concepts such as metric spaces, though nbd systems, limit points, sequential limits and closures. Historically, general topology was developed with an aim to extend the concept of continuity and dimension born in Euclidean spaces. Historically, metric spaces developed by F. Hausdorff (1869–1942) in his book **Grundzüge der Mengenlehre** published in 1914 are considered as a **milestone of point set topology**, where this branch of topology sets out its journey. The subject topology is very powerful and beautiful, as it provides various key tools to solve problems in almost all areas of mathematics such as algebra, analysis, geometry, knot and graph theories, differential equations and many other areas. The main objective of analysis or topology is to study continuity of functions.

### 8.6.3   Topological Spaces and Their Continuous Functions

The notion of metric spaces is not adequate to develop many mathematical concepts such as continuity of a function, specially, which are developed by using metrics in metric spaces. In the early twentieth century, a more general space was defined, called a topological space. The main objective of analysis or topology is to study continuity of functions. The creation of intuitive set theory by G. Cantor (1815–1897) leads to an abstract concept of continuity. General topology starts with sets admitting some cohesive properties leading to the concept of continuity of functions and addresses the basic set-theoretic definitions and constructions used in topology. It formulates the basic concepts used in all other branches of topology which are the concepts of continuity, compactness, connectedness, etc., thereby it establishes the foundational aspects of topology and investigates properties of topological spaces and concepts inherent to topological spaces. This aspect of topology formed a firm foundation through the work of Felix Hausdorff (1869–1942) Maurice Fréchet and others during 1900–1910. Early development of general topology during 1900–1930 witnessed several concepts such as metric spaces, nbd systems, limit points, sequential limits and closures. Sets enjoying such properties are called topological spaces. The set theory created by Georg Cantor (1845–1918) around 1880 has a great influence in topology, which is both enormous and decisive. The cohesive properties defining the concept of continuity were born through the concepts of limits as well as through a family of subsets in an axiomatic framework by introducing the concept of open or closed sets, where a notion of nearness is defined without any distance function or a metric. This axiomatic approach leads to the concepts of topological spaces and their continuous maps.

### 8.6.4   Geometry of Continuous Mappings of Segments, Circles and Disks and Spheres

This subsection conveys some early theorems obtained as continuous images of segments, circles and disks and spheres in the Euclidean space $\mathbf{R}^n$ for $n = 1, 2, 3$ displaying their geometry, which stimulated investigation for their generalization. For example, some basic results are given below:

**Theorem 8.6.1**  *Every continuous mapping of a closed interval $[a, b] \subset \mathbf{R}$ into itself has at least one fixed point.*

**Theorem 8.6.2**  *Every continuous mapping of a circle into a line $L$ sends some pair of diametrically opposite points onto the same point of $L$.*

**Theorem 8.6.3**  *A subspace $J$ of $\mathbf{R}^2$ homeomorphic to the circle separates $\mathbf{R}^2$ into two regions with $J$ their common boundary.*

**Theorem 8.6.4** *If* $f: \mathbf{D}^2 \to \mathbf{R}^2$ *be a continuous map from a disk* $\mathbf{D}^2$ *into itself, then* $f$ *has a fixed point.*

**Theorem 8.6.5** *For every continuous map* $f: S^2 \to \mathbf{R}^2$, *there exists some pair* $\{x, -x\}$ *of antipodal points of* $S^2$ *such that they have the same image point.*

### 8.6.5  Product of Topology and Tychonöff Topology

The Cartesian product $U \times V$ of two finite intervals U and V in $\mathbf{R}$ is an open rectangle in $\mathbf{R}^2$. The open rectangles form an open base for the natural topology on $\mathbf{R}^2$, which called a product topology on $\mathbf{R}^2$. This technique is borrowed for construction of any finite product topology. M. Fréchet first studied a finite product of abstract topological spaces in 1910. Construction of the product space of an arbitrary family of topological spaces is performed in a similar way. Tychonöff topology defined by Andrey Tychonöff (1906–1993) in 1930 with the help of product topology for any family (possibly, infinite) of topological spaces, which gives a generalization of the product topology for a finite family of topological spaces.

### 8.6.6  Quotient Topology and Quotient Spaces

The concept of quotient spaces was introduced by Robert L. Moore in 1925 and was also independently by Pavel Alexandroff (1896–1982). The concept of quotient spaces or identification spaces presents the mathematical version of a geometric process to obtain new geometric objects by several methods to obtain quotient spaces described in this section. Many interesting topological spaces can be constructed from a simple topological space by identifying some subset (or points) of $X$. For example, a circle is obtained by gluing together the end points of a closed line segment.

### 8.6.7  Möbius Band and Klein Bottle

Möbius Band and Klein bottle are important geometrical objects used in topology. Möbius band (Möbius strip) is named after A. F. Möbius (1790–1868), and the Klein bottle is named after F. Klein (1849–1925). Historically, from the square the Möbius band was constructed by Möbius in 1858 and the Klein bottle was constructed by Klein in 1882 **by identification methods,**

## 8.6.8   Separation Axioms

The topological spaces are studied by imposing certain conditions, called **separation axioms** on these spaces in terms of their points and open sets, initially used by P.S. Alexandroff (1896–1982) and H. Hopf (1894–1971), especially, where there is no concept of distance. These additional axioms are needed, because the defining axioms for a topological space are extremely general and are too weak to study them in depth. These separation axioms are natural restrictions on topological structure to make the structure nearer to metrizable structure. These axioms facilitate to classify topological spaces and provide enough supply of continuous functions which are linked to open sets.

Many important topological properties can be characterized with the help of separation axioms by distributing the open sets. Motivation of separation axioms was born through the observation that any two points in a metric space are separated if they have a strictly positive distance. But there exist many topological spaces satisfying a set of certain conditions in addition to the axioms defining topological spaces which can recover many significant properties of metric spaces lost to arbitrary topological spaces. Such spaces $X$ are important objects in topology as many important topological properties can be characterized with the help of separation axioms by distributing the open sets in the space $X$ and imposing natural conditions on $X$ such that $X$ behaves like a metric space. Several separation axioms are known, but this book studies only $T_i$-axioms for $i = 0, 1, 2, 3, 4, 5$ and the corresponding topological spaces, called $T_i$-spaces.

   (i) The concepts of $T_0$-axiom and $T_0$-spaces were introduced by Andrey Kolmogorov (1903–1987) around 1930.
   (ii) The concepts of $T_1$-axiom and $T_1$-spaces were introduced by Fréchet) (1878–1973) in 1906. But some authors say that the concept of $T_1$-spaces was given by Frigyes Riesz (1880–1956) in 1907.
   (iii) The concepts of $T_2$-axiom and $T_2$-spaces (commonly known as Hausdorff spaces) were introduced by Felix Hausdorff (1868–1942) in 1914.
   (iv) The concepts of $T_3$-axiom and $T_3$-spaces (commonly known as regular spaces) were introduced by Leopold Vietoris (1891–2002) in 1921.
   (v) The concepts of $T_4$-axiom and $T_4$-spaces (commonly known as normal spaces) were introduced by H. Tietze (1880–1964) in 1923 and also independently by Pavel Alexandroff (1896–1982) and Pavel Urysohn (1898–1924) in 1929.

## 8.6.9   Real-Valued Continuous Functions

Real-valued continuous functions or, simply, real functions play a central role in topology and analysis. For example, a deep result of Chap. 5 proves that two compact Hausdorff spaces $X$ and $Y$ are homeomorphic iff the corresponding rings $C(X, \mathbf{R})$ and $C(Y, \mathbf{R})$ of real-valued continuous functions are isomorphic. This result characterizes compact Hausdorff spaces in terms of algebras and recovers the topology of $X$ from the ring structure of $C(X, \mathbf{R})$. Moreover, this chapter characterizes compact Hausdorff spaces in terms of algebras. For more applications of the real-valued continuous functions see Sect. 8.6.11.

### 8.6.10 Urysohn Function and Urysohn Lemma for Metric Spaces

Urysohn's Lemma for metric spaces, named after P. S. Urysohn ( 1998–1924), is a very significant result in metric spaces with wide applications. It provides a rich supply of continuous functions, known as Urysohn function. For its generalization, see Sect. 8.6.11.

### 8.6.11 Separation by Real-Valued Continuous Functions and Urysohn Lemma

A nonconstant real-valued continuous function is not always defined on an arbitrary topological space. But on normal spaces, in particular, on metric spaces, there always exist nonconstant real-valued continuous functions by Urysohn lemma, named after P. S. Urysohn (1998–1924), which is an outstanding result characterizing normal spaces by real-valued continuous functions. His approach by using dyadic rational numbers is studied in th Chap. 6, is more general and different from Urysohn lemma for metric spaces discussed in Chap. 2.

### 8.6.12 Tietze Extension Theorem

**Urysohn lemma** is used to prove **Tietze extension theorem** saying that a topological space $(X, \tau)$ is normal iff for every closed set $Y$ in $(X, \tau)$ and every continuous map $f : Y \to \mathbf{I}$ has a continuous extension over $X$. Tietze Extension theorem is named after Heinerich Franz Friedrich Tietze (1880–1964). Historically, he proved a theorem in 1915, for a metric space. Urysohn published its general version in an article in 1925. Tietze also made significant contribution to topology and introduced Tietze transformation between presentations of groups. The general version of Tietze extension theorem as given in Corollary 6.5.3 was first time found in an article published by Urysohn in 1925.

**Corollary 8.6.6** (Tietze extension theorem) *If $A$ is a closed subspace of a metric space $X$, then every continuous map $f : X \to \mathbf{I}$ has a continuous extension over $X$.*

## 8.6.13   Countability and Separability Axioms

Topologies defined on an arbitrary set are weak in the sense that they fail to invite their deep study until certain additional condition or conditions are imposed on them. With this aim, the concepts of first and second countable spaces and also separable spaces by an axiomatic approach were introduced. **Historically**, two axioms of countability were formulated by F. Hausdorff (1868–1942) in 1914, and the concept of separability was introduced by M. Fréchet in 1906. They do not arise from the study of calculus and analysis in a natural way. They arise through a deep study of topology. The axiom of first countability arose through the study of convergent sequences. For example, a first countable topological space $X$ is Hausdorff iff every convergent sequence in $X$ has a unique limit in $X$. The classical Urysohn metrization theorem which asserts that every topological space satisfying the second axioms of countability and regularity can be embedded in a metric space , which implies that such a topological space is metrizable. This theorem gives a sufficient condition of metrizability of a topological space. This chapter also studies Lindelöf spaces from the viewpoint of countability and separability. Separability of topological spaces is a topological property, but it is not a hereditary property

## 8.6.14   Compactness Property

Compactness property dates back to Heine–Borel property of $\mathbf{R}$ in particular and $\mathbf{R}^n$ in general. Heinrich Heine (1821–1881) introduced the concept of finite subcovering in 1872 in his work on uniformly continuous functions. Emile Borel (1871–1956) proved a result in 1894 asserting that every open covering of a closed interval has a finite subcovering. This result is considered the beginning of the concept of compactness. A more general result was published by Henri Lebesgue (1875–1941) in 1898 and by Arthur Schönflies (1853–1928) in 1900. Heine–Borel theorem says that a subset of $\mathbf{R}$ is compact iff is closed and bounded. This theorem characterizes compactness in the setting of the real line space $\mathbf{R}$ in terms of its bounded and closed subsets. Another characterization of compactness in terms of closed sets having finite intersection property was given by Leopold Vietoris (1891–2002) in 1921.

## 8.6.15   Stone–Čech Compactification

This subsection proves Stone–Cech compactification theorem is a basic result in topology. This theorem named after M. H. Stone (1902–1989) and E. Čech (1893–1960) asserts that every completely regular space $X$ is embeddable as a dense subspace in a specified compact Hausdorrf space $\beta(X)$, which has an important property that every bounded real-valued function continuous on $X$ has a unique extension to a bounded continuous real-valued function on $\beta(X)$. Stone published a paper (Stone 1948) on compactification of topological spaces.

### 8.6.16 Lebesgue Lemma and Lebesgue Number

This subsection studies Lebesgue lemma and Lebesgue number named after H. Lebesgue (1875–1941) for an open covering of a compact metric space establishing a relation between such a space and Lebesgue number and proves Lebesgue lemma providing a technical result on open covering of a compact metric space. Given an open covering $\mathcal{F} = \{U_\alpha : \alpha \in \mathbf{A}\}$ of a compact metric space $X$, there exists a real number $\delta > 0$ (called **Lebesgue number** of $\mathcal{F} = \{U_\alpha\}$) such that every open ball $B_x(\epsilon)$ *in* $X$ for some $\epsilon > 0$ is contained in at least one open set $\{U_\alpha\} \in \mathcal{F}$. The concept of a Lebesgue number stems from Lebesgue's work on measure theory, starting with his thesis in 1902. **The existence of Lebesgue number is guaranteed in Lemma** 8.6.7, which is proved by Lebesgue and named after him.

**Lemma 8.6.7** (Lebesgue) *Let $X$ be a compact metric space. Given an open covering $\mathcal{F} = \{U_\alpha : \alpha \in \mathbf{A}\}$ of $X$, there exists a real number $\delta > 0$ (called **Lebesgue number** of $\mathcal{F}$) such that any subset $Y$ of $X$ of diameter $diam(Y) < \delta$ is contained in some member of $\mathcal{F}$, i.e., whenever $Y \subset X$ and $diam(Y) < \delta$, then $Y \subset U_\alpha$ for some $\alpha \in \mathbf{A}$.*

### 8.6.17 Paracompact Spaces

The concept of paracompactness of topological spaces was introduced by Jean Dieudonné in 1944. Paracompact spaces include regular and normal spaces. There are many important topological spaces which are not compact, but they are paracompact. For example, the real number space $\mathbf{R}$ is paracompact, but it is not compact. Paracompact spaces represent a special class of topological spaces having localization of its compactness. The significance of paracompactness is the assertion of the existence of partition of unity. The concept of paracompactness is closely related to that of metrizability of topological spaces. The former concept is sometimes applied in an easier way to study metrizable spaces.

Paracompactness is an important tool to study some problems in algebraic topology and geometry such as homotopy classification of vector bundles over paracompact spaces (see **Basic Topology: Volume 3** of the present series of books). Its other importance lies in the results that the class of paracompact spaces contains the compact Hausdorff spaces and metrizable spaces. By Stone's theorem, every metrizable space is paracompact. Its converse is true in the sense that every paracompact locally metrizable space is metrizable (see Nagata–Smirnov theorem). For a paracompact topological space $(X, \tau)$, a locally finite covering of $(X, \tau)$ always exists.

### 8.6.18    Baire Spaces

Baire spaces named after René-Louis Baire (1874–1932) form an important family of topological space. This family includes complete metric spaces, compact Hausdorff spaces and also locally compact Hausdorff spaces. A Baire space $X$ given in Definition 8.6.8 cannot be expressed as a countable union of closed sets with empty interior in $X$.

**Definition 8.6.8** A topological space $(X, \tau)$ is said to be a **Baire space** if given any countable family $\{X_n\}$ of closed sets of $X$ with each $X_i$ having empty interior in $X$, their union $\bigcup X_n$ has also empty interior in $X$. **Equivalently**, a topological space $(X, \tau)$ is said to be a Baire space if intersection of every countable family of open dense sets in $X$ is dense.

### 8.6.19    Connectedness Property

The concept of connectedness of some subsets of the Euclidean line $\mathbf{R}^2$ was given by Camille Jordan (1838–1922) in 1914. A systematic study of connected topological spaces was inaugurated by Felix Hausdorff in his book Grundzüge der Mengenlehre published in 1914. Again the "intermediate value theorem" asserts that if $f : [a, b] \to \mathbf{R}$ is continuous and $r \in \mathbf{R}$ lies between $f(a)$ and $f(b)$, then there is a point $\alpha \in [a, b]$ such that $f(\alpha) = r$. What is the generalization of this theorem in topology? The idea generalizing **"intermediate value theorem"** depends not only on the property of continuity of $f$ but also depends on a special property of $[a, b]$, called **connectedness**, which is also a topological property different from compactness property. Brower fixed-point theorem for dimension 1 saying that every continuous map

$$f : \mathbf{I} \to \mathbf{I}$$

has a fixed point is closely related to Bolzano theorem asserting that if

$$f : [a, b] \to \mathbf{R}$$

is a continuous function such that $f(a)f(b) < 0$, then there exists a point $x \in [a, b]$ such that $f(x) = 0$.

### 8.6.20    Compactification of Topological Spaces

Considering the importance of compact topological spaces, a noncompact space $X$ may be extended to a compact space $CX$ containing $X$ as a everywhere dense subspace of $CX$. Then the space $CX$ is called a compactification of $X$. If $CX$ is of the form $CX = X \cup \{\infty\}$, where $\{\infty\}$ is a point not isolated in $CX$, then $CX$ is called Alexandrov one-point compactification of $X$. One-point compactification of a noncompact space $X$ is characterized by **Alexandrov's theorem** in terms of locally compactness. Another compactification theorem, known as **(Stone–Čech compactification)**, says that if $X$ is a completely regular space, then there exists a compact Hausdorff space $\beta(X)$ such that $X$ is a dense subspace of $\beta(X)$ and every bounded continuous function $f \colon X \to \mathbf{R}$ has a unique extension to a bounded continuous function

$$\tilde{f} \colon \beta(X) \to \mathbf{R}.$$

### 8.6.21  Space-Filling Curve

A continuous map defined on a closed interval of the real line space whose image is a two-dimensional region in the Euclidean plane $\mathbf{R}^2$ is called a space-filling curve. There are several versions of space-filling curve theorem. The common version is that there exists a surjective continuous map

$$f \colon \mathbf{I} \to \mathbf{I}^2.$$

G. Cantor (1845–1918) proved in 1877 that the interval $\mathbf{I} = [0, 1]$ and the square $\mathbf{I}^2$ have the same number of points by establishing a bijective correspondence between the sets $\mathbf{I}$ and $\mathbf{I}^2$, i.e., $\mathbf{I} \sim \mathbf{I}^2$. He remarked that the dimension is not a set-theoretic concept. The difference of their dimensions involves topology. G. Peano (1858–1932) showed again in 1890 that there is a continuous function on $\mathbf{I}$ such that its image is a two-dimensional region, such as a square or a triangle in the Euclidean plane $\mathbf{R}^2$, called a space-filling curve or Peano curve named after him.

### 8.6.22  The Gelfand–Kolmogoroff Theorem

**The Gelfand–Kolmogoroff theorem**is a basic theorem in topology. It studies the rings $\mathcal{C}(X, \mathbf{R})$ of real-valued continuous functions for compact Hausdorff spaces $X$. This theorem asserts that two compact Hausdorff spaces $X$ and $Y$ are homeomorphic iff the corresponding rings $\mathcal{C}(X, \mathbf{R})$ and $\mathcal{C}(Y, \mathbf{R})$ are isomorphic. This deep result recovers the topology of $X$ from the ring structure of $\mathcal{C}(X, \mathbf{R})$.

### 8.6.23  Ascoli's Theorem

Ascoli's theorem characterizes compact subsets of a certain class of function spaces. This theorem is named after Giulio Ascoli (1843–1896). He introduced the concept of equicontinuity in 1884, which is one of the fundamental concepts in the theory of real functions.

## 8.7   Topology of Metric Spaces

Following the concept of (abstract) metric spaces introduced by M. Fréchet (1878–1973) in 1906, several topologists studied the topology of metric spaces and proved many fruitful results developing the subject topology. Every metric space $(X, d)$ induces a topology $\tau_d$, and equivalent metrics induce the same topology on $X$. From the topological viewpoint, metric spaces have many properties. Some of them studied in this book are stated in this subsection.

(i) **(Bolzano–Weierstrass)** Let $(X, d)$ be a compact metric space and $A$ be a compact subset of $X$. Then every sequence in $A$ has a convergent subsequence.

(ii) Every compact subset of a metric space is closed and bounded.

(iii) Every continuous map from a compact metric space to any metric space is uniformly continuous.

(iv) **(Lebesque covering lemma)** Given a compact metric space $(X, d)$ and an open covering $\{U_i\}$ of $X$, There exists a positive real number $\delta$ such that for every subset $A$ of $X$ with its diameter $< \delta$, there is an index $i$ such that $A \subset U_i$.

(v) **(Arzela–Ascoli Theorem)** Let $\mathbf{F} = \mathbf{R}$ or $\mathbf{C}$ and $\mathcal{C}(X, \mathbf{F})$ endowed with sup norm metric, a subset $\mathcal{B}$ of $\mathcal{C}(X, \mathbf{F})$ is compact iff $\mathcal{B}$ is bounded, closed and equicontinuous.

(vi) **(Cantor's intersection theorem)** Given a compact metric space $(X, d)$ and a family of closed subsets $\{A_n\}_{n \in \mathbb{N}}$ with $A_{n+1} \subset A_n$, $\forall n$ and diam $(A_n) \to 0$, then

$$\bigcap_{n=1}^{\infty} A_n$$

consists of exactly one element of $X$.

(vii) **(Baire category theorem)** Let $(X, d)$ be a complete metric space.

(i) If $\{U_n\}$ is sequence of open dense subsets of $X$, then

$$\bigcap_{n=1}^{\infty} U_n \neq \emptyset.$$

(ii) If $\{A_n\}$ is sequence of nonempty closed subsets of $X$ with $X = \bigcup_{n=1}^{\infty} A_n$, then exists at least one $A_n$ having nonempty interior.

(viii) **(Banach contraction theorem )** Let $(X, d)$ be a complete metric space and $f ; X \to X$ be a contraction. Then $f$ has a unique fixed point. This theorem proved by the Polish mathematician Stefan Banach (1892–1945) is named

after him. It applies to prove Picard's theorem on the existence of solutions of a
differential equation, which is named after the French mathematician Charles
Emile Picard (1856–1941).

(ix) **(Uniform boundedness principle)** Let $(X, d)$ be a complete metric space and
$\mathcal{F}$ be a family of real-valued continuous functions on $X$ having the property
that for every point $x \in X$, there exists a constant $C_x$ such that

$$|f(x)| \leq C_x, \ \forall f \in \mathcal{F}.$$

Then there exist a nonempty open set $U \subset U$ and a constant $C$ such that

$$|f(x)| \leq C, \ \forall x \in X \text{ and } \ \forall f \in \mathcal{F}.$$

***Remark 8.7.1*** The present book series "Basic Topology" consisting of three volumes
is actually three textbooks on different fields of topology.

(i) **Volume 1** considers the general properties of topological spaces and their mappings. This chapter addresses historical notes with an emphasis on development
of metric spaces and general topology.

(ii) **Volume 2** links with topological structure with other structures in a compatible
way to study topological groups, topological vector spaces, smooth manifolds,
fiber spaces, covering spaces and Lie group and Lie algebra. A historical note
on the texts of this volume is available in Chap. 5 of this volume.

(iii) **Volume 3** considers the problems of converting topological and geometrical
problems to algebraic one in a functorial way for better chance for solution. The
foundation of this useful idea was laid by Henri Poincaré (1854–1912) in his
land-marking "Analysis situs", Paris, 1895, through his invention of fundamental group and homology theory, which are topological invariants inaugurating
the subject "algebraic topology". Volume 3 also studies low-dimensional topology, manifolds and topology of fiber bundles by using algebraic topology. A
historical note on the texts of this volume is available in Chap. 7 of this volume.

# Additional Reading

Adhikari, A., Adhikari, M.R.: Basic Topology, Volume 2: Topological and Groups, Toology of
    Manifolds and Lie Groups. Springer, India (2022)
Adhikari, M.R.: Basic Topology. Volume 3: Algebraic Topology and Topology of Fiber Bundles.
    Springer, India (2022)
Adhikari, M.R.: Basic Algebraic Topology and its Applications. Springer, India (2016)
Adhikari, M.R., Adhikari, A.: Basic Modern Algebra with Applications. Springer, New Delhi, New
    York, Heidelberg (2014)
Alexandrov, P.S.: Introduction to Set Theory and General Topology. Moscow (1979)
James, I.M.: Handbook of the History of General Topology, vol. 3, pp. 809–834. Kluwer Academic
    Publishers (2001)

Borisovich, Y.U., Bliznyakov, N., Izrailevich, Y.A., Fomenko, T.: Introduction to Topology. Mir
    Publishers, Moscow (1985)
Bredon, G.E.: Topology and Geometry. Springer-Verlag, New York (1993)
CS Chinn, W.G., Steenrod. N.E.: First Concepts of Topology. Math Association of America (1966)
Lefschetz, S.: Introduction to Topology. Princeton University Press, New York (1949)
MacLane, S.: Categories for the Working Mathematician. Springer-Verlagn, New York (1971)
Poincaré, H.: Papers on Topology: Analysis Situs and its Five Supplements. Translated by Still-well.
    J. Hist. Math. **37**; Amer. Math. Soc (2010)
Stone, A.H.: Paracompactness and product spaces. Bull. Amer. Math. Soc. **54**, 977–982 (1948)
Stone, M.H.: Compactication of topological spaces. Ann. Soc. Pol. Math. **21**, 153–160 (1948)

# Index

**A**

Absolute value, 39
Accumulation point, 40
Affine space, 209
Arzela–Ascoli theorem, 496

**B**

Baire category theorem, 295
Baire space, 294
Banach contraction theorem, 98
Banach space, 89
Base, 140
Basis, 140
Bolzano theorem, 333
Bolzano weierstrass theorem, 449
Boundary point, 135
Bounded, 70
Box topology, 474
Brouwer Fixed Point Theorem for
    Dimension 1, 373

**C**

Cantor set, 17
Cantor space, 217
Cantor's intersection theorem, 77
Cardinal number, 21
Cartesian product, 16
Cauchy sequence, 45
Characteristic function, 400
Choice function, 17
Clopen set, 132
Closure of a set, 133
Cluster point, 40
Cofinite topology, 126
Compactly generated, 291
Compact open topology, 350

Compact set, 273
Complement, 4
Complete, 75
Completely normal, 247
Complete metric space, 294
Completeness property, 6
Component, 337
Connected component, 337
Connectedness, 325
Connected space, 326
Continuous, 48
Contraction, 371
Convergence of a sequence, 395
Countable compact, 323
Countable open base, 448
Countable space, 455
Countable subcover, 449

**D**

De Morgan's rules, 4
Dense, 65
Denseness property, 7
Density axiom, 6
Denumerable, 20
Derived set, 40, 130
Diameter, 114
Difference set, 4
Disconnectedness, 325
Discrete metric, 55
Discrete topology, 126
Distance, 69
Distance function, 51
Distributive laws, 28

**E**

Embedding theorem, 434

Printed in the United States
by Baker & Taylor Publisher Services